D0875639

CHEMISTRY OF HYPERVALENT COMPOUNDS

CHEMISTRY OF HYPERVALENT COMPOUNDS

Edited by

Kin-ya Akiba
Department of Chemistry
Faculty of Science
Hiroshima University
Higashi-Hiroshima 739-8526, Japan

NEW YORK / CHICHESTER / WEINHEIM / BRISBANE / SINGAPORE / TORONTO

This book is printed on acid-free paper. 

Published simultaneously in Canada.

Library of Congress Cataloging-in-Publication Data:

Chemistry of hypervalent compounds / edited by Kin-ya Akiba.
p. cm.
Includes index.
ISBN 0-471-24019-2 (acid-free paper)
1. Hypervalence (Theoretical chemistry) 2. Organic compounds.
3. Organometallic compounds. I. Akiba, Kin 'ya, 1936–
QD255.4.C48 1998
541.2′24—dc21 98-6578
CIP

Printed in the United States of America

10 9 8 7 6 5 4 3 2 1

To Former Professor James C. Martin of the University of Illinois, Urbana-Champaign, and Vanderbilt University
And to my wife Yoko

CONTENTS

FOREWORD

This book describes recent and up-to-date aspects of the chemistry of hypervalent compounds, mainly dealing with hypervalent organic compounds. I have read the original manuscript and I have found the quality of each chapter to be excellent and well written by outstanding experts in the field. To me, it seems good to include extra topics such as hypervalent lithium ions and modern theoretical aspects of hypervalent compounds as a sequel to the review article by J. I. Musher in 1969, which laid the theoretical basis for hypervalent compounds. However, these are not essential for making good use of this book.

When Professor Akiba asked me to write the Foreword to this book in Inchon, Korea, in 1996, where the 13th IUPAC Conference on Physical Organic Chemistry was held, he told me that his initial idea was to include two more chapters dealing with the above-mentioned subjects. He explained that there had been several discussions on the terminology of hypervalence(y) among theoretical chemists and others and that he had been concerned about this. I understood the situation because I have made a significant contribution to the chemistry of hypervalent bismuth compounds. I encouraged him to publish this book, telling him that it is timely and necessary to present the scope of the chemistry of hypervalent compounds to chemists in general, because the field has been growing rapidly recently. The term *hypervalence* is accepted and used worldwide. I said, "Do not worry, theoretical interpretations come later and are not essential here for using hypervalent compounds in organic synthesis." Akiba explains this situation briefly in Chapter 1.

N-X-*L* designation, where formally assignable valence electrons (*N*) of the central atom X are used in directly bonding a number of ligands (*L*), is a useful and convenient formalism.

I remember well the visit of Professor Akiba to College Station, Texas, on the way to Albany, New York, for ICHAC 2 in 1989. He was interested in the chemistry of hypervalent organic molecules, and he was not specialized in any one element. In Inchon, he clearly stated that he has been seeking "what is essentially new for hypervalent organic compounds" and has been choosing the correct elements to investigate. In this way new types of reactions and the preparation of new types of organic compounds, which cannot be realized for simple C, H, N, O compounds, can be accomplished. Thus, he has chosen S, P, Si, Sn, As, Sb, Bi, Te, and recently some transition metals for comparison. Hence, I am convinced that he is the best choice as editor for this book.

I am sure that this book will be read widely by organic and inorganic chemists who wish to update their own knowledge and to find new ideas for future research. The chemistry presented here is new and rapidly developing between the two fields. All branches of modern chemistry are greatly indebted to Professor Akiba for his initiative in editing this well-written and essential volume.

Derek H. R. Barton

February 1998
College Station, Texas
(*Deceased on March 16, 1998*)

PREFACE

The chemistry of hypervalent compounds has been acquiring wider and deeper attention, both by organic chemists and by inorganic chemists, as a new and rapidly developing area between the two fields during the past 20 years. The idea of hypervalent organic molecules stemmed from the dream of H. Staudinger and G. Wittig in the first half of the twentieth century to challenge the valence theory and to prepare nitrogen and phosphorus compounds bearing five homopolar bonds, which would necessarily violate the beliefs of most chemists of that period. Staudinger attempted to realize this idea using dialkylzincs as nucleophiles, but in vain (*Helv. Chim. Acta.* **2**, 1919). After about 30 years, Wittig independently reached the same idea as Staudinger and succeeded in preparing Ph_5P (1949), Ph_5Sb, and Ph_4Te (1952) by using PhLi. Wittig did not determine the structure of these compounds and did not use (and probably did not know) the term hypervalence. However, a theoretical interpretation of the structure of $K^+[I—I—I]^-$ and its analogs was proposed based on the three-centre–four-electron (3c–4e) bond using molecular orbital theory in 1951. The 3c–4e bond has been called a hypervalent bond and represents the apical bonds of PF_5, Ph_5P, and related compounds. The concept of hypervalent molecules and the area of chemistry it involves were made clear in the review article by J. I. Musher, which laid the theoretical basis for hypervalent compounds (*Angew. Chem. Intern. Ed. Engl.*, **8**, 1969).

During independent and active experimental research for more than 30 years on organic chemistry of silicon, phosphorus, and sulfur, several new, unique but fundamental features of structures and reaction patterns have been found to occur commonly to organic compounds of these elements. Among them, fluorosiliconates, phosphoranes, sulfonium and sulfuranes, iodonium and related iodoso compounds, and pentacoordinate bismuth compounds have found unique synthetic utilities. Investigation of better synthetic conditions and mechanisms of stereocontrol revealed the presence of hypervalent (higher coordinate) species as crucial intermediates, and some of them have been prepared stable and used as synthetic reagents. Synthetic utilities of phosphoranes, phosphonates, and sulfonium ylides have amply been reviewed already. Synthetic utilities of bismuth compounds were reviewed in *Tetrahedron* (D. H. R. Barton *et al.*, **44**, 1988), *Chem. Rev.* (J.-P. Finet, **89**, 1989), and *Heteroatom Chemistry* (D. H. R. Barton, Chapter 5, Ed. by E. Block, VCH, 1990). In this book, synthetic applications of hypercoordinate silicon species are elegantly presented (Chapter 5) and those of hypervalent organoiodinanes are summarized by the most active researcher in that field (Chapter 12).

By taking a closer look at the above-mentioned species, one can extract unique but common features of these species which occur in virtually all kinds of heavier main group elements. Those species can be categorized as hypervalent compounds. However, this situation is still not generally common knowledge to chemists, and the majority of them, in my opinion, are still reluctant to recognize the generality of these characters and prevalence of hypervalent compounds, because chemistry is fundamentally based on the character of each individual element.

General aspects of structure and reactivity of hypervalent organic compounds are overviewed at the beginning of this book (Chapter 2). Structure and reactivity of silicon species (Chapter 4), phosphorus compounds (Chapter 6), sulfuranes (Chapter 7), chalcogen compounds (Se, Te: Chapter 8), compounds bearing the $-XF_5$ group (S, Se, Te: Chapter 10), polycoordinate iodine compounds (Chapter 11), and xenon compounds (Chapter 13) are smartly presented. The static and dynamic structure of hypervalent Sb(III), Sn(II), and Ge(II) halides (Chapter 3) and the ligand-coupling reactions of Group 15 and 16 elements are summarized concisely (Chapter 9).

Rapid and great advances in theoretical calculations due to development and use of computers in chemistry, bonding and reaction patterns of hypervalent molecules have been the target of active discussion for many years. It is now essentially agreed that the contribution of *d* orbitals is not essential for *sp* elements to form hypervalent compounds.

Fundamental features of hypervalent molecules of heavier main group elements cannot be realized and experienced while we deal with typical and simple molecules of C, H, N, and O. It can be recognized that there are fundamental and formal similarities in structures and reactivity between hypervalent molecules and organometallic compounds in several spects. Detailed and more advanced studies in this direction are important for future development of chemistry. Naturally, it is assumed that readers of this book have studied organic chemistry at the undergraduate level.

A revised definition of hypervalent molecules is presented in Chapter 1. This definition can include all the hypervalent molecules and can give fundamental scope of the area:

> hypervalent compounds are main group element (*sp* element: Group 1,2, and 13–18) compounds that contain a number (N) of more than an octet of formally assignable electrons in a valence shell of the central atom (X) in directly bonding a number (L) of ligands (substituents).

The N-X-L designation is a convenient and useful formalism, especially to describe hypervalent molecules. The concept and terminology of hypervalence have been accepted and used widely and are gaining more popularity as an easily understandable and useful formalism.

Here, my sincere thanks is expressed to Professor Howel, Department of Chemistry, Sophia University, Tokyo, who kindly read and annotated the English of all the manuscripts, especially those of Japanese authors.

I wish to dedicate this book to my wife Yoko and to Professor James C. Martin, who had retired from Vanderbilt University after a bright career as Professor of Chemistry at the University of Illinois at Urbana–Champaign. Professor Martin pioneered this field experimentally and prepared a variety of hypervalent organic compounds. I learned a lot from him and remember and appreciate heartily his very kind friendship. Finally, but not least, I heartily express my sincere gratitude to Professor Derek Barton who kindly wrote the Foreword to this book.

Japanese newspapers reported that Professor Sir Derck H. R. Barton, Nobel Prize Laureate, passed away suddenly on March 16, 1998. I sincerely add here that I am heartily reminded of the kind and friendly encouragement I received throughout my contact with him.

KIN-YA AKIBA

February, 1998
Higashi-Hiroshima, Japan

CONTRIBUTORS

KIN-YA AKIBA, Department of Chemistry, Faculty of Science, Hiroshima University, Higashi-Hiroshima 739, Japan

VALERY K. BREL, Institute of Physiologically Active Compounds, Russian Academy of Science, 142432 Chernogolovka, Moscow Region, Russia

CLAUDE CHUIT, Laboratoire de Chimie Moléculaire et Organisation du Solide, UMR 5637 CNRS, Université Montpellier II, Sciences et Techniques du Languedoc, Place E. Bataillon, 34-095 Montpellier Cedex 5, France

ROBERT J. P. CORRIU, Laboratoire de Chimie Moléculaire et Organisation du Solide, UMR 5637 CNRS, Université Montpellier II, Sciences et Techniques du Languedoc, Place E. Bataillon, 34-095 Montpellier Cedex 5, France

JÓZEF DRABOWICZ, Center of Molecular and Macromolecular Studies, Polish Academy of Sciences, Department of Organic Sulfur Compounds, 90-363 Łódź, Poland

NAOMICHI FURUKAWA, Department of Chemistry, University of Tsukuba, 1-1 Tennoudai, Tsukuba 305, Japan

TAKAYUKI KAWASHIMA, Department of Chemistry, Graduate School of Science, The University of Tokyo, 7-3-1 Hongo, Bunkyo-ku, Tokyo 113, Japan

MITSUO KIRA, Department of Chemistry, Graduate School of Science, Tohoku University, Sendai 980-77, Japan

DIETER LENTZ, Institüt für Anorganische und Analytische Chemie der Freien Universität Berlin, Fabeckstrasse 34-36, 14195 Berlin, Germany

MASAHITO OCHIAI, Faculty of Pharmaceutical Sciences, University of Tokushima, 1078 Shomachi, Tokushima 770, Japan

TSUTOMU OKUDA, Department of Chemistry, Faculty of Science, Hiroshima University, Higashi-Hiroshima 739, Japan

CATHERINE REYE, Laboratoire de Chimie Moléculaire et Organisation du Solide, UMR 5637 CNRS, Université Montpellier II, Sciences et

Techniques du Languedoc, Place E. Bataillon, 34-095 Montpellier Cedex 5, France

SOICHI SATO, Department of Chemistry, University of Tsukuba, 1-1 Tennoudai, Tsukuba 305, Japan

KONRAD SEPPELT, Institüt für Anorganische und Analytische Chemie der Freien Universität Berlin, Fabeckstrasse 34-36, 14195 Berlin, Germany

PETER J. STANG, Department of Chemistry, The University of Utah, Salt Lake City, UT 84112, USA

KOJI YAMADA, Department of Chemistry, Faculty of Science, Hiroshima University, Higashi-Hiroshima 739, Japan

YOHSUKE YAMAMOTO, Department of Chemistry, Faculty of Science, Hiroshima University, Higashi-Hiroshima 739, Japan

NIKOLAI S. ZEFIROV, Institute of Physiologically Active Compounds, Russian Academy of Science, 142432 Chernogolovka, Moscow Region, Russia

LUO CHENG ZHANG, Photodynamics Research Center, The Institute of Physical and Chemical Research (RIKEN), 19-1399 Koeji, Nagamachi, Aoba-ku, Sendai 980, Japan

VIKTOR ZHDANKIN, Department of Chemistry, University of Minnesota-Duluth, Duluth, MN 55812, USA

CHEMISTRY OF HYPERVALENT COMPOUNDS

1 Hypervalent Compounds

KIN-YA AKIBA
Department of Chemistry, Faculty of Science, Hiroshima University, Higashi-Hiroshima, Japan

1.1 DEFINITION OF HYPERVALENT COMPOUNDS

The concept of hypervalent molecules was established by J. I. Musher in 1969: they are ions or molecules of the elements of Groups 15–18 bearing more electrons than the octet (nine or more) within a valence shell.[1]

There are essentially two ways to hold electrons beyond the octet within a valence shell: (1) make up a dsp^3 or d^2sp^3 orbital by hybridization using higher-lying *d* orbitals or (2) make up highly ionic (50% or more) orbitals revising (modifying) the basic idea of Lewis that a bond is formed by a localized pair of two electrons.

Pauling admitted both methods and explained that there are two kinds of hypervalent molecules, one where the contribution of (1) is predominant (e.g., ICl_2^-, I_3^-) and one where (2) is predominant (e.g., PCl_5, SF_6).

Pimentel and Rundle, independently in 1951, laid the basis for new developments in this area by proposing the idea of a three-center–four-electron (3c–4e) bond, employing molecular orbital theory.[2] According to the fundamental description of a 3c–4e bond, one pair of bonding electrons is delocalized to the two ligands (substituents), resulting in the charge distribution of almost −0.5 charge on each ligand and almost +1.0 charge on the central atom.

This established the possibility of (2). The idea of a 3c–4e bond, however, had not been accepted so easily and it gradually came to be used only when *d* orbital hybridization could apparently not be supported. Then, owing to the progress in computing and also to the efforts of Kutzelnigg and co-workers,[3] the idea of a 3c–4e bond has become supported and is presently accepted.

Schleyer and co-workers recently investigated the bonding of hypervalent molecules theoretically and published a paper where they included historical aspects of research on hypervalent bonding.[4] Accordingly, they concluded that

Chemistry of Hypervalent Compounds, edited by Kin-ya Akiba.
ISBN 0-471-24019-2

method (1) employing dsp^3 or d^2sp^3 is not at all correct but rather misleading, and that the contribution of *d* orbitals is not at all of fundamental importance for accepting electrons beyond the octet for hypervalent molecules.

Even after Schleyer's work, there have been a number of theoretical investigations on hypervalent bonding (molecules) and method (2) has been established. Because *d* orbitals cannot be responsible to hold extra electrons (the energy gap between n(*sp*) and n(*d*) is too large for *sp* elements), the number of orbitals is deficient; hence a 3c–4e bond is an electron-rich bond and the nonbonding molecular orbital (NBMO) becomes the highest occupied molecular orbital (HOMO). In a 3c–4e bond, extra electrons are distributed on ligands (substituents) and the number of pairs of effective electrons in a valence shell of the central atom is less than four and consequently does not exceed that of a Lewis octet. Based on this analysis, there have been some discussions about whether or not it is correct (appropriate) to use the term hyper*valency*(*e*), because the valence shell of the central atom is not expanded by hybridization and hence there is no violation of the octet rule.[5] However, there are some reports based on theoretical investigations on the structure of transition metal hydrides, realizing the importance of the 3c–4e bond, that state that most transition metal complexes are hypervalent![6] There are also a number of theoretical works on highly coordinated lithium cations.[7] At present, both the terms hypervalency(e) and hypercoordination are used.

Hence, the author consulted the keywords of CAS online for 1981–95. There are five keywords related to hypervalency: hypervalence, hypervalency, hypervalent, hypervalently, and hypervalents. The number of references including at least one of the five (some include more than one) was 783; 394 such papers were published during 1990–94. Also, there were three keywords related to hypercoordination: hypercoordinate, hypercoordinated, and hypercoordination. The number of references including at least one of the three (some include more than one) was 51; 36 such papers were published during 1990–94. Moreover, the term *hypervalent compounds* is listed in the Reviewer Expertise Survey of the *Journal of Organic Chemistry*, just like hormones, β-lactams, peroxides, porphyrins, and so on under the heading "Compounds."

Thus, it can be taken for granted that hypervalency and its related terms have been accepted in chemical society. Within hypervalent compounds, compounds of Group 14 and also elements of Group 1 or 2 as a central atom are presently included.

Terms related to higher coordination were not consulted. The concept of coordination is loose and ambiguous and does not include the idea of direction in itself, hence it is not appropriate to describe the chemistry of this area because a hypervalent bond (3c–4e bond) is linear and has a large enough bond energy (typically, the bond order is one-half of the corresponding single bond) compared to coordination. It is clear that bonding–coordination–interaction each have their own concept and definition in their fundamental or intrinsic states; however, the borderlines between them are not clear and they cannot be separated so simply and can even be continuous.

Now it is appropriate and necessary to define the term "hypervalent compounds" as a useful and formal concept as follows.

Hypervalent compounds are main group element (*sp* element: Groups 1, 2, 13–18) compounds that contain a number (*N*) of formally assignable electrons of more than the octet in a valence shell directly associated with the central atom (X) in directly bonding a number (*L*) of ligands (substituents). The designation *N*-X-*L* is conveniently used to describe hypervalent molecules.[8]

It is now clear that as far as we are dealing with fundamental organic chemistry, essential structures of molecules are based only on linear (*sp*), triangular (sp^2), and tetrahedral (sp^3) shapes; however, trigonal bipyramidal (TBP) and octahedral (Oh) shapes become fundamental for hypervalent compounds (Figure 1.1). Hence, there is an apparent similarity in shape between hypervalent compounds and organotransition metal compounds.

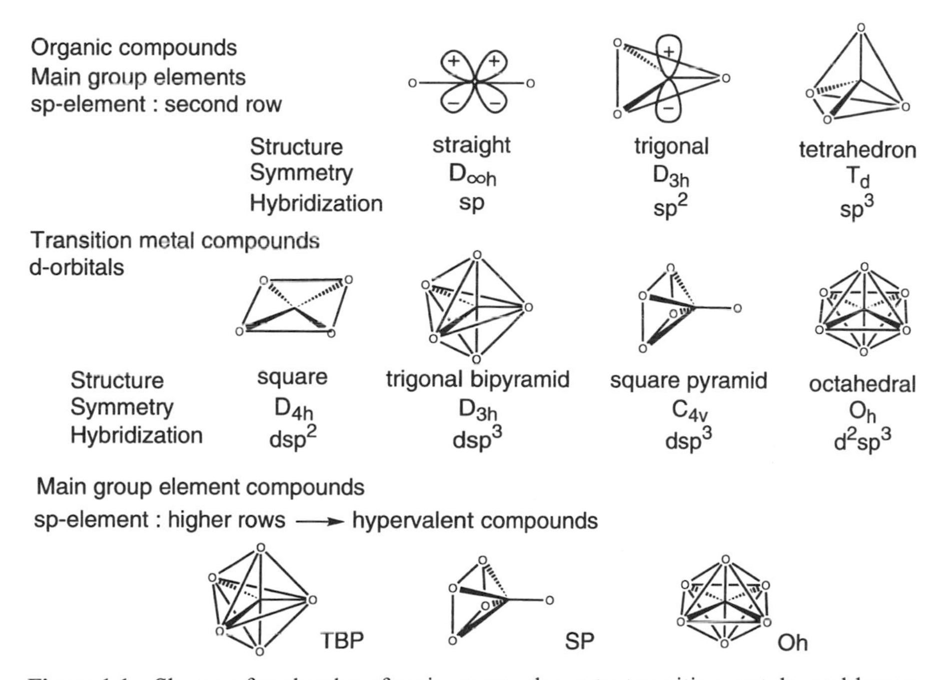

Figure 1.1 Shapes of molecules of main group elements, transition metals, and hypervalent compounds.

1.2 THREE-CENTER–FOUR-ELECTRON BOND (HYPERVALENT BOND)

The 3c–4e bond model obtained by the molecular orbital method was first applied to F—H—F^-, Cl—I—Cl^-, I_3^-, and I_5^-.[2] It was explained that I_3^- ion can be linear and the I—I bond length is longer by 10% compared with that of diiodine (I_2). A molecular orbital model for F_3^-, which should be the most unstable hypervalent compound, obtained by the TZP/ACCD (SCF) method is shown in Figure 1.2. The F—F bond length was calculated to be 1.70 Å which was 20% longer than that of difluorine (F_2: 1.412 Å). The two substituent fluorines bore −0.51 charges for each and the central fluorine had +0.03 charges. The hypervalent molecule of F_3^- obtained stabilization energy by 11 kcalmol^{-1} compared with the system of [$F_2 + F^-$].[9]

It is the apical bond of a pentacoordinate trigonal bipyramidal molecule where a hypervalent bond is most typically seen. The apical bond of the molecular orbital of PF_5 is shown in Figure 1.3. It is composed of one 3*pz* orbital of phosphorus and two 2*pz* orbitals of fluorine. The two 2*pz* orbitals of fluorine (Figure 1.3(1)) cannot overlap with the 3*pz* orbital of phosphorus and result in a nonbonding orbital (Ψ_n) and this is HOMO. The two 2*pz* orbitals of fluorine (Figure 1.3(2)) and the 3*pz* orbital of phosphorus can overlap and yield a bonding orbital (Ψ_b) and an antibonding orbital (Ψ_a). The molecule of PF_5 (10-P-5) is composed of six electrons of three P—F bonds in the *xy* plane and the four electrons of a 3c–4e bond. The apical bond (1.577 Å) is longer than the equatorial one (1.534 Å), and the minus charge is strongly localized on the two apical fluorines. The apical bond is polarized and weaker than the equatorial one. When there is a certain degree of contribution of the *d* orbital of the central atom to the 3c–4e bond, the nonbonding orbital will acquire some bonding character.

In order to make up a 3c–4e bond of pentacoordinate molecules experimentally, four methods can be proposed (Scheme 1.1). (1) Add two free radicals to coordinate with a pair of unshared electrons in a *p* orbital. (2) Add two pairs of

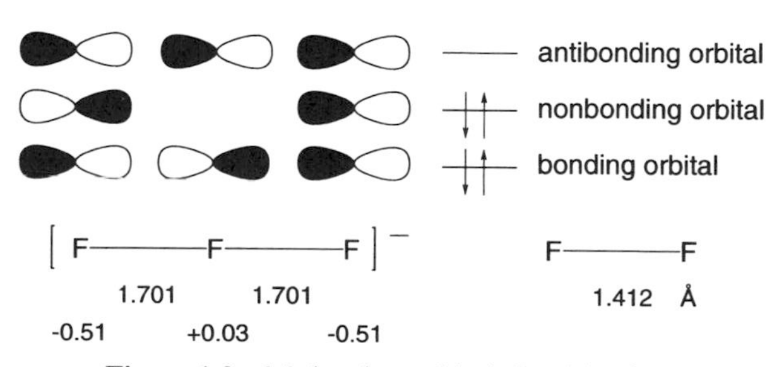

Figure 1.2 Molecular orbital (3c–4e) of F_3^-.

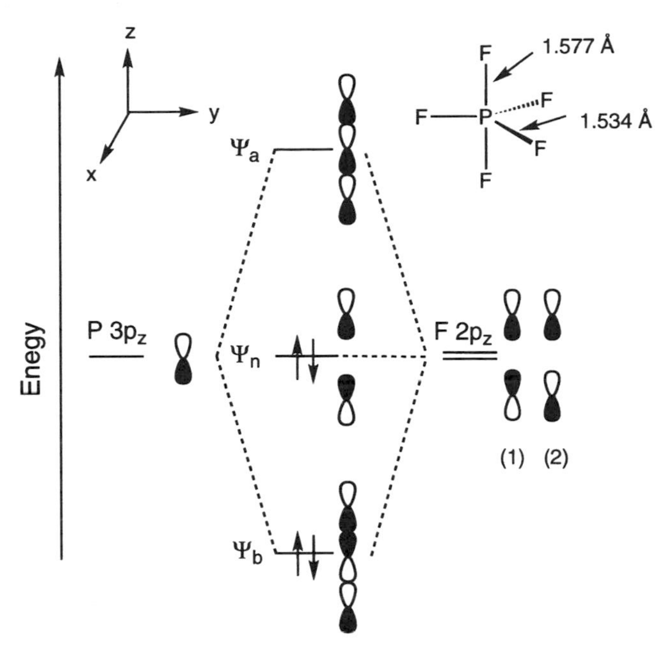

Figure 1.3 Molecular orbital (3c–4e) of apical bond of PF_5.

unshared electrons (nucleophiles) to coordinate to a vacant *p* orbital. (3) Add a pair of unshared electrons to coordinate with the σ^* orbital of a Z—X bond in a cationic molecule (e.g., sulphonium, phosphonium etc.). (4) Add a pair of unshared electrons to the σ^* orbital of a Z—X bond of a neutral molecule (e.g., silicon and tin compounds). When X is a carbon, method (4) corresponds to the transition state (TS) of S_N2 reaction. This is why hypervalent compounds are sometimes called molecules of frozen transition state.

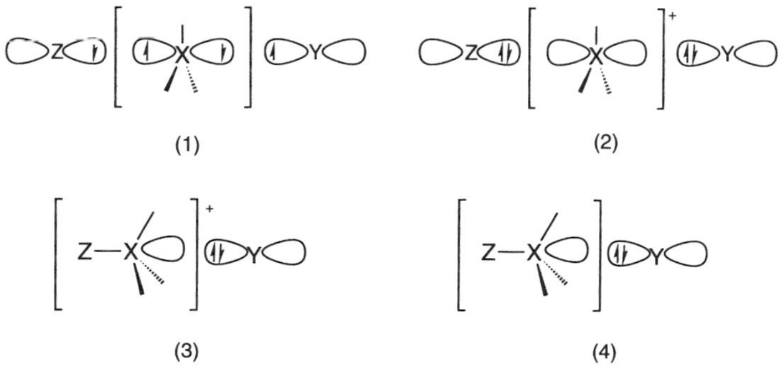

(1) (2)

(3) (4)

Scheme 1.1

One of the typical methods to form a hexacoordinate hypervalent molecule (12-X-6) is to add two pairs of unshared electrons (nucleophiles) to a vacant orbital which appears at the TS of edge inversion, thus forming a 3c–4e bond.[10] By bending two pairs of two single bonds of C_3 symmetry (tetrahedral molecule) in order to change into planar configuration, the character of the single bonds changes and a vacant *p*-type orbital appears perpendicularly to the plane, thus allowing one to make up approximately three 3c–4e bonds at the central atom by coordination of two nucleophiles. Scheme 1.2 shows the mechanism of edge inversion with that of vertex inversion.

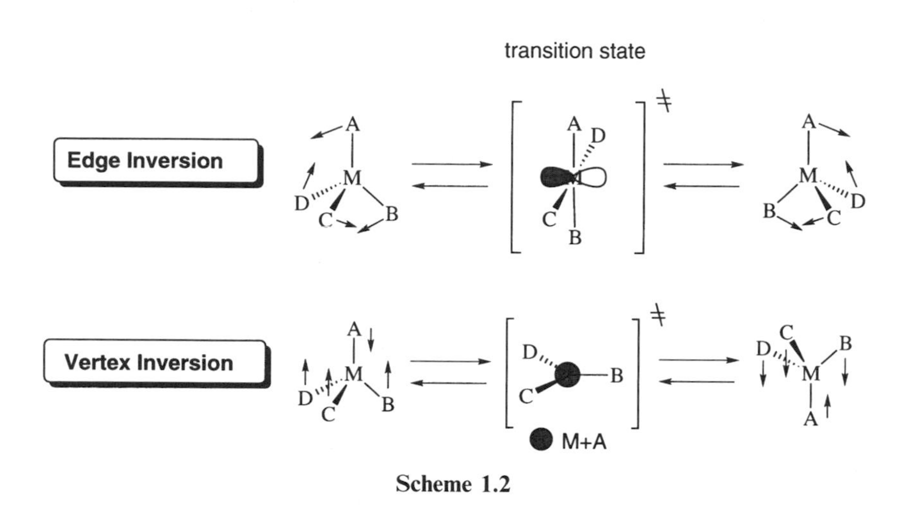

Scheme 1.2

1.3 WHAT RESULTS ARE ESSENTIALLY NEW FOR HYPERVALENT COMPOUNDS?

In order to answer the question of what results are essentially new (novel) for hypervalent compounds, we need a concrete standpoint as reference. The reference is the well-established structure and reactivity of organic compounds of C, H, N, O, and halogens.

Based on a vast amount of experimental results, the following characterics can be extracted and summarized as essentially different and unusual characters for main group element compunds below the third row of the periodic table: (1) Long chain compounds (polymers) connected by single bonds are unstable, unlike carbon compounds. (2) Unsaturated bonds containing main group elements are unstable. (3) Hypervalent compounds exist stably for almost all the main group elements below the third row.

Then, the next question is what are the essential characteristics of hypervalent compounds? These can be summarized as follows:

1. The structures of five-coordinate compounds (10-X-5 and 10-X-4 bearing a pair of unshared electrons) are essentially trigonal bipyramid (TBP), regardless of the kind of the central atom and the kind of formal charge (plus, neutral, or minus). An apical bond is a 3c–4e bond and an equatorial bond is an sp^2 bond.
2. The structures of six-coordinate compounds (12-X-6 and 12-X-5 bearing a pair of unshared electrons) are essentially octahedral (Oh), regardless of the kind of the central atom and the kind of formal charge (plus, neutral, or minus). It can be said approximately that essentially three equivalent 3c–4e bonds are formed.
3. For five-coordinate molecules, intramolecular positional isomerization takes place rapidly (Berry pseudorotation or turnstile rotation) but for six-coordinate molecules, it is very slow (Bailar twist). These correspond to fluxional molecules of organometallic compounds.
4. Ligand exchange reactions (LER) and ligand coupling reactions (LCR) take place for hypervalent compounds of 10-X-5 and 10-X-4. These are formally similar to LER and reductive elimination of organometallic compounds.
5. The stereochemistry of nucleophilic substitution of four-coordinate main group compounds is complex, unlike that of the S_N2 reaction at carbon. Both retention and inversion occur due to the stability of intermediate 10-X-5 species where rapid intramolecular positional isomerization can take place.
6. Edge inversion can be the mechanism of inversion for main group element compounds of higher rows. This is also applicable for 10-X-3 compounds bearing lone pair electrons. This enables the inversion to proceed via an essentially different mechanism from that of ammonia (vertex inversion). This invokes the appearance of a hypervalent bond at the TS.
7. A hypervalent bond (attractive interaction) can be formed by accepting unshared electron pairs of a main group element (heteroatom) by an X^+—Z bond and even by a neutral X—Z bond by using a lower-lying σ^* orbital of the X—Z bond.

Based on an understanding of items (2) and (6) above, porphrins (12-X-6) of main group elements were prepared.[11] The formation of a hypervalent bond was exemplified by 1,5-transannular bond formation between heteroatoms of eight-membered ring. The latter idea has been extended to the chemistry of atranes.[12] Moreover, contributions of the hypervalent bond have been suggested for bioorganic molecules recently. As an example, a bonding interaction (formation of hypervalent bonding) between a selenium atom (II) and an unshared pair of electrons of oxygen has been shown to be one apparent factor in maintaining the shape of bioorganic macromolecules.[13]

In Chapter 2, characteristics and reactions of hypervalent compounds are overviewed, which are classified according to the number of coordination (substituents).

REFERENCES

1. J. I. Musher, *Angew. Chem. Intern. Ed. Engl.*, **8**, 54 (1969).
2. (a) G. C. Pimentel, *J. Chem. Phys.*, **19,** 446 (1951); (b) R. J. Hackand, R. E. Rundle, *J. Am. Chem. Soc.*, **73**, 4321 (1951).
3. W. Kutzelnigg, *Angew. Chem. Intern. Ed. Engl.*, **23,** 272 (1984) and references therein.
4. A. E. Reed, P. v. R. Schleyer, *J. Am. Chem. Soc.*, **112**, 1434 (1990) and references therein.
5. R. J. Gillespie, E. A. Robinson, *Inorg. Chem.*, **34**, 978 (1995).
6. (a) C. R. Landis, T. Cleveland, T. K. Firman, *J. Am. Chem. Soc.*, **117,** 1859 (1995); (b) T. Cleaveland, C. R. Landis, *J. Am. Chem. Soc.*, **118**, 6020 (1996); (c) C. R. Landis, T. K. Firman, D. M. Root, T. Cleaveland, *J. Am. Chem. Soc.*, **120**, 1842 (1998).
7. (a) P. v. R. Schleyer, *Pure and Appl. Chem.*, **56,** 151 (1984); (b) H. Kudo, M. Hashimoto, K. Yokoyama, C. H. Wu, A. E. Dorigo, F. M. Bickelhaupt, P. v. R. Schleyer, *J. Phys. Chem.*, **99**, 6477 (1995).
8. C. W. Perkins, J. C. Martin, A. J. Arduengo, III, W. Lau, A. Alegria, J. K. Koci, *J. Am. Chem. Soc.*, **102**, 7753 (1980).
9. P. A. Cahill, C. E. Dykstra, J. C. Martin, *J. Am. Chem. Soc.,* **107,** 6359 (1985).
10. (a) D. A. Dixon, A. J. Arduengo, III, *J. Am. Chem. Soc.*, **109**, 338 (1987); (b) D. A. Dixon, A. J. Arduengo, III, *J. Chem. Phys.*, **91**, 3195 (1987); (c) Y. Yamamoto, X. Chen, S. Kojima, K. Ohdoi, M. Kitano, Y. Doi, K.-y. Akiba, *J. Am. Chem. Soc.*, **117**, 3922 (1995).
11. (a) Y. Yamamoto, R. Nadano, M. Itagaki, K.-y. Akiba, *J. Am. Chem. Soc.*, **117**, 8287 (1995); (b) K.-y. Akiba, Y. Onzuka, M. Itagaki, H. Hirota, Y. Yamamoto, *Organometallics,* **13,** 2800 (1994).
12. (a) K.-y. Akiba, K. Takee, Y. Shimizu, K. Ohkata, *J. Am. Chem. Soc.*, **108,** 6320 (1986); (b) J. G. Verkade, *Acc. Chem. Res.*, **26**, 483 (1993).
13. (a) F. T. Burling, B. M. Goldstein, *J. Am. Chem. Soc.*, **114,** 2313 (1992); (b) D. H. R. Barton, M. B. Hall, Z. Lin, S. I. Parekh, J. Reibenspies, *J. Am. Chem. Soc.*, **115**, 5056 (1993).

2 Structure and Reactivity of Hypervalent Organic Compounds: General Aspects

KIN-YA AKIBA
Department of Chemistry, Faculty of Science, Hiroshima University, Higashi-Hiroshima, Japan

2.1 INTRODUCTION

As is explained in Chapter 1, the *N*-X-*L* designation is quite useful and convenient to classify the structure of hypervalent compounds; typical examples of reported hypervalent compounds are classified here according to this designation and their structures are shown in Figures 2.1–2.7. According to the *N*-X-*L* designation, multiple bonds are written as polarized single bonds; this is applied for compounds **6**, **11**, **16**, **19**, **26**, and **29**, regardless of the actual character of the bonds. Here general aspects of structure and characteristics of hypervalent compounds are described.

2.2 THREE-COORDINATE HYPERVALENT ORGANIC COMPOUNDS

Typical examples of three-coordinate hypervalent compounds are classified as 10-X-3 type compounds. The fundamental electronic structure is shown by the valence bond method in Figure 2.8. The central atom bears two unshared electron pairs and one M—A bond (sp^2 bond) in a plane and the 3c–4e bond is perpendicular to the plane, which consists of apical bonds (B—M—B) in order to keep two ligands.

The 3c–4e bond is an essential feature for structure and stability of 10-X-3 compounds. Such a bond is stabilized when electron-withdrawing ligands (substituents) occupy the apical position, but the bond between M and B becomes longer and weaker according to the increase of electronegativity of the

Chemistry of Hypervalent Compounds, edited by Kin-ya Akiba.
ISBN 0-471-24019-2

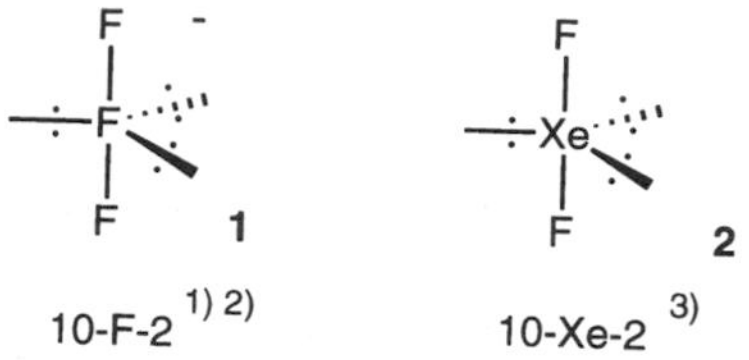

1
2
10-F-2 1) 2)
10-Xe-2 3)

Figure 2.1 Two-coordinate hypervalent compounds.[1–3]

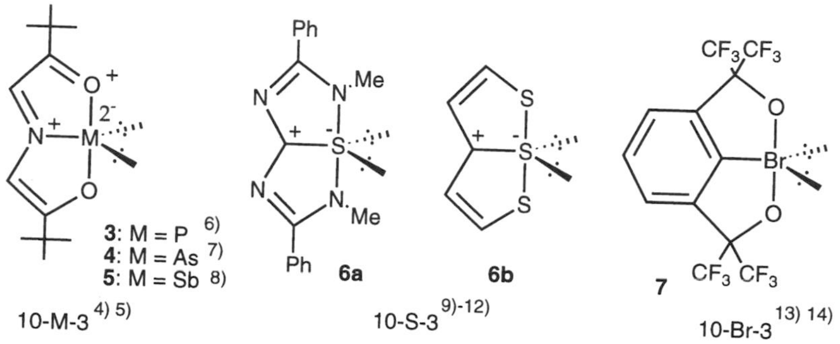

3: M = P 6)
4: M = As 7)
5: M = Sb 8)
10-M-3 4) 5)
6a
6b
10-S-3 9)-12)
7
10-Br-3 13) 14)

Figure 2.2 Three-coordinate hypervalent compounds[4–14]

Et4N+
9a : X = O-
9b : X = Me
9c : X = MeO
9d : X = Cl
TfO-
8
10-Bi-4 15)-17)
10-S-4 18)-21)
10
10-I-4 22) 23)
11
10-Xe-4 24)
12
12-Xe-4 25)

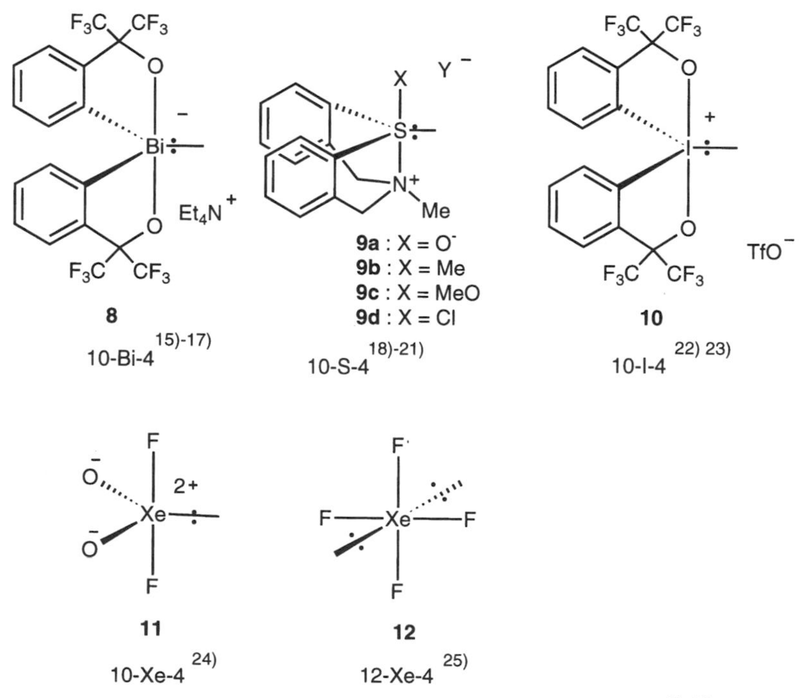

Figure 2.3 Four-coordinate hypervalent compounds.[15–25]

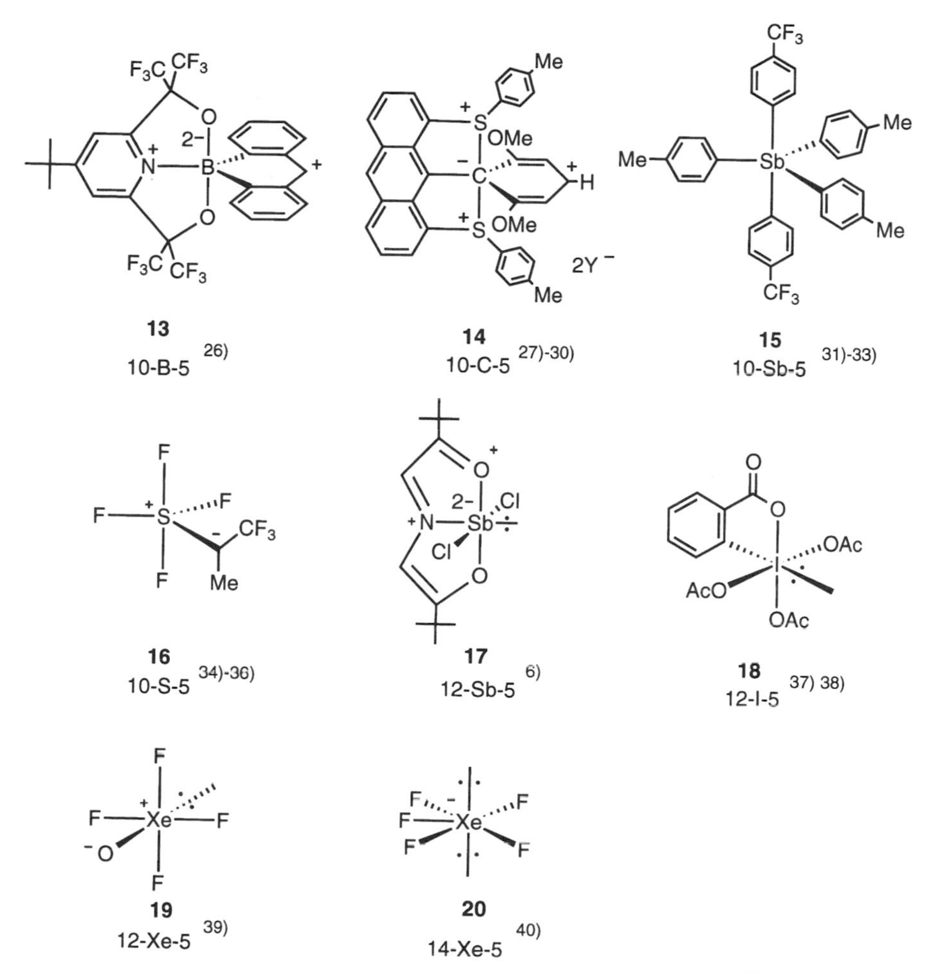

Figure 2.4 Five-coordinate hypervalent compounds.[26–40]

substituent (B). These characteristics, if based on superficial understanding, seem to be contradictory to the stability of the molecule. One of the best examples to show this peculiar character is the investigation of the bond switch of 10-S-3 sulfuranes.[62,63]

For the reaction of thiadiazolines and activated acetylenes, it was shown that the reaction path became totally different according to changes of the kind of substituent and of the position of heteroatoms in the five-membered ring. For example, 5-imino-1,2,4-thiadiazoline (**33**) reacts with activated acetylenes to give the corresponding 2-aminothiadiazole and a nitrile by addition-elimination; however, 2-iminothiazoline (**34**) reacts with the acetylenes to give simple addition products[64,65] (Scheme 2.1).

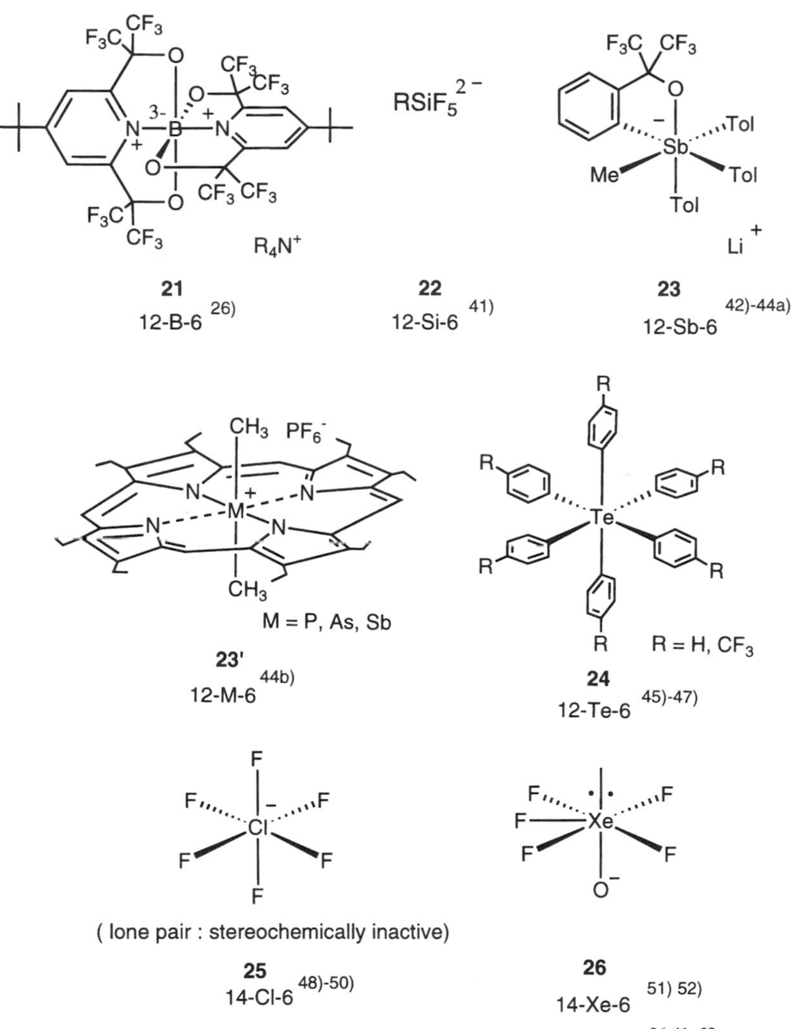

Figure 2.5 Six-coordinate hypervalent compounds.[26,41–52]

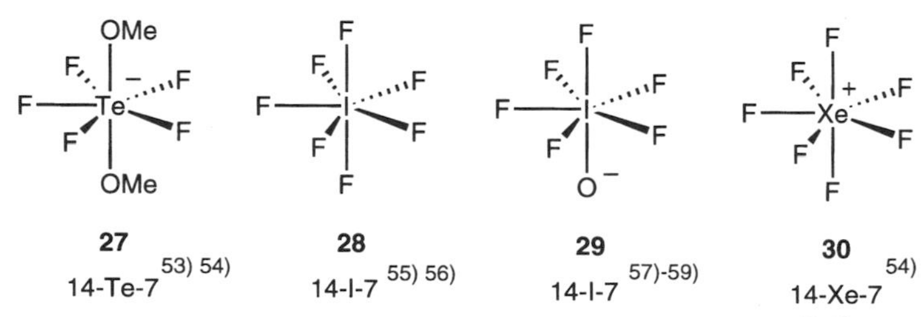

Figure 2.6 Seven-coordinate hypervalent compounds.[53–59]

F
2−
Te
−
I

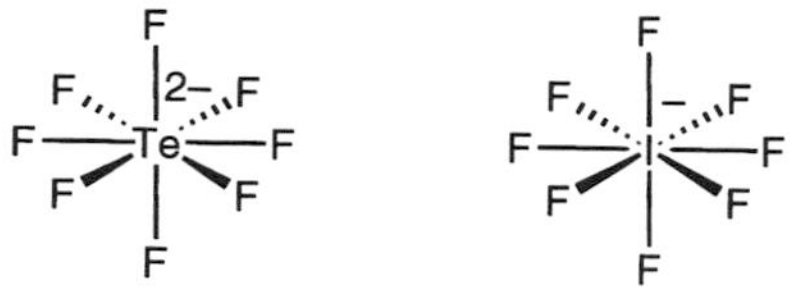

31 16-Te-8 [58)]

32 16-I-8 [58) 60) 61)]

Figure 2.7 Eight-coordinate hypervalent compounds.[58,60,61]

B
A
M
B

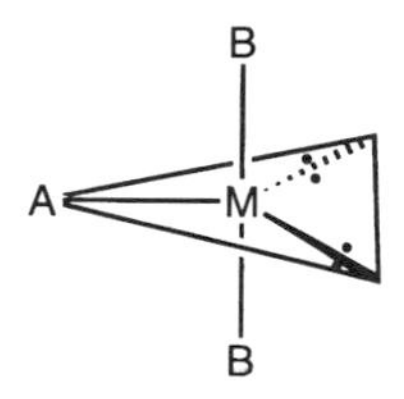

Figure 2.8 Description of 10-M-3 species.

33
A
34
B
C
R¹CN + R²NH
(X = CO₂Et, COPh)

Scheme 2.1

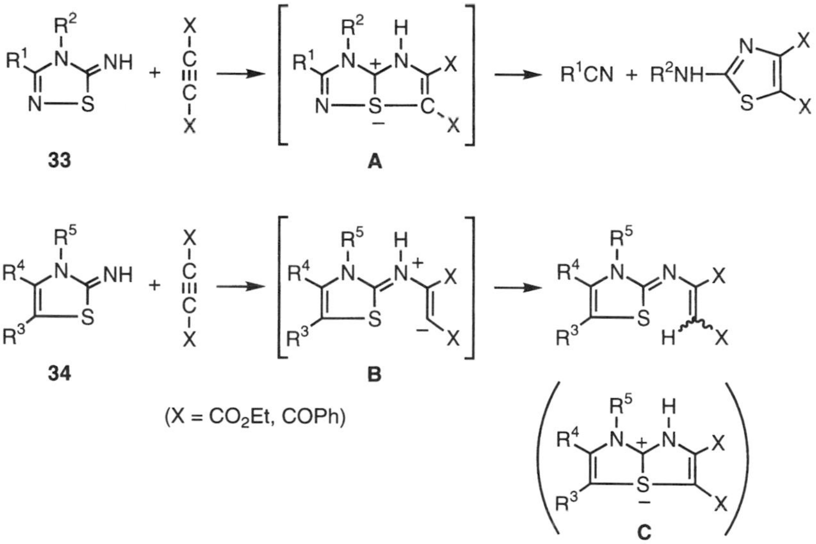

This result can be understood by the difference in stability of the 3c–4e bond assumed in the intermediate (A or C). In the latter case (C), the 3c–4e bond consists of C—S—C and in the former (A), it is made of N—S—C. The 3c–4e bond in the former (A) containing the more electronegative nitrogen atom should be more stable compared with C, hence the reaction through A is preferred. The reaction path involving C may surely be energetically unfavorable, thus the reaction proceeded through the zwitterion B to give an addition product. Then, 4-substituted 5-imino-1,3,4-thiadiazolines (**35** and **36**) were prepared (Scheme 2.2). It is designed to pass the intermediate (D or E) where the 3c–4e bond, which consists of C—S—C, should be generated during the reaction. The electron-withdrawing ability of the carbon at 2 is adjusted by a phenyl or a benzoyl group. The reaction of **35** with activated acetylenes gave simply the addition product. However, **36** bearing a benzoyl group at C-2 reacted with the same acetylenes in a competitive manner to afford a simple addition product via a zwitterion intermediate (F) and addition-elimination product via a 10-S-3 sulfurane intermediate (E). The ratio of the competition was shown to depend very much on polarity of solvents. Such competition was observed even in alcohols, thus certifying that the two paths are really competitive.[66]

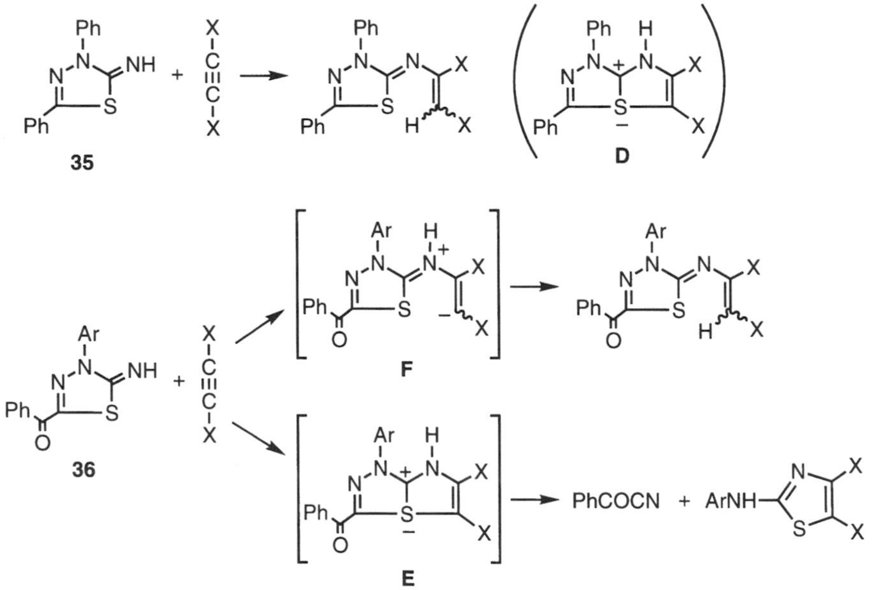

Ph N N NH S Ph + X C C X → Ph N N N X S Ph H X (Ph H N N + N S − Ph X X) 35 D

Ar N N NH S Ph O + X C C X → [Ar H N N N+ X S − Ph O X] F → Ar N N N X S Ph O H X

→ [Ar H N N + N X S − Ph O X] E → PhCOCN + ArNH—(N S X X) 36

Scheme 2.2

A reaction of **33** with nitriles in the presence of aluminum trichloride or with the corresponding imidates afforded **37**, where ring transformation took place induced by prototropy to give an aromatic thiadiazole ring (Scheme 2.3).[67-76] Bond-switching equilibration, that is, ring transformation equilibrium, was observed when the same skeleton (**38α** and **38β**) was obtained by ring transformation.[77] These kinds of ring transformation could not be observed for the corresponding oxygen analog.

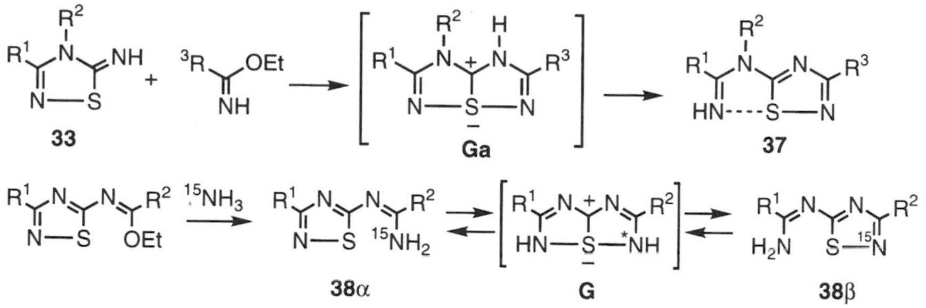

Scheme 2.3

The rate of ring transformation was rapid in solution and neither **38α** nor **38β** could be separated, but it was too slow to measure by ^{1}H NMR. The rate of ring transformation of symmetrical **38** ($R^1 = R^2 = Me$) was enormously accelerated by acid and that of monoprotonated species was measured at low temperatures (Scheme 2.4). Activation parameters were $\Delta G^{\ddagger}_{273} = 14.8 \pm 0.2\,\text{kcal mol}^{-1}$, $\Delta H^{\ddagger} = 11.6 \pm 1.0\,\text{kcal mol}^{-1}$, $\Delta S^{\ddagger} = -12 \pm 3\,\text{eu}$, $k_{273} = 9\,\text{sec}^{-1}$.[78]

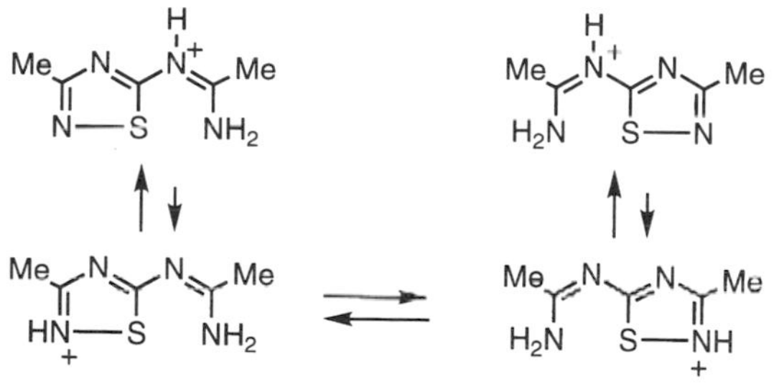

Scheme 2.4

By employing silyl group shift and de-sililation, 5-(2-aminovinyl)isothiazole (**40α**), which has ^{15}N at the vinyl group, was prepared in a pure form (Scheme 2.5).[79, 80] When **40α** was heated in C_6D_6 and DMSO-d_6 at 60°C, the equilibrium mixture of **40α** and **40β** was obtained, where hypervalent sulfirane H was invoked as an intermediate. The rate of equilibration from **40α** to **40β** followed exactly the first-order rate in the presence of equilibration, the rate was faster in benzene-d_6 than in DMSO-d_6. Deuterium, which was initially on the nitrogen, was gradually distributed to the vinyl and isothiazolyl hydrogen. Based on these findings, it was concluded that the ring transformation of **40** is effected by sigmatropy of hydrogen through a symmetrical intermediate H as 10-S-3 sulfurane where the hypervalent bond is not so strongly polarized according to the symmetrical structure. Kinetic parameters of equilibration of **40** ($Ar^1 = Ar^2 = p\text{-}Cl_6H_4$) in benzene-$d_6$ were also obtained as $\Delta G^{\ddagger}_{298} = 24.4 \pm 0.8$, $\Delta H^{\ddagger} = 12.2 \pm 0.2\,\text{kcal mol}^{-1}$, $\Delta S^{\ddagger} = -4.10 \pm 0.5\,\text{eu}$, and $k_{298} = 6.21 \times 10^{-6}\,\text{sec}^{-1}$.

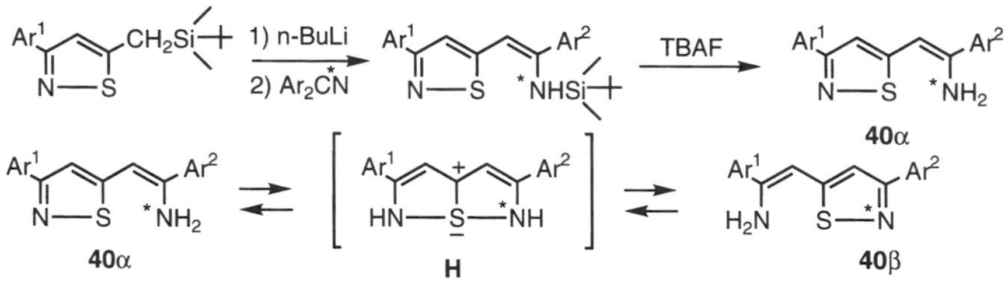

Scheme 2.5

From these results, the flexibility and weakness of the 3c–4e bond of 10-S-3 sulfuranes were exemplified.

In order to determine the bond energy of the hypervalent N—S—N bond, sulfuranes fused with pyrimidine rings (**41**) were prepared and the structures were determined to be planar by X-ray analysis (Scheme 2.6).[63, 81-84]

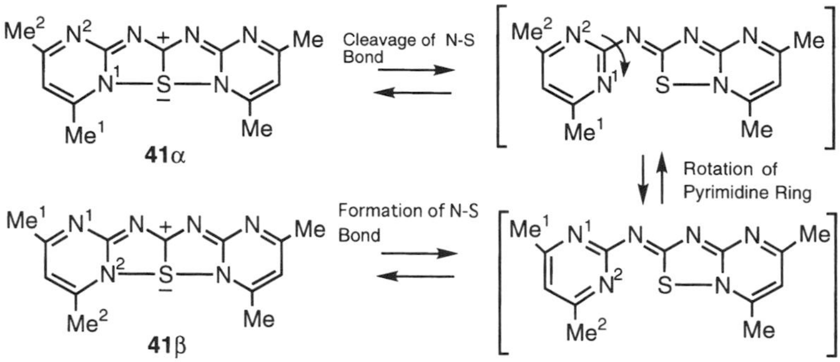

Scheme 2.6

As there are two nitrogens in a pyrimidine ring, a new hypervalent N^2—S bond can be generated by the rotation of the pyrimidine ring when the original N^1—S bond is cleaved. The bond energy of the hypervalent N—S—N bond can be approximated by the rotation energy of the pyrimidine ring, which can be measured by the exchange energy of the two methyl groups (Me^1 and Me^2). By using this system, the exchange energy of the two methyl groups can be measured separately for unsymmetrically substituted pyrimidine rings. As is shown in Scheme 2.7, the exchange energy for Me^1 and Me^2 corresponds to the bond dissociation energy of N^1–S and that of Me^3 and Me^4 to that of S—N^3. Kinetic parameters obtained by variable temperature measurement of the rates are shown in Table 2.1. The rate of exchange of Me^3 and Me^4 on

TABLE 2.1 Activation Parameters of S—N Bond Cleavage of 41 and 42 in CD_2Cl_2

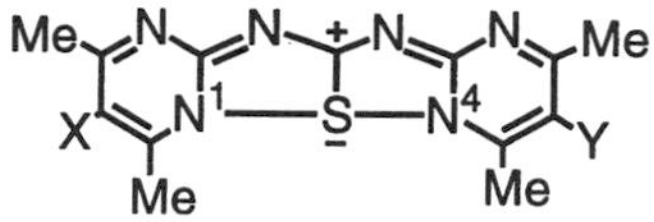

	X	Y		T_c (°C)	$\Delta G^{\ddagger}_{T_c}$ (kcal mol^{-1})	$\Delta G^{\ddagger}_{298}$ (kcal mol^{-1})	$\Delta H^{\ddagger}$ (kcal mol)$^{-1}$	$\Delta S^{\ddagger}$ (eu)
41	H	H		45	16.7	16.6	15.9 ± 1.1	-2.4 ± 3.4
42	H	Br	H-side	75	18.2	18.3	18.8 ± 0.3	$1.8 + 1.0$
			Br-side	24	15.6	15.5	16.1 ± 1.3	2.2 ± 0.9

the pyrimidine ring bearing an electron-withdrawing group (Br), that is, exchange between **42α** and **42β**, is apparently faster and easier than that of Me^1 and Me^2 on the pyrimidine ring with hydrogen, that is, exchange between **42α** and **42γ** (Scheme 2.7). When we compare $\Delta G^{\ddagger}_{298}$ of both processes, that for **42α** and **42β** is 15.5 kcal mol^{-1} and that for **42α** and **42γ** is 18.3 kcal mol^{-1}, the difference of $\Delta G^{\ddagger}_{298}$ is 2.8 kcal mol^{-1}. The average value (16.9 kcal mol^{-1}) is almost identical to that of the symmetrical molecule (**41**). The result clearly shows that the labile character of the hypervalent bond is deeply influenced ($\Delta G^{\ddagger}_{293} = 2.8$ kcal mol^{-1}) by a slight change of the electronegativity of the apical ligands (N^1 and N^3), which was effected by substitution by a bromine for a hydrogen.

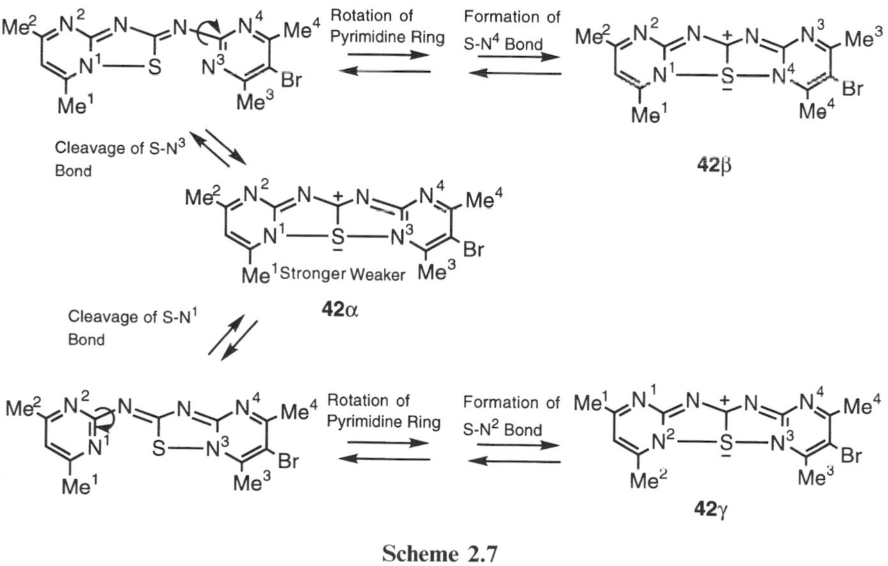

Scheme 2.7

Although a variety of stable 10-S-3 sulfuranes had been prepared and their chemistry has been described, including thiathiophthenes, stable 10-P-3 compounds were first prepared by Culley and Arduengo in 1984[6] (Scheme 2.8).

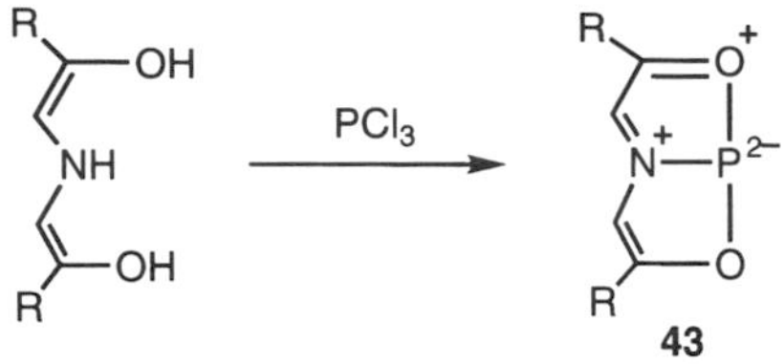

Scheme 2.8

The stability of this system is mainly based on two factors: (1) the molecule (**43**) is maintained neutral as a whole by the dicationic nature of the ligand, and (2) the cationic nature of the two apical oxygens strongly stabilizes the 3c–4e bond of 10-P-3. Arduengo and Dixon investigated electronic states of their 10-P-3 (**43**) system intensively by *ab initio* calculation. First they concluded that the electronic state of **43** is most accurately shown as **43a** where two pairs of unshared electrons reside on the central phosphorus atom (Figure 2.9). These unshared electrons can be reorganized also as one (σ-type) in the molecular plane and the other out of plane (π-type), the latter being perpendicular to the former. The planarity of the molecule can be considered maintained by the contribution of the π-type electron pairs. In the phosphorus compound, the electron density of the π-type pairs is decreased by the π-type conjugation to the ligand; hence the stability of the planar structure is much less for 10-P-3 compounds compared to that of 10-As-3 and 10-Sb-3. When transition metal complexes with vacant orbitals coordinate to one of the electron pairs of the 10-M-3 system, the 10-P-3 system cannot maintain the planarity and the structure changes into 8-P-4 (ordinary phosphonium salt), whereas the 10-M-3 systems of arsenic and antimony can maintain the 10-M-4 structure, keeping a pair of electrons still on the central atom (Scheme 2.9).

When a saturated tridentate ligand is employed, planar structure **44b** can no longer be stable and the molecule **44** becomes pyramidal (8-P-3: **44a**) (Scheme 2.10). This also shows that the planar structure of 10-P-3 is maintained based on a quite delicate balance of conjugation of π-type orbitals of the 10-P-3 to the ligand and the stability of 10-P-3 configuration, which is supported by the rather strong hypervalent bond of O—P—O.[85]

Based on the theoretical investigation of electronic states of **43** and **44**, it was proposed that the T-shaped structure of 8-P-3 compounds (**43c** and **44b**) can be the transition state of inversion (Scheme 2.10). This new mechanism of inversion is called edge inversion. It is shown in Scheme 2.11 together with the well-known vertex inversion.[86, 87]

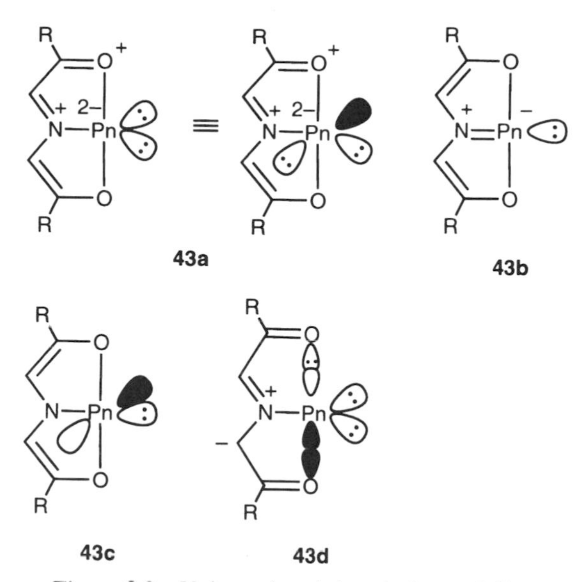

Figure 2.9 Valence bond descriptions of **43**.

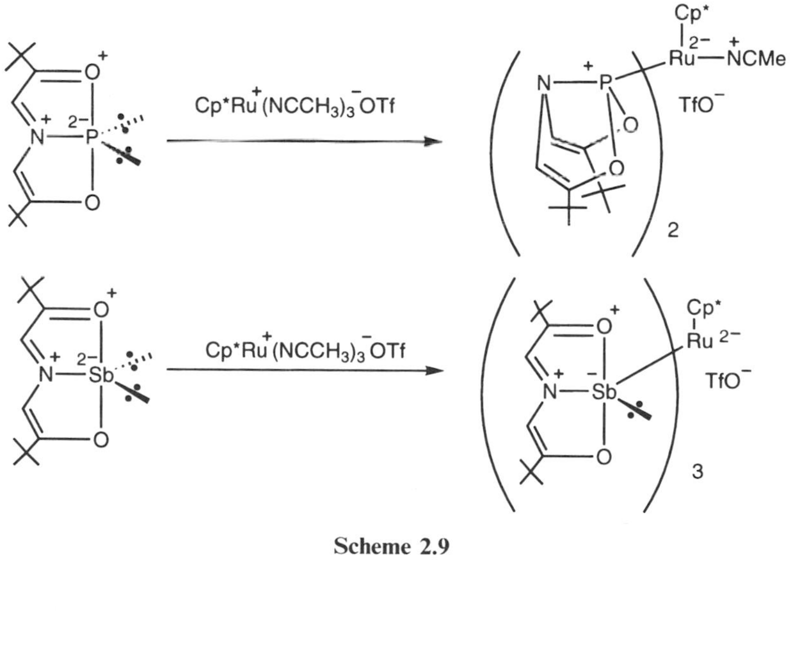

$Cp^{*}Ru^{+}(NCCH_3)_3\,OTf^{-}$

$Cp^{*}Ru^{+}(NCCH_3)_3\,OTf^{-}$

Scheme 2.9

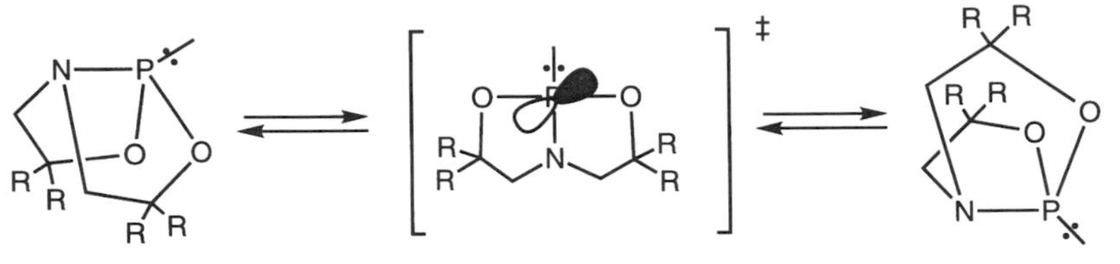

44a 44b 44a

Scheme 2.10

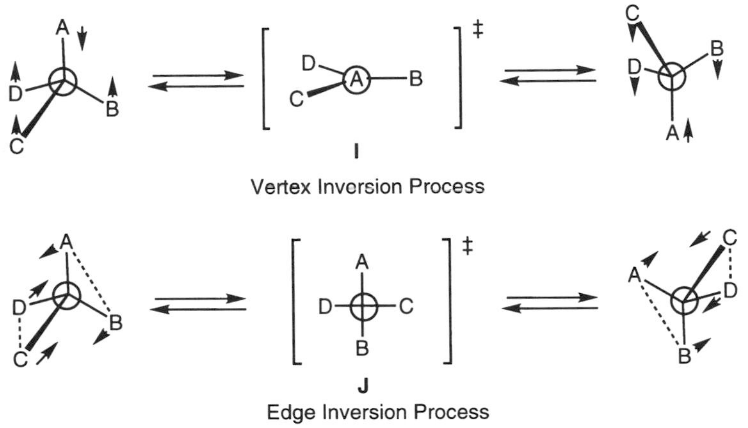

Vertex Inversion Process

Edge Inversion Process

Scheme 2.11

The energy barrier for edge inversion was calculated in detail by an *ab initio* method. Results predicted that edge inversion can be preferred to vertex inversion for higher row (heavy) main group element compounds. It is also concluded that the energy for edge inversion is decreased compared with that of vertex inversion according to (1) the increase of the size and of the electron-donating ability of the central atom, and (2) the decrease of the size and the increase of the electronegativity of substituents at the central atom (Table 2.2).[87,88,89]

Arduengo and co-workers experimentally showed that activation parameters for inversion of **44** (R, R = spiro(3,7)adamantyl) in toluene were $\Delta H^{\ddagger} = 23.4 \pm 1.6\,\text{kcal mol}^{-1}$, $\Delta S^{\ddagger} = -1.2 \pm 4.2$ eu, which should be intramolecular because the activation entropy was nearly zero. The value of $\Delta H^{\ddagger}$ is clearly less than the vertex inversion barrier of $34.4\,\text{kcal mol}^{-1}$ for PH_3.[85] The inversion of this 8-P-3 compound bearing a saturated tridentate ligand should take place by edge inversion through a T-shaped intermediate (or TS: **44b**) according to the molecular design (Scheme 2.10).

The possibility of edge inversion was investigated recently by using a less sterically restricted model where competition between edge inversion and

TABLE 2.2 Free Energy of Activation of Vertex Inversion and Edge Inversion Obtained by *Ab initio* Calculation (kcal mol^{-1})

	Vertex	Edge
PH_3	35.0,[a] 34.7[b]	15.9[a]
AsH_3	41.3,[a] 39.7[b]	142.2[a]
SbH_3	42.8,[a] 44.9[b]	112.2[a]
BiH_3	60.5,[b]	
PF_3	85.3,[a]	53.8,[a] 52.4[b]
AsF_3	66.3,[a]	46.3,[a] 45.7[b]
SbF_3	57.9,[a]	38.7,[a] 37.6[b]
BiF_3		33.5,[b]

[a]Reference 88 (MP2).
[b]Reference 89 (MP2).

vertex inversion can take place. Bismuth and antimony compounds (**45** and **46**) with a Martin ligand (hexafluorocumyl alcohol ligand) were prepared[90,91] (Scheme 2.12). There are two reasons to use a Martin ligand: (1) the ligand can stabilize organobismuth compounds, which are usually not so stable thermally; (2) the two CF_3 groups are anisochromous to the central atom. Thus, the inversion at the central atom can be observed as the exchange of the two CF_3 groups. However, the coalescence by the exchange of CF_3 group of bismuth compound (**45**) could not be detected by ^{19}F NMR in non-nucleophilic solvents up to 150°C.

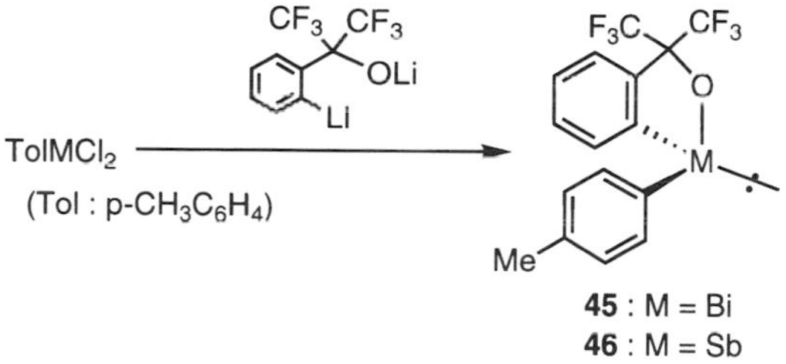

TolMCl$_2$ (Tol : p-CH$_3$C$_6$H$_4$) → **45** : M = Bi; **46** : M = Sb

Scheme 2.12

Another important aspect of edge inversion, that is, the appearance of vacant *p*-type orbital at the transition state perpendicularly to the molecular plane, was checked by the possibility of external nucleophile(s) to coordinate with the vacant *p*-type orbital of TS, thus lowering the energy of the TS (Scheme 2.13). Dixon and Arduengo predict that considerable stabilization of the TS will be obtained also by the second coordination of a nucleophile even after a large stabilization by the first one.[92]

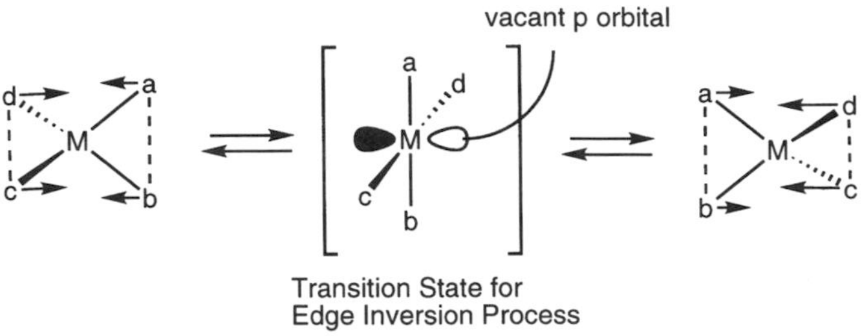

vacant p orbital

Transition State for
Edge Inversion Process

Scheme 2.13

Akiba and co-workers prepared bismuth (III) (**47** and **48**) and antimony (III) compounds (**49** and **50**), each bearing one or two dimethylaminomethyl groups in the molecule which can coordinate to the central atom (Figure 2.10). The structures of **47–50** were determined by X-ray crystallography and the rates of inversion were measured in several solvents. The energy of inversion in non-nucleophilic solvents decreased dramatically according to the increase of the number of intramolecular nucleophilic ligands (**45 > 47 > 48).** Also that in pyridine for **45** and **47** decreased significantly from that in non-nucleophilic solvents, but 2,6-dimethylpyridine could not show this accelerating effect as expected based on steric hindrance (Table 2.3).

The strong coordination of an intramolecular ligand and also of intermolecular nucleophilic solvents to the TS was supported by the fact that the activation entropies for the inversion of **45** and **47** were large minus values (Scheme 2.14).[90]

TABLE 2.3 Activation Parameters of Inversion of 45, 47, 48, and 50 in Various Solvents

Compound	Solvent	T_c(°C)	$\Delta G^{\ddagger}_{T_c}$ (kcal mol^{-1})	$\Delta H^{\ddagger}$ (kcal mol^{-1})	$\Delta S^{\ddagger}$ (eu)
45	*o*-Dichlorobenzene	> 175[a]	> 21		
	Pyridine-d_5	110[a]	18.0	9.0 ± 0.1	−23.5 ± 0.4
47	Toluene-d_8	125[b]	20.5	12.8 ± 0.4	−18.8 ± 1.0
	Pyridine-d_5	40[a]	14.6	7.1 ± 0.2	−23.6 ± 0.9
	2,6-Dimethylpyridine	170[a]	20.6	12.4 ± 0.2	−18.6 ± 0.5
48	Dichloromethane-d_2	< −90[a]	< 8		
50	Dichloromethane-d_2	−55[a]	9.4	9.3	−1.7
	Dichloromethane-d_2:pyridine-d_5 (9:1)	−36[a]	10.4		

[a] By exchange of CF_3.
[b] By exchange of CH_2.

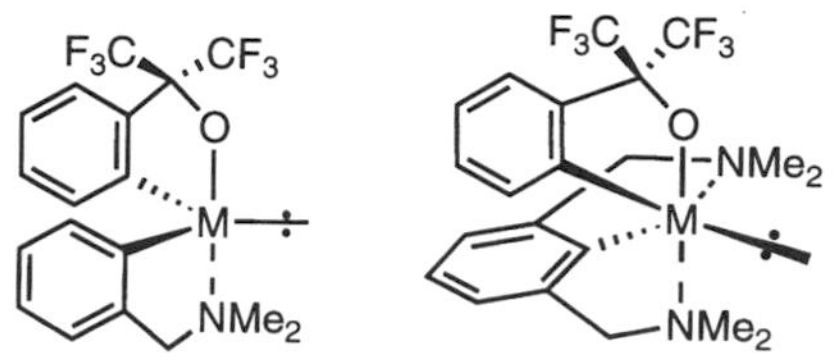

Figure 2.10 Coordinated Bi (III) and Sb (III) compounds for edge inversion.

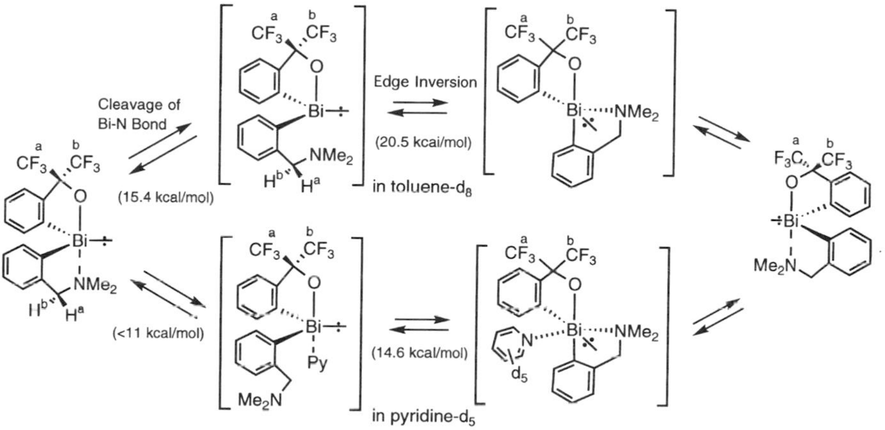

Scheme 2.14

The inversion of bismuth compound **48** was too rapid to measure by ^{19}F NMR even at −90°C; hence variable-temperature NMR measurements were carried out for the corresponding antimony compound **50**. As expected, the rates of inversion of **50** were much slower than those of **48**; however, the activation free energy ($\Delta G^{\ddagger}_{Tc}$) was very small, 9–10 kcal mol^{-1} (Table 2.3). Two facts for **50** are essentially different from those for **45** and **47**: that the activation entropy was nearly zero and that there was no effect of nucleophilic solvent (pyridine) on the rate. These strongly support the proposal that the inversion at the antimony of **50** is effected by the intramolecular coordination of the two dimethylamino groups to the vacant *p*-type orbital, thus generating a typical 3c–4e bond at the TS. The bond dissociation energy of the Sb—N bond of **50** could be obtained by the exchange of the two anisochromous methyl groups of NMe_2 to be 12.0 kcal mol^{-1}. This is clearly much larger than that of inversion; hence it is certain that the two Sb—N bonds are not dissociated at all during the inversion[91] (Scheme 2.15).

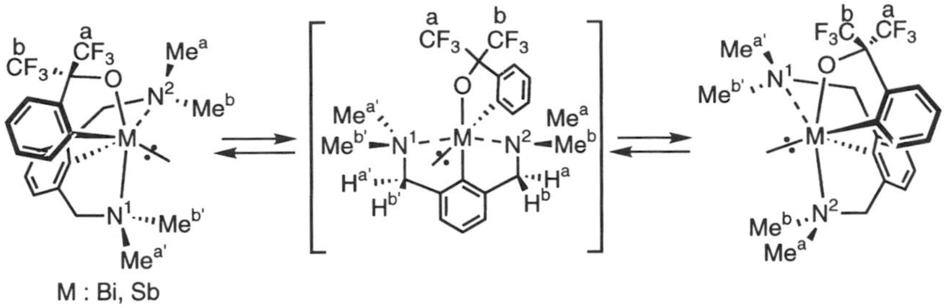

Scheme 2.15

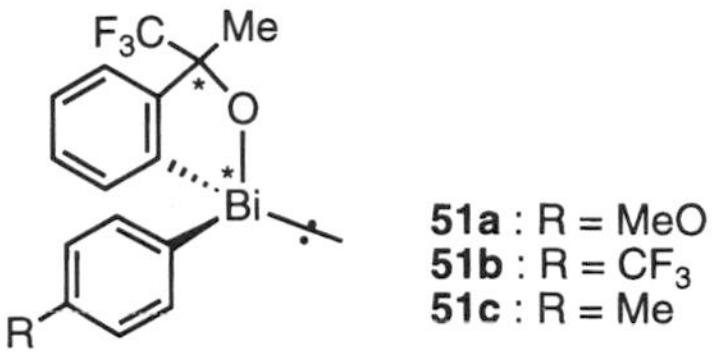

51a : R = MeO
51b : R = CF_3
51c : R = Me

Figure 2.11 Diastereomers of *p*-substituted phenylbismuth (III) compounds.

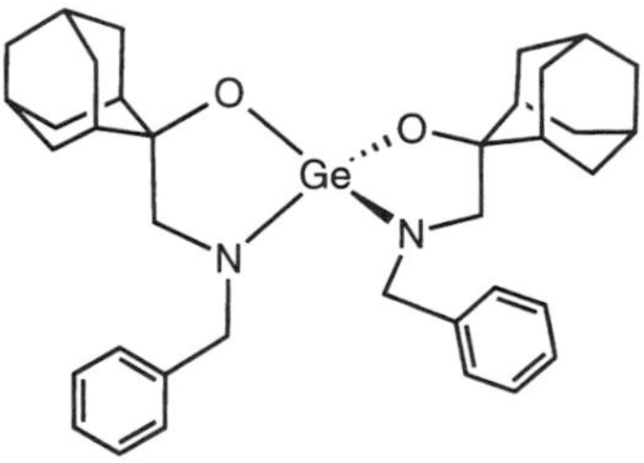

52

Figure 2.12 Germanium (IV) compound for edge inversion.

The essential results, that is, that the above-mentioned inversion was very much accelerated by the coordination of two nucleophilic substituents and that the rate of inversion of bismuth compound **48** was much faster than that of antimony compound **50**, are consistent with the prediction of edge inversion mechanism.

As the inversion of **45** was too slow to measure by the coalescence of the two CF_3 groups, one of the CF_3 groups was replaced by a CH_3 group in order to have a chiral carbon in the ligand (**51**). Thus diasteremers of **51** were separated and the barriers of inversion were obtained from the rates of equilibration of the diastereomers (Figure 2.11). The inversion rates of **51** in 1,2-dichloroethane were faster when the *p*-substituent of the benzene ring is π-donative or σ-attractive, compared to the methyl group (**51a** > **51b** > **51c**). This is also consistent with what is expected from the edge inversion mechanism.

Arduengo and co-workers have already investigated the inversion at the germanium atom of **52** and concluded that the inversion proceeds by an edge inversion mechanism[93] (Figure 2.12). Thus there is a good possibility for the edge inversion mechanism to have a very wide applicability.

2.3. FOUR-COORDINATE HYPERVALENT ORGANIC COMPOUNDS

Typical examples of four-coordinate hypervalent compounds are 10-X-4-type compounds. The fundamental structure of 10-X-4 is a pseudotrigonal bypyramid and the electronic state is shown by the valence bond method in Figure 2.13. One pair of unshared electrons and two M—A bonds (equatorial bond with sp^2 character) are in a plane and the 3c–4e bond is perpendicular to the plane, which consists of apical bonds in order to keep two ligands (B—M—B).

In order to explore the possibility of hypervalent bond formation between sulfur and nitrogen, formation of ammonio-sulfurane (10-S-4) was investigated using **53** and **54** as models (Figure 2.14). No detectable interaction could be observed between the sulfonium and the amino groups by ^{1}H NMR;[94] however, formation of a 3c–4e bond between the sulfur and the nitrogen was theoretically supported for the system of [SH_3^+ and $NH_{3]}$.[96b]

Dibenzothiazocine (**56**) was chosen as a skeleton to form a 3c 4e bond by 1,5-transannular interaction between the sulfur and the nitrogen, which should be induced by the oxidation of the sulfur. Compounds **9b–d** were prepared from the sulfoxide (**9a**). The conformation of the sulfide (**56**) was a mixture of BC (boat–chair) and TB (twist–boat) or BB (boat–boat)[95]; however, that of sulfoxide (**9a**) and sulfonium salts (**9b–d**) was confined only to TB (or BB); thus the formation of ammonio-sulfurane (**9a–d**) was verified. The S—N bond length became shorter according to the increase of electon-withdrawing ability of the substituent (X) at the sulfur (Table 2.4).[96–99]

The electronic effect of the substituent (X) was transmitted to the N-Me group according to the inductive effect of the substituent (X: σ_I) and it was found that the chemical shift of ^{1}II , ^{13}C and ^{15}N of the N-methyl group shifted down field according to σ_I value of the substituent (X). A coupling constant

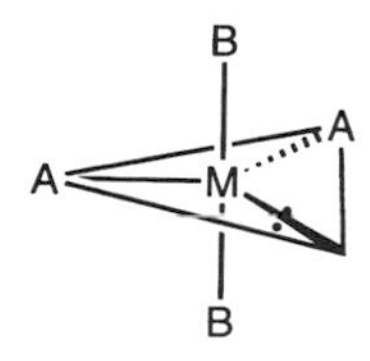

Figure 2.13 Description of 10-M-4 species.

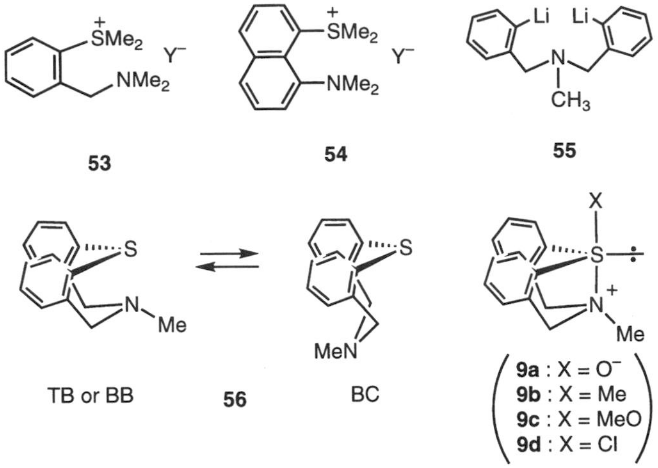

Figure 2.14 Direct interaction between the sulfonium and the amino groups: formation of a hypervalent bond.

TABLE 2.4 Bond Length of N—S and Chemical Shifts of ^{1}H, ^{13}C of N-Me of 9a–9d

Compound	X	N—S Bond length (Å)	Chemical Shift of N-Me ^{1}H NMR (δ, ppm)	(CD_3CN) ^{13}C NMR
9a	O^-	2.609		
9b	Me	2.466	2.53	39.8
9c	MeO	2.206	2.77	41.7
9d	Cl	2.091	3.13	43.8

($^{2}J^{13}C^{15}N = 4.8$ Hz; $\delta(^{15}N) = 23.7$) was determined between ^{15}N and $^{13}CH_3$-S in CD_3CN, thus showing the existence of bonding between the sulfur and the nitrogen even in the weak transannular interaction in **9b**. The rate of methyl transfer of **9b** to pyridine-d_5 was very much slower than that of diphenylmethylsulfonium salt (**57**); it was found to be about 1/1300[100,101] (Scheme 2.16). This is consistent with the above facts, which show that an electron is withdrawn from the nitrogen to the sulfur.

The tridentate ligand (**55**) is applicable to form and stabilize hypervalent compounds by transannular bond formation (**59, 60, 61**)[102–104] (Figure 2.15). Sulfur and oxygen are also used to effect transannular bond formation instead of nitrogen.[105–107]

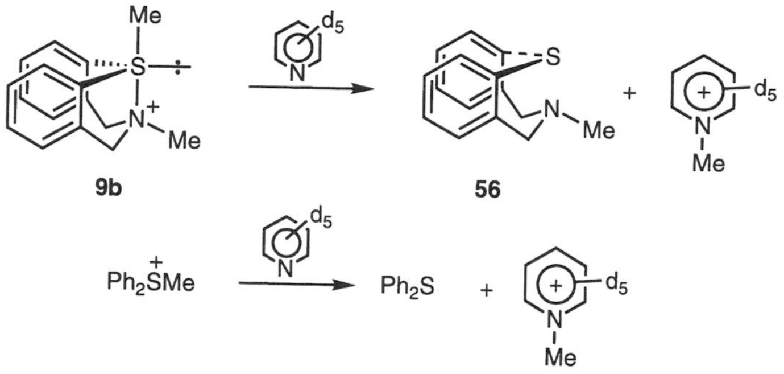

Scheme 2.16

Furukawa and Fujihara extended this kind of idea to Se and Te systems[108–110] (Figure 2.16). A variety of interesting multicenter interactions were found and will be described in Chapter 8.

The central atom of 10-M-4 can be optically active. This possibility had been recognized and the first synthesis of stable optically active sulfurane was reported by Martin and Balthazor in 1977. They prepared optically active chlorosulfurane **67** from optically active sulfoxide **66**[111] (Scheme 2.17).

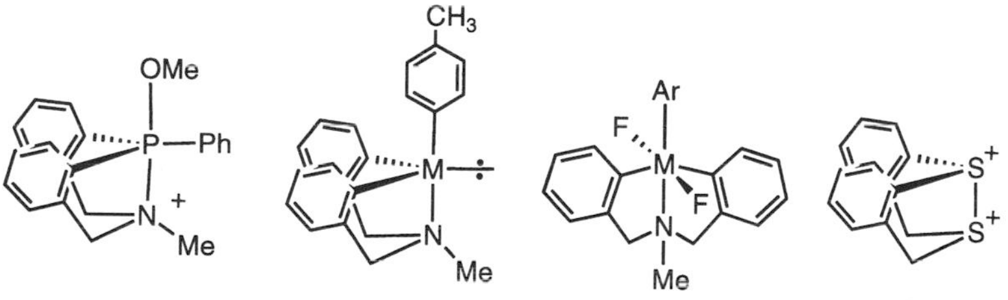

59 **60** M=Sb, Bi **61** M=Sb, Bi **62**

Figure 2.15 Formation of a hypervalent bond by transannular interaction for Group 15 elements.

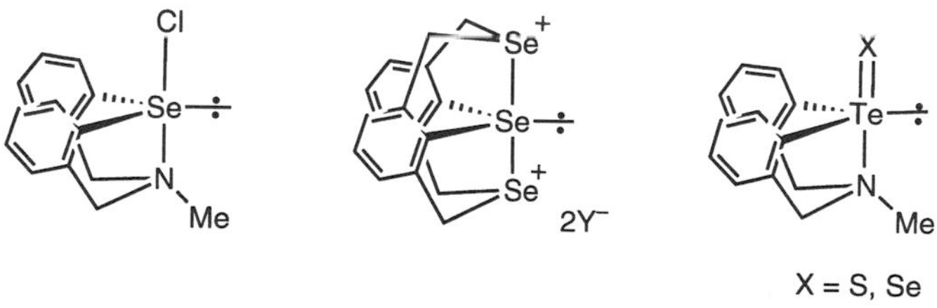

63 **64** **65**

Figure 2.16 Formation of a hypervalent bond by transannular interaction for Group 16 elements.

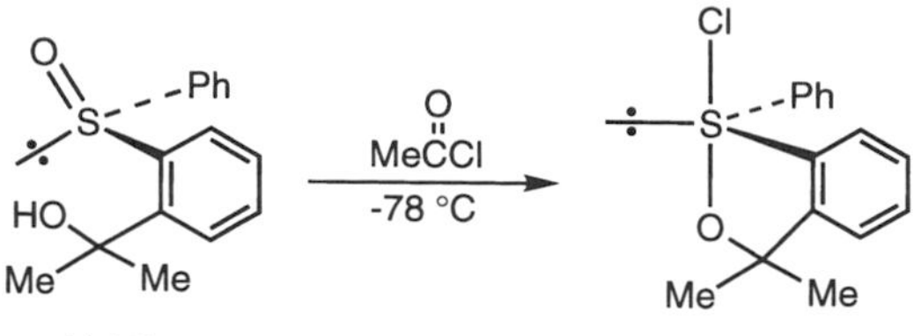

(-)-(S)-**66**

(+)-**67**

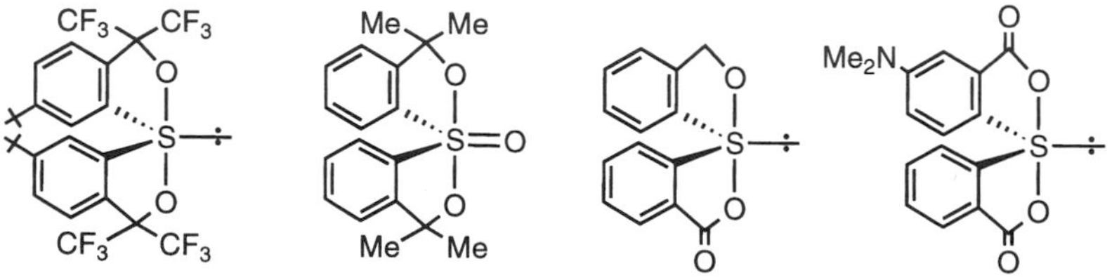

68 **69** **70** **71**

Scheme 2.17

Recently, optically active spirosulfuranes (**68–71**) were prepared by Drabowicz and co-workers[20,112,113] and by Kapovits.[114]

Another interesting aspect of 10-M-4 species is the ligand coupling reaction (LCR). Barton and co-workers reported ligand exchange reaction (LER) and LCR of tetraphenyltellurium in 1977.[115]

Oae's group reported LCR of 2-pyridylsulfurane generated *in situ* from the reaction of sulfoxide and Grignard reagents (Scheme 2.18). They showed that the LCR proceeded with retention of configuration of the migrating group and that the coupling took place between apical and equatorial ligands selectively.[116–118] These will find some synthetic applications.

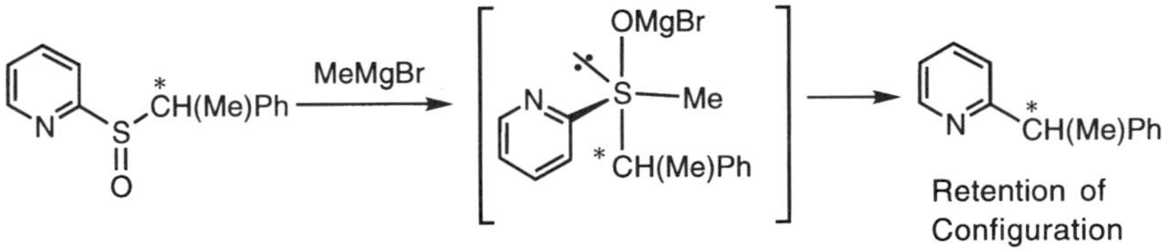

Retention of Configuration

Scheme 2.18

A recent *ab initio* calculation using MH_4 as model compounds (M = S, Se Te) predicted that LCR of these selectively occurs with apical–equatorial ligands through strongly ionic transition state.[119] More detailed discussions will be presented in Chapter 9.

2.4. FIVE-COORDINATE HYPERVALENT ORGANIC COMPOUNDS

The typical structure of five-coordinate hypervalent molecules is the trigonal bipyramid (TBP); there are also molecules with square pyramid (SP) or rectangular pyramid (RP) structures (Figure 2.17). The TBP structure is obtained by replacing a pair of unshared electrons of 10-X-4 with a substituent (ligand). Three M—A bonds are in a plane consisting of equatorial bonds (sp^2 type); the 3c–4e bond is perpendicular to the plane and consists of apical bonds, which are longer and weaker than equatorial ones.

In the SP structure, four bonds (basal bonds), connecting four substituents at the corners of a square and the central atom, are longer and weaker than the bond (apical bond) connecting a substituent at the top and the central atom.[120] That is, the apical bond in TBP and the basal bond in SP are hypervalent bonds (essentially 3c–4e bonds); thus electron-withdrawing substituents prefer apical and basal positions, respectively. In five-coordinate molecules, the TBP structure is the most popular and prevails over the others. The tendency of more electron-withdrawing substituents to prefer apical positions in TBP is called apicophilicity.

Holmes summarized factors leading to the preference for pyramidal structures (SP) for 10-P-5 phosphoranes, based on X-ray structural analyses of a number of phosphoranes.[33,121,122] According to his analysis, 10-P-5 compounds tend to approach a pyramidal structure when a spiro structure bearing two five- (or four-) membered rings is attained by using bidentate ligands that have the same kind of element (for example, oxygen) bonding to the phosphorus and also when the remaining monodentate ligand is sterically large and small in electronegativity.

There are examples of pyramidal structure even for acyclic compounds, which do not have the above-mentioned factors. The most well-known example is Ph_5Sb. The structure of Ph_5Sb was analyzed by X-ray crystallography in 1968 and found to be almost a square pyramid[123] and some discussions followed about the structure.[124] Then, the structure of Ph_5Sb containing 0.5 molecule of cyclohexane was analyzed and found to have the TBP structure.[125]

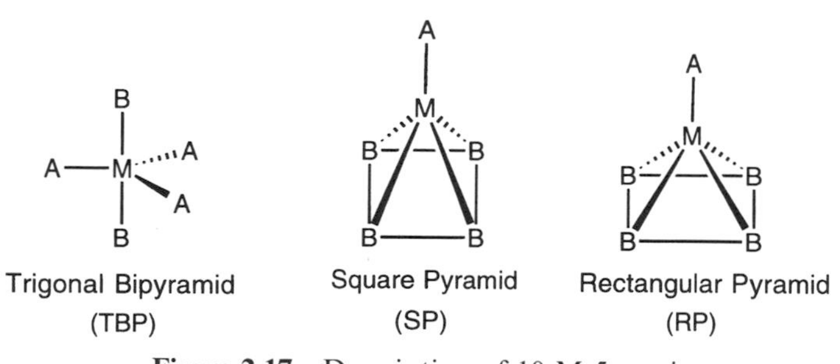

Figure 2.17 Description of 10-M-5 species.

There was no interaction between Ph_5Sb and cyclohexane; moreover, the structure of $(p\text{-}CH_3C_6H_4)_5Sb$ was reported to be TBP.[126] From these facts, the pyramidal structure of Ph_5Sb without cyclohexane can be understood as being due to the packing effect in crystals. This conclusion implies that the energy difference between TBP and pyramidal structures is very small.

The positional exchange between apical and equatorial in TBP molecules is explained as taking place through SP structure by the Berry pseudorotation mechanism[127] (Scheme 2.19). This exchange proceeds with very low activation energy. Two mechanisms of positional exchange (isomerization), namely Berry pseudorotation[127, 128] and turnstile rotation,[129 130] are shown in Scheme 2.19.

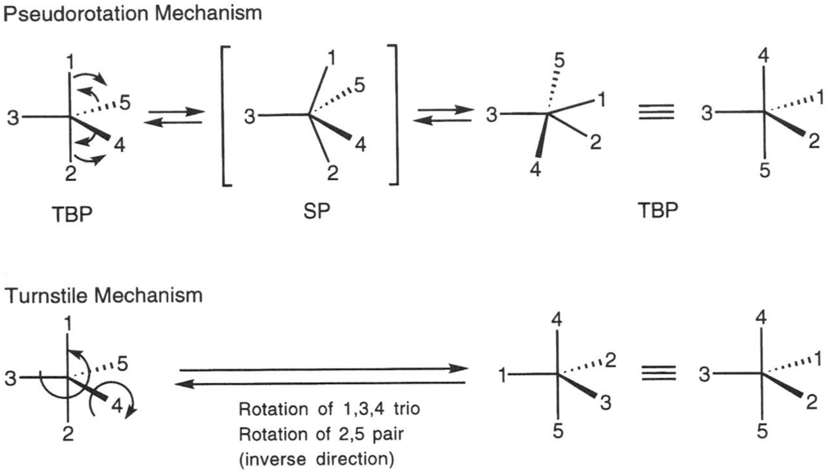

Scheme 2.19

Energy barriers of Berry pseudorotation have recently been calculated by a variety of methods. According to one of the recent *ab initio* calculations for neutral models of PH_5, AsH_5, SbH_5, and BiH_5, the barrier was 1.9, 2.1, 2.3, and 1.9 kcal mol^{-1}, respectively,[131] and for anionic models of PH_4^-, AsH_4^-, SbH_4^-, and BiH_4^-, it was 4.5, 6.1, 2.1, and 7.2 kcal mol^{-1}, respectively.[132] The pseudorotation barrier of SMe_4, $SeMe_4$, and $TeMe_4$ was also calculated to be 6.0, 5.9, and 0.5 kcal mol^{-1}, respectively,[133,134] and MH_4 (M = S, Se, Te) is said to be a little bit more stable in pseudosquare (SP) than in TBP structures.[135,136] The pseudorotation barrier for acyclic 10-M-5 species should be very low, as mentioned above; however, the pseudorotation path can be divided into a low-energy path and a high-energy path by attaching a five-membered ring, using an appropriate bidentate ligand (Scheme 2.20). Monodentate ligands can interconvert their relative positions by two pseudorotations in which the five-membered ring can stay on the apical–equatorial plane (low-energy path). In order to invert the chirality of the central atom, it is mandatory to pass through an intermediate (or TS) where the five-membered ring comes onto the equatorial–equatorial plane. In this conformation, the angle of the five-membered ring should be enlarged from 90° to 120° (in a typical case), thus becoming energetically unfavorable; hence the pseudorotation path involving equatorial–equatorial TS for a five-membered ring is a high-energy path. Actually, the tolyl groups of monocyclic antimony compound **73** isomerize very rapidly by a low-energy path, and two kinds of methyl groups (two equatorial and one apical) were observed at −80°C (Scheme 2.20). The energy barrier was obtained as 9.1 kcal mol^{-1}.[137] However, the exchange of the two CF_3 groups of **72** could not be observed even at 150°C and the barrier should be larger than 21 kcal mol^{-1}.[138]

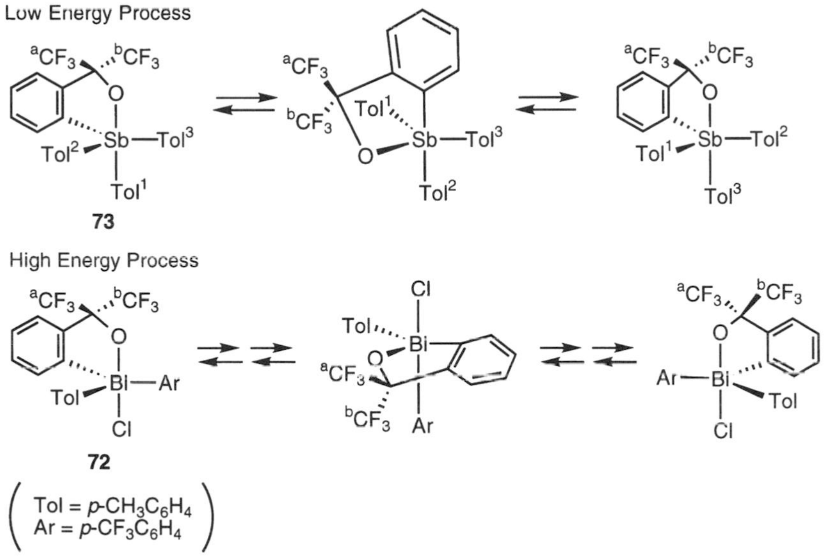

Scheme 2.20

The energy barriers of inversion of 10-M-5 of group 15 elements bearing two biphenylene rings (**74–76**) were experimentally obtained as 15.8 for phosphorane, 15.4 for arsorane, and 11.6 kcal mol^{-1} for stiborane, thus showing that the barrier decreases according to the increase of the row of the element of the same group[139–141] (Figure 2.18).

Recently, it has been reported that the chirality of spirocyclic 10-P-5 compounds can remain stable even at room temperature by employing two Martin ligands. The diastereomer **77** having an *l*-menthyl group as a chiral auxiliary was separated easily by recrystallization from methanol at room temperature. After determining the exact structure of both diastereomers by X-ray analysis, each isomer was reduced to a pair of enantiomers **78**,which have chirality only at the phosphorus and are stable at room temperature (Scheme 2.21). Chiral purity (100%) of the enantiomer was checked by converting each to a Mosher ester.[142]

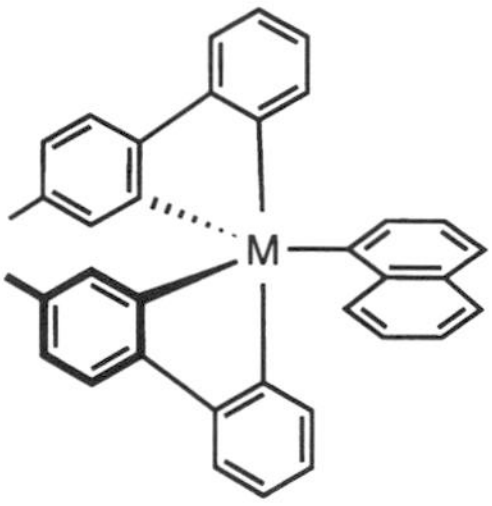

74 : M = P
75 : M = As
76 : M = Sb

Figure 2.18 Optically active 10-M-5 species.

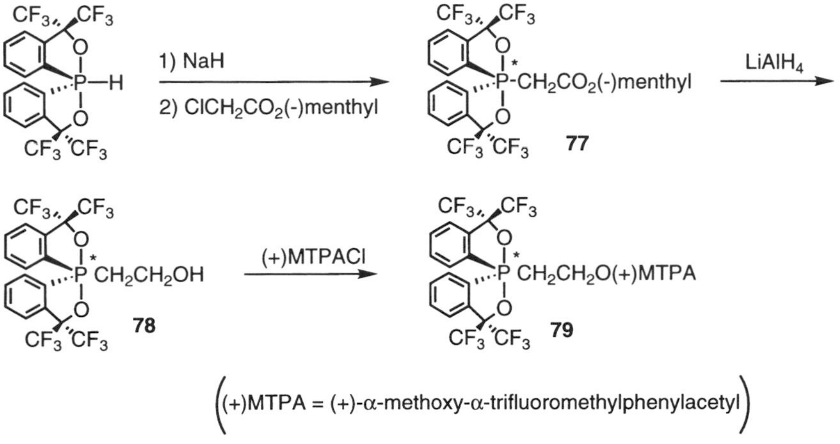

1) NaH
2) $ClCH_2CO_2$(-)menthyl
P-CH_2CO_2(-)menthyl
77
$LiAlH_4$
CH_2CH_2OH
78
(+)MTPACl
CH_2CH_2O(+)MTPA
79

((+)MTPA = (+)-α-methoxy-α-trifluoromethylphenylacetyl)

Scheme 2.21

Similarly, diastereomers (**80**) were separated and the structure of each was determined by X-ray crystallography. By treating each **80** with methyllithium, a pair of enantimers of 10-P-5 hydrophosphorane (**81**) was obtained pure at room temperature (Scheme 2.22). **81** has chirality only at the phosphorus. The alkylation of the hydrophosphorane in the presence of DBU proceeded with complete retention of the stereochemistry.[143]

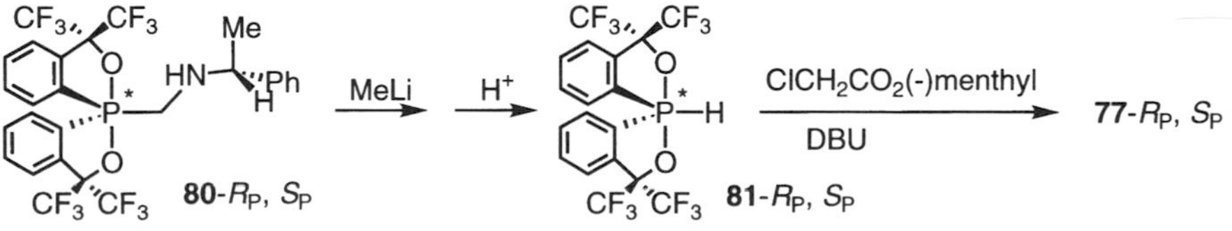

80-R_P, S_P
MeLi
H^+
81-R_P, S_P
$ClCH_2CO_2$(-)menthyl
DBU
77-R_P, S_P

Scheme 2.22

Furthermore, stereochemically stable diastereomers of 10-P-5 (**82**) and 10-Sb-5 (**83**) were isolated by using a modified Martin ligand which has one methyl group instead of CF_3[144,145a] (Figure 2.19). The energy barrier of isomerization between diastereomers of each **83** (high-energy path of pseudorotation) was measured and the results show that the barrier becomes lower when the monodentate substituent becomes more electronegative (Table 2.5).

This is interpreted as shown in Scheme 2.23. Intermediate K is the highest energy intermediate to effect the inversion at the antimony atom; hence intermediate K is more stabilized and the barrier is lowered according to the increase of apicophilicity of the monodentate substituent R. It is also shown that the apicophilicity of a transition metal ligand such as $FeCp(CO)_2$ **84** is much smaller than that of monodentate substituent R in **83** (Table 2.5).[146]

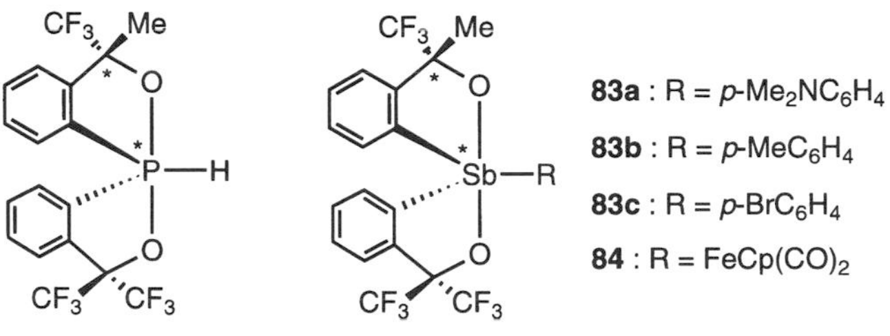

82

83a : R = *p*-$Me_2NC_6H_4$
83b : R = *p*-MeC_6H_4
83c : R = *p*-BrC_6H_4
84 : R = $FeCp(CO)_2$

Figure 2.19 Diastereomers of 10-M-5 species.

TABLE 2.5 Activation Parameters of Isomerization of 83 in *p*-Xylene (from Minor to Major)

Compound	R	$\Delta G^{\ddagger}_{298}$ (kcal mol^{-1})	$\Delta H^{\ddagger}$ (kcal mol^{-1})	$\Delta S^{\ddagger}$ (eu)
83a	*p*-$Me_2NC_6H_4$	28.2	26.1 ± 0.3	−7.0 ± 0.9
83b	*p*-MeC_6H_4	27.7	25.0 ± 1.0	−8.3 ± 2.9
83c	*p*-BrC_6H_4	27.2	24.6 ± 6.5	−8.4 ± 18.5
84	$FeCp(CO)_2$	30.3	29.1	−3.1

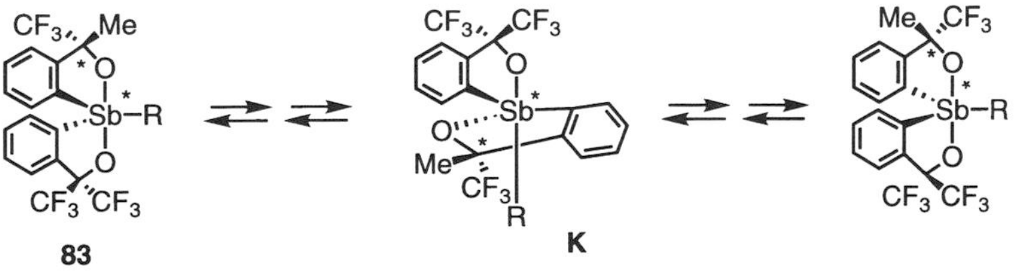

83 **K**

R: **a** = *p*-$Me_2NC_6H_4$, **b** = *p*-MeC_6H_4, **c** = *p*-BrC_6H_4, **84** = $FeCp(Co)_2$

Scheme 2.23

The rate of inversion of **83** was found to be very much accelerated by donor solvents like pyridine and *N*,*N*-dimethylaminoacetamide (Table 2.6) and also by the presence of intramolecular coordinating ligands (**85–87**). This can be interpreted as shown in Scheme 2.24. The highest energy intermediate L (or TS) for the inversion in the absence of any donor is stabilized by the coordination of a donor, because the angle strain of equatorial–equatorial five-membered ring in L can be relieved by the possible change of the angle of five-membered ligand from 120° to 90° in M. The activation free energy of the isomerization of **85–87** decreases from 26.5 to 21.6 kcal mol^{-1} by intramolecular coordination. The accelerating effect of donors on stereoisomerization of 10-M-5 species is a novel and interesting phenomenon.[145a, 145b]

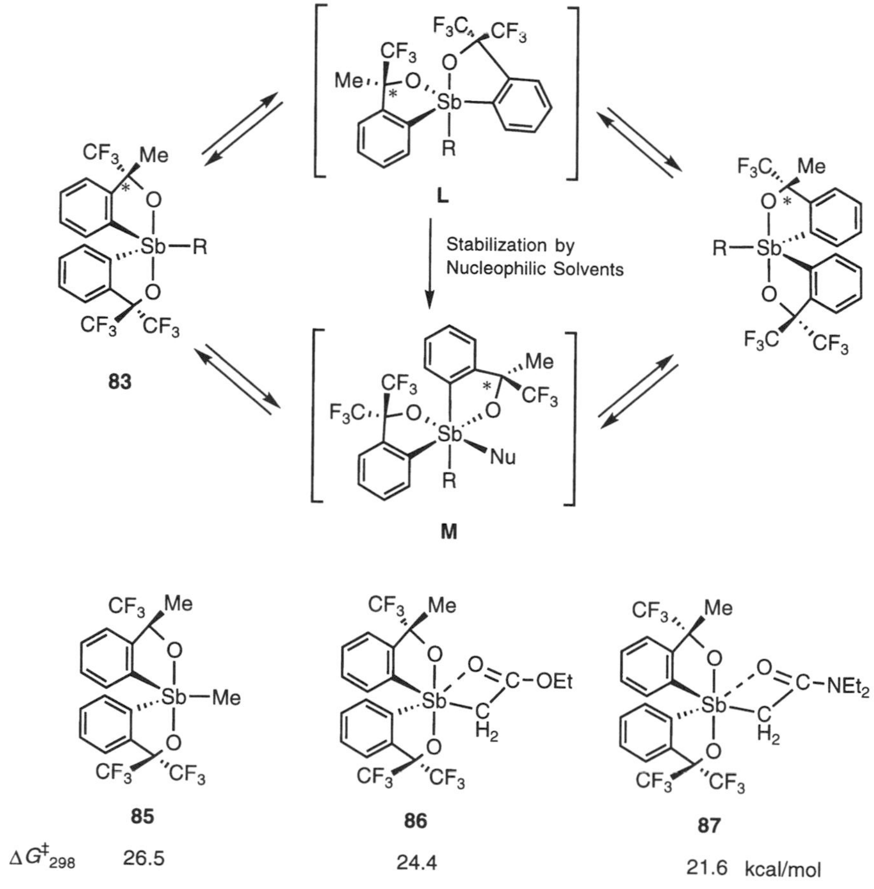

Scheme 2.24

TABLE 2.6 Solvent Effect of Isomerization of 83b (from Minor to Major)

Solvent	E_T^N	DN	$\Delta G^{\ddagger}_{298}$ (kcal mol^{-1})	$\Delta H^{\ddagger}$ (kcal mol^{-1})	$\Delta S^{\ddagger}$ (eu)
n-Octane	0.012	0.00	27.9	26.2 ± 2.0	-6.3 ± 5.6
p-Xylene	0.074	(0.13)	27.7	25.0 ± 1.0	-8.3 ± 2.9
1,2-Dichloroethane	0.327	0.00	26.8	23.6 ± 0.5	-11.1 ± 1.5
THF	0.207	0.52	25.7	21.8 ± 0.9	-13.4 ± 2.8
Pyridine	0.302	0.85	24.3	17.1 ± 1.0	-24.1 ± 3.1
Dimethylacetamide	0.401	0.72	23.0	18.1 ± 1.3	16.4 ± 4.1

It has recently been reported that a 10-P-5 spirophosphorane with an apical-carbon–equatorial-oxygen ring (*cis* isomer: **88**) was isolated and characterized together with the corresponding apical-oxygen–equatorial-carbon ring stereoisomer (*trans* isomer: **89**). The activation enthalpy of inversion at the phosphorus of **88**, that is, one-step Berry psudorotation, $\Psi_{n\text{-butyl}}$, was measured to be 10 kcal mol^{-1} and that for conversion from **88** to **89** was about 21.8 kcal mol^{-1} (Scheme 2.25).[147]

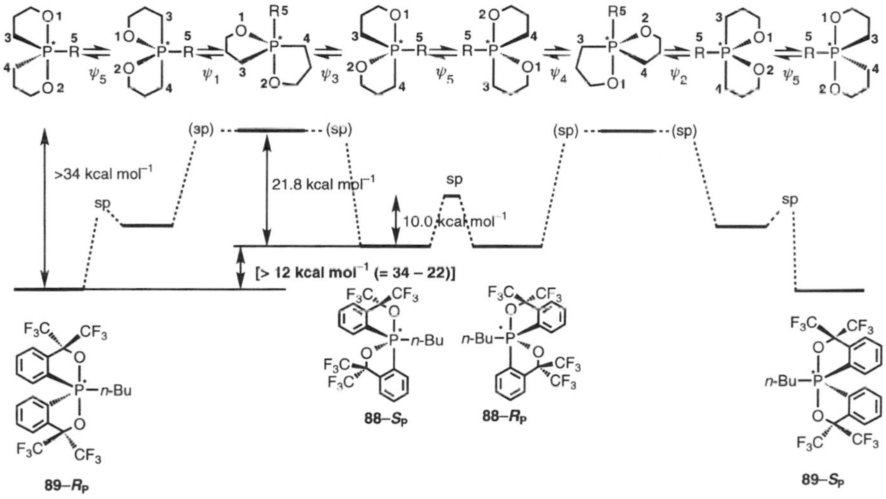

Scheme 2.25

One of the most fundamental and important reactions of 10-M-5 compounds is the ligand-coupling reaction (LCR), that is, a thermal reductive coupling reaction like $PH_5 \rightarrow PH_3 + H_2$, which is formally analogous to reductive elimination of organotransition metal compounds. A theoretical calculation on this problem using the above-mentioned model was first made by Hoffmann in 1972,[120] and was followed by some advanced calculations.[148,149] The essential conclusion obtained for TBP molecules was that apical–apical or equatorial–equatorial coupling is symmetry-allowed and apical–equatorial coupling is symmetry-forbidden. For an SP molecule, apical–basal or *trans*–basal coupling is allowed but *cis*–basal coupling is forbidden. Recently, theoretical investigations on the selectivity of the LCR by advanced *ab initio* methods were reported on 10-M-5 species using MH_5 as a model (M = P, As, Sb, Bi).[131,150a,b] According to them, for MH_5 of P, As, and Sb, equatorial–equatorial coupling is energetically favored, but for BiH_5, apical–equatorial coupling (which is symmetry-forbidden) is energetically favored.

Experimental investigations on this fundamental problem have been quite scarce.[151–153] This is because it is very difficult to find suitable systems that meet the following criteria: (1) the reaction should be concerted and irreversible; (2) it is mandatory to prepare pure acyclic 10-M-5 compounds bearing electronically different but sterically similar (preferably almost identical) substituents. Furthermore, there is inevitably rapid intramolecular pseudorotation, which should proceed with much less activation energy than the LCR; ligand-exchange reactions (LERs) among those 10-M-5 species may take place before the LCR occurs. Experimental investigations on LCRs will be described in Chapter 9 including LCRs of $Tol_nAr_{5-n}Sb$. It is proposed that apical–apical coupling is a favored process obtained by the memory effect[152] (Scheme 2.26). There is still much to be done in order to understand correctly the LCR mechanism that actually occurs in 10-M-4 and 10-M-5 species.

The LCR of 10-Bi-5, especially Ph_3BiXY, has been shown to be synthetically useful by Barton and Finet, where copper reagents like $Cu(acac)_2$ are used as catalysts. Phenylation of enolate type carbon by this method is a valuable one.[154]

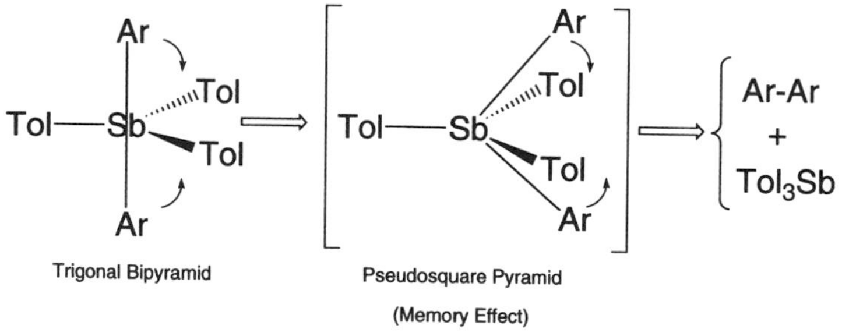

Trigonal Bipyramid

Pseudosquare Pyramid (Memory Effect)

$Ar\text{-}Ar + Tol_3Sb$

Scheme 2.26

2.5 SIX-COORDINATE HYPERVALENT ORGANIC COMPOUNDS

The typical structure for six-coordinate hypervalent molecules is octahedral (Oh). According to Musher's classification, six-coordinate compounds belong to hypervalent II type. The six bonds are equivalent and these are composed of three equivalent 3c–4e bonds; each 3c–4e bond should be stronger than that of an apical bond of the corresponding TBP species.[155, 156] As an example, let us compare the structures of Ph_4Te and Ph_6Te. For Ph_4Te, the average bond length of equatorial bonds is 2.13 Å and that for apical bonds is 2.30 Å.[157] For Ph_6Te, the average bond length is 2.228 Å, each is almost identical; for Ar_6Te ($Ar = p\text{-}CF_3C_6H_4$), the average C—Te length is 2.232 Å.[158] A ligand coupling reaction of Ph_4Te takes place to give Ph_2Te and Ph_2 at ca. 100°C in benzene,[115] but Ph_6Te is thermally stable up to the melting point (312°C).

Ab initio calculation of bond dissociation energy of $Te(CH_3)_n$ ($n = 2, 4, 6$) was carried out recently. It was shown that BDE of the apical bond of Te(CH3)4 is the weakest as expected, but surprisingly, that of $Te(CH_3)_6$ is stronger than $Te(CH_3)_2$[159] (Table 2.7).

Although the six bonds in the Oh structure should be equivalent, a considerable electronic effect of one substituent (ligand) to the one in its *trans* position can be expected because the fundamental nature of the bond is based on the 3c–4e bond. When ArylLi ($Ar = p\text{-}CF_3C_6H)_4$ is added to five-coordinate antimony compound bearing a Martin ligand (**90**), a six-coordinate complex (**91**) is obtained quantitatively in solution (Scheme 2.27).[42] At low temperature, **91a** was obtained selectively, probably effected by coordination of the lithium cation to the oxygen of the ligand. When TolLi was added to **92** under the same conditions, a mixture of **91a–c** was obtained at the beginning, where **91c** was the major species. The ratio of **91c** to (**91a** + **91b**), 70:30, should correspond to the equilibrium ratio of **92a** to **92b** under very rapid Berry pseudorotation, reflecting the apicophilicity of the Ar group compared to that of the Tol group.

TABLE 2.7 Calculated Bond Lengths and Bond Dissociation Energies of $Te(CH_3)_n$ ($n = 2, 4, 6$)

	C–Te Bond Length (Å)	Bond Dissociation Energy (kcal mol^{-1})	
		MP2/ECP	QCISDT(T)/ECP
$Te(CH_3)_2$	2.15 [2.142(5)][a]	58.1	55.8
$Te(CH_3)_4$	2.15 [2.140(8)][a] eq 2.25 [2.265(9)][a] ap	46.2[b]	42.0[b]
$Te(CH_3)_6$	2.18 [2.19(3)][a]	61.6	63.1

[a]Values in parentheses are values observed from the literature.
[b]BDE should be that of apical bond.
Source: K.-y. Akiba, section lecture at ICCST-7 at Aachen, July, 1997; M. Minoura, N. Yamashita, and K.-y. Akiba, unpublished results.

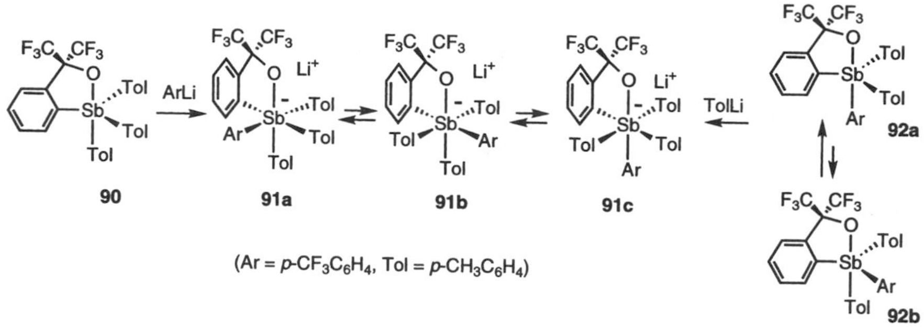

(Ar = p-$CF_3C_6H_4$, Tol = p-$CH_3C_6H_4$)

Scheme 2.27

When a mixture of **91** obtained by each method was allowed to stand at room temperature for equilibration, it became apparent that the ratio of **91c** (6–11) in equilibrium was shown to be much less than that expected by statistical ratio (**91a**:**91b**:**91c** = 50:25:25), although complete equilibrium was not attained (Tables 2.8 and 2.9).

This can be explained if an electron-withdrawing substituent is not preferred to the *trans* position of the oxygen in Martin ligand because the oxygen is strongly electronegative. This was supported by *ab initio* calculation on simplified models (**93, 94**) that a hydrogen *trans* to fluorine or oxygen bears a smaller negative charge compared to what *cis* hydrogens bear (Figure 2.20).

TABLE 2.8 Isomerization of 91 Prepared from 90 and ArLi *In Situ* (25°C)

	Time				
91	10 min	3 h	17 h	36 h	66 h
91a	98	94	90	83	74
91b	1	5	8	12	20
91c		1	2	5	6

TABLE 2.9 Isomerization of 91 Prepared from 92 and TolLi *In Situ* (25°C)

	Time				
91	0 min	5 min	32 min	8 h	66 h
91a	17	13	24	31	51
91b	13	27	37	45	38
91c	70	60	39	24	11

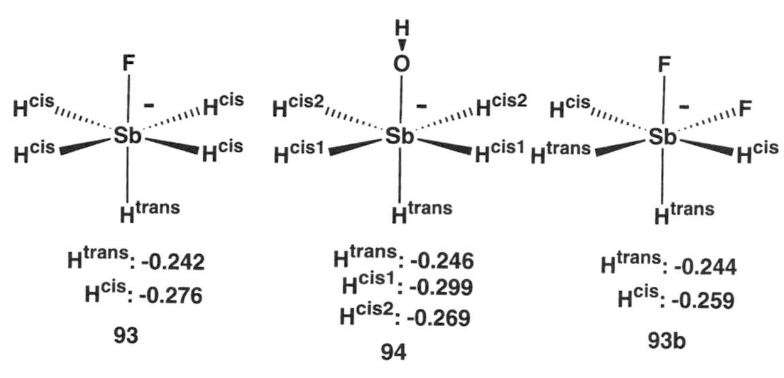

Figure 2.20 Electron density of hydrogen atoms of SbH_5F^-, $SbH_5(OH)^-$, $SbH_4F_2^-$ (by calculation).

Rapid equilibration was observed for the ate complex (**23**). The mechanism of isomerization of **23** was determined as follows: (1) the rate of isomerization (1.9×10^{-4} sec^{-1} at $-20°C$ was 10 times faster than the rate (1.8×10^{-5} $L\,mol^{-1}$) of protonolysis by ethanol ($1\ mol\ L^{-1}$); (2) the rate of isomerization is greatly retarded by addition of crown-15 and HMPA. Accordingly, it was concluded that the isomerization proceeds by the cleavage of the Sb—O bond induced by a lithium cation, followed by pseudorotation of 10-Sb-5 generated *in situ*, and does not contain the cleavage of the Sb—C bond generating TolLi *in situ*.[42]

There are only a few examples of investigation of the isomerization mechanism of six-coordinate compounds. It was shown that six-coordinate difluorotellurorane (**95-*trans***) isomerized to **95-*cis*** by a nondissociative intramolecular mechanism[160] (Scheme 2.28). However, six-coordinate difluorosulfurane (**96**) was stable for 254 hours at 238°C under basic conditions but it isomerized rapidly by addition of a catalytic amount of a strong Lewis acid (SbF_5), which cleaved one of the S—F bonds, generating a cationic intermediate.[161] The size and the charge of the central atom are dominating factors in determining the mechanism of isomerization.

Another interesting aspect of six-coordinate compounds is the formation of porphyrins. Porphrins can be designated as 12-M-6 and have been known for Group 14 elements as neutral molecules, for Group 15 elements as monocationic molecules; however, Group 16 and 17 elements have not been incorporated into porphyrin rings yet.

Group 14 element porphyrins (**97**) of Si, Ge, and Sn have been prepared and their photochemical trapping processes of oxygen are established, but they are not usually stable for handling in air.[162] Group 15 element porphyrins were prepared for phosphorus as the smallest element that can be incorporated into the ring,[163] arsenic, to settle the controversy concerning on its preparation,[164]

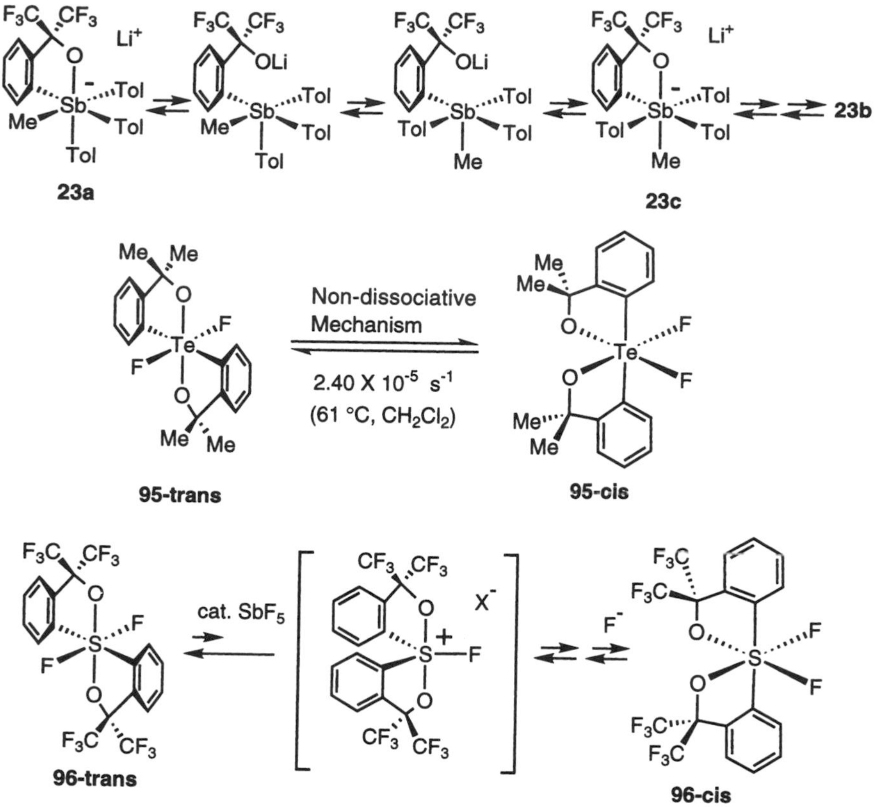

Scheme 2.28

and antimony, as metalloidal porphyrins.[165] For this purpose, synthetic methods were developed recently to prepare a variety of Group 15 porphyrins (**98**) which bear carbon–element bond(s) (Figure 2.21).

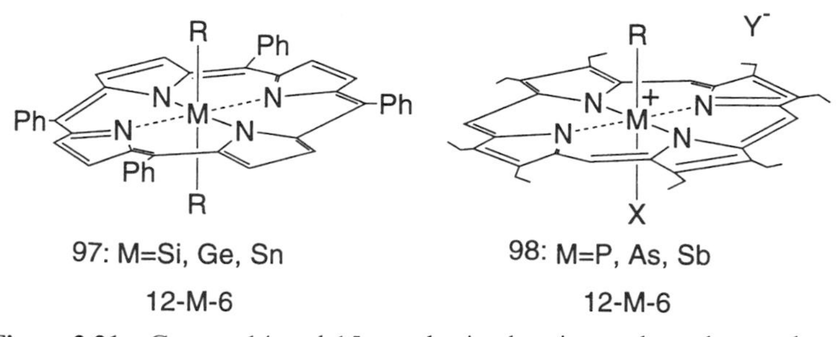

Figure 2.21 Groups 14 and 15 porphyrins bearing carbon-element bonds.

2.6 SEVEN- AND EIGHT-COORDINATE HYPERVALENT COMPOUNDS

Examples of seven-coordinate and eight-coordinate compounds consisting of sigma bonds are limited; however, Seppelt and Christe have been exploring this region of chemistry independently. A summary of recent research is given briefly (see Figures 2.22 and 2.23).

The structure of seven-coordinate compounds has been disclosed to be a pentagonal bipyramid. It consists of five bonds forming a plane pentagon (equatorial bonds) and two bonds perpendicular to the plane (apical bonds) (Figure 2.22). It is shown that the equatorial bond is longer and weaker than the apical bond, which is in contrast to the character of pentacoordinate compounds. An apical bond has a strong character of a covalent bond using the hybridized *sp* orbital of the central atom; equatorial bonds are of six-center ten-electron bond types and the electron densities of equatorial bond positions are larger than the apical positions.[57]

Accordingly, a pair of unshared electrons and an oxide anion are selectively placed at apical positions and electron-withdrawing substituents like fluorine are placed selectively at equatorial positions (**20** of Figure 2.4, **26** of Figure 2.5, **27** and **29** of Figure 2.6). When the central atom is not large enough to hold seven ligands, the formally residual one pair of unshared electrons becomes stereochemically inactive. For example, the unshared pair of electrons of **25** of Figure 2.5, ClF_6^-,[48] and BrF_6^-,[49] are stereochemically inactive but that of IF_6^- [50] is stereochemically active.

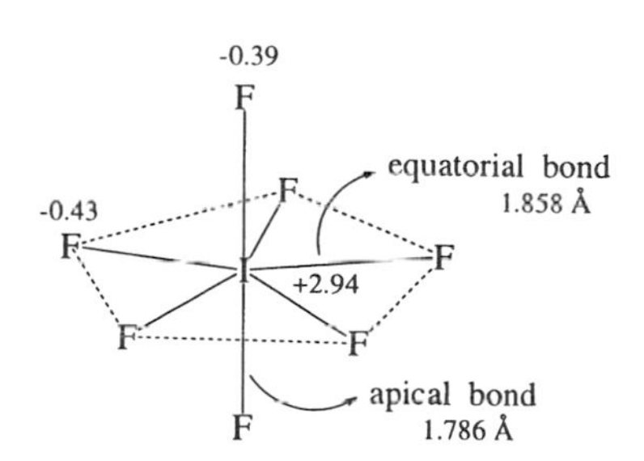

Figure 2.22 Bond distance and electron density of IF_7 (by calculation).

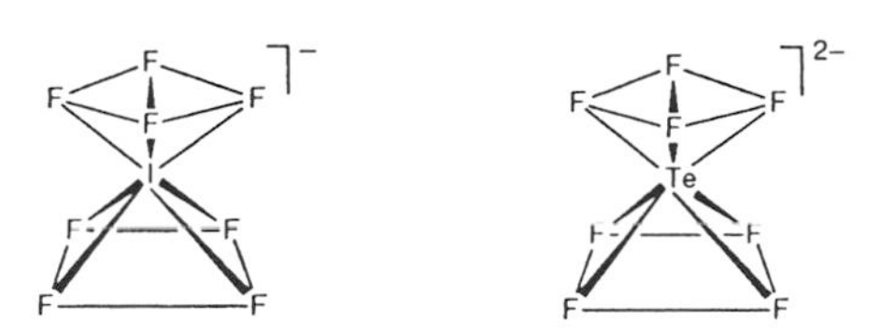

Figure 2.23 Structure of IF_8^- and TeF_8^{2-}.

Examples of eight-coordinate molecules are quite rare; however, structures of TeF_8^- and IF_8^{2-} have been reported as antiprism structures[58,60,61] (Figure 2.23).

REFERENCES

1. P. A. Cahill, C. E. Dykstra, J. C. Martin, *J. Am. Chem. Soc.*, **107**, 6359 (1985).
2. B. S. Ault, A. L. Andrews, *Inorg. Chem.*, **16**, 2024 (1977).
3. R. Filler, *Isr. J. Chem.*, **17**, 71 (1978).
4. A. J. Arduengo, III, C. A. Stewart, F. Davidson, D. A. Dixon, J. Y. Becker, S. A. Culley, M. B. Mizen, *J. Am. Chem. Soc.*, **109**, 627 (1987).
5. A. J. Arduengo, III, D. A. Dixon, *Heteroatom Chemistry*, E. Block, Ed. (VCH, New York, 1990), Chapter 3, p. 47.
6. S. A. Culley, A. J. Arduengo, III, *J. Am. Chem. Soc.*, **106**, 1164 (1984).
7. S. A. Culley, A. J. Arduengo, III, *J. Am. Chem. Soc.*, **107**, 1089 (1985).
8. C. A. Stewart, R. L. Harlow, A. J. Arduengo, III, *J. Am. Chem. Soc.*, **107**, 5543 (1985).
9. Y. Yamamoto, K.-y. Akiba, *Heterocycles*, **13**, 297 (1979).
10. F. Iwasaki, S. Yoshida, S. Kakume, T. Watanabe, M. Yasui, *J. Mol. Struct.*, **352/353**, 203 (1995).

For reviews on 10-S-3 sulfuranes [11,12]:

11. R. Gleiter, R. Gygax, *Top. Curr. Chem.*, **63**, 49 (1976).
12. N. Lozac'h, *Adv. Heterocycl. Chem.*, **13**, 161 (1971).
13. T. T. Nguyen, S. R. Wilson, J. C. Martin, *J. Am. Chem. Soc.*, **108**, 3803 (1986).
14. T. T. Nguyen, J. C. Martin, *J. Am. Chem. Soc.*, **102**, 7382 (1980).
15. X. Chen, Y. Yamamoto, K.-y. Akiba, *Heteroatom Chem.*, **6**, 293 (1995).
16. X. Chen, Y. Yamamoto, K.-y. Akiba, S. Yoshida, M. Yasui, F. Iwasaki, *Tetrahedron Lett.*, **33**, 6653 (1992).

For a review on 10-P-4 phosphoranide anions [17]:

17. K. B. Dillon, *Chem. Rev.*, **94**, 1441 (1994).
18. K.-y. Akiba, K. Takee, Y. Shimizu, K. Ohkata, *J. Am. Chem. Soc.*, **108**, 6320 (1986).
19. K.-y. Akiba, K. Takee, K. Ohkata, F. Iwasaki, *J. Am. Chem. Soc.*, **105**, 6955 (1983).

For reviews on 10-S-4 sulfuranes [20,21]:

20. J. Drabowicz, P. Lyzwa, M. Mikolajczyk, *Supplement S: The Chemistry of Sulphur-Containing Functional Groups*, S. Patai, Z. Rapppoport, Eds (John Wiley & Sons, 1993), Chapter 15, p. 799.
21. R. A. Hayes, J. C. Martin, *Organic Sulfur Chemistry. Theoretical and Experimental Advances*, F. Bernardi, I. G. Csizmadia, A. Mangani, Eds (Elsevier, 1985), Chapter 8, p. 408.
22. D. B. Dess, J. C. Martin, *J. Am. Chem. Soc.*, **104**, 902 (1982).

For a review on hypervalent iodinanes [23]:

23. T. T. Nguyen, J. C. Martin, *Comprehensive Heterocyclic Chemistry*, (1984) Vol.1, Chapter 1.19, p. 563.
24. S. W. Peterson, R. D. Willett, J. L. Houston, *J. Chem. Phys.*, **59**, 453 (1973).
25. N. Bartlett, F. O. Sladky, *J. Am. Chem. Soc.*, **90**, 5316 (1968).
26. D. Y. Lee, J. C. Martin, *J. Am. Chem. Soc.*, **106**, 5745 (1984).
27. T. R. Forbus, J. C. Martin, *J. Am. Chem. Soc.*, **101**, 5057 (1979).
28. T. R. Forbus, J. C. Martin, *Heteroatom Chem.*, **4**, 113 (1993).
29. T. R. Forbus, J. C. Martin, *Heteroatom Chem.*, **4**, 129 (1993).
30. T. R. Forbus, J. C. Martin, *Heteroatom Chem.*, **4**, 137 (1993).
31. K.-y. AKiba, G. Schroder, Y. Mimura, M. Watanabe, A. Hasuoka, Y. Yamamoto, unpublished results.

For reviews on hypervalent phosphoranes [32,33]:

32. R. Burgada, R. Setton, *The Chemistry of Organophosphorus Compounds*, S. Patai and F. R. Hartley, Eds (John Wiley & Sons, 1994), Chapter 3, p. 185.
33. R. R. Holmes, *Pentacoordinated Phosphorus* (American Chemical Society, 1980).
34. J. Buschmann, T. Koritsanszky, R. Kuschel, P. Luger, K. Seppelt, J. *Am. Chem. Soc.*, **113**, 233 (1991).
35. B. Potter, G. Kleemann, K. Seppelt, Chem. Ber., **117**, 3255 (1984).
36. K. Seppelt, *Angew. Chem. Int. Ed. Engl.*, **30**, 361 (1991).
37. D. B. Dess, J. C. Martin, *J. Am. Chem. Soc.*, **113**, 7277 (1991).
38. D. B. Dess, J. C. Martin, *J. Org. Chem.*, **48**, 4155 (1983).
39. J. H. Holloway, V. Kaucic, D. Martin-Rovet, D. R. Russell, G. J. Schrobilgen, H. Selig, *Inorg. Chem.*, **24**, 678 (1985).
40. K. O. Christe, E. C. Curtis, D. A. Dixon, H. P. Mercier, J. C. P. Sanders, G. J. Schrobilgen, *J. Am. Chem. Soc.*, **113**, 3351 (1991).
41. K. Tamao, J. Yoshida, M. Takahashi, H. Yamamoto, T. Kakui, H. Matsumoto, A. Kurita, M. Kumada, *J. Am. Chem. Soc.*, **100**, 290 (1978).
42. Y. Yamamoto, H. Fujikawa, H. Fujishima, K.-y. Akiba, *J. Am. Chem. Soc.*, **111**, 2276 (1989).
43. Y. Yamamoto, K. Ohdoi, X. Chen, M. Kitano, K.-y. Akiba, *Organometallics*, **12**, 3297 (1993).

44a. S. Wallenhauer, K. Seppelt, *Inorg. Chem.*, **34**, 116 (1995).

44b. Y. Yamamoto, R. Nadano, M. Itagaki, K.-y. Akiba, *J. Am. Chem. Soc.*, **117**, 8287 (1995).

45. M. Minoura, T. Sagami, K.-y. Akiba, unpublished results.
46. K. Seppelt, private communication.
47. L. Ahmed, J. A. Morrison, *J. Am. Chem. Soc.*, **112**, 7411 (1990).
48. K. O. Christe, W. W. Wilson, R. V. Chirakal, J. C. P. Sanders, G. J. Schrobilgen, *Inorg. Chem.*, **29**, 3506 (1990).
49. K. O. Christe, W. W. Wilson, *Inorg. Chem.*, **28**, 3275 (1989).
50. A. R. Mahjoub, K. Seppelt, *Angew. Chem. Int. Ed.*, **30**, 323 (1991).

51. K. O. Christe, D. A. Dixon, J. C. P. Sanders, G. J. Schrobilgen, S. S. Tsai, *Inorg. Chem.*, **34**, 1868 (1995).
52. A. Ellern, K. Seppelt, *Angew. Chem. Int. Ed. Engl.*, **34**, 1586 (1995).
53. A. R. Mahjoub, T. Drews, K. Seppelt, *Angew. Chem. Int. Ed. Engl.*, **31**, 1036 (1992).
54. K. O. Christe, D. A. Dixon, J. C. P. Sanders, G. J. Schrobilgen, W. W. Wilson, *J. Am. Chem. Soc.*, **115**, 9461 (1993).
55. K. O. Christe, E. C. Curtis, D. A. Dixon, *J. Am. Chem. Soc.*, **115**, 1520 (1993).
56. O. Ruff, R. Klein, *Z. Anorg. Chem.*, **201**, 245 (1931).
57. K. O. Christe, D. A. Dixon, A. R. Mahjoub, H. P. A. Mercier, J. C. P. Sanders, K. Seppelt, G. J. Schrobilgen, W. W. Wilson, *J. Am. Chem. Soc.*, **115**, 2696 (1993).
58. K. O. Christe, J. C. P. Sanders, G. J. Schrobilgen, W. W. Wilson, *J. Chem. Soc. Chem. Commun.*, 837 (1991).
59. A. R. Mahjoub, K. Seppelt, *J.. Chem. Soc. Chem. Commun.*, 840 (1991).
60. A. R. Mahjoub, K. Seppelt, *Angew. Chem. Int. Ed. Engl.*, **30**, 876 (1991) .
61. C. J. Adams, *Inorg. Nucl. Chem. Lett.*, **10**, 831 (1974).
62. K.-y. Akiba, *J. Chem. Soc. Jpn.*, 1130 (1987).
63. K. Ohkata, K.-y. Akiba, *Synth. Org. Chem. Jpn.*, **53**, 687 (1995).
64. K.-y. Akiba, M. Ochiumi, T. Tsuchiya, N. Inamoto, *Tetrahedron Lett.*, 459, (1975).
65. K.-y. Akiba, T. Tsuchiya, N. Inamoto, *Tetrahedron Lett.*, 1877 (1976).
66. Y. Yamamoto, T. Tsuchiya, M. Ochiumi, S. Arai, N. Inamoto, K.-y. Akiba, *Bull. Chem. Soc. Jpn.*, **62**, 211 (1989).
67. K.-y. Akiba, T. Tsuchiya, M. Ochiumi, N. Inamoto, *Tetrahedron Lett.*, 455, (1975).
68. K.-y. Akiba, T. Tsuchiya, N. Inamoto, K. Onuma, N. Nagashima, A. Nakamura, *Chem. Lett.*, 723 (1976).
69. K.-y. Akiba, T. Tsuchiya, N. Inamoto, K. Yamada, H. Tanaka, H. Kawazura, *Tetrahedron Lett.*, 3819 (1976).
70. K.-y. Akiba, S. Arai, F. Iwasaki, *Tetrahedron. Lett.,* 4117 (1978).
71. K.-y. Akiba, S. Arai, N. Inamoto, K. Yamada, H. Tanaka, H. Kawazura, *Chem. Lett.*, 1415 (1978).
72. K.-y. Akiba, S. Arai, T. Tsuchiya, Y. Yamamoto, F. Iwasaki, *Angew. Chem. Int. Ed. Engl.,* **18**, 166 (1979).
73. K.-y. Akiba, T. Kobayashi, S. Arai, *J. Am. Chem. Soc.,* **101**, 5857 (1979).
74. Y. Yamamoto, K.-y. Akiba, *Heterocycles*, **13**, 297 (1979).
75. K.-y. Akiba, A. Noda, K. Ohkata, T. Akiyama, Y. Murata, Y. Yamamoto, *Heterocycles*, **15**, 1155 (1981).
76. K.-y. Akiba, A. Noda, K. Ohkata, *Phosphorus and Sulfur*, **16**, 111 (1983).
77. Y. Yamamoto, K.-y. Akiba, *J. Am. Chem. Soc.*, **106**, 2713 (1984).
78. Y. Yamamoto, K.-y. Akiba, *Bull. Chem. Soc. Jpn.*, **62**, 479 (1989).
79. K.-y. Akiba, K. Kashiwagi, Y. Ohyama, Y. Yamamoto, K. Ohkata, *J. Am. Chem. Soc.*, **107**, 2721 (1985).

80. K. Ohkata, Y. Ohyama, Y. Watanabe, K.-y. Akiba, *Tetrahedron Lett.*, **25**, 4561 (1984).
81. K.-y. Akiba, M. Ohsugi, H. Iwasaki, K. Ohkata, *J. Am. Chem. Soc.*, **110**, 5576 (1988).
82. K. Ohkata, M. Ohsugi, K. Yamamoto, M. Ohsawa, K.-y. Akiba, *J. Am. Chem. Soc.*, **118**, 6355 (1996).
83. K. Ohkata, M. Ohsugi, T. Kuwaki, K. Yamamoto, K.-y. Akiba, *Tetrahedron Lett.*, **31**, 1605 (1990).
84. K. Ohkata, K. Yamamoto, M. Ohsugi, M. Ohsawa, K.-y. Akiba, *Heterocycles*, **37**, 1707 (1994).
85. A. J. Arduengo, III, D. A. Dixon, D. C. Roe, *J. Am. Chem. Soc.*, **108**, 6821 (1986).
86. D. A. Dixon, A. J. Arduengo, III, T. Fukunaga, *J. Am. Chem. Soc.*, **108**, 2461 (1986).
87. D. A. Dixon, A. J. Arduengo, III, *J. Chem. Phys.*, **91**, 3195 (1987).
88. D. A. Dixon, A. J. Arduengo, III, T. Fukunaga, *J. Am. Chem. Soc.*, **109**, 338 (1987).
89. J. Moc, K. Morokuma, *Inorg. Chem.*, **33**, 551 (1994).
90. Y. Yamamoto, X. Chen, K.-y. Akiba, *J. Am. Chem. Soc.*, **114**, 7906 (1992).
91. Y. Yamamoto, X. Chen, S. Kojima, K. Ohdoi, M. Kitano, Y. Doi, K.-y. Akiba, *J. Am. Chem. Soc.*, **117**, 3922 (1995).
92. D. A. Dixon, A. J. Arduengo, III, *Intern. J. Quantum Chem., Symp.*, **22**, 85 (1988).
93. A. J. Arduengo, III, D. A. Dixon, D. C. Roe, M. Kline, *J. Am. Chem. Soc.*, **110**, 4437 (1988).
94. Y. Ohara, K.-y. Akiba, N. Inamoto, *Bull. Chem. Soc. Jpn.*, **56**, 1508 (1983).
95. R. P. Gellatly, W. D. Ollis, I. O. Sutherland, *J. Chem. Soc. Perkin Trans. I*, 913 (1976).
96. (a) K.-y. Akiba, K. Takee, K. Ohkata, F. Iwasaki, *J. Am. Chem. Soc.*, **105**, 6965 (1983); (b) K. Morokuma, M. Hanamura, K.-y. Akiba, *Chem. Lett.*, 1557 (1984).
97. K.-y. Akiba, K. Takee, Y. Shimizu, K. Ohkata, *J. Am. Chem. Soc.*, **108**, 6320 (1986).
98. F. Iwasaki, K.-y. Akiba, *Acta Cryst.*, **B41**, 445 (1985).
99. K. Ohkata, K. Takee, K.-y. Akiba, *Bull. Chem. Soc. Jpn.*, **58**, 1946 (1985).
100. K. Ohkata, K. Takee, K.-y. Akiba, *Tetrahedron Lett.*, **24**, 4859 (1983).
101. K. Ohkata, M. Ohnishi, K. Yoshinaga, K.-y. Akiba, J. C. Rongione, J. C. Martin, *J. Am. Chem. Soc.*, **113**, 9270 (1991).
102. K.-y. Akiba, K. Okada, K. Ohkata, *Tetrahedron Lett.*, **27**, 5221 (1986).
103. K. Ohkata, M. Ohnishi, K.-y. Akiba, *Tetrahedron Lett.*, **29**, 5401 (1988).
104. K. Ohkata, S. Takemoto, M. Ohnishi, K.-y. Akiba, *Tetrahedron Lett.*, **30**, 4841 (1989).
105. K.-y. Akiba, K. Okada, K. Ohkata, *Tetrahedron Lett.*, **26**, 4491 (1985).
106. K.-y. Akiba, K. Okada, K. Maruyama, K. Ohkata, *Tetrahedron Lett.*, **27**, 3257 (1986).
107. K. Ohkata, K. Okada, K.-y. Akiba, *Heteroatom Chem.*, **6**, 145 (1995).

108. H. Fujihara, H. Mima, M. Ikemori, N. Furukawa, *J. Am. Chem. Soc.*, **113**, 6337 (1991).
109. H. Fujihara, H. Mima, T. Erata, N. Furukawa, *J. Am. Chem. Soc.*, **114**, 3117 (1992).
110. H. Fujihara, T. Uehara, N. Furukawa, *J. Am. Chem. Soc.*, **117**, 6388 (1995).
111. J. C. Martin, T. M. Balthazor, *J. Am. Chem. Soc.*, **99**, 152 (1977).
112. J. Drabowicz, J. C. Martin, *Tetrahedron Assymmetry*, **4**, 297 (1993).
113. J. Drabowicz, J. C. Martin, *Phosphorus, Sulfur, and Silicon*, **74**, 439 (1993).
114. I. Kapovits, *Phosphorus, Sulfur, and Silicon*, **58**, 39 (1991).
115. D. H. R. Barton, S. A. Glover, S. V. Ley, *J. Chem. Soc. Chem. Commun.*, 266 (1977); S. A. Glover, *J. Chem. Soc. Perkin Trans. I*, 1338 (1980).
116. S. Oae, T. Kawai, N. Furukawa, T*etrahedron Lett.*, **25**, 69 (1984); S. Oae, T. Kawai, N. Furukawa, and F. Iwasaki, *J. Chem. Soc. Perkin Trans. II*, 405 (1987).
117. S. Oae, Y. Uchida, *Acc. Chem. Res.*, **24**, 202 (1991).
118. S. Oae, N. Furukawa, *Adv. Heterocycl. Chem.*, **48**, 1 (1990).
119. J. Moc, A. E. Dorigo, K. Morokuma, *Chem. Phys. Lett.*, **204**, 65 (1993).
120. R. Hoffmann, J. M. Howell, E. L. Muetterties, *J. Am. Chem. Soc.*, **94**, 3047 (1972).
121. R. R. Holmes, *Acc. Chem. Res.*, **12**, 257 (1979).
122. R. K. Brown, R. O. Day, S. Husebye, R. R. Holmes, *Inorg. Chem.,* **17**, 3276 (1978).
123. A. L. Beauchamp, M. J. Bennett, F. A. Cotton, *J. Am. Chem. Soc.*, **90**, 6675 (1968).
124. F. A. Cotton, G. Wilkinson, in *Advanced Inorganic Chemistry*, 5th ed. (Wiley, New York, 1988), p. 28
125. C. Brabant, B. Blanck, A. Beauchamp, *J. Organometal. Chem.*, **82**, 231 (1974).
126. C. Brabant, J. Hubert, A. L. Beauchamp, *Can. J. Chem.*, **51**, 2952 (1973).
127. R. S. Berry, *J. Chem. Phys.*, **32**, 933 (1960).
128. M. Kumada, K. Tamao, *Kagaku (Jpn),* **25**, 957, 1036 (1970).
129. P. Gillespie, P. Hoffmann, H. Klusacek, D. Marquarding, S. Pfohl, F. Ramirez, E. A. Tsolis, I. Ugi, *Angew. Chem. Int. Ed. Engl.,* **10**, 687 (1971).
130. T. Kawashima, K.-y. Akiba, *Kagaku no Ryoiki (Jpn)*, **29**, 804 (1975).
131. J. Moc, K. Morokuma, *J. Am. Chem. Soc.*, **117**, 11790 (1995).
132. J. Moc, K. Morokuma, *Inorg. Chem.*, **33**, 551 (1994).
133. J. E. Fowler, H. F. Schaefer, III, *J. Am. Chem. Soc.*, **116**, 9596 (1994).
134. C. J. Marsden, B. A. Smart, *Organometallics*, **14**, 5399 (1995).
135. J. Moc, A. E. Dorigo, K. Morokuma, *Chem. Phys. Lett*, **204**, 65 (1993).
136. Y. Yoshioka, J. D. Goddard, H. F. Schaefer, III, *J. Chem. Phys.*, **74**, 1855 (1981).
137. S. Kojima, K.-y. Akiba, unpublished results.
138. X. Chen, K. Ohdoi, Y. Yamamoto, K.-y. Akiba, *Organometallics*, **12**, 1857 (1993).
139. D. Hellwinkel, B. Knaebe, *Phosphorus*, **2**, 129 (1972).
140. D. Hellwinkel, M. Bach, *Naturwissenschaften*, **56**, 214 (1969).
141. D. Hellwinkel, W. Lindner, *Chem. Ber.*, **109**, 1497 (1976).

142. S. Kojima, K. Kajiyama, K.-y. Akiba, *Tetrahedron Lett.*, **35**, 7037 (1994).
143. S. Kojima, K. Kajiyama, K.-y. Akiba, *Bull. Chem. Soc. Jpn.*, **68**, 1785 (1995).
144. S. Kojima, M. Nakamoto, K. Kajiyama, K.-y. Akiba, *Tetrahedron Lett.*, **36**, 2261 (1995).
145. (a) S. Kojima, Y. Doi, M. Okuda, K.-y. Akiba, *Organometallics*, **14**, 1928 (1995); (b) S. Kojima, M. Okuda, K.-y. Akiba, unpublished results.
146. Y. Yamamoto, M. Okazaki, Y. Wakisaka, K.-y. Akiba, *Organometallics*, **14**, 3364 (1995).
147. S. Kojima, K. Kajiyama, M. Nakamoto, K.-y. Akiba, *J. Am. Chem. Soc.*, **118**, 12866 (1996) and unpublished results.
148. J. M. Howell, *J. Am. Chem. Soc.*, **99**, 7447 (1977).
149. W. Kutzelnigg, J. Wasilewski, *J. Am. Chem. Soc.*, **104**, 953 (1982).
150 (a) P. Kolandaivel, R. Kumaresan, *J. Mol. Struct.*, **337**, 225 (1995); (b) M. Minoura, S. Nagase, K.-y. Akiba, unpublished results.
151. K.-y. Akiba, T. Okinaka, M. Nakatani, Y. Yamamoto, *Tetrahedron Lett.*, **28**, 3367 (1987).
152. K.-y. Akiba, *Pure Appl. Chem.*, **68**, 837 (1996).
153. K. Shen, W. E. McEwen, A. P. Wolf, *J. Am. Chem. Soc.*, **91**, 1283 (1969).
154. J.-P. Finet, *Chem. Rev.*, **89**, 1487 (1989).
155. J. I. Musher, *Angew. Chem. Int. Ed. Engl.*, **8**, 54 (1969).
156. S. Tamagakli, S. Oae, *Kagaku no Ryoiki (Jpn)*, **31**, 117, 218 (1977).
157. C. S. Smith, J.-S. Lee, D. D. Titus, R. F. Ziolo, *Organometallics*, **1**, 350 (1982).
158. M. Minoura, T. Sagami, K.-y. Akiba, C. Modrakowski, A. Sudau, K. Seppelt, S. Wallenbauer, *Angew. Chem. Int. Ed. Engl.*, **36**, 2660 (1996).
159. K.-y. Akiba, lecture at ICCST-7 at Aachen, July, 1997; M. Minoura, N. Yamashita, K.-y. Akiba, unpublished results.
160. R. S. Michalak, J. C. Martin, *J. Am. Chem. Soc.*, **106**, 7529 (1984).
161. R. S. Michalak, S. R. Wilson, J. C. Martin, *J. Am. Chem. Soc.*, **104**, 1683 (1982).
162. C. Cloutour, D. Lafargue, J. A. Richards, J.-C. Pommier, *J. Organomet. Chem.*, **137**, 157 (1977).
163. Y. Yamamoto, R. Nadano, M. Itagaki, K.-y. Akiba, *J. Am. Chem. Soc.*, **117**, 8287 (1996).
164. W. Satoh, R. Nadano, Y. Yamamoto, K.-y. Akiba, *J. Chem. Soc. Chem. Commun.*, 2451 (1996); W. Satoh, R. Nadano, G. Yamamoto, Y. Yamamoto, K.-y. Akiba, *Organometallics*, **16**, 3664 (1997); G. Yamamoto, R. Nadano, W. Satoh, Y. Yamamoto, K.-y. Akiba, *J. Chem. Soc. Chem. Commun.*, 1325 (1997).
165. K.-y. Akiba, Y. Onzuka, M. Itagaki, H. Hirota, Y. Yamamoto, *Organometallics*, **13**, 2800 (1994).

3 Characteristic Properties of Hypervalent Compounds: Static and Dynamic Structures of Sb(III), Sn(II), and Ge(II) Halides

KOJI YAMADA AND TSUTOMU OKUDA
Department of Chemistry, Faculty of Science, Hiroshima University, Higashi-Hiroshima, Japan

3.1 INTRODUCTION

For complexes with main group elements as central atoms, the mutual influence of ligands is quite different from that of the transition elements. This is because the number of available orbitals is deficient compared with that of electrons from the donors. Musher proposed the idea of a "hypervalent bond" for these orbital-deficient molecules.[1] The hypervalent bond is distinguished from the normal covalent bonds by the fact that there is only one bonding orbital with two atoms located at the *trans* positions. Musher classified the hypervalent bond into two categories: "hypervalent bond I" and "hypervalent bond II", depending whether the central metal ion has an ns^2 lone pair or no ns^2 lone pair. The mutual influence of the ligands in the octahedral environment was also studied theoretically for hypervalent bonds I and II by Shustorovich and Buslaev.[2] According to them, a strong donor causes a weakening of the *trans* bond in hypervalent bond I. Hypervalent bond II, in contrast, experiences, in the case of a strong donor, a strengthening of the *trans* bond at the expense of weakening of the *cis* bond. These theoretical results agree with the experimental data and permit a number of predictions to be made.

Alcock has summarized the crystal structures of the compounds containing main group elements having an *s*-electron lone pair and has investigated especially the intramolecular and intermolecular distances in these crystals. In his review an approximately linear arrangement Y—A···X was also discussed in

Chemistry of Hypervalent Compounds, edited by Kin-ya Akiba.
ISBN 0-471-24019-2

detail; in this, the short intermolecular interaction A···X was termed a "secondary bond" in contrast to the primary Y—A bond located at the *trans* position.[3]

In this chapter we will describe the static and dynamic structures of the compounds having hypervalent bond I, that is, the bonding of the low oxidation state of a main group metal having an *s*-electron lone pair. The most simple and one-dimensional example of hypervalent bond I is a bonding of the I_3^- anion. Pimentel proposed the three-center–four-electron (3c–4e) bond model for this anion using the simple molecular orbitals.[4] Since three atomic p_x orbitals of iodine atoms along the molecular axis are available for the bonding, the following three molecular orbitals can be formed:

$$\begin{aligned} \sigma_2^* &= p_x - (p_x' + p_x'') \\ \sigma_N &= p_x' - p_x'' \\ \sigma_1 &= p_x + (p_x' + p_x'') \end{aligned} \tag{3.1}$$

where p_x, p_x', and p_x'' are atomic p orbitals of the central and two terminal atoms, respectively. If the I_3^- anion is symmetric, there are occupied bonding orbital σ_1 and nonbonding orbital σ_N, and an unoccupied antibonding orbital σ_2^*, resulting in a weak bond between each pair of atoms. However, if the I_3^- anion is asymmetric, the bonding orbital is concentrated between the I—I bond and the nonbonding one is concentrated on the other side, I···I. The structure of the I_3^- anion is an interesting example of the 3c–4e bond. For example, if the counter cation is bulky, such as $(C_6H_5)_4As$ or $(CH_3)_4N$, a symmetric anion appears, whereas if the cation is small and has large polarization force, an asymmetric anion appears. Two extreme structures of the I_3^- anions, symmetric and asymmetric, were reported for $(C_6H_5)_4AsI_3$ (I—I: 2.90 Å)[5] and for NH_4I_3 (I—I: 2.79 Å and I···I: 3.11 Å).[6] The chemical bonding in these trihalides was reviewed by Nakamura and Kubo by means of nuclear quadrupole resonance spectroscopy.[7] Quite similar situations have been reported for the hydrogen bonding systems such as O—H···O and F—H···F, for which Pimentel proposed first the idea of the 3c–4e bond.[8]

In order to explain the structural variety and many phase transitions of the hypervalent compounds having octahedral coordination, the 3c–4e bond must be introduced into the three orthogonal directions. One simple example using this model is a structural variety of the BX_3^- (B = Ge(II) and Sn(II)) anion, for which four different types of the coordination appear as a result of the symmetry change of the 3c–4e bonds, as will be described in Section 3.4.

In the subsequent sections we will describe the static and dynamic structure of the coordination compounds of Sb(III), Ge(II), and Sn(II). First, we will show a phase transition of $C_6H_5NHSbBr_4$ as a characteristic example of phase transition having a hypervalent bond, and then we will discuss the structural variety of the halocomplexes of Sb(III). Second, we will describe the successive phase transitions of trihalogenogermanate(II) salts, for which drastic increases of the ionic conductivities were observed, associated with the transitions to the cubic perovskite phases. Finally, we will discuss the metallic and semiconducting properties of Sn(II) halides on the basis of their structures and solid-state nuclear magnetic resonance (NMR) spectroscopy.

3.2 PHASE TRANSITION OF $C_5H_5NHSbBr_4$ HAVING A HYPERVALENT BOND

According to the valence-shell electron-pair repulsion (VSEPR) model,[9] the structure of the $SbBr_4^-$ anion can be described as a trigonal bipyramid having a lone pair of electrons at the equatorial position. However, in order to describe the structure of a complex ion containing a heavy main metal such as Sb or Bi, the hypervalent model is simpler and more versatile than VSEPR because we can ignore the hybridization using *s* or *d* orbitals. According to the hypervalent bonding model, the $SbBr_4^-$ anion can be described as consisting of one symmetric 3c–4e bond (axial Br—Sb—Br) and two asymmetric 3c–4e bonds (equatorial Br—Sb···Br), as shown in Figure 3.1. Furthermore, this hypervalent model can predict the structural change associated with a phase transition. In order to show the characteristic properties of the hypervalent bond, we will first describe a phase transition of pyridinium tetrabromoantimonate(III), $C_5H_5NHSbBr_4$.[10]

$C_5H_5NHSbBr_4$ shows a second-order phase transition at 253K; one phase is designated as Phase I and the other as Phase II, from the high-temperature side. The extremely large splitting of the ^{81}Br nuclear quadrupole resonance

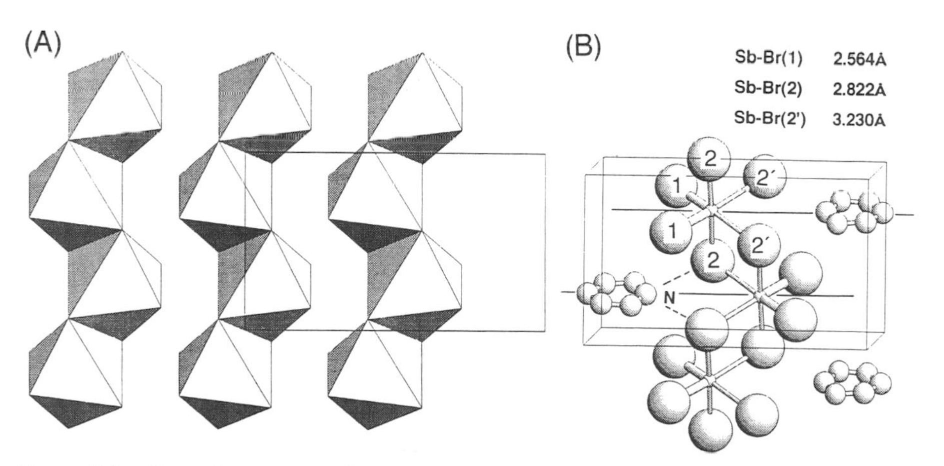

Figure 3.1 Crystal structure of $C_5H_5NHSbBr_4$ (room temperature phase). (A) Edge sharing $SbBr_6$ octahedra along the crystal *c*-axis. (B) Possible hydrogen bonds between N—H and axial Br atoms.

(NQR) was observed below 253K, as shown in Figure 3.2. The ^{81}Br NQR lines at around 120 and 60 MHz above T_{tr} could be assigned easily to equatorial Br atoms and to axial ones, respectively. Since the halogen NQR frequency is proportional to its covalent character, the bond order of the Sb–Br (axial) is nearly one-half of the Sb–Br (equatorial) bond. The extremely large splitting of the ^{81}Br NQR below T_{tr} suggests a deformation of the axial hypervalent bond, that is, $[\text{Br—Sb—Br}]^- \rightarrow \text{Br—Sb}\cdots\text{Br}^-$. Furthermore, it is interesting to note that the sum of the two NQR frequencies at the axial positions remains nearly constant even below T_{tr} and is nearly equal to that of the equatorial one. In order to understand this phase transition in static and dynamic aspects, Rietveld analysis of the X-ray diffraction at 116K and the temperature dependence of ^{2}H NMR have been performed.[11,12]

The crystal structure of $C_5H_5NHSbBr_4$ above T_{tr} was first determined by DeHaven and Jacobson using a single crystal.[13] According to them, it belongs to a monoclinic system ($a = 11.824$ Å, $b = 13.040$ Å, $c = 7.703$ Å, and $\beta = 93.89°$) with space group $C2/c$ and the $SbBr_4^-$ anions can be described as the edge-sharing octahedra shown in Figure 3.1. It should be emphasized

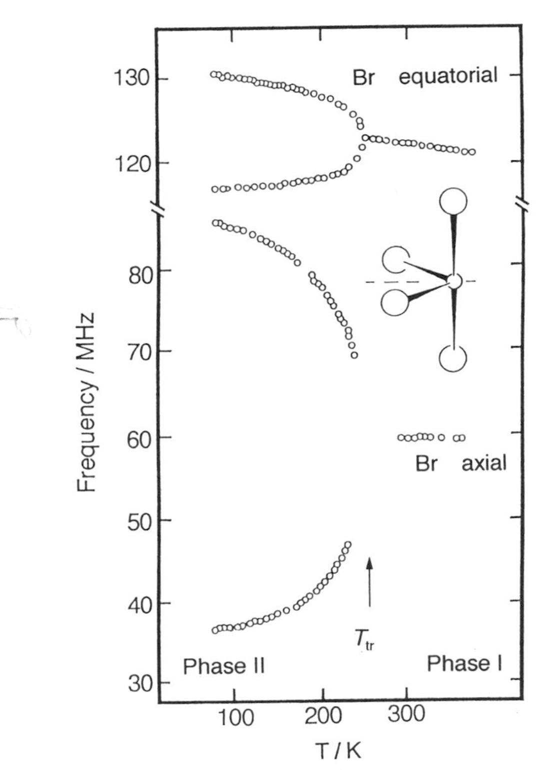

Figure 3.2 Temperature dependence of the ^{81}Br NQR frequencies for $C_5H_5NHSbBr_4$.

that both the nitrogen and antimony atoms are located on the twofold axis of the monoclinic system, that is, the two axial Sb—Br bonds are crystallographically equivalent. The possible hydrogen bonds between the axial Br atom and N—H bond are shown symmetrically about the twofold axis. Figure 3.3 shows the final plots of the Rietveld refinements at 116K and 297K. The powder pattern at 116K could be indexed as a triclinic system with space group $P\bar{1}$ (No. 2). The unit cell vectors ($\boldsymbol{a}'$, $\boldsymbol{b}'$, and $\boldsymbol{c}'$) in Phase II are roughly expressed as

$$\begin{aligned} \boldsymbol{a}' &= (1/2)(\boldsymbol{a} + \boldsymbol{b}) \\ \boldsymbol{b}' &= (1/2)(\boldsymbol{a} - \boldsymbol{b}) \\ \boldsymbol{c}' &= \boldsymbol{c} \end{aligned} \tag{3.2}$$

where $\boldsymbol{a}$, $\boldsymbol{b}$, and $\boldsymbol{c}$ denote the unit cell vectors at Phase I. Only the inversion symmetry is maintained in Phase II and the unit cell volume decreases to around half of Phase I. Figure 3.4 shows the anion structure together with the Sb—Br bond lengths. From these bond lengths, the anion structure in Phase II can be described more properly as $SbBr_3{\cdot}Br^-$ and the structure is consistent with the ^{81}Br NQR spectrum at Phase II. This structure appears

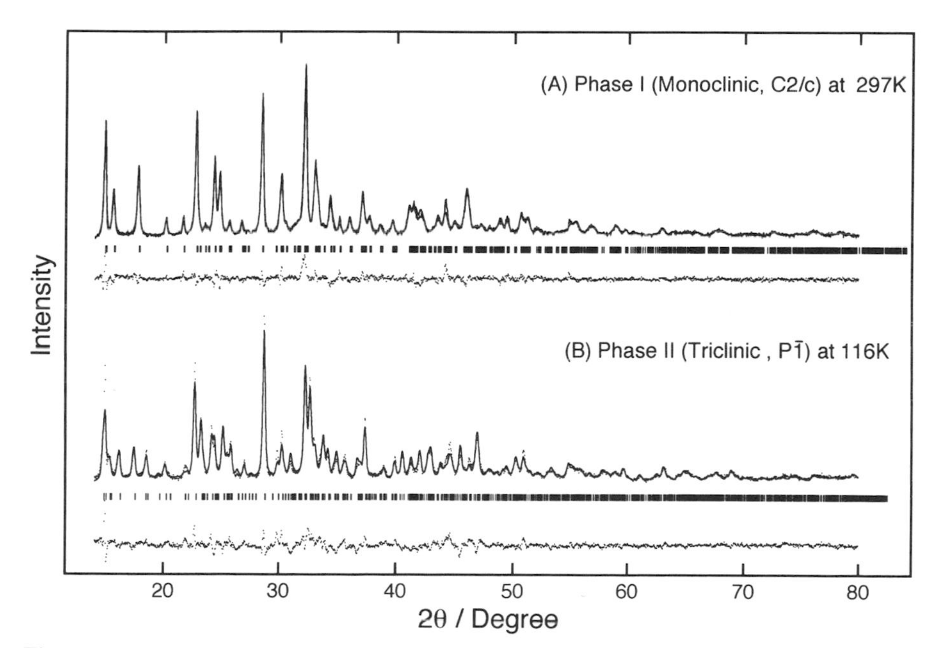

Figure 3.3 Final plot of the Rietveld analysis XRD data for $C_5H_5NHSbBr_4$. Solid lines and dots are calculated and observed patterns, respectively. The differences between them are shown at the lower portions.

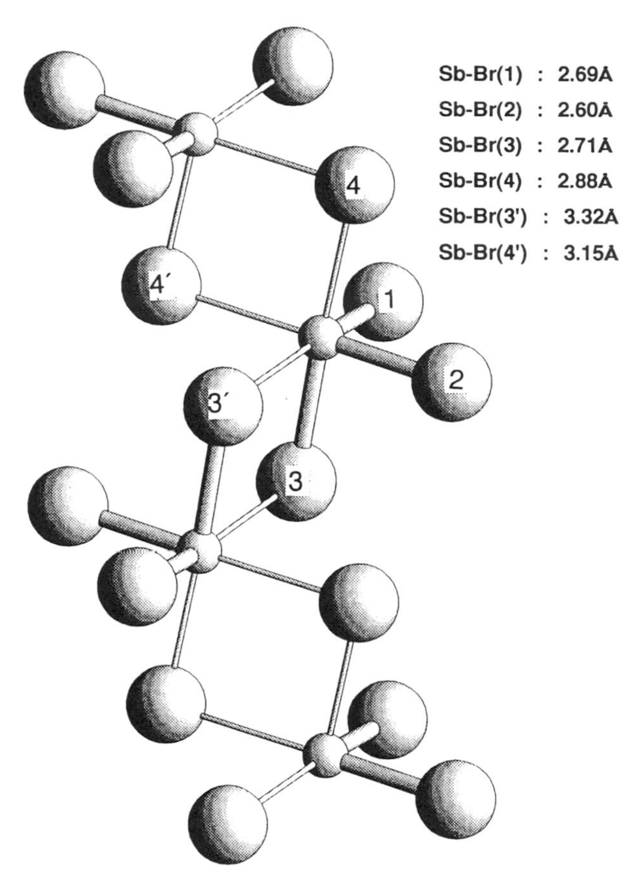

Figure 3.4 Anion structure in $C_5H_5NHSbBr_4$ at 116K. The deformation of the Br(3)—Sb—Br(4) bond is clearly seen from the bond length.

as a result of the symmetry lowering of the *trans* Br—Sb—Br bond, similar to the asymmetric I_3^- ion in NH_4I_3.[6]

Since a hydrogen bond between $C_5H_5NH^+$ and Br^- may contribute to the stabilization of the ionic form, the motion of the pyridinium ion was investigated as a function of temperature by means of broadline ^2H NMR using a deuterated sample, $C_5{}^2H_5NHSbBr_4$. ^2H NMR is a useful technique to investigate molecular motions because the quadrupole coupling constant reflects the motion containing probe nuclei. If the molecular motion is slower than $10^3\,sec^{-1}$ or faster than $10^7\,sec^{-1}$, the observed spectrum of the ^2H NMR can be simulated easily using e^2Qq_{zz}/h and η. In the limit of the fast motion, the motional averaged efg tensor having $\boldsymbol{q}'_{ij}$ $(i, j = x, y, z)$ components is expressed as a sum of the different orientations weighted with the equilibrium distribution:

$$\boldsymbol{q}'_{ij} = \Sigma \boldsymbol{P}(k)\boldsymbol{q}_{ij}(k) \tag{3.3}$$

where $\boldsymbol{P}(k)$ and $\boldsymbol{q}_{ij}(k)$ are the occupation probability and the tensor components of the electric field gradient at the kth site, respectively. The resultant $\boldsymbol{q}'_{ij}$ is diagonalized and then three new components, $\boldsymbol{q}'_{ii}$, and its orientation are obtained. Figure 3.5 plots three experimentally determined components of the quadrupole coupling constant (e^2Qq_{ii}/h) against temperature, together with some spectra at selective temperatures. Although the z and x components decrease continuously with increasing temperature, the y component remains constant over the whole temperature range studied. The corresponding motional mode is a librational motion about an axis normal to the plane. In the limit of fast motion, the motionally averaged efg components, q'_{ii} ($ii = xx, yy, zz$), are expressed using a root-mean-square (rms) amplitude defined by θ as

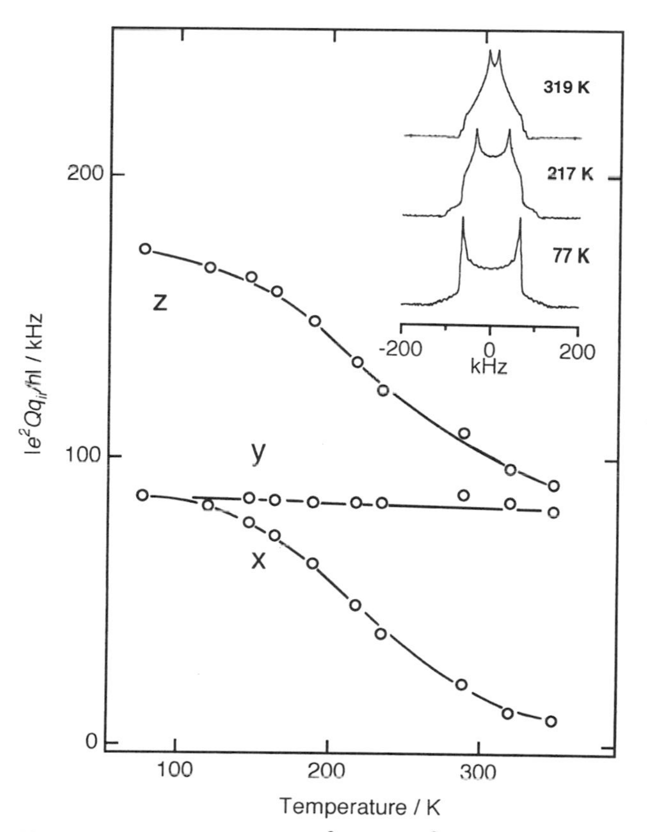

Figure 3.5 Temperature dependence of e^2Qq_{ii}/h of ^{2}H NMR together with the spectra at selected temperatures.

$$
\begin{aligned}
q'_{xx} &= q_{xx}\cos^2\theta + q_{zz}\sin^2\theta = (1/2)q_{zz}(3\sin^2\theta - 1) \\
q'_{yy} &= q_{yy} = -(1/2)q_{zz} \\
q'_{zz} &= q_{xx}\sin^2\theta + q_{zz}\cos^2\theta = (1/2)q_{zz}(3\cos^2\theta - 1)
\end{aligned}
\tag{3.4}
$$

where q_{ii} are efg components at the rigid lattice. Figure 3.6 plots the θ values estimated from Eq. 3.4 against temperature. Although no discontinuity was observed at T_{tr}, a quite large amplitude of the librational motion of the pyridinium ring is seen in Phase I. With increasing amplitude of the motion, the hydrogen bond is supposed to be weakened and the asymmetric *trans* Br—Sb···Br bond is gradually transformed into a symmetric form. Although θ reaches around 30° at Phase I, the isotropic reorientation about the pseudo C_6 axis like that found for the pyridinium cation[14] or isoelectronic benzene[15] was not observed even at 350K. This suggests that the motion of the pyridinium cation in Phase I is still restricted by the two hydrogen bonds formed between N—H and axial Br atoms, as shown in Figure 3.1. However, two polymorphic forms were reported for the chloride analog at room temperature, depending upon the preparation conditions.[16] Two different structures of anion correspond to the structures in Phase I and II of the bromide analog.

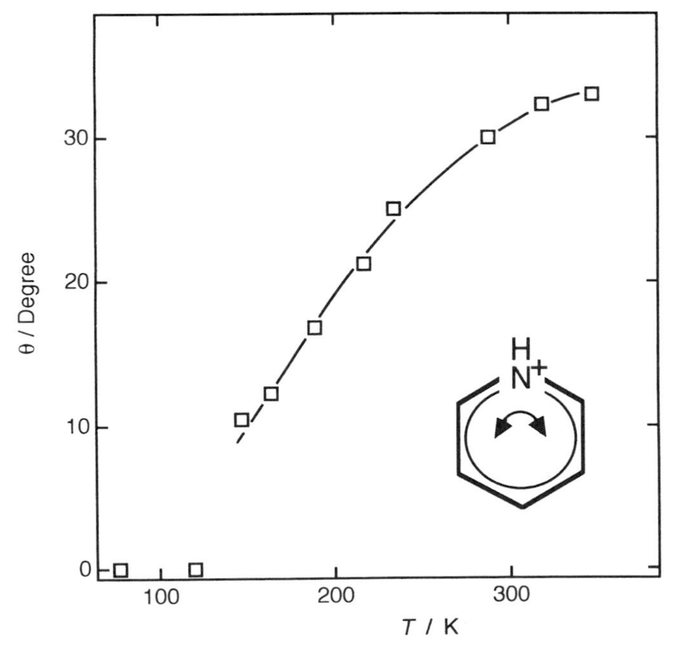

Figure 3.6 Root-mean-square amplitude of the librational oscillation against temperature.

3.3 STRUCTURE AND BONDING OF HALOGENOANTIMONATE (III) COMPLEXES

As is apparent from the examples stated above, a symmetric 3c–4e bond easily changes to an asymmetric one, depending upon the electrostatic interactions in the lattice. Similar deformation of the *trans* X—Sb—X bond is also induced by the replacement of the counter cation; the deformation leads to the structural variety of the halocomplexes of Sb(III). Many salts with general formula A_{x-3} $SbX_x (x = 4, 4.5, 5, 5.5, 6)$ have been synthesized and studied by X-ray diffraction, NQR spectroscopy, and dielectric measurement. We will show here one example of the structural variety of an isolated $SbBr_6{}^{3-}$ anion because of its simple structure, having no bridging atom. Since ^{81}Br NQR frequency is proportional to the covalent character of the Sb—Br bond, Table 3.1 summarizes NQR frequencies for a series of hexabromoatimonates together with their crystallographic parameters. The structures of $SbBr_6{}^{3-}$ anions are divided into two extreme types: a nearly regular octahedral anion found in $(CH_3)_2NH_2$, $(n\text{-}C_3H_7)_2NH_2$, or $(n\text{-}C_4H_9)NH_2$, and a remarkably distorted anion along the C_3 axis found in $(C_2H_5)_2NH_2$ salt.[17–19] In the former regular octahedral anion, three symmetric Br—Sb—Br bonds are formed. In the latter distorted octahedron, there are three short Sb—Br and three long Sb···Br bonds, that is, $SbBr_6{}^{3-}$ anion has three asymmetric Br—Sb···Br bonds.

In order to examine the nature of the X—Sb···X bond, the length of the Sb—X bond is plotted against that of the *trans* Sb···X bond. A good correlation is found between them, as shown in Figure 3.7. According to Pauling,[20] the bond length (d_{Sb-X}) is expressed using the bond order as

$$d_{Sb-X} = d_0 - a\log(n) \tag{3.5}$$

where n is the bond order, d_0 is the bond length corresponding to $n = 1$, and a is a numerical constant. However, the bond order assigned to the *trans* Sb···X is assumed to be $1 - n$, because the sum of the two halogen NQR frequencies assigned to the X—Sb···X bond is held almost constant in spite of the distortion. Then, the *trans* Sb···X bond length is expressed as

$$d_{Sb\cdots X} = d_0 - a\log(1 - n) \tag{3.6}$$

where $d_{Sb\cdots X}$ is the bond length assigned to Sb···X. By eliminating n from these equations, d_0 and a were determined to be 2.34 Å and 0.95 Å for the Cl—Sb···Cl bond and 2.46 Å and 1.11 Å for the Br—Sb···Br bond by using 27 chloroantimonates and 12 bromoantimonates, respectively. The solid curves in Figure 3.7 show the correlations that are reproduced using these parameters. This figure represents the characteristic feature of the 3c–4e bond, that is, the strong Sb—X bond leads to a weakening of the *trans* Sb···X bond or *vice versa*. Therefore, the structural variety of the anion or its structural deformation with temperature arise mainly from the symmetry change of the

TABLE 3.1 ^{81}**Br NQR Frequencies and Crystallographic Data for** $[(C_nH_{2n+1})_2NH_2]_3SbBr_6$ ($n = 1, 2, 3, 4$)

Cation	Crystal Parameter	Sb—Br or Sb···Br Bond Length (Å)	Point Symmetry of $SbBr_6^{3-}$ Anion	^{81}Br NQR at 298K (MHz)	Reference
		2.81		67.24	
$(CH_3)_2NH_2$[a]	$R\bar{3}$, $z = 12$,	2.81	$\bar{1}$	66.47	
	$a = 29.234$ Å, $c = 8.442$ Å	2.84		61.52	18
		2.82	$\bar{3}$	63.84	
$(C_2H_5)NH_2$	$R3c$, $z = 6$	2.622	3	105.25	
	$a = 15.183$ Å, $c = 20.126$ Å	3.061		30.97	17
$(n\text{-}C_3H_7)_2NH_2$	$R\bar{3}c$, $z = 6$	2.794	$\bar{3}$	65.10	
	$a = 13.667$ Å, $c = 31.566$ Å				17
$(n\text{-}C_4H_9)_2NH_2$[b]	$R3c$, $z = 6$		3	67.78	
	$a = 14.556$ Å, $c = 32.992$ Å			65.80	

[a]Crystallographic parameters were determined by Rietveld analysis on the basis of the isomorphous Bi analog.[19] Two crystallographically different anions having $\bar{1}$ and $\bar{3}$ point symmetry exist in the lattice.
[b]Crystallographic parameters were determined by Rietveld analysis of the XRD data.

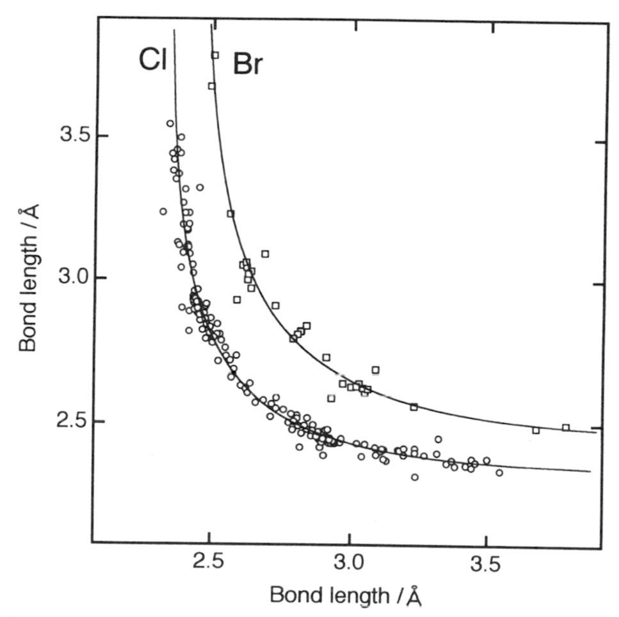

Figure 3.7 Correlation between bond lengths for a series of $A_{x-3}SbX_x$. Sb—X length is plotted against its *trans* Sb···X length. The solid curves were calculated according to Eqs (3.5) and (3.6) by the elimination of n.

Br—Sb···Br bonds. The stereochemical activity of the s-electron lone pair is a result of the deformation.

Many ferroelectric compounds containing halogenoantimonate(III)[21,22] or halogenobismuthate(III)[23,24] anions have been synthesized by R. Jakubas. These ferroelectric properties are partly due to the large polarizability of the complex anion having a hypervalent bond and partly due to the dynamics of the cationic sublattice.

3.4 STRUCTURAL CLASSIFICATION OF THE BX_3^- ANION IN ABX_3

Many crystalline compounds such as ASn_2X_5, $ASnX_3$, A_2SnX_4, and A_4SnX_6 have been reported in the case of B = Sn(II). Among these double salts, we have studied mainly the structures and properties of ABX_3 (A = alkali metal and alkylammonium, B = Ge, and Sn, X = Cl, Br, and I) compounds because of their interesting electrical properties. Taking the hypervalent bond in the octahedral coordination into account, four model structures appear for the BX_3^- anion.[25] Figure 3.8 shows these four structures. For simplicity, the long interactions X···B are not drawn in this figure. These coordination models result from whether the three *trans* X—B—X bonds are symmetric or not, in a similar manner to the deformation of the Sb(III) complexes. Model A appears in a

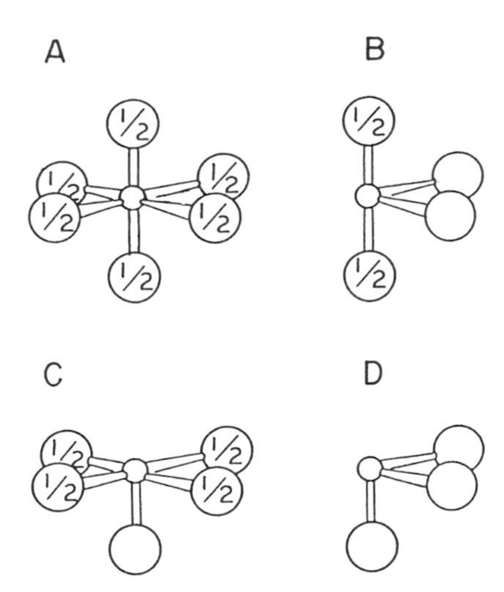

Figure 3.8 Structural models of BX_3^- (B = Ge(II), Sn(II), and Pb(II)) anion based upon the hypervalent bond model. 1/2 indicates the bridging halogen.

cubic or in a slightly distorted perovskite structure. Several prototype structures for the ABX_3 appear, depending upon not only these models, but also upon the types of the network of bridging halogens. Furthermore, many phase transitions associated with the structural changes of the anion have been reported. For example, a structural change from model D to C was suggested for $KSnBr_3{\cdot}H_2O$ or $NH_4SnBr_3{\cdot}H_2O$ with increasing temperature by means of a ^{81}Br NQR study.[26] Many crystals containing the isolated BX_3^- anions (model D) undergo phase transitions to cubic perovskites having regular BX_6 octahedra.[27,28] However, in the case of the reconstructive phase transition of $CsSnI_3$, the anionic sublattice changes from a edge-sharing octahedron to a perovskite type and at the same time the coordination around the Sn(II) changes from C to A.[29,30] This structural change results in an increase of the conductivity up to the extent of metals.

As was stated above, many phase transitions to the cubic perovskites have been found for $ASnX_3$ and $AGeX_3$ at the highest temperature phases. It is particularly interesting to note that two different types of conductivities, electronic and ionic, have been found for the cubic or slightly distorted perovskite phases. Electronic conductivity was found for Sn(II) compounds such as $CsSnI_3$, $CH_3NH_3SnI_3$, $CsSnBr_3$, and $CsSnCl_3$, and the conductivity varies from a metallic order for the iodides to a semiconducting one for the chlorides. High ionic conductivity, however, has been found for the Ge(II) compounds such as $CH_3NH_3GeCl_3$ and $(CH_3)_4NGeCl_3$ in their cubic phases.[28,31] Both electronic and ionic conductivities are closely related to the hypervalent nature of the central metal, that is, these properties depend upon whether the 3c–4e bonds in the perovskite lattice are formed symmetrically or formed asymmetrically in a disordered state (B· · ·X—B ↔ B—X· · ·B). First we will discuss the structures and the mechanisms of the ionic conductivity that have been found in the trihalogenogermanate salts.

3.5 NEW TYPE OF CHLORIDE ION CONDUCTOR HAVING HYPERVALENT BOND

Figure 3.9 shows the temperature dependence of the conductivity determined by a complex impedance method in a frequency range from 100 Hz to 100 kHz. There is no pronounced increase in conductivity for $CsGeCl_3$ at the transition temperature from the trigonal to the cubic phase. However, abrupt increases in conductivity were observed for $CH_3NH_3GeCl_3$ and $(CH_3)_4NGeCl_3$ just below the phase transitions to their cubic phases. In order to clarify the origin of the high conductivity at the cubic phases, X-ray diffraction (XRD) patterns were measured as a function of temperature. Figure 3.10 shows the final plots of the Rietveld refinements for $CH_3NH_3GeCl_3$ at three different phases. The simple

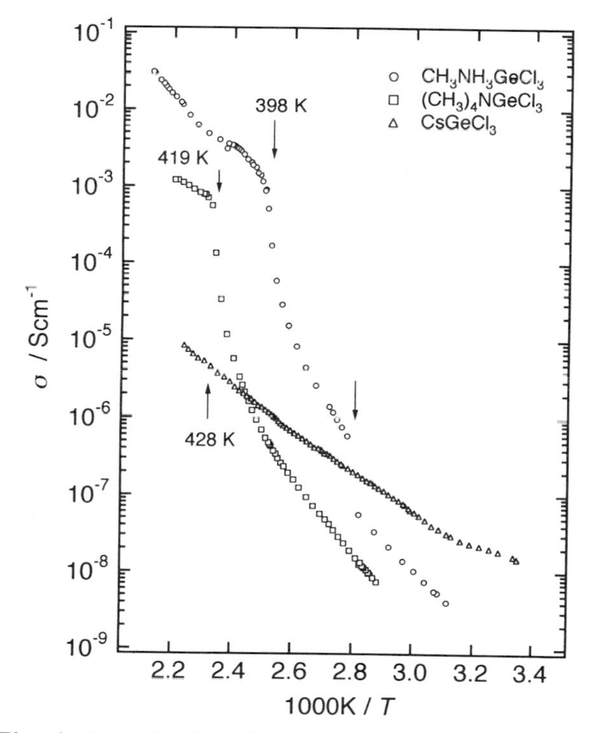

Figure 3.9 Electrical conductivity for $AGeCl_3$ (M = Cs, CH_3NH_3, and $(CH_3)_4N$) against inverse temperature. Arrows indicate phase transitions.

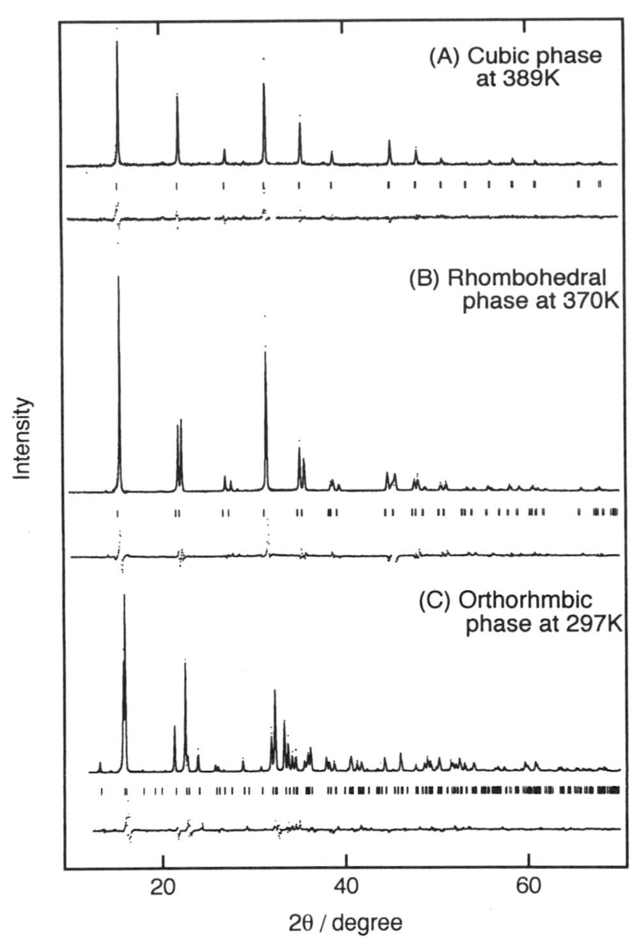

Figure 3.10 Final plot of the Rietveld analysis of $CH_3NH_3GeCl_3$. Solid lines and dots are calculated and observed patterns, respectively. The differences between them are shown at the lower portions.

powder pattern at 389K suggests that this phase belongs to a cubic system having a perovskite structure. The perovskite structure, however, is not an ideal but a disordered one which has not been confirmed for the cubic perovskite so far. A quite similar disorder was more clearly seen in $(CH_3)_4NGeCl_3$ on the Fourier map, because the separation between disordered sites increased with increasing the cubic lattice constant. With decrease of the cationic radius the disordered sites merged into one site and indeed no disorder was found for $CsGeCl_3$. Figure 3.11 shows the structures of $(CH_3)_4NGeCl_3$ at Phase I (440K) and II (294K) along the crystal *a*-axes. From the comparison between these structures, it was proved that the disorder at the Cl site leads to a cubic phase.

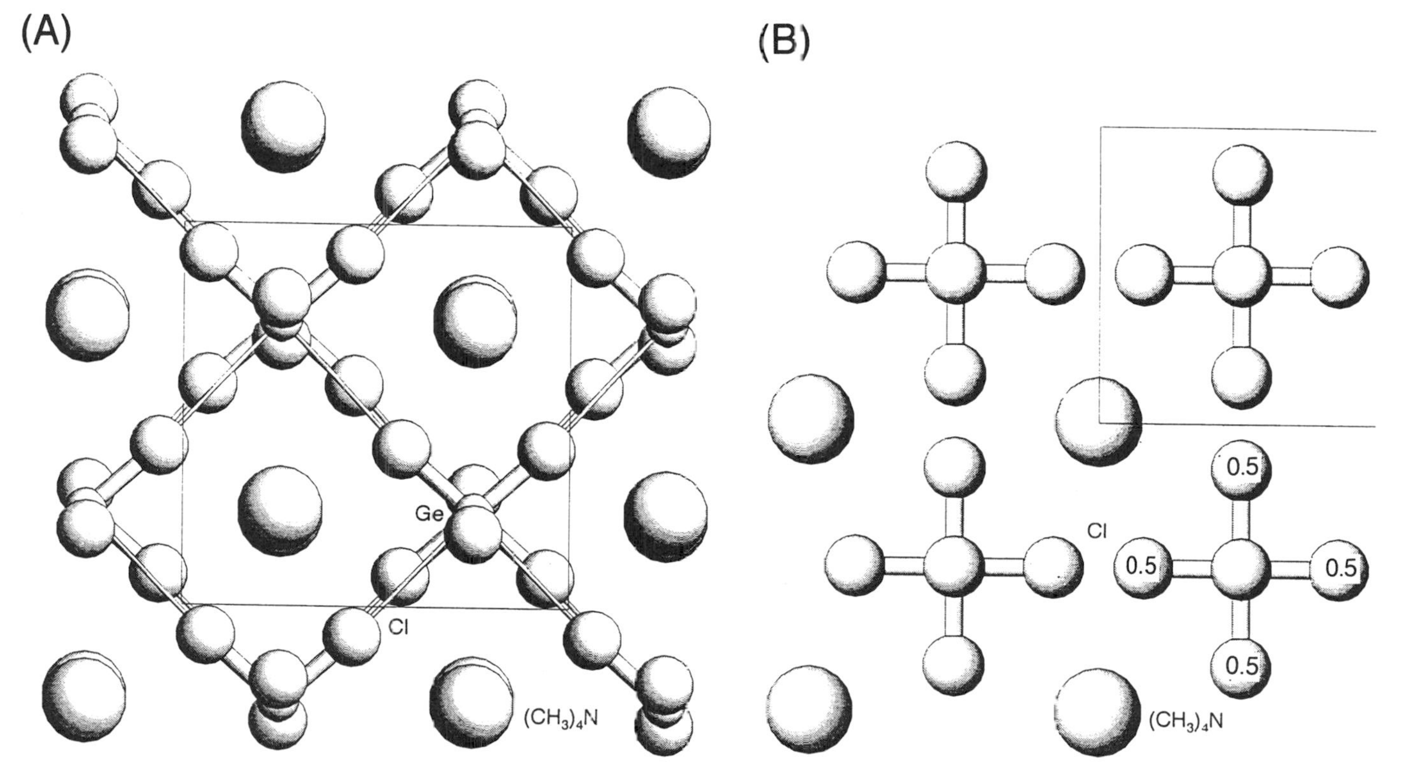

Figure 3.11 Structure of $(CH_3)_4NGeCl_3$ at (A) Phase II (orthorhombic), (B) Phase I (cubic). The lattice constants of Phase II are approximately $a \approx 2a_c$, $b \approx \sqrt{2}a_c$ and $c \approx 2a_c$, where a_c is a cubic lattice constant.

It is particularly interesting that this type of order–disorder transition reflects the characteristic feature of the hypervalent bond. In these cubic perovskite phases, a dynamically disordered state Ge—Cl···Ge ↔ Ge···Cl—Ge appears instead of a symmetric bridge Ge—Cl—Ge, which exists in an ideal perovskite structure. This bond-switching disorder appears along three orthogonal directions in the cubic lattice. Therefore, at every instant a discrete $GeCl_3^-$ anion exists in the cubic phase due to the characteristic *trans* effect of the hypervalent bond.

According to the ^{35}Cl NQR study for a series of $AGeCl_3$ compounds, the quadrupole coupling constant (e^2Qq/h) assigned to the $GeCl_3^-$ anion is around 28 MHz. Therefore, it is impossible to detect ^{35}Cl NMR for the powdered sample because of the over-broadening of the central line due to the second-order quadrupole effect. As shown in Figure 3.12, however, the ^{35}Cl NMR signal could be detected above approximately 364K, and the signal intensity increased with increasing temperature. This NMR behavior suggests that the carrier of the electric conductivity is a chloride ion and that the correlation

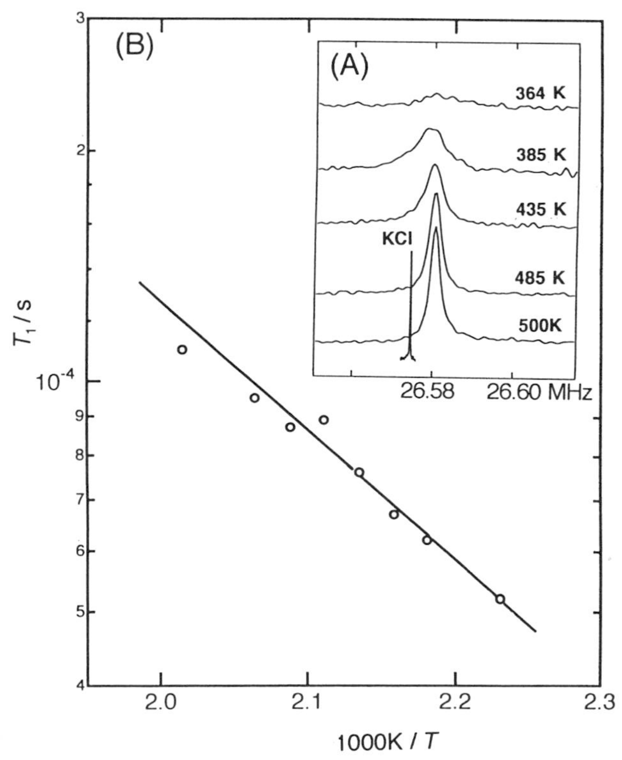

Figure 3.12 ^{35}Cl NMR spectra of $CH_3NH_3GeCl_3$ at selected temperatures (A) and the temperature dependence of the spin–lattice relaxation times (B).

frequency ($1/\tau_c$) of the motion is much faster than e^2Qq/h, that is, $1/\tau_c \gg e^2Qq/h$, where τ_c is the correlation time of the translational diffusion. Since the principal axis of the e^2Qq/h at the Cl site is parallel to the *trans* Ge—Cl···Ge bond and there are three possible directions which are mutually orthogonal, the time-averaged e^2Qq/h vanishes if the condition stated above can be satisfied. Figure 3.12(B) shows the temperature dependence of the spin–lattice relaxation time T_1 observed at 26.58 MHz (6.37 T). The relaxation is governed by the quadrupole interaction because of its large interaction. The T_1 values increased with increasing temperature according to a fast-motion limit of the BPP formula ($\tau_c\omega \ll 1$, ω: NMR Larmor frequency), as was expected. Then the relaxation rate for a ^{35}Cl nucleus ($\boldsymbol{I} = 3/2$), $1/T_1$, is given like in the liquid as

$$1/T_1 = (1/10)(e^2Qq/h)^2\tau_c \tag{3.7}$$

From the log(T_1) plot against $1/T$, the activation energy for the translational diffusion was calculated to be 28 ± 1 kJ mol^{-1}. Ionic conductivity can be estimated from the following equation:

$$\sigma = Nel/(kT\tau_c) \tag{3.8}$$

where N is the concentration of the conduction ion, e the charge on the Cl atom, and l the length of an elementary jump. Using τ_c evaluated from Eq. 3.7, σ was estimated to be around 0.2 S cm^{-1} (at 450K), which is about one order larger than the conductivity measurement. The discrepancy probably arises from the numerical constant in Eq. 3.7, which we could not determine experimentally for the cubic phase.

Thus we have proposed the following mechanism for the high ionic conductivity. The mechanism consists of two steps: a translational displacement of the Cl ion between two 6f sites (Wyckoff notation of space group *Pm*3*m*), Ge—Cl···Ge $\leftrightarrow$ Ge···Cl—Ge, and a reorientation of the $GeCl_3^-$ anion about the C_3 axis. The latter reorientational mode of the $GeCl_3^-$ anion was found by means of the ^{35}Cl NQR technique even in the lower temperature phase, at which a pyramidal anion could be recognized.[28,31]

Quite similar temperature behavior of the conductivity was also confirmed for $(CH_3)_4NGeBr_3$ and $(CH_3)_3NHGeBr_3$. In the case of $(CH_3)_4NGeBr_3$ a disordered Br site was clearly recognized on the Fourier map, similar to that found for the chloride analog.[32] From the conductivity measurements for a series of $AGeX_3$, it became apparent that the conductivity at the cubic phase shows a maximum at a suitable cation size. This suggests that the height of the potential barrier may increase with increasing distance between a pair of disordered sites.

Figure 3.13 plots the Ge—Cl bond length against that of the Ge···Cl bond located at the *trans* position. A good correlation between them is found similar to the Sb(III) complexes shown in Figure 3.7. This figure includes all available data, not only low- but also high-temperature phases. It becomes apparent that

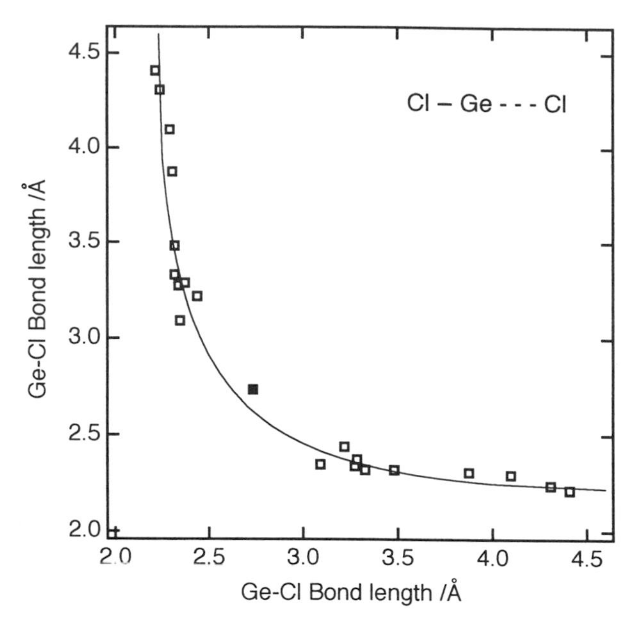

Figure 3.13 Correlation between bond lengths for a series of $AGeCl_3$. Ge—X bond length is plotted against its *trans* Ge···X length located at the *trans* position. The solid curves were calculated according to Eqs 3.5 and 3.6 by the elimination of n ($d_0 = 2.19$ Å, $a = 1.60$ Å). ■ = Cubic phase of $CsGeCl_3$.

the strong Ge—Cl bond leads to a weakening of the *trans* Ge···Cl bond and that the same correlation is kept at the disordered phase. Furthermore, Cs^+ is an ideal size for the cubic perovskite structure, and a certain amount of disorder or a large thermal anisotropy along the bond may be introduced if the cation is larger than Cs^+.

3.6 METALLIC OR SEMICONDUCTING PROPERTY OF $ASnX_3$ (X = I AND Br)

Table 3.2 summarizes the physical properties and spectroscopic parameters for crystalline $ASnBr_3$ (A = Cs and alkylammonium) in order of cationic size. $CsSnBr_3$ and $CH_3NH_3SnBr_3$ belong to cubic perovskite structures at 298K and are black and red, respectively.[25] In spite of the phase transition at 290K, the conductivity of $CsSnBr_3$ remains high over the temperature range studied. However, $CH_3NH_3SnBr_3$ showed a first-order phase transition at 229K, below which a sudden decrease of the conductivity[33,34] and a large anisotropy of the chemical shift (^{119}Sn NMR) were confirmed. In the case of a large cation such as A = $(CH_3)_2NH_2$ or $(CH_3)_3NH$, however, the color of the powdered sample is white and an isolated $SnBr_3^-$ anion is recognized in a

TABLE 3.2 Physical Properties and Spectroscopic Parameters for $ASnBr_3$

A	Color	Conductivity at 298K ($S\,cm^{-1}$)	Mean ^{81}Br NQR Frequency at 77K (MHz)	^{119}Sn NMR at 77K			*trans* Br—Sn—Br Bond Length (Å)
				$\delta_{iso}{}^{a}$ (ppm)	$\Delta\delta^{b}$ (ppm)	η^{c}	
Cs	Black	4×10^{1}	64.41	420^{d}			$2.904, 2.904^{f}$
CH_3NH_3	Red	2×10^{-5}	65.73 (298K)	0^{e}	0	0	$2.951, 2.951^{f}$
CH_3NH_3	Yellow(77K)		69.09	96	424	0.11	
$(CH_3)_2NH_2$	White		76.25	260	630	0.13	$2.684, 3.514^{g}$
$(CH_3)_3NH$	White		79.79	283	660	0.23	$2.665, 3.764^{g}$

[a] Isotropic chemical shift, $\delta_{iso} = (\delta_{11} + \delta_{22} + \delta_{33})/3$.

[b] Anisotropic parameter of chemical shift, $\Delta\delta = \delta_{33} - (\delta_{11} + \delta_{22})/2$.

[c] Asymmetry parameter of chemical shift, $\eta = (\delta_{22} - \delta_{11})/(\delta_{33} - \delta_{iso})$

[d] Knight shift.

[e] ^{119}Sn NMR spectra of $CH_3NH_3SnBr_3$ at 298K was used as a chemical shift standard for these Sn(II) halides because its sharp singlet has short relaxation time.

[f] Cubic perovskite structure.

[g] Mean values of three directions.

distorted octahedron.[35,36] The bond lengths corresponding to Br—Sn···Br are summarized in the last column in Table 3.2. Although the $SnBr_6$ octahedron in $(CH_3)_2NH_2$ or $(CH_3)_3NH$ salt distorts considerably from the regular one, the anion sublattice maintains the perovskite structure in which linear bridges Sn—Br···Sn are formed three-dimensionally throughout the crystal. The structural change of the $SnBr_3^-$ anion with increasing the cationic radius was also confirmed by the increases of the ^{81}Br NQR frequency and the chemical shift of the ^{119}Sn NMR. The large anisotropy of the ^{119}Sn chemical shift over 660 ppm suggests that the valence electron configuration changes to a coordination of trigonal pyramid. It should be emphasized that the electronic conductivity and color also change drastically in accordance with this cation replacement. Much higher conductivity is expected by the replacement of the cation with a smaller one; however, the Cs^+ is the smallest to form the perovskite structure. In contrast to the Ge halides, we could not confirm the cubic phase with a disordered anionic sublattice, for which a high conductivity due to the bromide ion was expected.

Figure 3.14 shows the ^{119}Sn NMR spectra and the temperature dependence of their spin–lattice relaxation times, T_1, for $CsSnBr_3$ and $CH_3NH_3SnBr_3$. As is expected from the cubic perovskite structure, $CH_3NH_3SnBr_3$ shows a sharp singlet at 297K, whereas the spectrum at 77K exhibits a large anisotropy of the chemical shift, suggesting a strong distortion of the tin coordination from regular octahedron. $CsSnBr_3$, however, shows a remarkable paramagnetic shift of 420 ppm at 77K relative to the cubic $CH_3NH_3SnBr_3$. Furthermore, its relaxation rate ($1/T_1$) is very fast compared to that of the CH_3NH_3 salt and the rate is proportional to the absolute temperature below 290K ($T_1 \cdot T = 2.3\,\mathrm{s\,K}$). This temperature dependence suggests that the $CsSnBr_3$ is metallic, and the relaxation rate is governed by the electron in the conduction band. Therefore, the paramagnetic shift stated above can be assigned to the Knight shift, which is commonly observed for the metallic compounds. These metallic features on the NMR spectra were also observed for $CsSnI_3$ or $CH_3NH_3SnI_3$; however, no successful spectrum could be detected, due to the large line width.

Figure 3.15 shows the temperature dependence of the conductivity for $CsSnI_3$ and $CH_3NH_3SnI_3$ determined by a four-point probe method using a DC current.[29] Although a small tilting of the SnI_6 unit was recognized in $CH_3NH_3SnI_3$ at the low-temperature phase, the perovskite structure is maintained essentially over the whole temperature range. As expected from the structure, the conductivity also remains metallic over the whole temperature range studied. However, $CsSnI_3$ obtained from an aqueous solution is a needle-like yellow crystal having a semiconducting conductivity. With increasing temperature, the conductivity of $CsSnI_3$ increased 10^4 times at 425K, and the metallic conductivity was kept in the cooling process.[29] The DTA measurement on $CsSnI_3$ showed a first-order phase transition accompanied by a drastic color change to black at 425K. The phase transition was irreversible and of a reconstructive type. Figure 3.16 shows both structures at room temperature. The

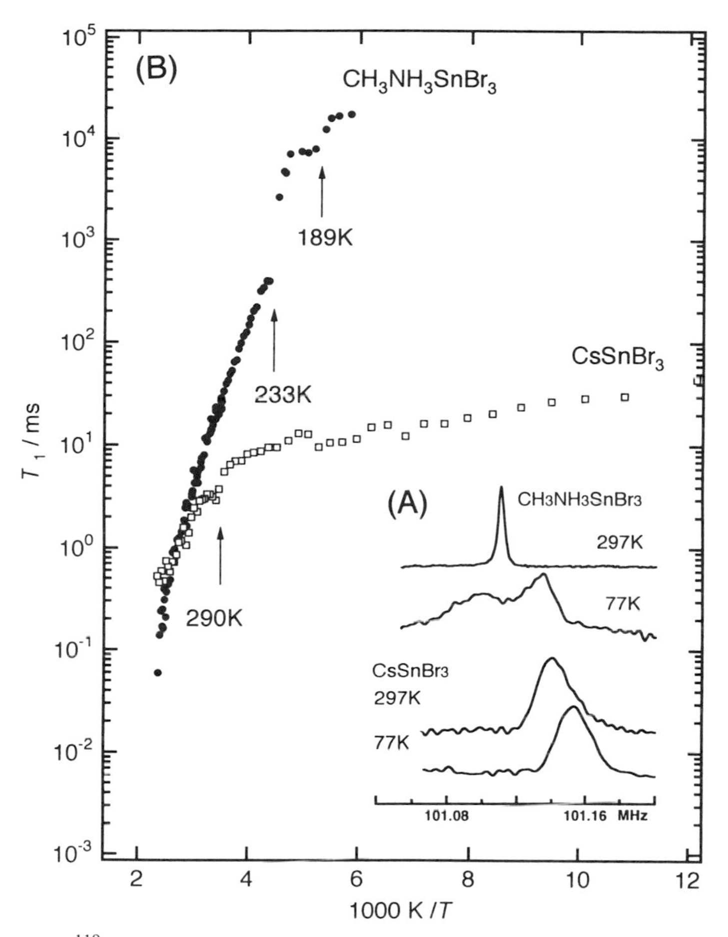

Figure 3.14 ^{119}Sn NMR spectra (A) and the temperature dependence of their spin–lattice relaxation times (B) for $CsSnBr_3$ and $CH_3NH_3SnBr_3$.

structure of the yellow phase was determined by X-ray diffraction using a single crystal.[37] That of the black phase, was determined by Rietveld analysis using a powdered sample, because this phase was unstable around the ambient temperature.[30] Although the black metallic phase showed successive phase transitions with decreasing temperature ($Pm3m \rightarrow P4/mbm \rightarrow Pnam$), SnI_6 remains almost a regular octahedron. The conductivity of $CsSnI_3$ (black phase) increased gradually with decreasing temperature down to 77K, and no anomalies were observed at the phase transition temperatures. The metallic conductivity may originate from the infinite and linear chains having almost the same Sn—I bond length; thereby the band structure is not affected by the tilt of the SnI_6 octahedra. In a similar manner to $ASnBr_3$ compounds, Cs^+ is the smallest ion that adopts a perovskite structure. A recent transport and optical study on

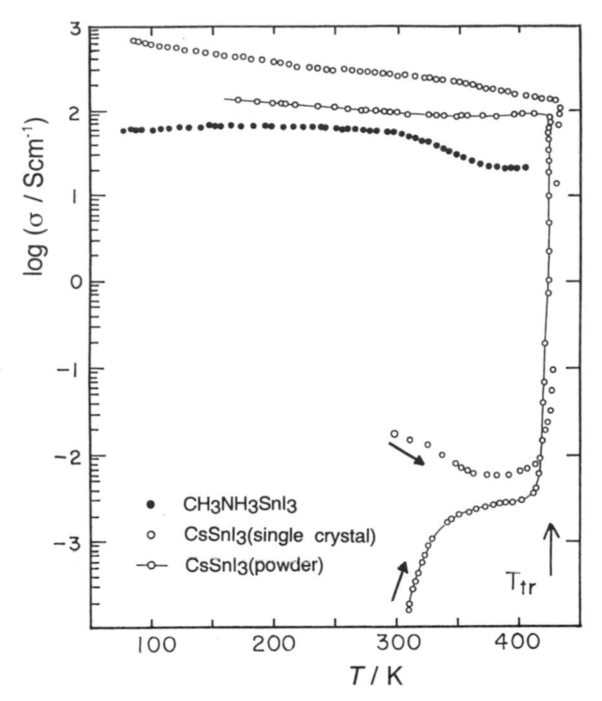

Figure 3.15 Temperature dependence of the electrical conductivities for $CsSnI_3$ and $CH_3NH_3SnI_3$.

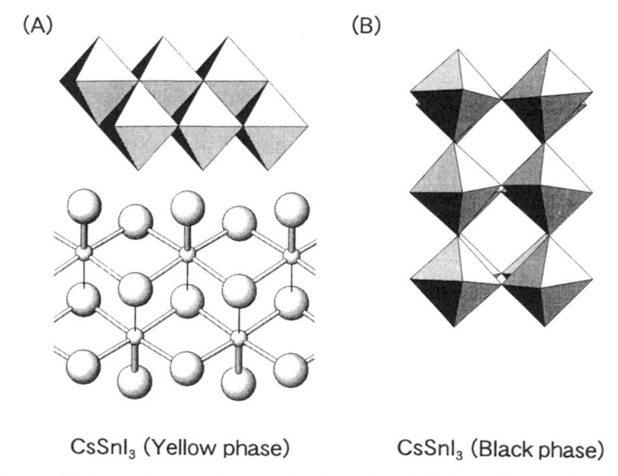

Figure 3.16 Two different anionic sublattices in $CsSnI_3$. (A) yellow phase; (B) black phase.

the perovskite $CH_3NH_3SnI_3$ confirmed that this unusual conducting halide was a low-carrier-density *p*-type metal in the temperature range 1.8–300K with no transition to the superconducting state in this temperature range.[38,39]

In order to understand the metallic properties of these Sn(II) compounds, a series of monomethylammonium salts substituting the central metal by Ge(II) or Pb(II) were synthesized. Figure 3.17 plots the conductivities for $CH_3NH_3BI_3$ (B = Ge, Sn, and Pb) against temperature. The structural changes accompanied by the successive phase transitions were determined by means of Rietveld analysis of the XRD data. Two transitions of $CH_3NH_3PbI_3$ were detected at 162K and 314K, and the highest temperature phase belongs to a cubic perovskite structure. With decreasing temperature it deforms to a tetragonal and further to a orthorhombic perovskite structure, as shown in Figure 3.18(A) and (B), respectively. A cubic $CH_3NH_3SnI_3$ undergoes a phase transition to a tetragonal system at 267K, which has a structure similar to the Pb analog shown in Figure 3.18(A). $CH_3NH_3GeI_3$, however, belongs to a rhombohedral system above 220K. No phase transition to a cubic phase has been confirmed in the temperature range studied. The rhombohedrally distorted structure is shown in Figure 3.18(C), in which an alternation of the short and long bonds (Ge—I···Ge—I) is formed linearly. On the basis of such structural inform-

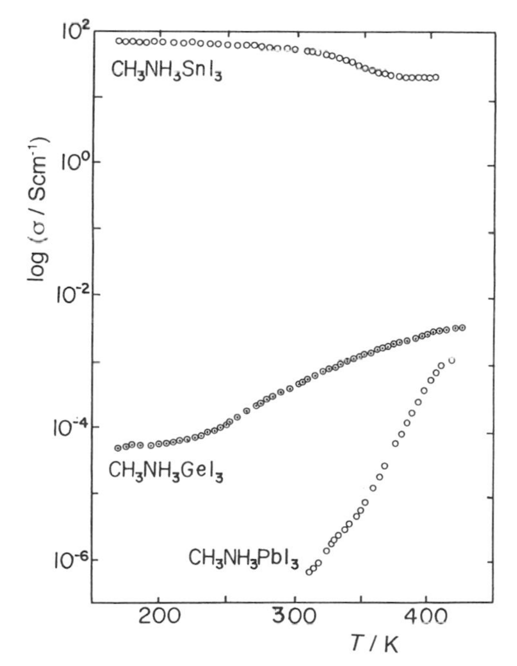

Figure 3.17 Temperature dependence of the electrical conductivities for $CH_3NH_3SnI_3$, $CH_3NH_3GeI_3$, and $CH_3NH_3PbI_3$.

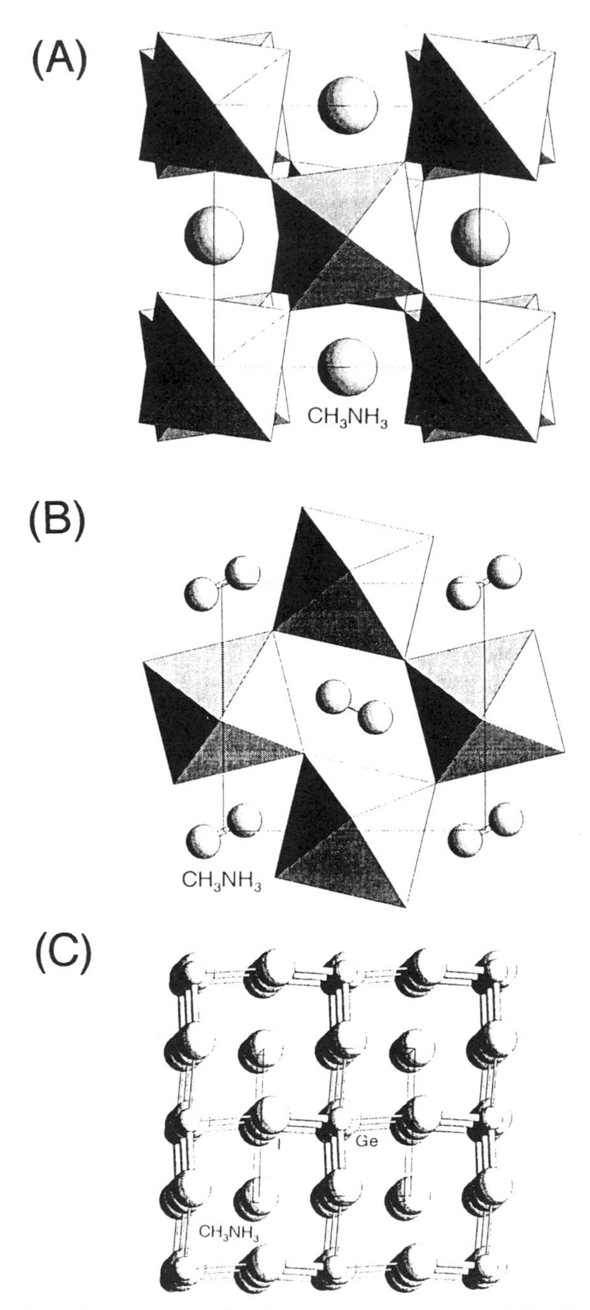

Figure 3.18 Crystal structures of (A) tetragonal phase of $CH_3NH_3PbI_3$ (294K), (B) orthorhombic phase of $CH_3NH_3PbI_3$ (116K), and (C) rhombohedral phase of $CH_3NH_3GeI_3$ (298K).

ation, the following empirical conclusions may be derived concerning the electronic properties of these perovskite halides. The metallic conductivity appears in the perovskite lattice in which the I—B—I bond is formed symmetrically. Thereby, the conductivity increases with decreasing lattice constant as long as there is no Peierls-type distortion, such as seen in $CH_3NH_3GeI_3$.

3.7 CONDUCTIVITY AND DYNAMIC STRUCTURE OF $ASnCl_3$ ($A = Cs$ AND CH_3NH_3)

In the previous sections we have described a high ionic conductivity for $AGeCl_3$ ($A = CH_3NH_3$ or $(CH_3)_4N$) and a metallic conductivity of $ASnI_3$ ($A = Cs$ or CH_3NH_3). From the standpoints of electronic property and structure, it is interesting to determine which type of conductivity appears for the trichlorostannate, $ASnCl_3$. Furthermore, it is important to establish the structural conditions for the electronic or ionic conductor.

Figure 3.19 shows the DTA curves for $CsSnCl_3$ and $CH_3NH_3SnCl_3$. An irreversible phase transition was detected for $CsSnCl_3$ at 379K, together with the color change from white to pale yellow. However, $CH_3NH_3SnCl_3$ showed complex phase transitions between 250–331K and a transition to a cubic perovskite phase at 463K.[40] The endothermic peak at 331K has a long tail at the

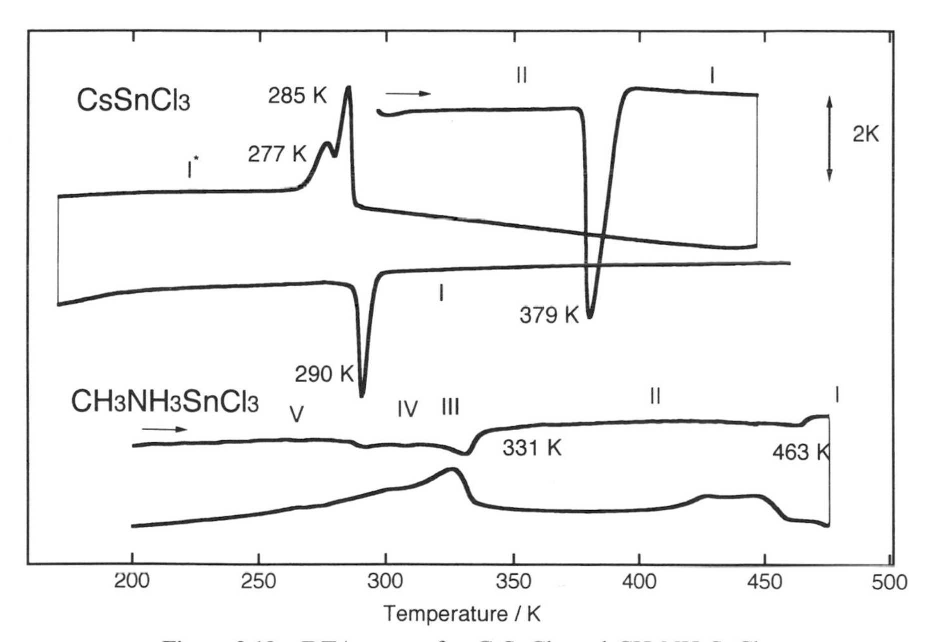

Figure 3.19 DTA curves for $CsSnCl_3$ and $CH_3NH_3SnCl_3$.

lower-temperature side, suggesting an order–disorder transition. Each phase was tentatively designated as Phase I, II, and so on from the high-temperature side. Figure 3.20 reproduces the temperature dependencies of the conductivities. The conductivity of $CH_3NH_3SnCl_3$ increases continuously with increasing temperature and there is no discontinuity at T_{tr}. However, a sudden increase in the conductivity was observed for $CsSnCl_3$ at T_{tr}. In the cooling process, however, the conductivity of $CsSnCl_3$ remained high because of the quenching of the high-temperature phase. After about 24 hours, the conductivity decreased to that of Phase II and a similar hysteresis loop could be reproduced again. In order to assign the charge carrier at the highly conducting phase, we measured the polarization effect using a DC current. However, no polarization effect was observed, in contrast to our expectation. These preliminary experiments suggested that the conductivity at the cubic phase was

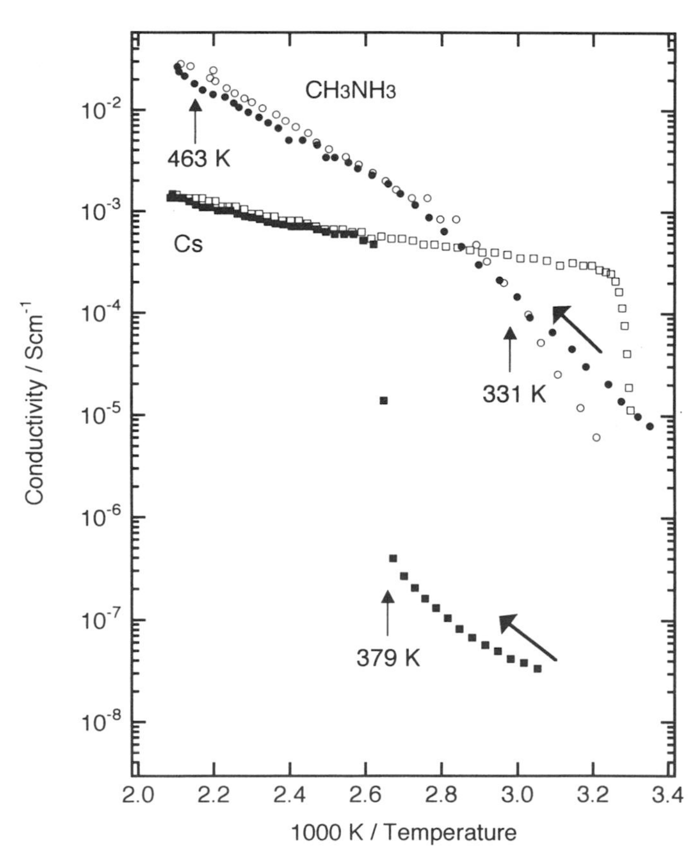

Figure 3.20 Temperature dependence of the conductivities for $CsSnCl_3$ and $CH_3NH_3SnCl_3$ determined by a complex impedance method.

mostly governed by the semiconducting property similar to that found for the cubic $CH_3NH_3SnBr_3$.

According to Poulsen and colleagues, $CsSnCl_3$ belongs to a monoclinic system in which Sn forms a strongly distorted octahedron having three short bonds (S—Cl; 2.50–2.55 Å) and three long bonds (Sn···Cl: 3.21–3.77 Å).[41] Above T_{tr} the XRD pattern changed to a cubic system having a perovskite structure, as was already reported by Sharma.[27] Figure 3.21 shows structures below and above the phase transition. These structures clearly show the reconstructive nature of the phase transition, and the large hysteresis of the conductivity is also consistent with the structural change. However, although the crystal symmetry of $CH_3NH_3SnCl_3$ decreased successively at the phase transitions, the Rietveld analysis at each phase suggested that the perovskite structure was essentially kept over the wide temperature range studied. The symmetric *trans* Cl—Sn—Cl bonds in the cubic phase deform asymmetrically below the rhombohedral phase. The structure at the rhombohedral phase is isomorphous with $CsGeBr_3$[42] or $CsGeCl_3$,[43] and a pyramidal $SnCl_3^-$ anion is recognized in a distorted octahedral geometry. The change of the slope in the conductivity versus $1/T$ at around 370K may suggest a gradual change of the electronic structure, due to the asymmetric deformation of the Cl—Sn—Cl bond stated above. The high ionic conductivity similar to $CH_3NH_3GeCl_3$ was not confirmed for $CH_3NH_3SnCl_3$ by our conductivity measurements because of the relatively high semiconducting conductivity. However, a large thermal anisotropy at the Cl site parallel to the Sn—Cl—Sn bond has become apparent for $CH_3NH_3SnCl_3$ from the Fourier map at the cubic phase. This finding suggests a possibility of the chloride ion conductivity together with a semiconducting property. Therefore, the dynamic aspects of $CH_3NH_3SnCl_3$ were studied by means of the broadline ^1H, ^2H, and ^{119}Sn NMR spectroscopies. ^1H and

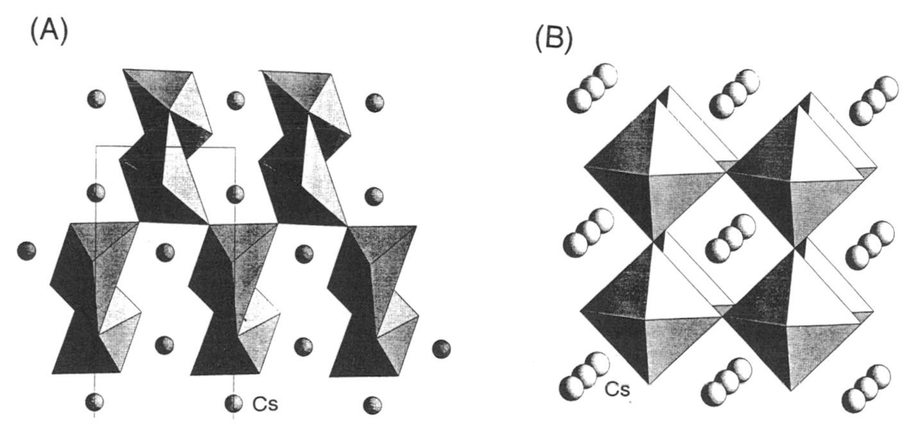

Figure 3.21 Crystal structure of $CsSnCl_3$. (A) Phase II (monoclinic); (B) Phase I (cubic perovskite).

^{2}H NMR spectra over the wide temperature range suggested that an isotropic reorientation of the cation was excited above the rhombohedral phase, but a translational diffusion of the cation could not be confirmed even in the cubic perovskite phase. These NMR observations suggested that the phase transition at 331K was an orientational order–disorder transition of the $CH_3NH_3^+$ ion. From the T_1 measurements of the ^{2}H NMR, the activation energy for the isotropic reorientation of the cation was determined to be 17.7 kJ mol^{-1} at the rhombohedral phase. In order to get some dynamic information about the chloride ion, the ^{119}Sn NMR spectrum and its relaxation time were investigated as a function of temperature. A superposition of a sharp and a broad component appears even in the rhombohedral phase, as shown in Figure 3.22. This is in contrast to the ^{119}Sn NMR in $CsSnCl_3$, in which the broad powder pattern changed to a sharp singlet at T_{tr}. This partially averaging effect of the chemical shift in $CH_3NH_3SnCl_3$ suggests the onset of the chloride ion diffusion. Mizusaki and co-workers reported the halide ionic conductivity for

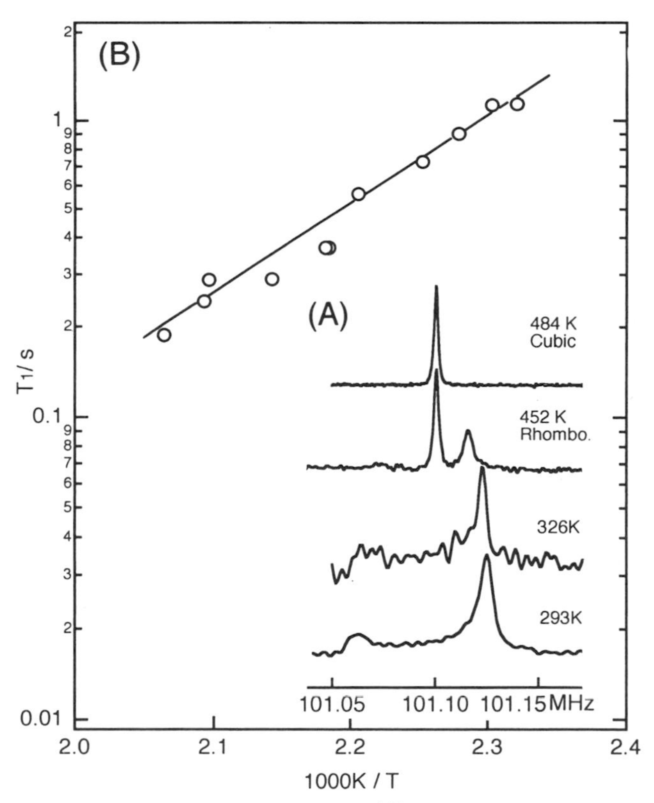

Figure 3.22 Temperature dependence of (A) ^{119}Sn NMR spectra and (B) spin–lattice relaxation times for $CH_3NH_3SnCl_3$.

$CsPbCl_3$ or $CsPbBr_3$.[44] According to them the ionic conduction was caused by the migration of halide ion vacancies. As the result of the vacancy diffusion, a time-averaged coordination around the Sn becomes regular octahedral, resulting in no anisotropy of the chemical shift. At the same time the interaction due to the time-dependent chemical shift affects the relaxation of the Zeeman energy. The form of the T_1 for ^{119}Sn is as follows;

$$1/T_1 = (2/15)\gamma^2 B_0^2\ \delta^2\ [\tau_c/(1+\omega^2\tau_c^2)] \tag{3.9}$$

where γ is a magnetogyric ratio, B_0 the magnetic field, and δ the chemical shift anisotropy, $\delta = \sigma_{33} - \sigma_{iso}$. Assuming an Arrhenius equation for τ_c and a slow-motion limit ($\omega\tau_c \gg 1$), the activation energy for the diffusion of chloride ions was determined to be 54 kJ mol^{-1}, which was much higher than that of the chloride ion conductor $CH_3NH_3GeCl_3$.

^{35}Cl NQR provides useful information not only about the chemical bond but also about the dynamic information. Figure 3.23 shows temperature dependence of NQR frequency and its spin–lattice relaxation time for $CsSnCl_3$ at Phase II. Three ^{35}Cl NQR frequencies agreed well with that reported by Scaife and co-workers.[45] Just below the phase transition at 379K, the relaxation rate, $1/T_1$, increases exponentially against temperature. The temperature dependence of the relaxation rate assigned to the ν_2 line could be satisfactorily reproduced by

$$1/T_1(\mathrm{sec}^{-1}) = (5.75\times10^{-5})T^{2.02} + (1.32\times10^{12})\exp(-62.8\,\mathrm{kJ\,mol^{-1}}/RT) \tag{3.10}$$

The first term represents a Raman process that is nearly proportional to T^2, as expected from the mechanism. The second term represents a contribution from the reorientation of the group containing probe nuclei. The activation energy, 62.8 kJ mol^{-1}, was assigned to the reorientation of $SnCl_3^-$ anions around its pseudo C_3 axis, because an isolated anion was recognized in Phase II. This activation energy is comparable to those determined for $AGeCl_3$ compounds.[28] Since the energy corresponds roughly to the sum of the three Cl···Sn interactions in the anionic sublattice, we can estimate the bond energy to be around 21 kJ mol^{-1} for the Cl···Sn interaction that appears in the asymmetric 3c–4e bond.

ACKNOWLEDGMENT

This work was supported by the Grant-in-Aid for Scientific Research on Priority Area of "Organic Unusual Valency" and "Solid State Ionics" from the Ministry of Education, Science and Culture, Japan.

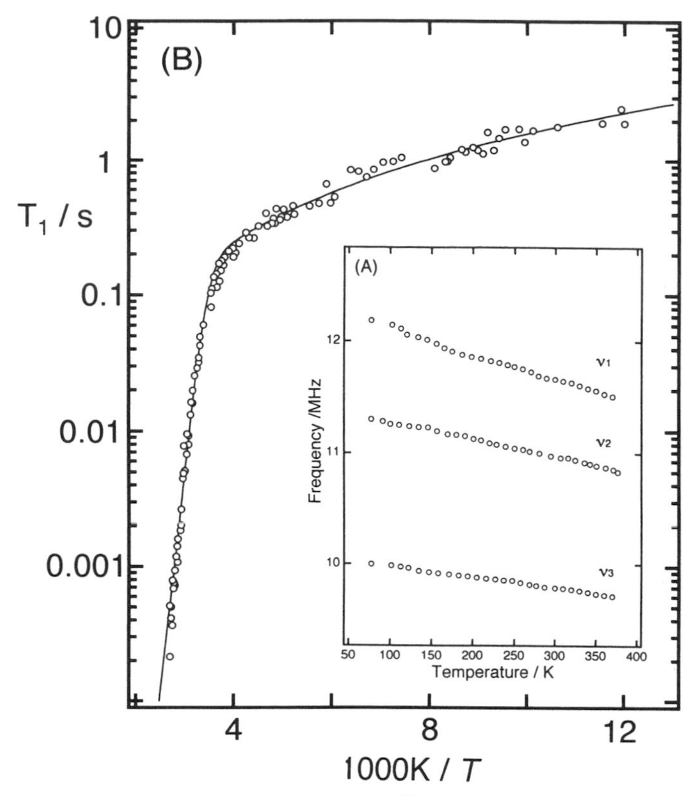

Figure 3.23 Temperature dependence of (A) ^{35}Cl NQR frequencies and (B) spin-lattice relaxation times for $CsSnCl_3$.

REFERENCES

1. J. I. Musher, *Angew. Chem. Int. Ed.*, **8**, 54 (1969).
2. E. M. Shustorovich, Yu. A. Buslaev, *Inorg. Chem.*, **15**, 1142 (1976).
3. N. W. Alcock, *Adv. Inorg. Chem. Radiochem.*, **15**, 1(1972).
4. G. C. Pimentel, R. D. Spratley, *Chemical Bonding* (Holden-Day, 1969), Chapter 7.
5. R. C. L. Mooney Stater, *Acta Crystallogr.*, **12**, 187 (1959).
6. G. H. Cheesman, A. J. T. Finney, *Acta Crystallogr., sect. B,* **26**, 904 (1970).
7. D. Nakamura, M. Kubo, in *Advances in Nuclear Quadrupole Resonance Vol. 2*, J. A. S. Smith, Ed. (Heyden & Son, London, 1975), Chapter 4.
8. G. C. Pimentel, *J. Chem. Phys.*, **19**, 446 (1951).
9. R. J. Gillespie, I. Hargittai, *The VSEPR Model of Molecular Geometry* (Allyn and Bacon, Massachusetts, 1991), Chapter 5.

10. T. Okuda, K. Yamada, H. Ishihara, M. Hiura, S. Gima, H. Negita, *J. Chem. Soc. Chem. Commun.*, 979 (1981).
11. F. Izumi, in *The Rietveld method*, R. A. Young, Ed. (Oxford University Press, Oxford, 1993), Chapter 13. p. 236.
12. K. Yamada, T. Ohtani, S. Shirakawa, H. Ohki, T. Okuda, T. Kamiyama, H. Oikawa, *Z. Naturforsch.*, **51a**, 739 (1996).
13. P. W. DeHaven, R. A. Jacobson, *Struct. Comm.*, **5**, 31 (1976).
14. J. A. Ripmeester, *J. Chem. Phys.*, **85**, 747 (1986).
15. M. A. Altbach, Y. Hiyama, R. J. Wittebort, L. G. Butler, *Inorg. Chem.*, **29**, 741 (1990).
16. T. Okuda, Y. Aihara, N. Tanaka, K. Yamada, S. Ichiba, *J. Chem. Soc. Dalton Trans.*, 631 (1989).
17. T. Okuda, M. Hiura, E. Koshimizu, H. Ishihara, Y. Kushi, H. Negita, *Chem. Lett.*, 1321 (1982).
18. T. Okuda, Y. Kinoshita, H. Terao, K. Yamada, *Z. Naturforsch.*, **49a**,185 (1994).
19. W. G. McPherson, E. A. Meyers, *J. Phys. Chem.*, **72**, 3117 (1968).
20. L. Pauling, *The Nature of the Chemical Bond* (Cornell University Press, New York, 1960), Chapter 7.
21. R. Jakubas, *Solid State Commun.*, **60**, 389 (1986).
22. R. Jakubas, Z. Czapla, Z. Galewski, L. Sobczyk, *Ferroelectrics Lett.*, **5**, 143 (1988).
23. J. Mroz, R. Jakubas, *Ferroelectrics Lett.*, **11**, 53 (1990).
24. R, Jakubas, *Solid State Commun.*, **69**, 267 (1989).
25. K. Yamada, S. Nose, T. Umehara, T. Okuda, S. Ichiba, *Bull. Chem. Soc. Jpn.*, **61**, 4265 (1988).
26. K. Yamada, T. Hayashi, T. Umehara, T. Okuda, S. Ichiba, *Bull. Chem. Soc. Jpn.*, **60**, 4203 (1987).
27. S. Sharma, N. Weiden, A. Weiss, *Z. Naturforsch.* **46a**, 329 (1991).
28. K. Yamada, K. Isobe, T. Okuda, Y. Furukawa, *Z. Naturforsch.*, **49a**, 258 (1994).
29. K. Yamada, T. Matsui, T. Tsuritani, T. Okuda, S. Ichiba, *Z. Naturforsch.*, **45a**, 307 (1990).
30. K. Yamada, S. Funabiki, H. Horimoto, T. Matsui, T. Okuda, S. Ichiba, *Chem. Lett.*, **1991**, 801.
31. K. Yamada, K. Isobe, E. Tsuyama, T. Okuda, Y. Furukawa, *Solid State Ionics*, **79**, 152 (1995).
32. Y. Okuda, S. Gotou, T. Takahashi, H. Terao, K. Yamada, *Z. Naturforsch.*, **51a**, 686 (1996).
33. K. Yamada, H. Kawaguchi, T. Matsui, T. Okuda, S. Ichiba, *Bull. Chem. Soc. Jpn.*, **63**, 2521 (1990).
34. N. Onoda-Yamamuro, T. Matsuo, H. Suga, *J. Chem. Thermodynamics* , **23**, 987 (1991).
35. G. Thiele, B. G. Serr, *Z. Kristallogr.*, **211**, 47 (1996).
36. G. Thiele, B. G. Serr, *Z. Kristallogr.*, **211**, 46 (1996).
37. P. Mauersberger, F. Huber, *Acta Crystallogr., Sect. B*, **36**, 683 (1980).

38. D. B. Mitzi, C. A. Feild, Z. Schlesinger, R. B. Laibowitz, *J. Solid State Chem.*, **114**, 159 (1995).
39. D. B. Mitzi, C. A. Feild, W. T. A. Harrison, A. M. Guloy, *Nature*, **369**, 467 (1994).
40. K. Yamada, Y. Kuranaga, K. Ueda, S. Goto, T. Okuda, Y. Furukawa, *Bull. Chem. Soc. Jpn.*, **71**, 127 (1998).
41. R. Poulsen, S. E. Rasmussen, *Acta Chem. Scand.*, **24**, 150 (1970).
42. G. Thiele, H. W. Rotter, K. D. Schmidt, *Z. anorg. allg. Chem.*, **545**, 148 (1987).
43. A. N. Christensen, S. E. Rasmussen, *Acta Chem. Scand.*, **19,** 421 (1965).
44. J. Mizusaki, K. Arai, K. Fueki, *Solid State Ionics*, **11**, 203 (1983).
45. D. E. Scaife, P. F. Weller, W. G. Fisher, *J. Solid State Chem.*, **9**, 308 (1974).

4 Structure and Reactivity of Hypercoordinate Silicon Species

CLAUDE CHUIT, ROBERT J. P. CORRIU, AND CATHERINE REYE

Laboratoire de Chimie Moléculaire et Organisation du Solide, Université Montpellier II, Sciences et Techniques du Languedoc, Montpellier, France

4.1 INTRODUCTION

Silicon compounds with coordination number greater than four have been known since the beginning of the nineteenth century, when Gay-Lussac[1] and Davy[2] reported the formation of the adduct $SiF_4 \cdot 2NH_3$. Some years later, Berzelius[3] synthesized fluorosilicic acid and its salts, the hexafluorosilicates. About a century later the octahedral geometry of the $[SiF_6]^{2-}$ dianion was established.[4] The formation and structure of hypercoordinate silicon compounds continue to be an area of lively interest,[5–7] which has been regularly reviewed.[8–15] In recent years, interest in hypercoordinate silicon compounds has grown considerably, because of their unexpected reactivity. Interest in this field started with studies of nucleophilic activation at silicon. It was shown that some reactions (racemization of optically active silicon compounds, hydrolysis and alcoholysis of tetracoordinate chlorosilanes) are highly accelerated by nucleophiles.[16] These studies suggested that a pentacoordinate intermediate reacts with nucleophiles faster than does the starting tetracoordinate silane. The enhanced reactivity of pentacoordinate silicon compounds in nucleophilic substitutions was demonstrated experimentally[17] and was supported by calculations which show that the positive charge on the central atom is at least maintained[18] and may well be increased[19] by coordination of an additional ligand, even when that ligand is anionic (fluoride or hydride ion).

In this chapter we describe the preparation, structure, and reactivity of pentacoordinate silicon compounds, hexacoordinate silicon compounds, and silicon compounds with higher coordination numbers.

Chemistry of Hypervalent Compounds, edited by Kin-ya Akiba.
ISBN 0-471-24019-2

4.2 PENTACOORDINATE SILICON COMPOUNDS

4.2.1 Anionic Acyclic Complexes

4.2.1.1 Preparation, Structure, and Fluxional Behavior

4.2.1.1.1 Pentacoordinate Fluorosilicates In 1966, Müller and Dathe[20] prepared the first pentacoordinate fluorosilicate, $[MeSiF_4]^-$ $[Et_4N]^+$, by reaction of $MeSiF_3$ with $Et_4N^+F^-$. Soon after, Clark and Dixon[21] and Harris and Rudner[22] obtained the pentafluorosilicate anion $[SiF_5]^-$. The preparation of the tetrafluorosilicates was then developed by Klanberg and Muetterties.[23] They prepared also the diphenyltrifluorosilicate anion $[Ph_2SiF_3]^-$. Nuclear magnetic resonance (NMR)[23,24] and vibrational spectroscopic data[25,26] strongly suggest that these anions are pentacoordinated at silicon. Recently, $[Ph_3SiF_2]^-$,[27] $[Ph_2MeSiF_2]^-$,[27b] and $[PhMe_2SiF_2]^-$[27b] were prepared as their ammonium salts in good yields.

Addition of fluorine anions to organosilanes is also possible in the gas phase. Thus, in 1980, DePuy and co-workers,[28] by using a flowing afterglow system, added fluoride anions, generated *in situ*, to tetramethyl, tetravinyl, and tetramethoxysilane to give the corresponding pentacoordinate silicates $[R_4SiF]^-$ (R = Me, vinyl, MeO). Cyclic silicates **1** and **2** were also obtained by treatment of the corresponding silane with fluoride ion.[29] Following computational studies[30] this reaction was extended to the reaction of various anions with organosilanes Me_3SiX forming pentacoordinate silicates. Study of these complexes in the gas phase allows the exploration of their chemistry and fundamental properties in the absence of solvent and counteranions.[31,32]

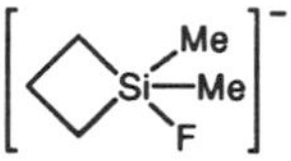

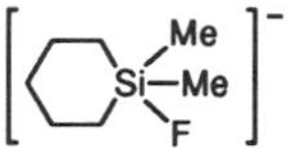

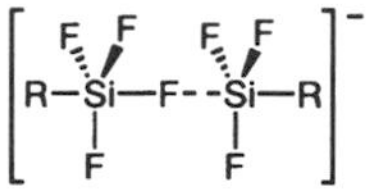

The crystallographic analysis of pentacoordinate silicates was first performed by Schomberg and Krebs[33,34] for $[SiF_5]^-$, $[PhSiF_4]^-$, and $[Ph_2SiF_3]^-$ and subsequently by Holmes and coworkers[35] for $[PhMeSiF_3]^-$ and $[\alpha NpPh_2SiF_2]^-$. The structure of these anions is close to trigonal bipyramidal, with the organic groups in equatorial positions. The length of an axial Si–F bond is always greater than the length of an equatorial Si—F bond. Moreover, there is a tendency for both lengths to increase with increasing steric bulk of the organic groups as well as with a diminishing number of electronegative atoms. Dynamic NMR studies of these silicates performed by Klanberg and Muetterties[23] suggested that these complexes are subject to more or less rapid intramolecular exchange of fluorine between axial and equatorial sites. Accurate studies are difficult however, the fluorosilicates being very hygroscopic and giving rise to impurities able to catalyze intermolecular fluorine ion exchange reactions.[24] NMR studies have shown that intermolecular fluorine exchange via the hexacoordinate intermediate **3**[36] occurs also on reaction of $[RSiF_4]^-$ anions with $RSiF_3$.

In 1986, Damrauer and Danahey[37] found a general way to prepare stable and non-hygroscopic pentacoordinate organofluorosilicates as their 18-crown-6 potassium salts. X-ray structure analysis of organotetrafluoro $[RSiF_4]^-$[38] and of organotrifluorosilicates $[R_2SiF_3]^-$[39–42] confirmed the trigonal pyramidal arrangement around the silicon atom. Dynamic NMR studies[37] provide further evidence of the intramolecular nature of the fluorine site exchange. These exchanges are extremely fast for $[SiF_5]^-$ and $[RSiF_4]^-$ anions, with the exception of the hindered silicate **4**, for which the fluorine site exchange occurs with an energy of 53.5 kJ mol^{-1}. This exchange is also fast for the five-membered silicates **5**,[41] **6a**,[39] and **7**.[43] The free energies of activation for the fluorine site exchange lie in the range of about 38 to 50 kJ mol^{-1} for acyclic fluorosilicates of type $[RR'SiF_3]^-$[42,44] and for the cyclic silicates **6b**[39] and **8**.[43] Recent calculations performed on $[SiH_5]^-$[45] and $[SiF_5]^-$[46] indicate that pseudorotation barriers for these anions are less than 12.5 kJ mol^{-1}.

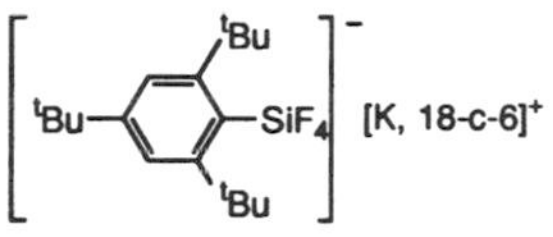

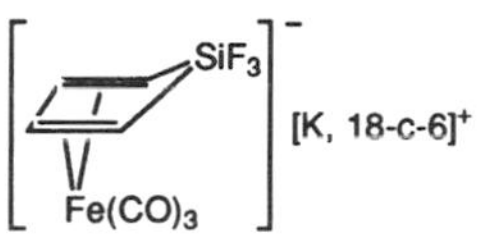

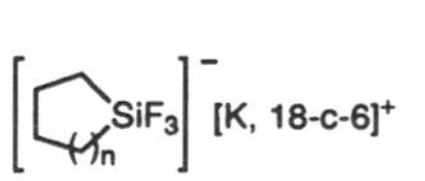

Recently, derivatives **9**,[38] **10**[47] and **11**[48] with two pentacoordinate silicon atoms have been prepared by addition of KF in the presence of 18-crown-6 to the corresponding tetracoordinate fluorosilanes. The X-ray structure analysis of **10**[47a] showed that the geometry around each silicon atom is similar to that found for $[Ph_2SiF_3]^-$ Me_4N^+.[34] Dynamic ^{13}C NMR studies have shown fluorine-ion exchange in solution, for which a concerted bimolecular exchange mechanism through a cyclophane-like transition state **12** was proposed.[47b]

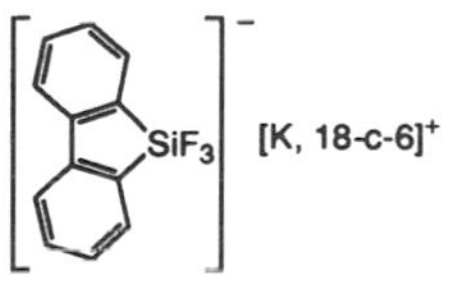

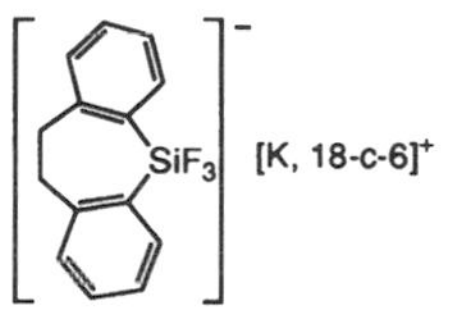

The X-ray structure determination of the pentacoordinate bis(silicates)**13a**–**13c**[49] showed the presence of a bent fluorine bridge between the two silicon atoms. The geometry about each silicon atom corresponds to a distorted trigonal bipyramid, with two fluorine atoms occupying the apical positions. The fluoride bridge is unsymmetric, with Si—F bond lengths of 1.898 Å and 2.065 Å

$[F_4SiCH_2CH_2SiF_4]^{2-}$ 2 $[K, 18\text{-c-}6]^+$

9

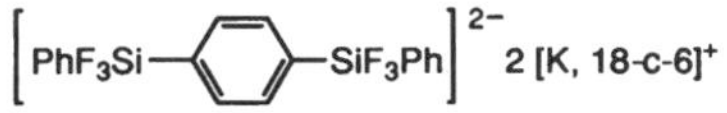

$[PhF_3Si-C_6H_4-SiF_3Ph]^{2-}$ 2 $[K, 18\text{-c-}6]^+$

10

$[EtF_3Si-C_6H_4-C_6H_4-SiF_3Et]^{2-}$ 2 $[K, 18\text{-c-}6]^+$

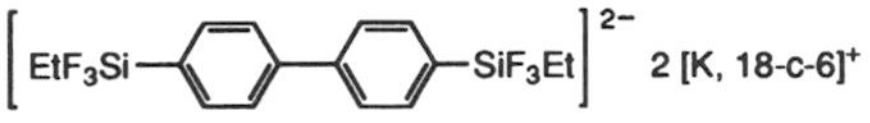

11

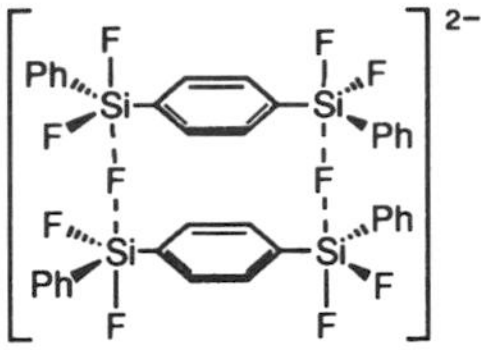

12

in **13a**, 1.700 and 2.369 Å in **13b**, and 1.805 and 2.090 Å in **13c**. These Si—F bonds are the longest found in pentacoordinate silicates. Dynamic ^{19}F NMR studies showed that all the fluorine ligands undergo intramolecular exchange between the two silicon atoms. These exchanges have been explained by the following consecutive processes: (1) exchange of the fluorine bridge between tetracoordinate and pentacoordinate silicon atoms, (2) flipping of the five-membered ring, (3) rotation about the Si—C bond, and (4) pseudorotation at pentacoordinate silicon centres.

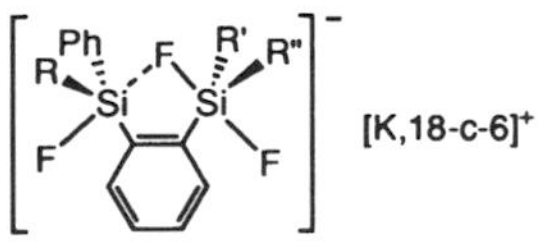

$[K, 18\text{-c-}6]^+$

13 a : R = R' = F, R" = Ph
b : R = Ph, R' = R" = F
c : R = R' = Ph, R" = F

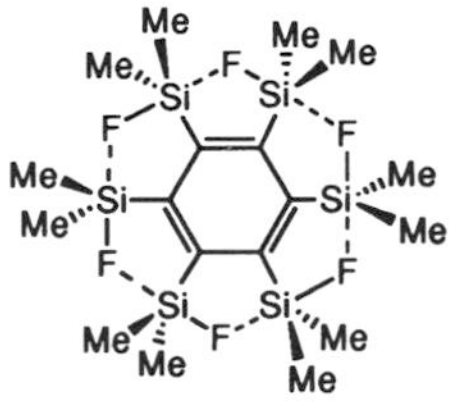

14

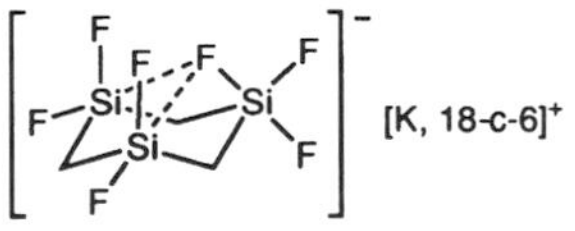

15 $[K, 18\text{-c-}6]^+$

Sakurai and co-workers[50] described recently the structure and dynamic NMR behavior of hexakis(fluorodimethylsilyl)benzene **14**, a neutral derivative. The X-ray structure determination showed that all the silicon atoms are pentacoordinated with a nearly trigonal pyramidal geometry, the average Si—F bond lengths being 1.68 and 2.39 Å. Dynamic NMR studies showed that at 328K there is rotation of the dimethylsilyl groups around the Si—aryl bonds and migration of the fluorine atoms throughout the ring, the fluorine atoms moving like a "merry-go-round."

Another example of migration of a fluorine anion between silicon atoms was detected in compound **15**, which contains three silicon atoms.[51] The X-ray structure analysis showed that one fluorine anion is not well positioned, and dynamic NMR studies have shown that the fluorine atoms exchange very rapidly at room temperature. A mechanism was proposed in which rapid transfer of the bridging fluorine ion between two silicon atoms occurs simultaneously with conformational equilibrium of the six-membered ring.

The pentacoordinate fluorosilicate F_4Si^{-}–$(CH_2)_3NH_3^{+}$, in which both anionic and cationic centres are in the same molecule, was first mentioned by Müller[52] in 1966. Recently these derivatives were extensively studied by Tacke and co-workers.[53,54] They are usually prepared by reaction of HF with the appropriate methoxysilane, but they can also be obtained by Si—phenyl bond cleavage (Eq. 4.1). X-ray structure analyses of these compounds showed that the silicon atom is pentacoordinated with a trigonal bipyramidal geometry, and dynamic NMR studies indicated that they undergo facile intramolecular fluorine exchange.[54b]

$$PhMe(MeO)_2SiCH_2N\langle\text{pyrrolidine}\rangle \xrightarrow{HF} MeF_3\overset{-}{Si}CH_2\overset{+}{N}H\langle\text{pyrrolidine}\rangle \quad (4.1)$$

4.2.1.1.2 Pentacoordinate Alkoxysilicates Stable pentacoordinate alkoxysilicates **16**,[55] **17**,[55] and **18**[56] have been prepared by potassium alkoxide addition to the corresponding alkoxysilane in the presence of 18-crown-6 ether. **16** has also been obtained by Si—phenyl bond cleavage of the organosilane **19**[55b] (Eq. 4.2). The formation of pentaalkoxysilicates $[Si(OR)_5]^-$ $[K, 18\text{-c-}6]^+$ resulting from the addition of RO^- K^+ to $Si(OR)_4$, in the presence of 18-crown-6 ether, was inferred from the upfield shift of the ^{29}Si NMR resonances.[56]

$[MeSi(OEt)_4]^-$ $[K, 18\text{-c-}6]^+$ **16**

$[Ph_nSi(OMe)_{5-n}]^-$ $[K, 18\text{-c-}6]^+$ **17**

$[PhSi(OCH_2CF_3)_4]^-$ $[K, 18\text{-c-}6]^+$ **18**

$$\underset{\mathbf{19}}{PhMeSi(OEt)_2} + EtO^- K^+ \xrightarrow[\text{18-crown-6}]{\text{Toluene}} \underset{\mathbf{16}}{[MeSi(OEt)_4]^- [K, 18\text{-c-}6]^+} \quad (4.2)$$

Reaction of RO^-K^+ with trialkoxysilanes $HSi(OR)_3$ affords the anionic pentaccordinate hydrosilicates $[HSi(OR)_4]^-$ K^+, **20** in good yields even in the absence of crown ether.[57,58] Potassium hydride reacts with trialkoxysilanes[59,60] (Eq. 4.3) to give a mixture of dihydrosilicates **20** and **21**, the **20**/**21** ratio depending on the nature of the R group. When R = Me or Ph, only **20** is formed, but with secondary R groups dihydrosilicates **21** are obtained in good yields. The formation of **20** was explained by the disproportionation of **21**. Dynamic ^1H NMR studies of hydrosilicates **20** failed to distinguish between substituents in axial and equatorial sites; rapid intramolecular exchange appears highly likely. However, with the dihydrosilicates **21** (R = ^iPr, ^sBu), it was inferred from NMR studies that in nonpolar solvants (benzene, toluene) one hydrogen is in an equatorial and the other in an axial position. In polar solvents (THF, DMF, or in the presence of 18-crown-6), an intramolecular process occurs in which the hydrogen atoms interchange axial-equatorial positions.[60]

$$HSi(OR)_3 \xrightarrow[THF]{KH} \underset{\mathbf{20}}{[HSi(OR)_4]^- K^+} + \underset{\mathbf{21}}{[H_2Si(OR)_3]^- K^+} \quad (4.3)$$

Hydrosilicate **20a** (R = Et) was not formed in the reaction of KH with $Si(OEt)_4$.[61] However, addition of KH to an equimolar mixture of Ph_3SiH and $Si(OEt)_4$ affords the complex **20a** (Eq. 4.4). It was suggested that KH adds to Ph_3SiH to give the transient complex $[Ph_3SiH_2]^-K^+$ which then transfers a hydride to $Si(OEt)_4$ to form the stable complex **20a**. Such a mechanism had previously been proposed to explain the racemization of the optically active organosilane αNpPhMeSiH catalyzed by KH or $LiAlH_4$[62] as well as the disproportionation of $PhSiH_3$ in the presence of KH or $LiAlH_4$.[63]

$$Ph_3SiH + KH + Si(OEt)_4 \xrightarrow{THF} \underset{\mathbf{20a}}{[HSi(OEt)_4]^- K^+} + Ph_3SiH \quad (4.4)$$

Squires and co-workers[64,65] prepared in the gas phase under flowing afterglow conditions the pentacoordinate silicon hydrides $[RSiH_4]^-$ by addition of H^- to $RSiH_3$ and $[SiH_5]^-$ by hydride transfer from $[RSiH_4]^-$ to SiH_4. Attempts to prepare $[SiH_5]^-$ directly by H^- addition to SiH_4 gave only $[SiH_3]^-$.[66]

4.2.1.1.3 Pentacoordinate Silicates with Five Si–C Bonds Using the flowing afterglow technique, Damrauer and co-workers[29a] showed in 1981 that allyl anion adds to fluorotrimethylsilane and to 1,1-dimethylsilacyclobutane to form adducts **22** and **23**, respectively. Complex **23** provided the first example of a pentacoordinate silicon anion containing five Si—C bonds. The reversible addition of $^nBu_4N^+CN^-$ to Me_3SiCN gave $[Me_3Si(CN)_2]^-$ $^nBu_4N^+$, which is the first isolable pentacoordinate silicate complex containing five Si—C bonds.[67]

$[Me_3Si(F)(CH_2CH{=}CH_2)]^-$ **22** $\qquad$ $[(CH_2)_3Si(Me)_2(CH_2CH{=}CH_2)]^-$ **23**

4.2.1.2 *Reactivity*

4.2.1.2.1 Pentacoordinate Fluoro and Alkoxysilicates Pentacoordinate silicates **24** and **25** react with nucleophiles such as $LiAlH_4$, RMgX, RLi and

MeONa to give the corresponding tetravalent silicon derivatives in good yields[55] (Schemes 4.1 and 4.2). They are more reactive than the corresponding tetracoordinate silanes toward hindered Grignard reagents. Thus the reaction of ^{t}BuMgBr or ^{i}PrMgBr is more than 100-fold faster with $[PhMeSiF_3]^-$ [K, 18-c-6]$^+$ than with $PhMeSiF_2$.[55b]

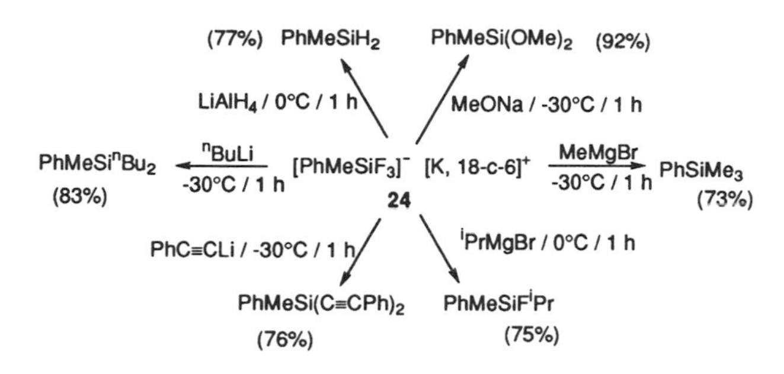

Scheme 4.1

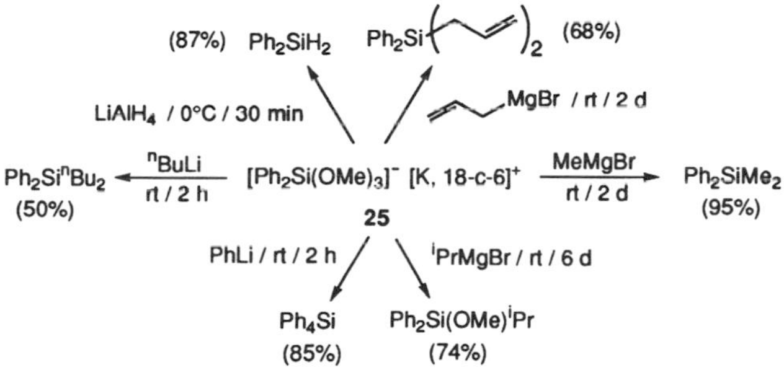

(87%) Ph_2SiH_2 $Ph_2Si(CH_2CH=CH_2)_2$ (68%)

LiAlH$_4$ / 0°C / 30 min; $CH_2=CHCH_2MgBr$ / rt / 2 d

$Ph_2Si^nBu_2$ (50%) ← ^{n}BuLi, rt / 2 h — $[Ph_2Si(OMe)_3]^-$ [K, 18-c-6]$^+$ **25** — MeMgBr, rt / 2 d → Ph_2SiMe_2 (95%)

PhLi / rt / 2 h; ^{i}PrMgBr / rt / 6 d

Ph_4Si (85%) $Ph_2Si(OMe)^iPr$ (74%)

Scheme 4.2

The high reactivity of pentaorganofluorosilicates towards nucleophiles was also shown in the study of the hydrolysis of dimesityldifluorosilane Mes_2SiF_2.[40] This compound does not react with water in refluxing acetonitrile. However, rapid hydrolysis occurs when tetraethylammonium fluoride hydrate is introduced. To confirm that $[Mes_2SiF_3]^-$, which was presumed to be formed, is very sensitive to hydrolysis, $[Mes_2SiF_3]^-$ [K,18-c-6]$^+$ was prepared and was found to be extensively hydrolysed after 5 min in aqueous acetone.

Recently, it was shown that cleavage of the Si–C bond occurs in the reaction of pentafluorosilicate **26** with NBS[68] (Eq. 4.5). The ease of the cleavage decreases with the nature of X in the following order: 4-$MeOC_6H_4$ > 4-MeC_6H_4 > C_6H_5 > 4-ClC_6H_4 > 4-$CF_3C_6H_4$. Similarly, cleavage of the Si—C bond was observed in the reaction of $[C_6F_5SiF_4]^-$ Me_4N^+ with various electrophiles (EtOD, Br_2, IF_5, ICN).[69]

$$[X\text{-}C_6H_4\text{-}SiF_3(C_6H_5)]^- [K, 18\text{-c-6}]^+ \xrightarrow[\text{THF, 10°C}]{\text{NBS}} X\text{-}C_6H_4\text{-}Br + C_6H_5\text{-}Br \tag{4.5}$$

(X = MeO, Me, Cl, CF_3)

26

4.2.1.2.2 Pentacoordinate Alkoxyhydrosilicates $[HSi(OEt)_4]^-K^+$ **20a** and $[H_2Si(O\textit{i}Pr)_3]^-K^+$ **21a** show a much greater variation in the nature of their reactions than does $HSi(OR)_3$. Thus they can behave as: electrophilic reagents, basic reagents, reducing reagents, and SET reagents (Schemes 4.3 and 4.4).[58,60]

In the presence of 18-crown-6 ether, the hydrosilicate **20a** reacts with EtOH with rapid evolution of dihydrogen to give the pentaalkoxysilicate $[Si(OEt)_5]^-$ $[K,18\text{-c-6}]^+$ [70] (Scheme 4.3). When the reaction was performed in thc absence of crown ether, tetraethoxysilane $Si(OEt)_4$ was obtained, presumably from the decomposition of the transient $[Si(OEt)_5]^-K^+$ complex, very unstable in the absence of crown ether. The dihydrosilicate **21a** reacted with one equivalent of 2-propanol to give quantitatively the stable pentacoordinate silicon derivative **20b**[70] (Scheme 4.4). This reaction was explained by a direct nucleophilic substitution at silicon involving the hexacoordinate silicon intermediate **27**.

The hydrolysis of $Si(OR)_4$ is the key reaction in the sol–gel process[71–74] for the preparation of silica. With a view to determining the possible rôle of pentacoordinate silicates in this reaction, the hydrolysis of $[Si(OR)_5]^-$ $[K,18\text{-c-6}]^+$ and of $[HSi(OR)_4]^-$ $[K,18\text{-c-6}]^+$ was studied[70] and compared to that of $Si(OR)_4$ and $HSi(OR)_3$. It was found that the hydrolysis of the pentacoordinate

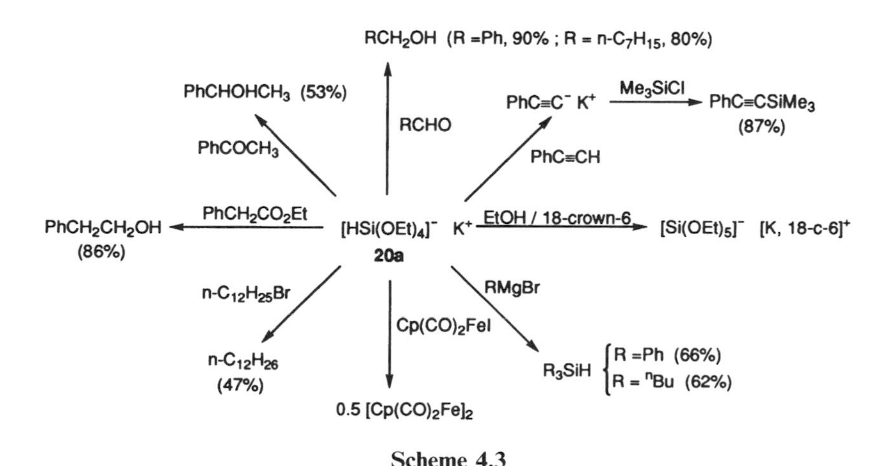

Scheme 4.3

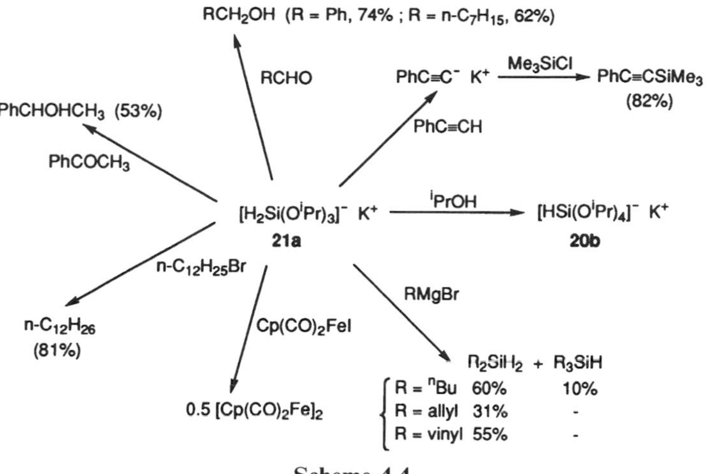

Scheme 4.4

silicates was much faster than those of the corresponding tetracoordinate silanes. The mechanism proposed involves the formation of the hexacoordinate intermediate **28**.

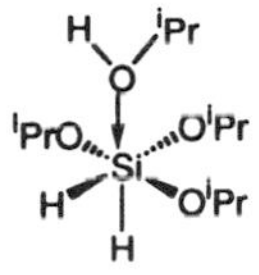

27

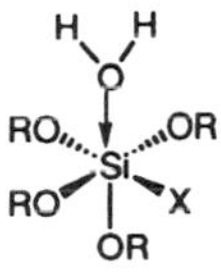

28 (X = H, OR)

4.2.1.2.3 Reactivity of Hydrosilicates in the Gas Phase The reactivity of [^{n}BuSiH$_4$]$^-$ was studied in the gas phase. This complex was shown to be a good reducing reagent,[66] as illustrated in Scheme 4.5.

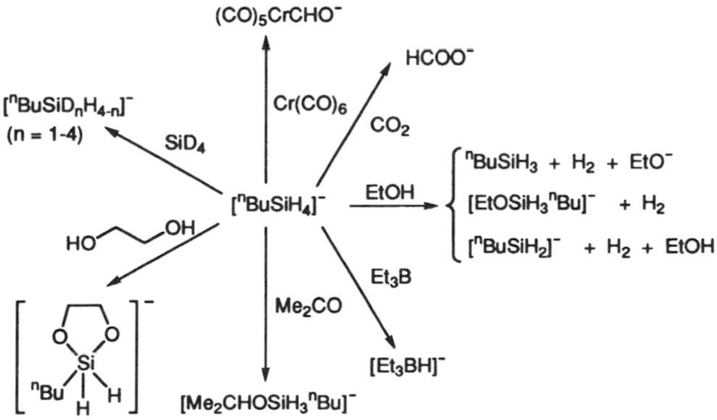

Scheme 4.5

4.2.2 Anionic Spirosilicates

4.2.2.1 Preparation In 1961, Müller and Heinrich[75] presented evidence for the formation of the two silicates **29** and **30**, isolated from the reaction mixture of the corresponding spirosilanes with lithium and sodium methoxides respectively. Later, Frye[76] found that simple amines were sufficiently basic to afford similar silicates (Eq. 4.6). This method requires strong nucleophiles for addition to spirosilanes derived from glycols or diphenols, but it is the only method up to now for the preparation of pentacoordinate fluorospirosilicates. Thus silicates **31**,[77] **32**,[78] **33**,[43] and **34**[79] were obtained by addition of fluoride to the corresponding spirosilanes. This method was also used for the preparation of the pentacoordinate alkoxysilicate **35**[80] (Eq. 4.7).

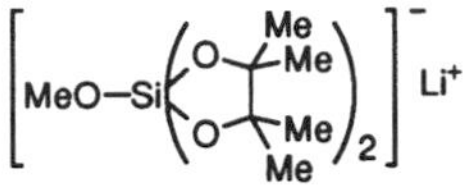

29

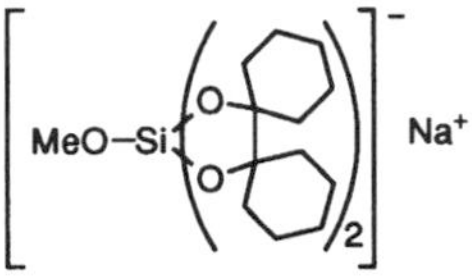

30

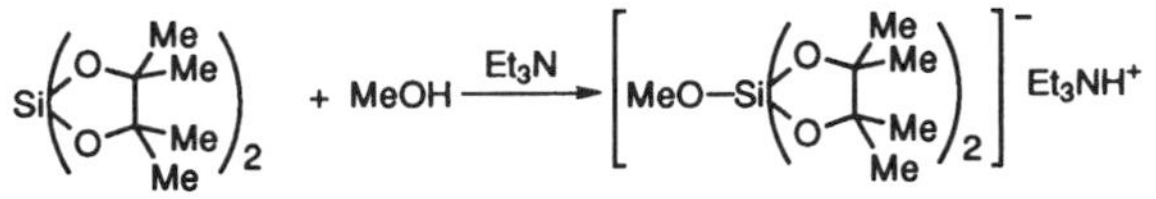

(4.6)

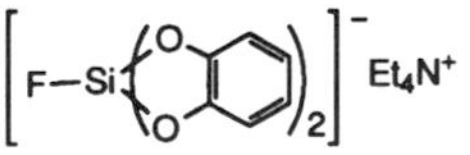

31

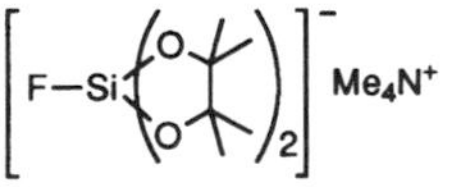

32

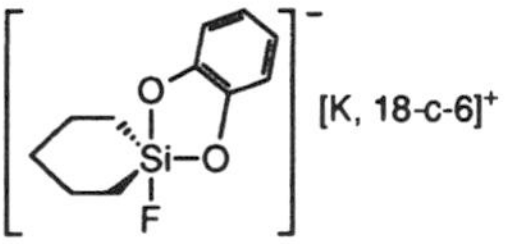

33

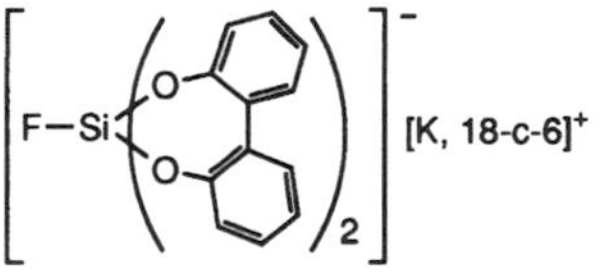

34

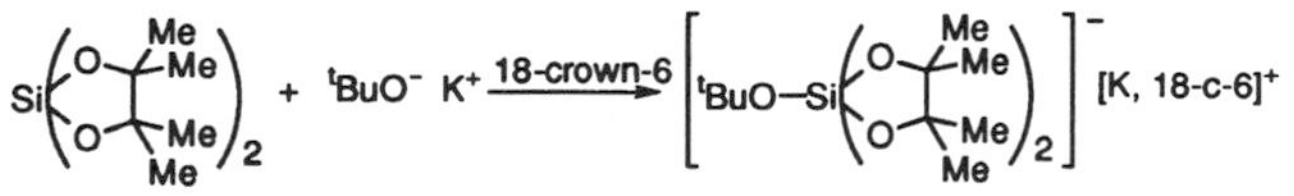

(4.7)

35

By contrast, the spirosilane **36** has exceptional ability to coordinate a further ligand including organic,[81] fluoride,[82] hydride,[83] and neutral donors.[84] The pentacoordinate silicon complexes **37**,[81] **38**,[82], **39**,[85] and **40**[83] were prepared in this way (Scheme 4.6).

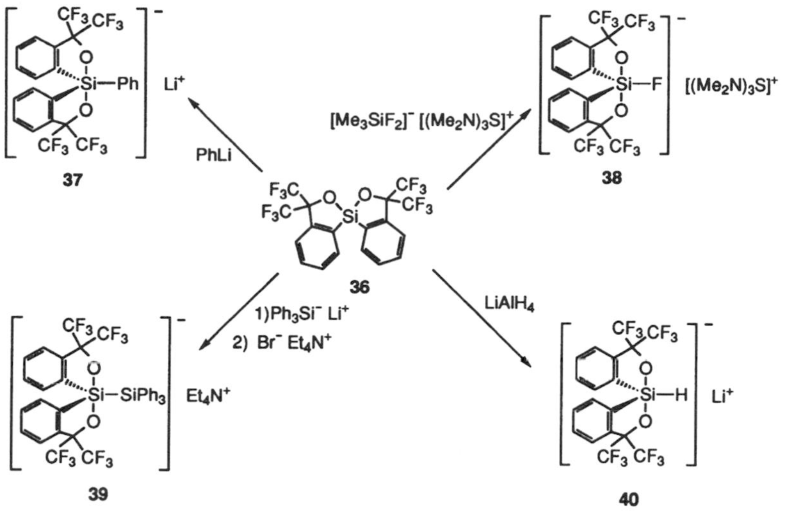

Scheme 4.6

In 1964, Frye[86] found that phenyltrimethoxysilane $PhSi(OMe)_3$ reacted with catechol in the presence of triethylamine to give quantitatively the pentacoordinate spirosilicate **41** (R = Ph, M = Et_3NH). This reaction is general; trialkoxysilanes $RSi(OMe)_3$ also react readily with catechol in the presence of various bases such as Me_4NOH[87], NaOMe,[88,89] or KOMe[87] (Eq. 4.8). Complexes **41** (M = Li) in which R = allyl[90] or H[91] were obtained from $RSiCl_3$ and lithium catecholate, but they could not be isolated. However, ammonium salts of the allylsilicate[87,92,93] prepared according to Eq. 4.8 are stable and were isolated as crystalline solids. Similar complexes with 1,2-diols instead of catechol as the ligand were prepared by the same reaction.[76] Pentacoordinate alkoxyspirosilicates were obtained by reacting 1,2-diols with tetraalkoxysilanes $Si(OR)_4$[76,80] (Eq. 4.9).

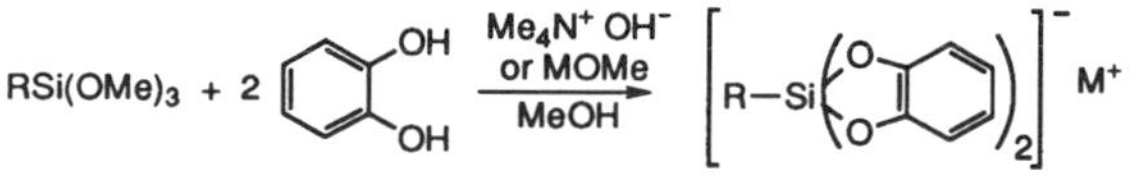

(4.8)

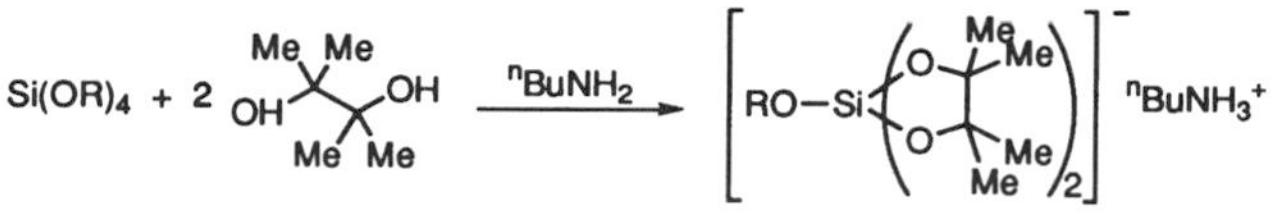

(4.9)

The spirosilicate **37**, obtained from the spirosilane **36** (Scheme 4.6) was also prepared by reaction of the dilithio derivative **42** with $PhSiCl_3$[94] (Eq. 4.10). The corresponding methyl complex was obtained by the same reaction starting from $MeSiCl_3$.

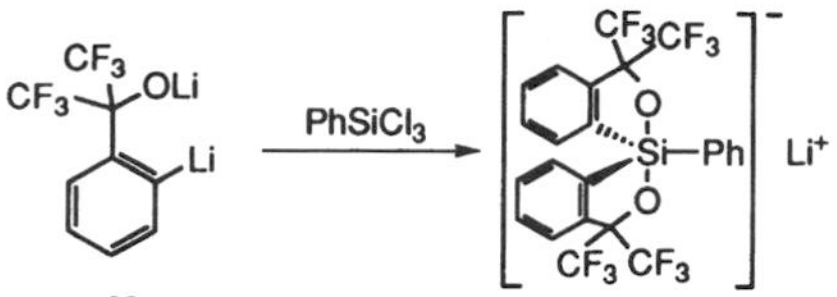

CF_3 CF_3 OLi Li → ($PhSiCl_3$) [CF_3CF_3 O Si–Ph O CF_3 CF_3]$^-$ Li^+

42 **37**

(4.10)

Compounds **43**[95] and **44**[96] with two pentacoordinate silicon atoms have been obtained according to the methods developed for the preparation of the monopentacoordinate silicon species, starting from the corresponding bis-trimethoxy and bis-triethoxysilanes respectively. Condensation of $PhSi(OEt)_3$ with 1,2,4,5-tetrahydroxybenzene[97] led to new polymers incorporating pentacoordinate silicon atoms (Eq. 4.11).

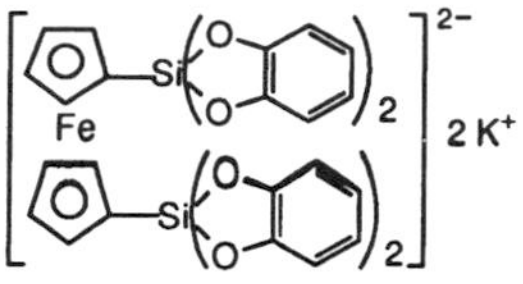

[Fe Si(O O)$_2$ Si(O O)$_2$]$^{2-}$ 2 K^+

43

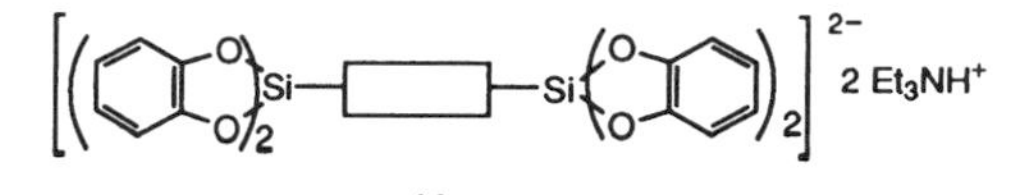

[(O O)$_2$Si—▭—Si(O O)$_2$]$^{2-}$ 2 Et_3NH^+

44

(▭ = , , , $-(CH_2)_n-$)

$PhSi(OEt)_3$ + HO OH HO OH → (NEt_3) [O O Et_3NH^+ Si Ph O O O O Et_3NH^+ Si Ph O O]$_n$

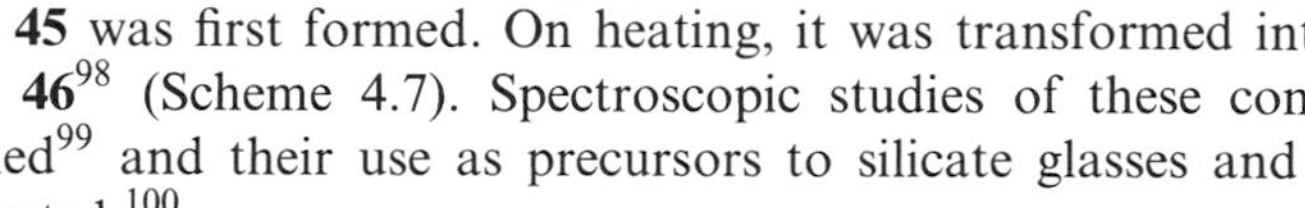

(4.11)

The reaction of SiO_2 with ethylene glycol in the presence of alkali metal hydroxides was studied thoroughly.[98] It was shown that the pentacoordinate silicate **45** was first formed. On heating, it was transformed into the dimeric species **46**[98] (Scheme 4.7). Spectroscopic studies of these compounds were peformed[99] and their use as precursors to silicate glasses and ceramics was investigated.[100]

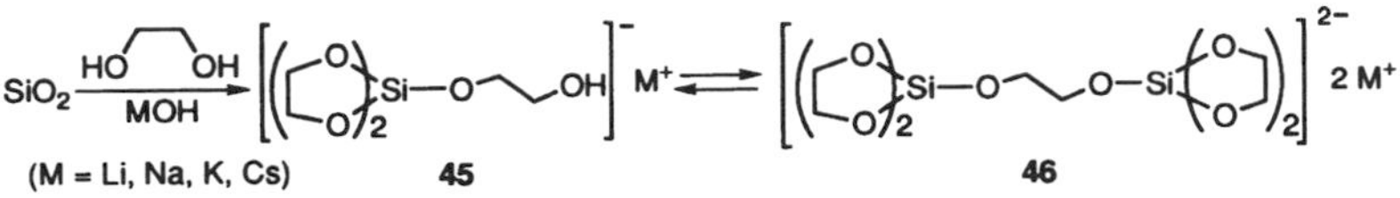

SiO_2 → (HO OH, MOH) [(O O)$_2$Si–O~OH]$^-$ M^+ ⇌ [(O O)$_2$Si–O~O–Si(O O)$_2$]$^{2-}$ 2 M^+

(M = Li, Na, K, Cs) **45** **46**

Scheme 4.7

Tacke and co-workers have extensively studied the formation of zwitterionic pentacoordinate spirosilicates. Silicates **47–49** were prepared by reaction of the corresponding trimethoxysilanes with 1,2-dihydroxynaphthalene,[101] catechol,[102] or glycolic acid.[103] These types of complexes could also be obtained by Si–C cleavage from a dimethoxysilane (Eq. 4.12).[104] These unusual Si—C cleavages were shown to be quite general in the preparation of many zwitterionic silicates.[105] The zwitterionic bis-spirosilicate **50** was prepared from 1,4-bis [(trimethoxysilyl)methyl]piperazine[106] following the same methodology.

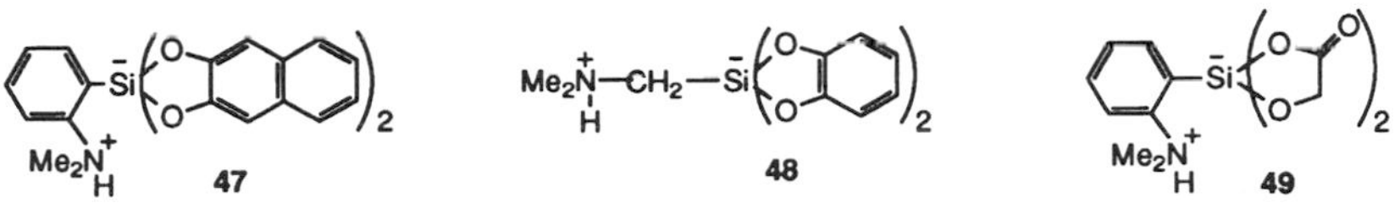

$Me_2\overset{+}{N}H$ **47** $Me_2\overset{+}{N}H-CH_2-\overset{-}{Si}$ **48** $Me_2\overset{+}{N}H$ **49**

$$RSi(OMe)_2-(CH_2)_n-N \xrightarrow{\text{naphthalenediol (OH, OH)}} \overset{+}{N}H(CH_2)_n-\overset{-}{Si}(O,O)_2 \qquad (4.12)$$

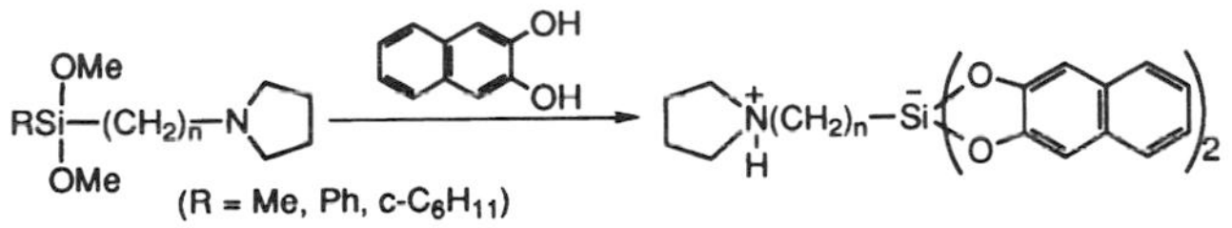

(R = Me, Ph, c-C_6H_{11})

$\overset{-}{Si}-CH_2-\overset{+}{N}H \quad \overset{+}{N}H-CH_2-\overset{-}{Si}$

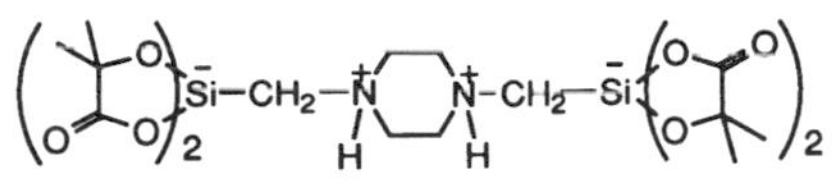

50

*4.2.2.2 **Structure and Fluxional Behavior*** In 1968, Boer and co-workers[107] showed that the geometry around the silicon atom in the pentacoordinate spirosilicate **51** is a slightly deformed trigonal bipyramid. Further determination concerning the structures of these spirosilicates carried out by Schomburg[78,108] and particularly by Holmes and co-workers[77,80,109–112] have shown that a wide range of geometries may exist, extending from the trigonal bipyramid (TBP) to the rectangular pyramid (RP) and distributed along the Berry pseudorotational coordinate.[113] The geometry of the silicate is a function of the nature of the monochelating ligand, a bulky group promoting the RP geometry (29.5% in **51**,[107] 63.8% in **52**,[110] and 91.4% in **53**[110]). The cation also influences the geometry of the silicate, especially primary, secondary, and tertiary ammonium cations which promote hydrogen bonding in the crystal lattice with one or more oxygen atoms of the cyclic ligand. Thus in compounds **54**,[109] **55**,[108] and **56**[111] the geometry is displaced by more than 50% towards RP and is nearly RP (97.6%) in **57**.[112] In compounds **35**,[80] **58**,[80] **59**,[80] the displacement toward RP from **35** (24%) to **58** (71%) is due to the number of hydrogen bondings.[80] The geometry of the fluorosilicates **31**[77] and **32**[78] (cations R_4N^+), which both crystallize as two discrete molecules, is displaced toward RP by about 50 and 70% in both cases. One possible reason for that may be the apicophilicity of the fluorine atom. The same variation of geometries from TBP to RP is also found in the zwitterionic spirosilicates. Thus in **60** the geometry around the silicon atom is nearly TBP, while it is nearly RP in **61**.[102] Here too the role of hydrogen bonding in the crystal is a factor promoting RP geometry.

Ph—Si(O₂C₆H₄)₂⁻ Me₄N⁺ **51**; ⁿBu—Si(O₂C₆H₄)₂⁻ Et₄N⁺ **52**; ᵗBu—Si(O₂C₆H₄)₂⁻ Et₄N⁺ **53**

Ph—Si(OCH₂CH₂O)₂⁻ RNH₃⁺ **54**; Me—Si(OCH₂CH₂O)₂⁻ RNH₃⁺ **55**; Ph—Si(O₂C₆H₄)₂⁻ Et₃NH⁺ **56**

Ph—Si(O₂C₁₀H₆)₂⁻ C₅H₅NH⁺ **57**; MeO—Si(pinacolato)₂⁻ ⁿBu₃NH⁺ **58**; EtO—Si(pinacolato)₂⁻ ⁿBu₃NH⁺ **59**

$Me_2\overset{+}{N}H$—CH_2—$\bar{Si}(O_2C_6H_4)_2$ **60**; pyrrolidinium-$\overset{+}{N}H$—CH_2CH_2—$\bar{Si}(O_2C_6Br_4)_2$ **61**

X-ray crystal structure analyses of compounds **37**,[81] **38**,[82] and **39**[85] (Scheme 4.6) as well as of compound **62**,[114] which is an intermediate of the Peterson reaction, have shown that in these four complexes the silicon atom is pentacoordinated in a somewhat distorted TBP. The preference for TBP geometry can be explained by the difference in electronegativities of oxygen atoms and the ring carbon of the ligand. In the TBP, carbon atoms occupy equatorial sites and oxygen atoms are in apical positions.

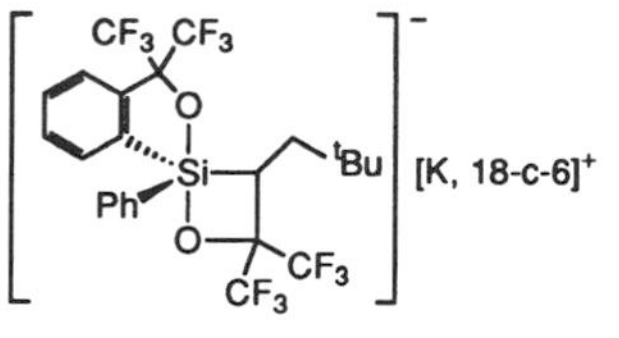

62

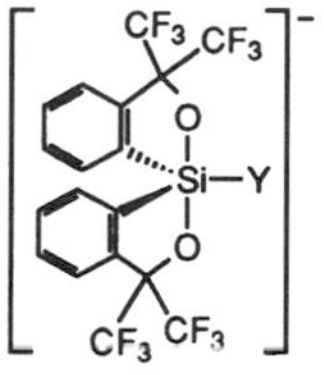

63

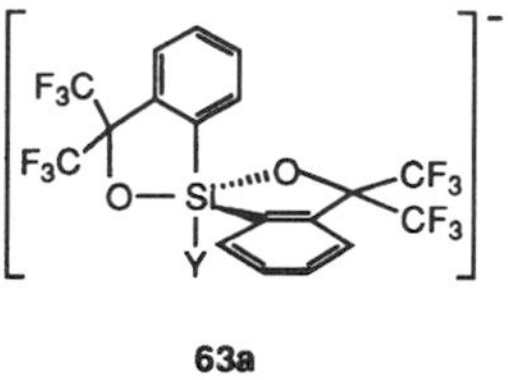

63a

The stereoisomerization of the bicyclic pentacoordinate silicates **63**[115] and the related inversion at silicon of the parent spirosilane **36** under the influence of weak nucleophiles,[116] have been extensively studied. The values of the free energy of activation for the inversion decrease as the electron-withdrawing effect of the substituent increases, from 119.5 kJ mol^{-1} for Y = ^{n}Bu to 73.1 kJ mol^{-1} for Y = F and 70.2 kJ mol^{-1} for Y = CN. Although an irregular mechanism cannot be completely excluded, a regular mechanism is more likely. The high energies for the isomerization process are probably due to the formation of the intermediate **63a**, in which one five-membered ring is in a diequatorial position.

4.2.2.3 ***Reactivity*** Pentaorganospirosilane **64** was found to be very reactive toward nucleophilic reagents such as RMgX, RLi, and metallic hydrides[88,89] (Scheme 4.8). Reaction with Grignard reagents containing a hydrogen atom in the β position, activated by Cp_2TiCl_2, gave monohydrogenosilanes with primary alkyl Grignard reagents, and dihydrogenosilanes with secondary and tertiary alkyl Grignard reagents.[117] Electrophilic BF_3 induced ligand exchange to give the corresponding borate in good yield.[118]

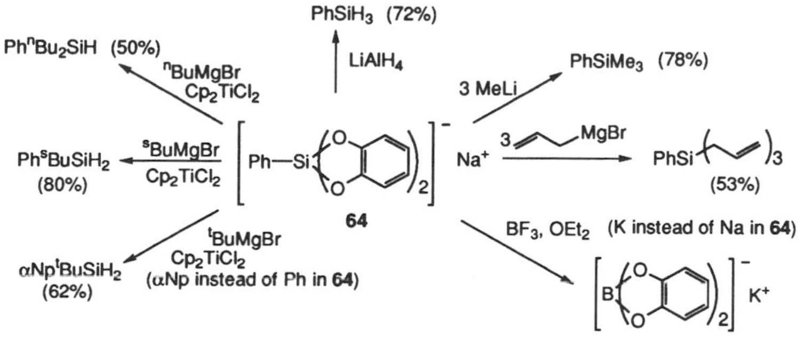

Scheme 4.8

Complexes **65** and **66**,[91] obtained by reaction of $HSiCl_3$ with the corresponding dilithium salts of catechol and of 2,2′-dihydroxybiphenyl, are not stable. However, in THF solution, they reduced quantitatively aldehydes and ketones (but not esters) at room temperature. Complex **40** is much less reactive and reduced only slowly aldehydes and ketones. However, the spirosilane **36** is a very efficient catalyst for this reaction.[83] Complexes **67** reacted with aldehydes to give homoallylic alcohols (Eq. 4.13), but there was no reaction with ketones.[93] The reactivity of complex **67c** was enhanced by performing the reaction in the presence of a nucleophile (KF, MeONa).[92]

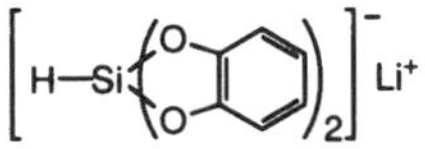

65 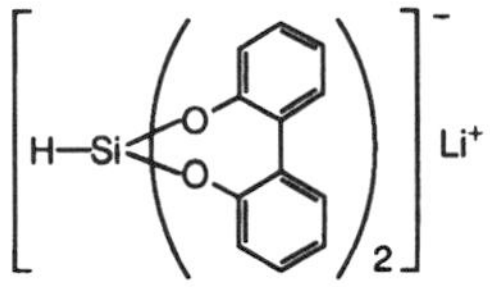 **66**

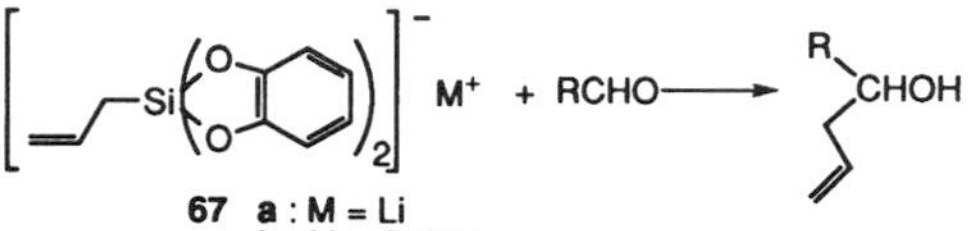

67 a : M = Li; b : M = Et_3NH; c : M = NMe_4 + RCHO → (4.13)

In order to study the mechanism of group transfer polymerization[119] (GTP) (Eq. 4.14), complex **68** was prepared by addition of the corresponding lithium enolate to the spirosilane **36**.[120] This complex, which is the first pentacoordinate silyl enolate described, was shown to polymerize methyl methacrylate without catalyst, thus indicating that the GTP reaction takes place via pentacoordinate intermediates.[120]

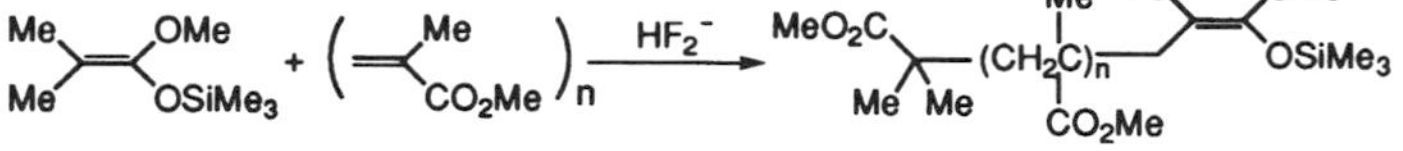

(4.14)

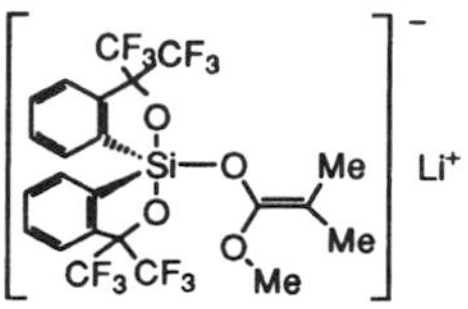

68

4.2.3 Intermolecular Donation by a Neutral Donor to an Organosilane

Neutral pentacoordinate silicon complexes can be formed by intermolecular coordination of neutral atoms to the silicon center.

4.2.3.1 Neutral Complexes X-ray crystal structure determination showed that dimethylsilylamine[121,122] exists as a pentamer **69**, in which the nitrogen atoms lie at the corner of a nearly regular pentagon whilst the silicon atoms are located midway along each side. The silicon atoms are pentacoordinated in a TBP geometry with N—Si distances of 1.98 Å, the hydrogen atoms occupying the equatorial position. In the same way, *N*(trifluorosilyl)trimethylphosphinimine and chlorosilyl-*N*,*N*-dimethylamine exist as dimers **70**[123] and **71**[124] in the solid state. In both compounds, the silicon atom is pentacoordinated with a TBP geometry. The coordinative N→Si bonds in **70** (1.857 Å) and in **71** (2.054 Å) are longer than the other N—Si bonds (1.736 Å in **70** and 1.814 Å in **71**).

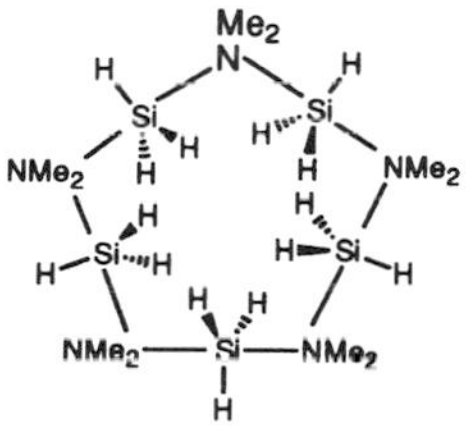

69 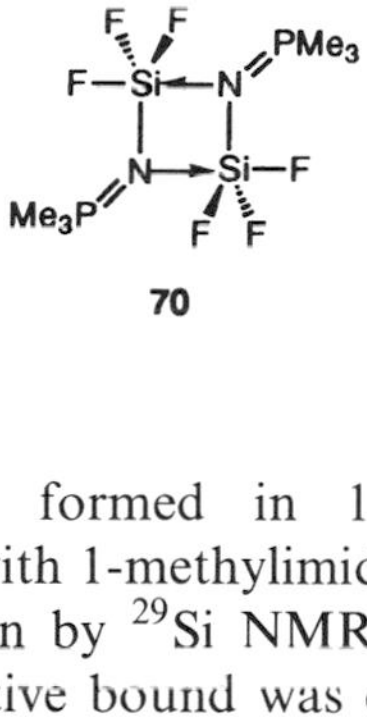 **70** 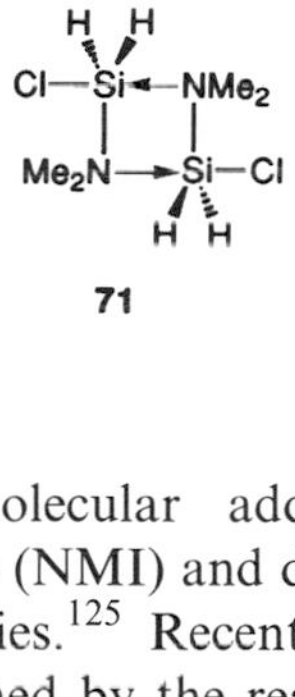 **71**

Neutral complexes are also formed in 1:1 molecular adducts of $MeHSi(OTf)_2$ and $PhHSi(OTf)_2$ with 1-methylimidazole (NMI) and dimethylpropyleneurea (DMPU), as shown by ^{29}Si NMR studies.[125] Recently, compound **73** with a C→Si coordinative bound was obtained by the reaction of the carbene **72** with $SiCl_4$[126] (Eq. 4.15). The X-ray structure analysis showed that the silicon is pentacoordinated with a TBP geometry. The two apical Si Cl bonds are longer (2.219 and 2.225 Å) than the equatorial bonds (2.045 and 2.070 Å). The C—Si coordinative bond is quite short (1.991 Å).

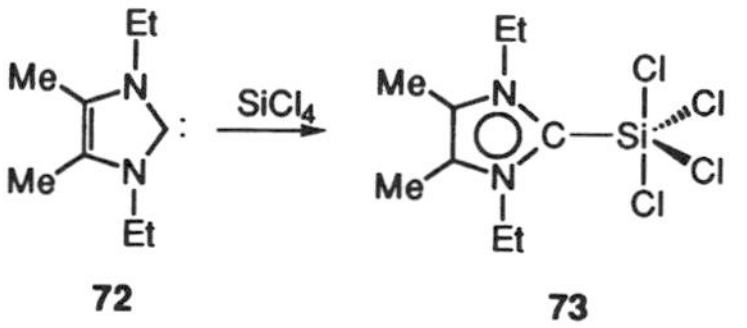

72 $\xrightarrow{SiCl_4}$ **73** (4.15)

Neutral donors are also able to stabilize silylene derivatives. Thus Zybill and co-workers[127–129] prepared silylene–iron **74** and silylene–chromium **75** complexes stabilized by one molecule of HMPA by reacting R_2SiCl_2 with

$Na_2Fe(CO)_4$ and $NaCr(CO)_5$, respectively. The X-ray structure determinations of these compounds showed short O→Si coordinative bonds of 1.73–1.74 Å when R = Me or ^{t}BuO and 1.69 Å when R = Cl.

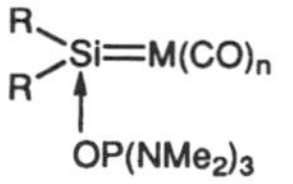

74 : M = Fe, n = 4
75 : M = Cr, n = 5

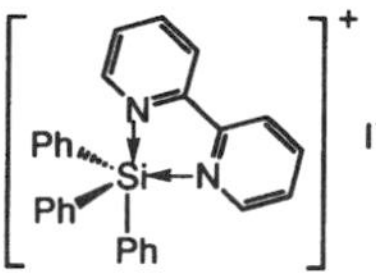

76

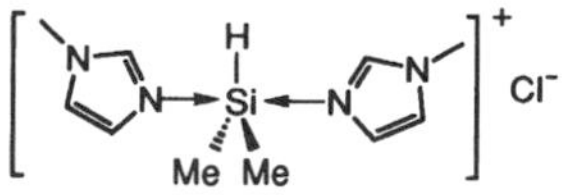

77

4.2.3.2 Cationic Complexes Hydrosilanes SiH_3X and $MeSiH_2X$ (X = Br, I) react with pyridine to give 1:2 adducts for which the ionic structures $[RSiH_2\ py_2]^+ X^-$ (R = H, Me; X = Br, I) were proposed.[130] An ionic structure has also been suggested[131] for the adduct **76** of iodotriphenylsilane with 2, 2′-bipyridine. The reaction of NMI with Me_2SiHCl gave the siliconium ion **77**,[132] the X-ray structure of which has revealed[133] that the silicon atom is pentacoordinated with TBP geometry and with two coordinative N → Si bonds of 2.034 and 2.005 Å.

4.2.4 N→Si Intramolecular Coordination to an Organosilane

Numerous neutral pentacoordinate silicon compounds have been prepared in which extra-coordination is obtained by intramolecular chelation of a donor atom to the silicon atom. Such chelation can be facilitated by the rigid geometry of the ligand (e.g., **78**[134]) but it is also possible with ligands for which there exists a favorable conformation allowing chelation (e.g., **79**[134]).

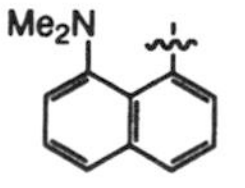

78

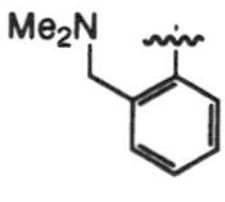

79

4.2.4.1 Neutral Complexes

4.2.4.1.1 Structure The X-ray crystal structure determinations for fluorosilicon compounds **80**, **81a** and **82–86** indicated that the silicon atoms are pentacoordinated with TBP geometries. The N—Si distances are shorter in **80**[135] (1.974 Å), **81a**[136] (1.969 Å), and **82**[137] (1.988 Å) than in **83**[138] (2.318 Å) and **84**[139] (2.436 Å). These distances do not depend on the nature of the ligand (rigid or flexible) in the cases of **80**, **81a**, and **82**.

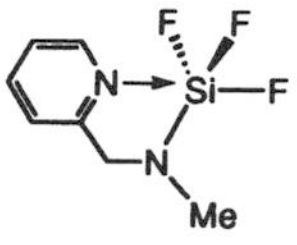

80

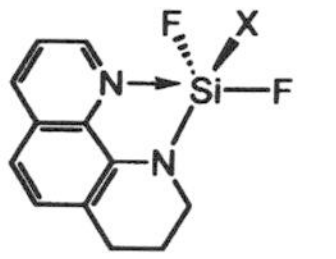

81 a : X = F
b : X = Cl

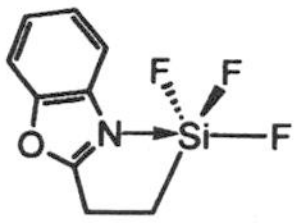

82

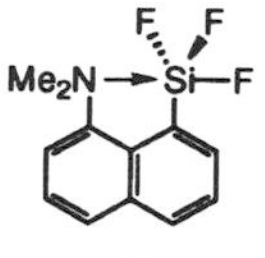

83

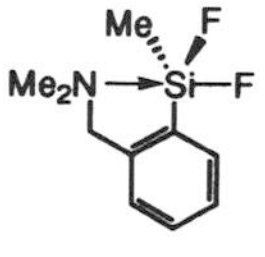

84

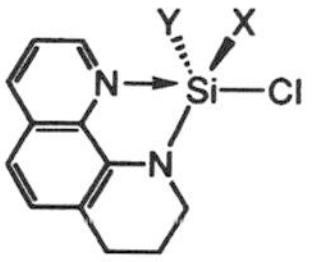

85 a : X = Y = Cl
b : X = Me, Y = Cl
c : X = Y = Me

The silicon atoms of chlorosilicon compounds **85a**[140], **85b**[141] and **85c**[141] are also pentacoordinated with TBP geometries. On going from **85a** to **85c**, the N—Si distances increase only slightly (due to the rigidity of the ligand) from 1.984 Å to 2.027 Å to 2.028 Å. More important deformations are found in compounds **86–88**, which incorporate flexible ligands. In **86**[142] the silicon atom is pentacoordinated with TBP geometry, the N—Si distance being 1.901 Å and the axial Si—Cl bond length 2.238 Å. In contrast, in **87**[142] the N–Si distance is a little shorter (1.898 Å), while the Si—Cl bond length is much longer (2.598 Å) so that the geometry around the silicon atom could be best described as a distorted tetrahedron capped by a chloride ion. The same geometry is found in **88**[143] with an even shorter N—Si bond (1.852 Å) and a longer $Si \leftarrow Cl^-$ interaction (2.679 Å). Such $N \rightarrow Si$ intramolecular interactions are also found in the four-membered chelates **89**[144] and **90**[145] in spite of the ring strain. **90** is the only pentacoordinate silanol described up to now. In this compound, the silicon atom has a distorted TBP geometry, the OH group occupying the axial position.

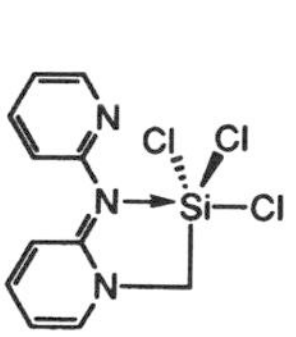

86

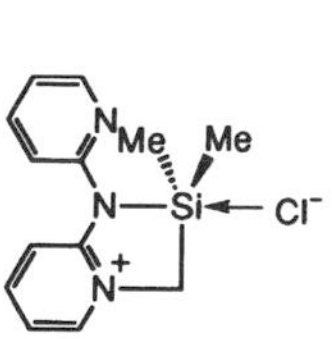

87

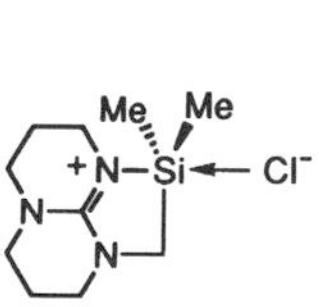

88

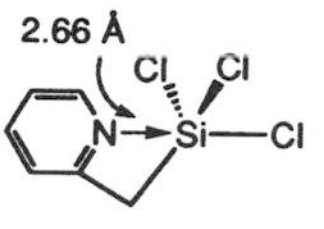

89

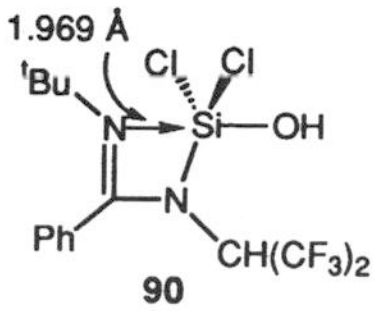

90

The X-ray structure determination of compound **91**[146] with two potentially chelating groups showed that only one intramolecular N→Si coordination occurs and that **91** has the same structure as that of the corresponding ammonium salt **92**,[147] the N–Si distance being, however, longer in **91** (2.221 Å) than in **92** (2.163 Å). A dynamic coordination–decoordination of the NMe_2 groups in compound **91** was observed by ^1H NMR spectroscopy. The NMe_2 groups displace each other rapidly at room temperature through a hexacoordinate transition state. The activation energy of this process was estimated to be $46.5\,kJ\,mol^{-1}$.[148] Recently, the X-ray structure analysis of compound **93**[149] showed that the silicon atom is pentacoordinated with an N–Si distance of 2.90 Å, the other NMe_2 group being noncoordinated (as in **91**). The geometry around the silicon atom is that of a somewhat distorted tetrahedron, monocapped by the NMe_2 group.

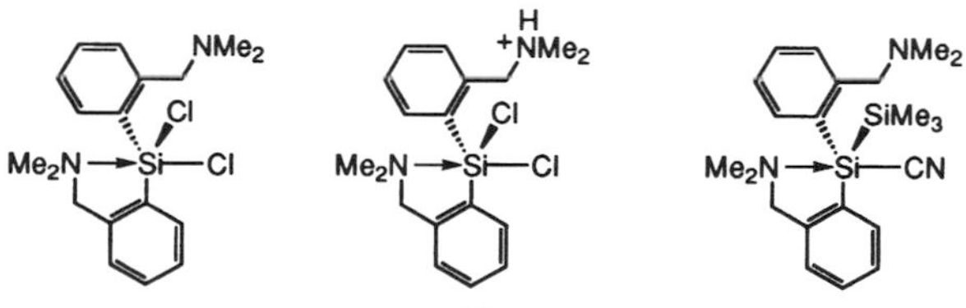

NMe2 Cl Me2N→Si—Cl
91

H +NMe2 Cl Me2N→Si—Cl
92

NMe2 SiMe3 Me2N→Si—CN
93

The occurrence of pentacoordination in the hydrosilanes is particularly interesting. The X-ray structures of **94**,[150] **95b**,[151] and **96a**[151] have shown significant N→Si coordination (2.44 Å, 2.66 Å, and 2.58 Å respectively), the geometry around the silicon center being that of a somewhat distorted TBP with the hydrogen atoms in equatorial sites. In **96b**,[152a] which contains a bulky R group, the N—Si distance is 2.790 Å and the naphthalene ring is highly distorted, the dihedral N—C-Si—C angle being 29.5°.

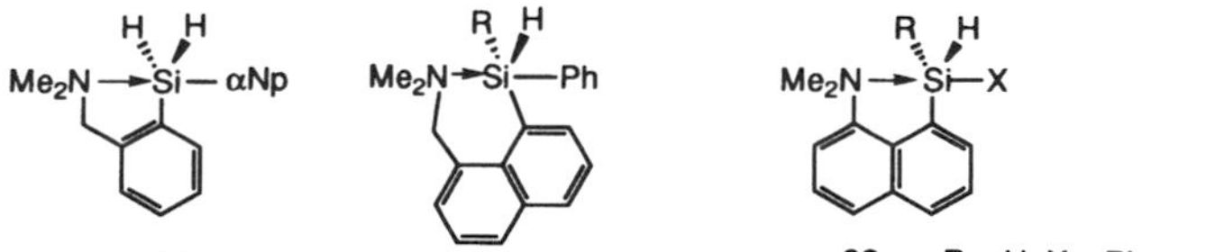

H H Me2N→Si—αNp
94

R H Me2N→Si—Ph
95 **a** : R = H
b : R = Me

R H Me2N→Si—X
96 **a** : R = H, X = Ph
b : R = Me, X = $SiPh_3$

Only a few compounds with an intramolecular N→Si coordination are known where the silicon atom does not bear polar groups. The pentacoordination around the silicon center for **97**[153] was inferred from the ^{29}Si NMR spectrum. In compounds **98**, the N→Si interactions are very weak (3.159 Å in **98a**[152b] and 2.969 Å in **98b**[152c]) and the naphthalene rings are highly distorted, the dihedral N–C-Si–C angle being 34.35° in **98a** and 31.32° in **98b**.

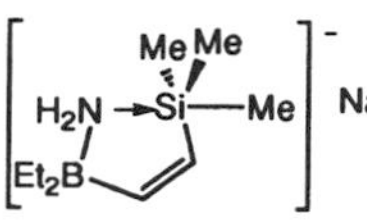

97

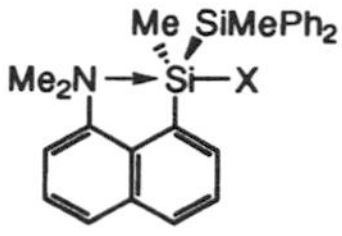

98 **a** : X = $SiMePh_2$
b : X = OEt

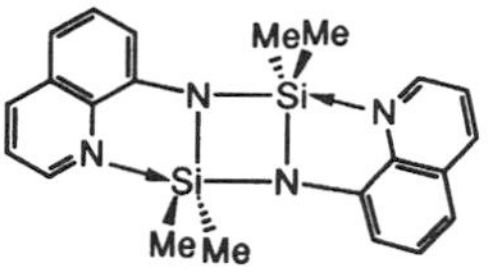

99

In compound **99**[154] the two silicon atoms are pentacoordinated with N—Si distances of 2.33 Å. The N and Si atoms are included in a four-membered ring and the geometries around the silicon atoms are those of distorted TBPs.

4.2.4.1.2 Fluxional Behavior Compounds in which the pentacoordination around the silicon atom is induced by intramolecular coordination show a more or less rapid exchange of the substituents on the silicon atom. This exchange can occur by a regular mechanism in which the coordination number remains unchanged throughout or by an irregular mechanism in which the ring is opened, allowing the silicon atom to become at least momentarily tetracoordinated.

Variable temperature ^{1}H NMR studies[155] of monofunctional and bifunctional derivatives **100a** (R = H, αNp) and **100b** ($X \neq H$) showed two signals for diastereotopic N-methyl groups at low temperature thus indicating intramolecular coordination. From the coalescence of these signals, the free energies of activation for N → Si cleavage were calculated to lie in the range 38.4–72 kJ mol^{-1}, the stability of the chelated form depending on X in the order: RO < H < F, SF < OAc, Cl, Br. ^{19}F NMR studies of fluorosilanes **100c**[156] established the apicophilicity of X relative to fluorine. Resonances due to axial fluorine are found at relatively low field, and an upfield shift indicates occupation of an equatorial site.[157] The experiments showed fluorine to be more apicophilic than H, OR, and NR_2, but less than Cl.[156] However, fluorine appears to be more apicophilic than chlorine in compound **81b**.[158]

Activation energies for site exchange have been determined for compounds **80**,[135] **84**,[139] and **100**.[159] In these cases, it is difficult to distinguish between the regular and the nonregular mechanism. However, this ambiguity can be eliminated by the incorporation of a chiral center in the molecule. Thus in compound **101**, the low-temperature ^{1}H NMR spectrum displays four separate methyl group resonances for the dimethylamino group, because of the chirality of both the benzylic carbon atom and the pentacoordinate silicon atom. As the temperature is raised, a first coalescence occurs, giving two resonances with a $\Delta G^{\ddagger}$ value of 39.3 kJ mol^{-1}. This arises from an intramolecular ligand exchange process. A second coalescence giving a singlet for the NMe_2 signal takes place with a $\Delta G^{\ddagger}$ of 49.3 kJ mol^{-1}, corresponding to Si—N bond opening accompanied by rotation and inversion at nitrogen.[160]

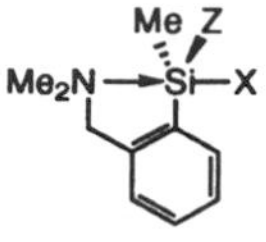

100 a : Z = R
b : Z = X
c : Z = F

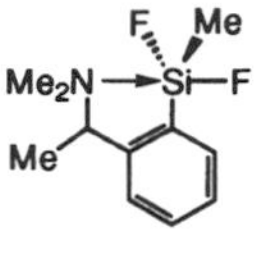

101

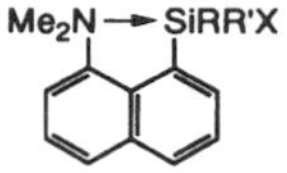

102 a : R = R' = X
b : R = Ph, R' = X
c : R = Ph, R' = Me

Low values of activation energies for site exchange were found for compounds bearing a rigid ligand as in **81**[136] ($\Delta H^{\ddagger} = 31.3\,\text{kJ mol}^{-1}$), **102a**[161,162] ($\Delta G^{\ddagger} < 29\,\text{kJ mol}^{-1}$ (X = H, OMe), $50.2\,\text{kJ mol}^{-1}$ (X = F)), and **102b**[162] ($\Delta G^{\ddagger} = 37.6\,\text{kJ mol}^{-1}$ (X = OMe), $46\,\text{kJ mol}^{-1}$ (X = Cl), $50.2\,\text{kJ mol}^{-1}$ (X = F)). These values were attributed to site exchange occurring by the regular mechanism. However, high values of activation energies for site exchange were found for compounds **102c**[162] ($\Delta G^{\ddagger} > 83\,\text{kJ mol}^{-1}$ (X = H, F, Cl, OMe)) and **103**[163] ($\Delta G^{\ddagger} = 63{-}79\,\text{kJ mol}^{-1}$ (X = H, F, Cl, Ph, OPh)). These values were attributed to site exchanges occurring by the non-regular mechanism.

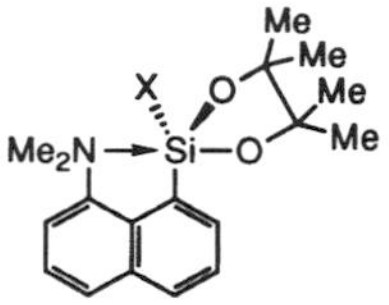

103

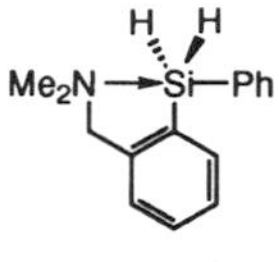

104

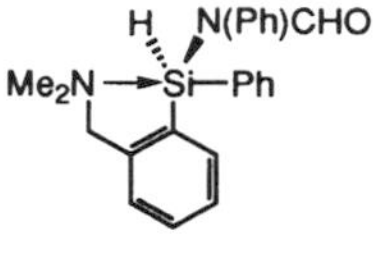

107

4.2.4.1.3 Reactivity The pentacoordinate silicon dihydrides **95a**, **96a**, and **104** are much more reactive than the corresponding tetracoordinate dihydrosilanes. They react at room temperature with alcohols,[164] diols,[165] acides,[166] and acyl chlorides[167] (Scheme 4.9). Hydrosilylation of phenylisothiocyanate[168] and arylcarbodiimides[169] at room temperature gives compounds **105** and **106** in good yields. In the same way, compound **107** is obtained by hydrosilylation of phenylisocyanate from silane **104**.[168] Reaction of α-hydroxyketones or 1,2-diketones with silane **108** gives selectively the erythro silane **109**, which can be reduced to the erythro diol[170] (Eq. 4.16).

A very fast hydrogen–chlorine exchange reaction occurs on reaction of silane **96a** with Me_3SiCl, PCl_3, or PCl_5[171] (Eq. 4.17). Monochloration takes place with Me_3SiCl, 1 molar equivalent of PCl_3 or 0.5 molar equivalent of PCl_5, while an excess of these last two reagents gives dichlorinated products.

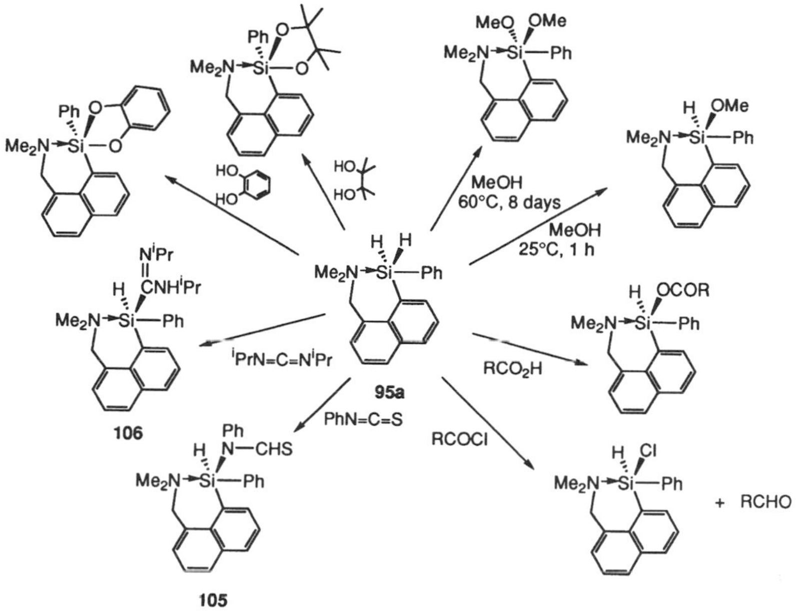

Scheme 4.9

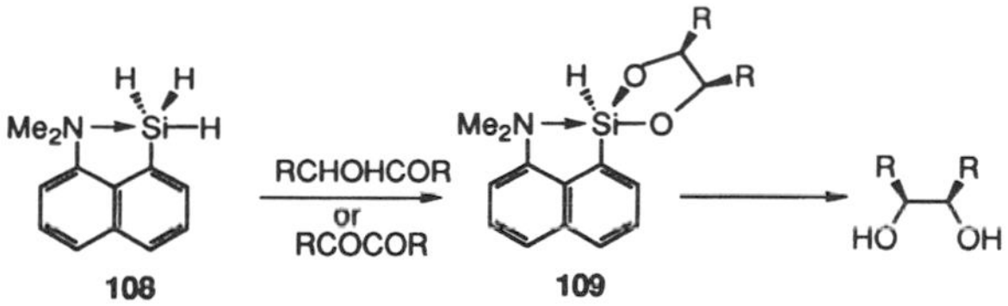

(4.16)

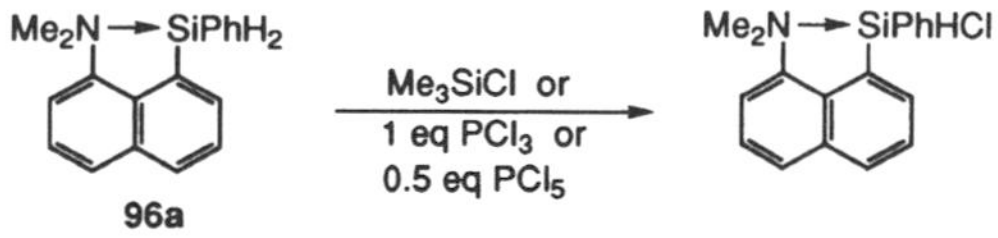

(4.17)

On heating complex **98b** at 90°C, facile α-elimination of $Ph_2MeSiOEt$ occurs, giving the silylene **110**,[152b] which was identified by trapping with 2,3-dimethyl-1,3-butadiene (Scheme 4.10). The same silylene can be obtained by Ni(0)-catalyzed degradation of silane **96b**.[152a]

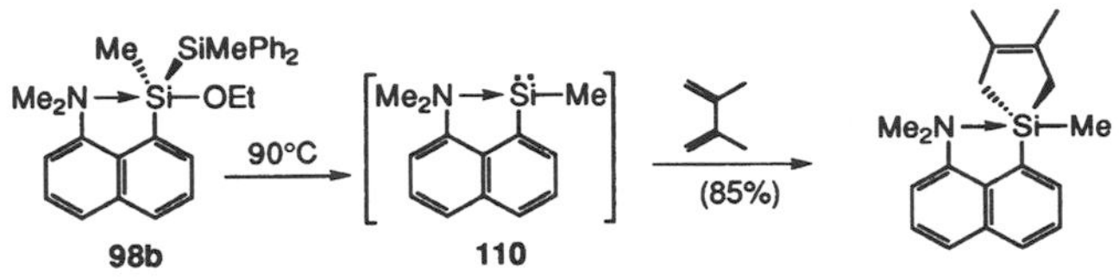

Scheme 4.10

4.2.4.2 *Neutral Donor-Stabilized Low-Coordinate Silicon Compounds*

4.2.4.2.1 Silanethione, Silaphosphene, and Silanimine Unstable silanone **111**,[172] silanethione **112a**,[173] and silaneselone **113**[173] were obtained according to the reactions indicated in Scheme 4.11. Silanethione **112a** was also obtained by reaction of CS_2 with silane **114**.[174] X-ray structure determination of silanethione **112b**[173] showed that the silicon atom has tetrahedral geometry with an Si—S bond (2.013 Å) shorter than an Si—S single bond (2.16 Å) and an N→Si coordinative bond of length 1.96 Å. Other stabilized low-coordinate silicon compounds including the silaphosphene **115** and the silanimine **116** were both prepared from the difluorosilane **117**[175] (Scheme 4.12). They are inert toward nucleophiles and electrophiles but are highly sensitive to oxygen and moisture, like silanethione and silaneselone.

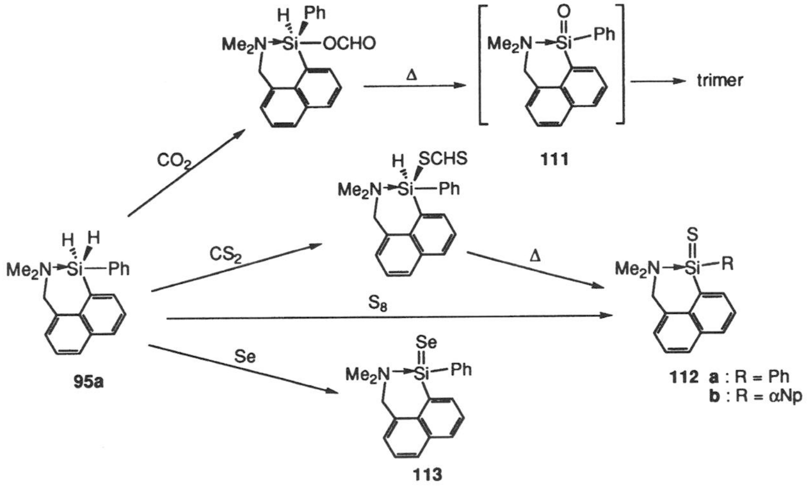

Scheme 4.11

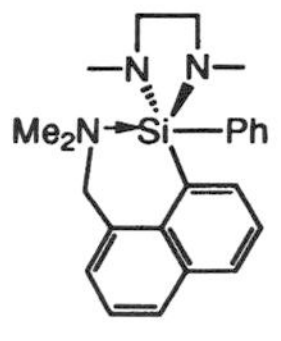

114

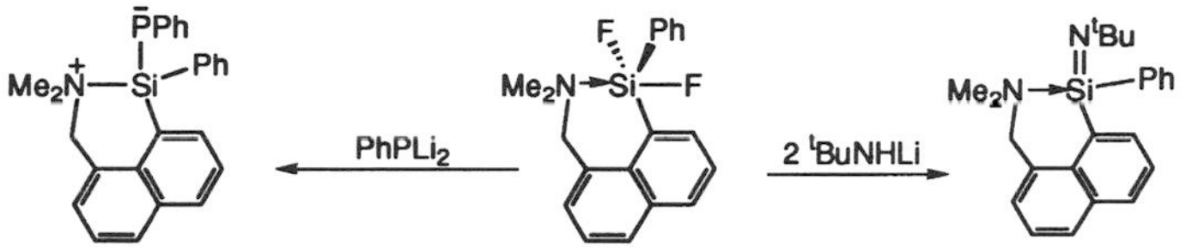

PhPLi$_2$ 2 ^{t}BuNHLi

115 117 116

Scheme 4.12

4.2.4.2.2 Transition Metal–Silylene Complexes Intramolecularly stabilized transition metal–silylene complexes **118** and **119** were prepared by reaction of hydrosilanes with the corresponding metal carbonyl in the presence or absence of UV irradiation.[146,176] Complexes **120** were obtained by reaction of the corresponding dichlorosilane with $Na_2[Cr(CO)_5]$[146] under UV irradiation.

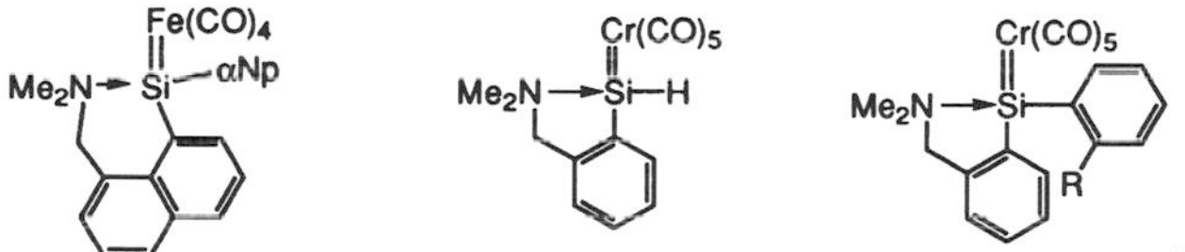

118 119 120 **a** : R = H **b** : R = CH_2NMe_2

The X-ray crystal structures of complexes **120a**,[146] **121**,[177] and **122**[178] showed that the silicon atoms have tetrahedral geometries with N—Si distances of 1.991 Å in **120a**, 1.962 Å in **121**, and 1.949 Å in **122**. In **123**[179] only one PPh_2 group is bonded to the silicon atom, with a P—Si distance of 2.380 Å. Dynamic coordination–decoordination of the chelating groups was inferred from NMR studies for compounds **120b**,[146] **123**,[179] **124**[178] and **125**.[178]

Complex **119** reacts with acetyl chloride, $Ph_3C^+BF_4^-$, and $Ph_3C^+TfO^-$ to give the corresponding chloro, fluoro, and trifluoromethanesulfonate complexes.[176b] It also reacts with organolithium derivatives to give the corresponding alkylated complexes in good yields[180] (Scheme 4.13).

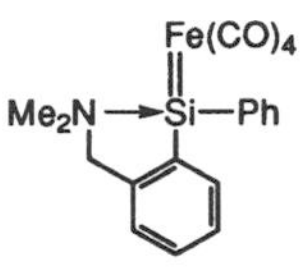

121 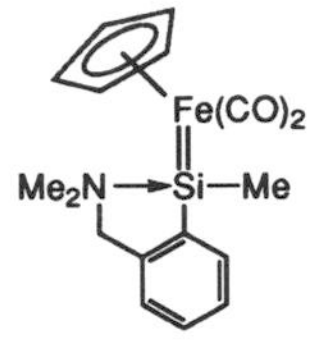 **122** 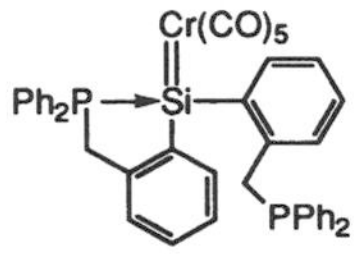 **123**

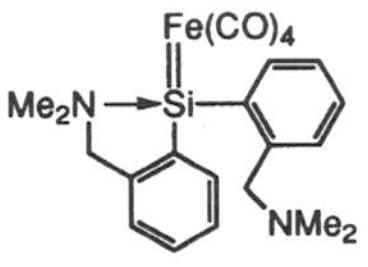

124 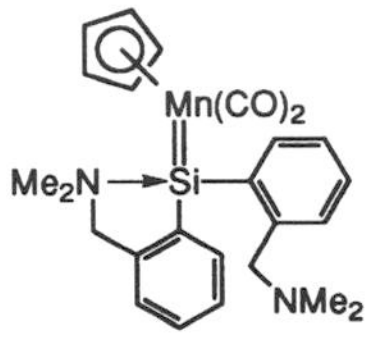 **125**

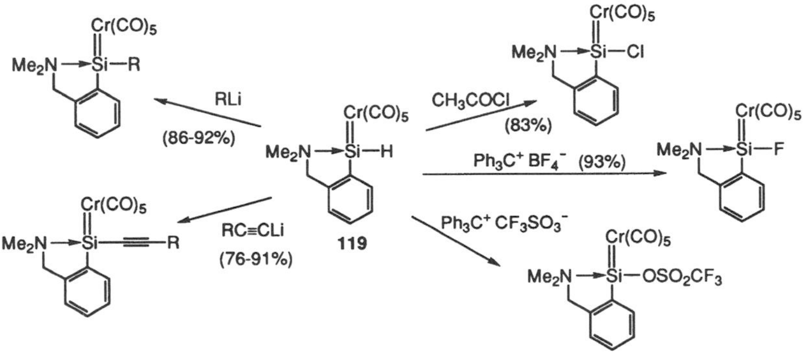

Scheme 4.13

4.2.4.2.3 Siliconium Ions Recently, siliconium ions **126a** with one Si—H bond were prepared by the reactions described in Scheme 4.14.[181] The structure of the related silyl cation with three Si—C bonds depends on the solvent as well as the counteranion.[182] A non-nucleophilic anion (tetra (3.5 bis-trifluoromethylphenyl)borate (TFPB)) allowed the formation of siliconium ion **126b**. With nucleophilic anions (Cl^-, Br^-, I^-, $CF_3SO_3^-$), siliconium ions **127** were obtained in CD_2Cl_2,[182] whereas four-coordinate silyl cations **128** were obtained in CD_3OD.[183] ^1H NMR studies have shown that coordination–decoordination of the two NMe_2 groups occurs in **127** and **128** in solution.

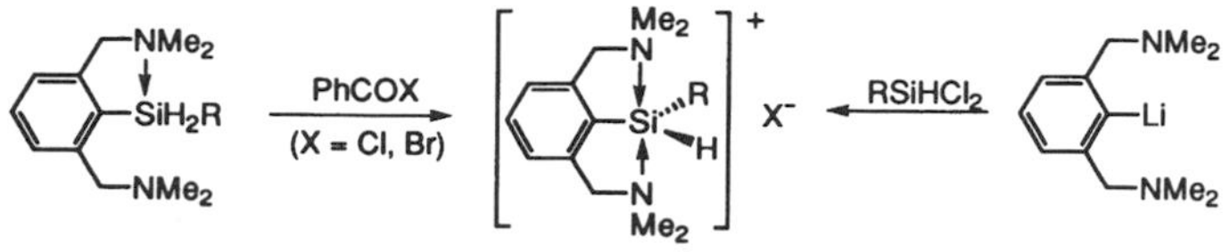

NMe_2 SiH_2R NMe_2 — PhCOX (X = Cl, Br) → [Me_2N, Si, R, H, N Me_2]$^+$ X^- ← $RSiHCl_2$ — NMe_2 Li NMe_2

126a

Scheme 4.14

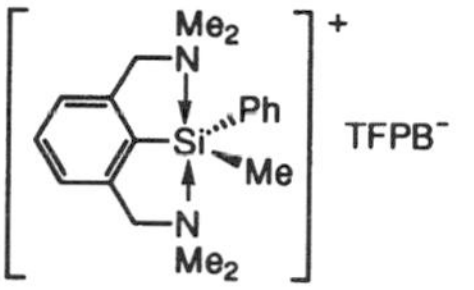

[Me_2N, Si, Ph, Me, N Me_2]$^+$ TFPB$^-$

126b

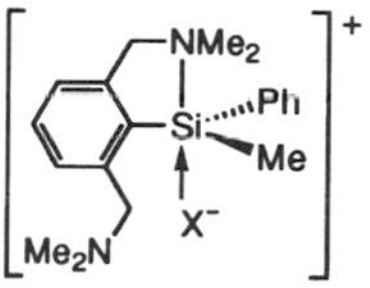

[NMe_2, Si, Ph, Me, X^-, Me_2N]$^+$

127

$^+NMe_2$ X^-, ^tBu, Si, Me, Me, Me_2N

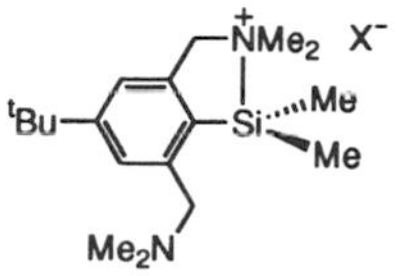

128

The X-ray structure analyses of **129**[184] (prepared by reaction of an excess of iodide with the corresponding dihydrosilane) and of **130**[185] and **131a**[185] (prepared by reaction of trimethylsilyl triflate with the corresponding dihydrosilanes) were performed recently. In these compounds, the silicon atom is pentacoordinated with slightly distorted TBP geometries. In **129** and **130**, the two NMe_2 groups are arranged *trans* to each other with N—Si distances of 2.05–2.08 Å. In **131a** the N and O atoms are *trans* to each other with N—Si and O—Si distances of 2.05 Å and 1.95 Å respectively.

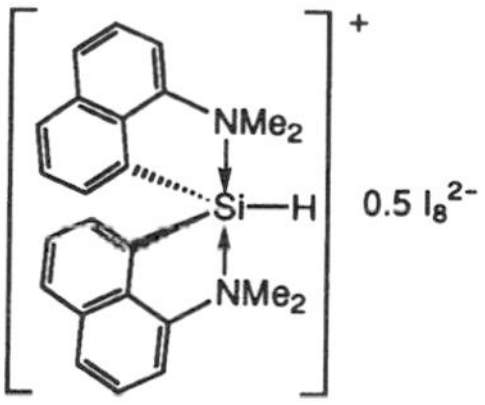

[NMe_2, Si—H, NMe_2]$^+$ 0.5 I_8^{2-}

129

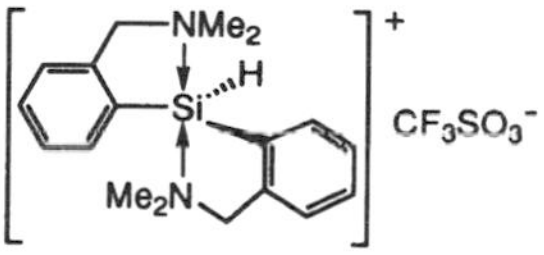

[NMe_2, Si, H, Me_2N]$^+$ $CF_3SO_3^-$

130

[Z, Si, Ph, H]$^+$ $^-OSO_2CF_3$

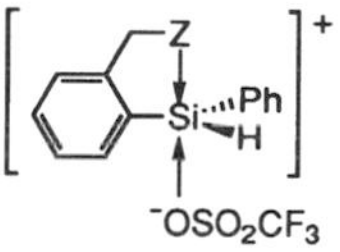

131 a : Z = Me_2N
b : Z = MeO

4.2.5 O→Si Intramolecular Coordination to an Organosilane

4.2.5.1 Neutral Complexes

4.2.5.1.1 Structure and Fluxional Behavior In 1970, Boer and van Remoortere[186] determined by X-ray crystallography the structure of compound **132**. They found that one silicon atom is pentacoordinated with a distorted TBP geometry, the axial sites being occupied by the two oxygen atoms. The coordinative O→Si bond is longer (2.613 Å) than the covalent O—Si bond

(1.670 Å). Very recently, compound **133**[187] has been prepared by reaction of $MeSiCl_3$ with pinacol. The X-ray structure of **133** showed that in this case too the geometry around the silicon atom is that of a distorted TBP, the coordinative Si→O bond being longer (2.736 Å) than the covalent ones (1.62–1.66 Å).

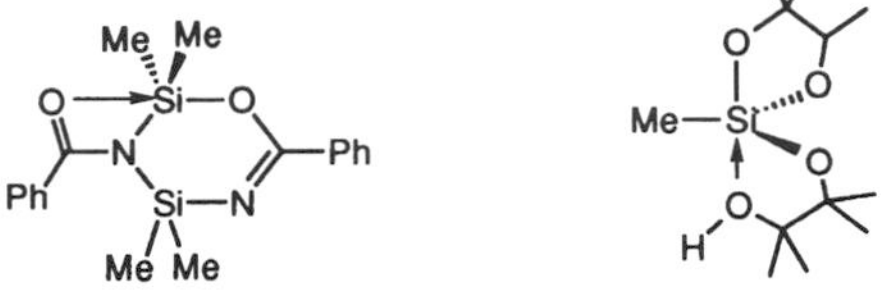

Me Me O Si O N Ph Ph Si N Me Me

132

O O Me Si O O H

133

Numerous compounds with Si—halogen bonds and intramolecular O→Si coordination have been described. The X-ray crystal structure determination of **134**[188] showed that the Si(1)—Cl(1) bond at pentacoordinate silicon is 15% longer (2.35 Å) than the Si(2)—Cl(2) bond at tetracoordinate silicon (2.05 Å). The geometry around the silicon atom is a slightly deformed TBP with an O→Si distance of 1.913 Å, which is much shorter than the O→Si coordinative bond in **132**. The O→Si coordinative bonds of compounds **135** (X = H, F, Cl, Br)[10,189] lie between 1.94 Å (X = Br) and 2.08 Å (X = Cl), and the ν(C=O) absorptions appear 80 to 100 cm^{-1} lower than those for the tetracoordinate silicon compounds. Dynamic ^{19}F NMR studies[190] showed fluorine exchange occurring with a $\Delta G^{\ddagger}$ value of 31.7 to 35.1 kJ mol^{-1} (at 188–198K). The pentacoordination of the silicon atom was inferred from dynamic ^{19}F NMR studies for compounds **136** (R = Me, Et, $PhCH_3$[191]), and from IR and NMR data for compound **137**.[192]

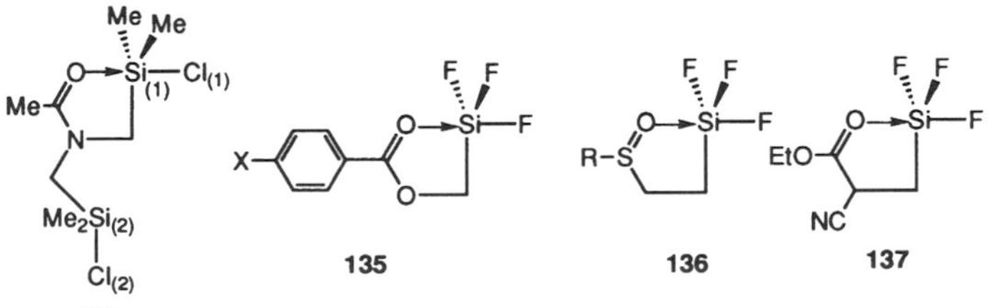

Me Me O Si(1) Cl(1) Me N $Me_2Si_{(2)}$ Cl(2)

134

F F X O Si F O

135

F F R S O Si F

136

F F EtO O Si F NC

137

The fact that in these complexes the length of the coordinative bond depends on the other equatorial ligand is particularly well illustrated for compounds **138–144**. Three groups of compounds can be distinguished. The first group includes compounds **138–140**. In **138**,[193,194] the coordinative O—Si distance is 2.395 Å, whereas the length of the Si–F bond is 1.652 Å, which is only 0.1 Å longer than an Si—F bond for tetrahedral Si. The geometry around the silicon atom is that of a tetrahedron monocapped by the oxygen atom. Similar geometries are found in compounds **139**[195] and **140**.[196] In the chloro compounds

141[193] and **142**[196] (second group), the O→Si coordinative bonds are shorter than in compounds **138–140**. In these cases, the geometry around the silicon atom corresponds to a slightly distorted TBP. In the third group, including compounds **143**[193,197] and **144**,[195] the O—Si distances are much more shorter than in the preceding compounds (1.749 Å in **143**), whereas the opposite X—Si bonds are very long (e.g., 3.743 Å for the Si—I distance in **143**). The geometries around the silicon atom in the derivatives **143** and **144** are those of inverted (compared to the first group) tetrahedrons monocapped by iodide or triflate ions. The changes in the O—Si lengths and in the geometries around the silicon atoms on going from **138** to **141** and **143** mimic the course of S_N2 substitution at silicon occurring with inversion of configuration. **138** represents the initial attack of the oxygen at the silicon, **141** the transition state, and **143** the final attack close to the "reaction product."[198]

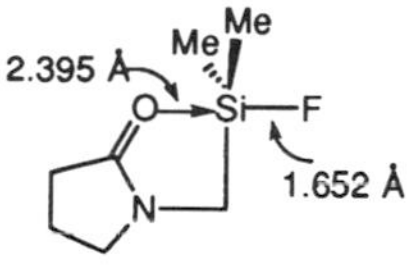

2.395 Å Me Me Si F 1.652 Å O N

138

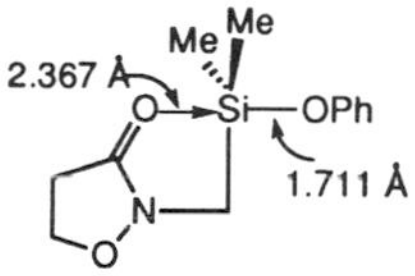

2.367 Å Me Me Si OPh 1.711 Å O N O

139

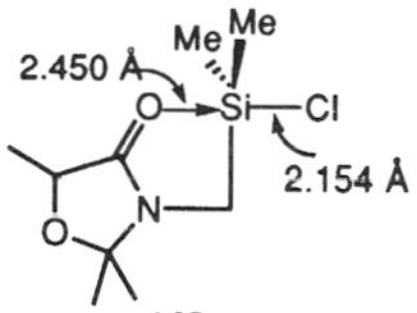

2.450 Å Me Me Si Cl 2.154 Å O N O

140

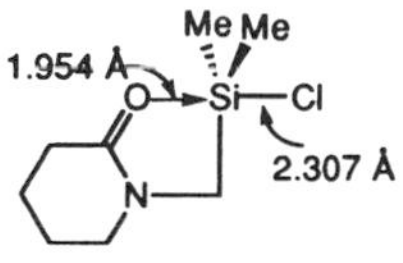

1.954 Å Me Me Si Cl 2.307 Å O N

141

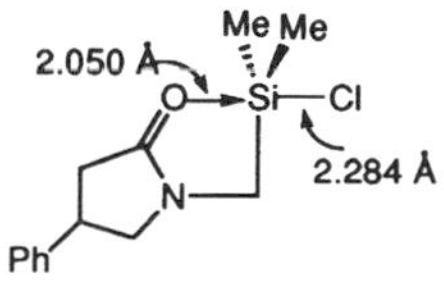

2.050 Å Me Me Si Cl 2.284 Å O N Ph

142

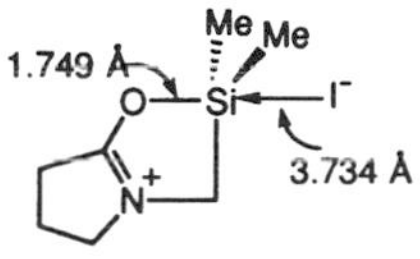

1.749 Å Me Me Si I^- 3.734 Å O N+

143

1.753 Å Me Me Si $^-OSO_2CF_3$ 2.785 Å O N+

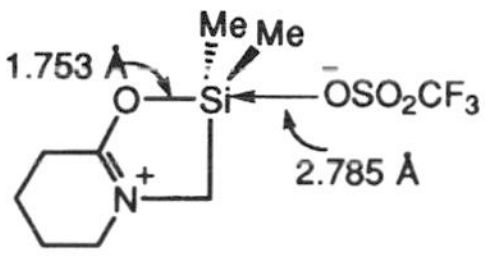

144

The same change of geometry due to the influence of the monochelating ligand on the O—Si distance was found in the bis-pentacoordinate silicon complexes **145**. However, the influence is less pronounced.[199] Similarly, the X-ray crystal structure determination of compounds **146a**[200] and **146b**[201] showed that the O→Si coordinative bond is shorter in **146a** (1.788 Å) than in **146b** (1.879 Å). Conversely, the Si—Cl bond is longer in **146a** (2.624 Å) than in **146b** (2.432 Å). The geometries around the silicon atom are best described as a monocapped (by Cl^-) tetrahedron in **146a** and as a TBP in **146b**).

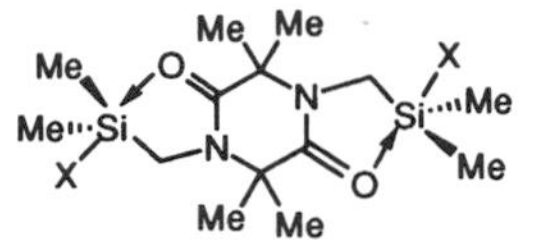

145 **a** : X = Cl
b : X = OSO_2CF_3

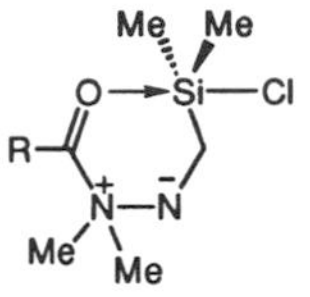

146 **a** : R = p-$CH_3OC_6H_4$
b : R = CF_3

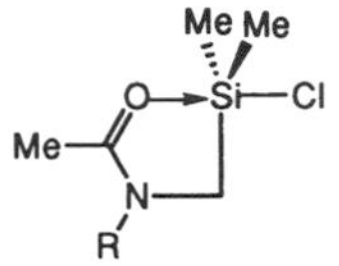

147 (R = Me, Ph)

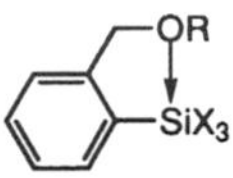

148 **a** : X = F
b : X = H

Recently,[202] it was inferred from NMR studies that compound **147** exists in solution at −90°C as two pentacoordinate species, one being formed by intramolecular O→Si coordination, the other being the dimer formed by two intermolecular O→Si coordinations. At higher temperatures their fast interconversion is observed.

A suitably situated RO group can give rise to pentacoordination of the silicon atom as in **148a**.[203] However, this group seems less favorable than the NMe_2 group as there is no coordination in the corresponding trihydrosilyl derivative **148b**.

4.2.5.1.2 Reactivity Complex **149** reacted with Grignard reagents,[204a] water,[196] alcohols,[195] cyclopentanone,[205] and $MeOSnEt_3$[204b] (Scheme 4.15).

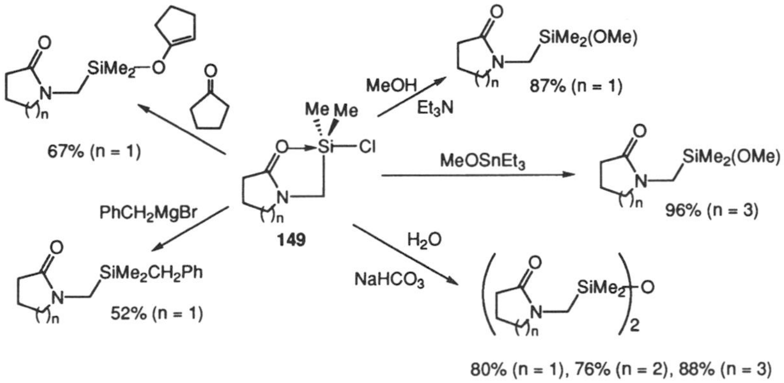

67% (n = 1)

MeOH, Et_3N: 87% (n = 1)

$MeOSnEt_3$: 96% (n = 3)

$PhCH_2MgBr$: 52% (n = 1)

H_2O, $NaHCO_3$: 80% (n = 1), 76% (n = 2), 88% (n = 3)

149

Scheme 4.15

Bassindale and Borbaruah[206] have studied the reactivity of complex **150** toward N-methylimidazole (Scheme 4.16). Reaction occurred at the pentacoordinate silicon atom when X = Cl and at the tetracoordinate one when X = F. This can be explained by the geometry around the silicon atoms. When X = Cl, the geometry at the pentacoordinate silicon atom is TBP[188] and this site is more reactive toward nucleophiles than is the tetrahedral silicon atom, whereas when X = F the geometry of the pentacoordinate silicon atom probably corresponds to a monocapped tetrahedron,[194] thus, for steric reasons, rendering this site less reactive than the tetracoordinate silicon toward NMI.

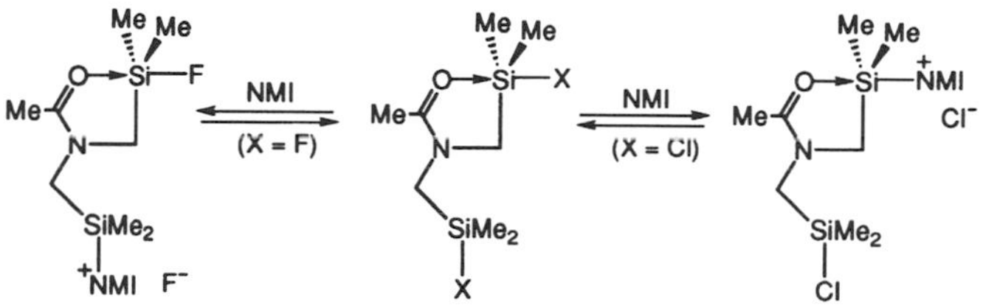

150

Scheme 4.16

4.2.5.2 ***Siliconium Ions*** Siliconium ions **151** and **152** were obtained by reaction of $RSiCl_3$ with two moles each of tropolone[207] and of 1,3-diketone.[208,209] The X-ray structure of the silyl cation **131b**[203] (obtained by reaction of trimethylsilyl triflate with the corresponding dihydrosilane) showed that the geometry around the silicon atom is very close to TBP. The bond length for Si—O(Me) is 2.03 Å and that for Si—O(SO_2CF_3) is 1.86 Å, the latter being shorter than the same distance in the silyl cation **131a**[185] (1.95 Å) with an N→Si coordinative bond.

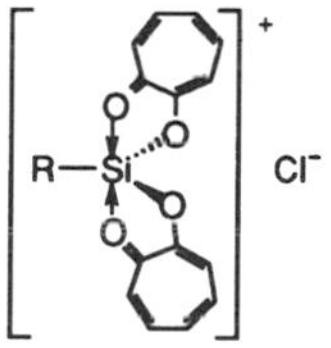

151

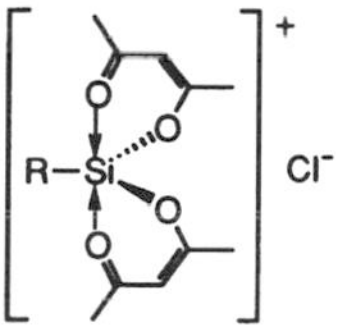

152

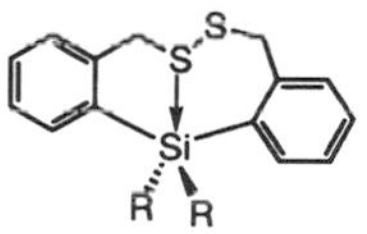

153 a : R = Me
b : R = Ph

4.2.6 S→Si and P→Si Intramolecular Coordination to an Organosilane

An S→Si intramolecular interaction was observed in compounds **153**.[210,211] The S—Si distances of the coordinated sulfur atom are 3.438 Å in **153a** and 3.372 Å in **153b**, whilst thc S—Si distances for the other sulfur atom are 4.190 Å in **153a** and 4.193 Å in **153b**. In both cases, the geometry around the silicon atom is that of a monocapped tetrahedron. Replacement of the sulfur atoms in **153b** by selenium atoms also gave a pentacoordinated silicon atom with the same geometry.[212] Recently, S→Si intramolecular interactions were also found in compounds **154**[213a] and **155**[213b], with coordinative S→Si distances of 3.04 Å and 3.11 Å in **154**, 3.292 Å in **155a**, and 3.074 Å in **155b**. In these compounds, the geometry around the silicon atom lies somewhere between that of a TBP and a monocapped tetrahedron.

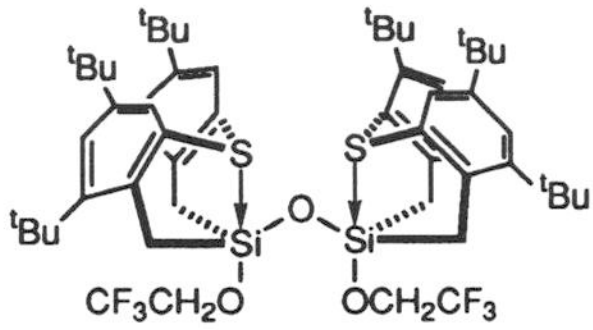

154

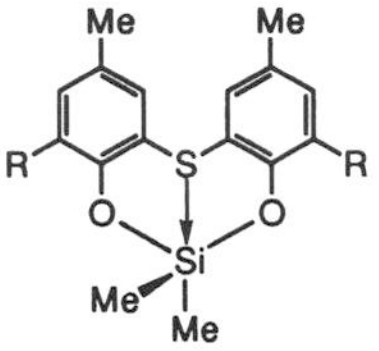

155 a : R = Me
b : R = ^tBu

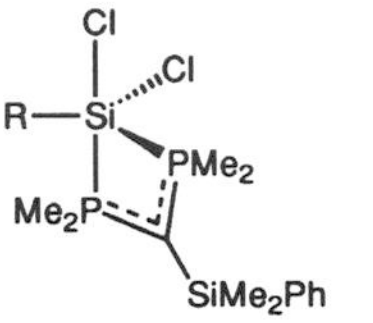

156 a : R = C_6H_5
b : R = $pMeC_6H_4$

Recently, pentacoordinate silicon compounds **156** with a phosphorus donor atom were prepared.[214] The X-ray structure of **156b**[214b] (obtained by reaction of $p\text{-}MeC_6H_4SiCl_3$ with $Li[C(PMe_2)_2(SiMe_3)]$) showed the pentacoordination of the silicon atom with P—Si distances in the range 2.30 to 2.40 Å. The geometry at the central silicon atom lies between that of a TBP and a tetragonal pyramid.

4.2.7 Silatranes and Azasilatranes

4.2.7.1 Silatranes Silatranes[215] are an important class of neutral pentacoordinate silicon compounds, in which pentacoordination occurs by N→Si intramolecular chelation in the apical position, three Si—O bonds lying in equatorial positions. These compounds, which exhibit interesting and unusual chemical properties as well as biological activity, have been extensively studied by Voronkov and co-workers.[216]

4.2.7.1.1 Preparation Silatranes were first prepared independently by Fineston[217] and by Frye and co-workers[218] by reaction of triethanolamine with alkoxysilanes $RSi(OMe)_3$ (Scheme 4.17). This reaction is the most versatile method for the preparation of silatranes. However, they can also be obtained from polyorganosilsesquioxanes[215] or polyorganyl hydrosiloxanes[215] (Scheme 4.17). 1-Organoxysilatranes can be prepared by reaction of tetraalkoxysilanes with an equimolar mixture of triethanolamine and an alcohol (ROH)[205], where R = alkyl, cycloalkyl, arylalkyl, or aryl (Eq. 4.18). Alkali catalysts accelerate the reaction and allow the preparation of 1-alkoxysilatranes bearing bulky R groups such as ^{i}Pr or ^{t}Bu. 1-Halosilatranes can be prepared from silatrane[219] by the reactions indicated in Scheme 4.18. The Si—H bond of 3,7,10-trimethylsilatrane was also cleaved by *N*-chloro and *N*-bromosuccinimide to give the corresponding 1-halo-3,7,10-trimethylsilatranes[220] (Eq. 4.19).

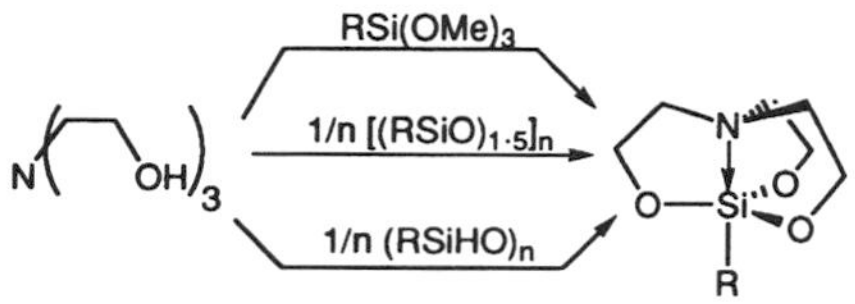

$N(CH_2CH_2OH)_3$ — $RSi(OMe)_3$ / $1/n\ [(RSiO)_{1.5}]_n$ / $1/n\ (RSiHO)_n$ → **157** (R)

Scheme 4.17

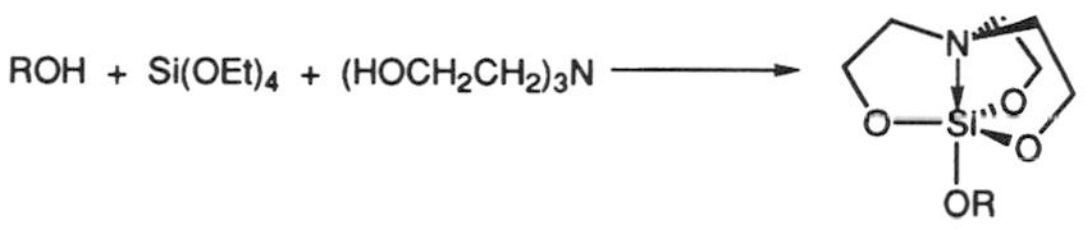

$$ROH + Si(OEt)_4 + (HOCH_2CH_2)_3N \longrightarrow \text{silatrane (OR)} \quad (4.18)$$

Ph_3CCl → (Cl); Ph_3CBr → (Br); CF_3I / Et_3N → (F); $C_3F_7I / h\nu$ → (I); from silatrane (H)

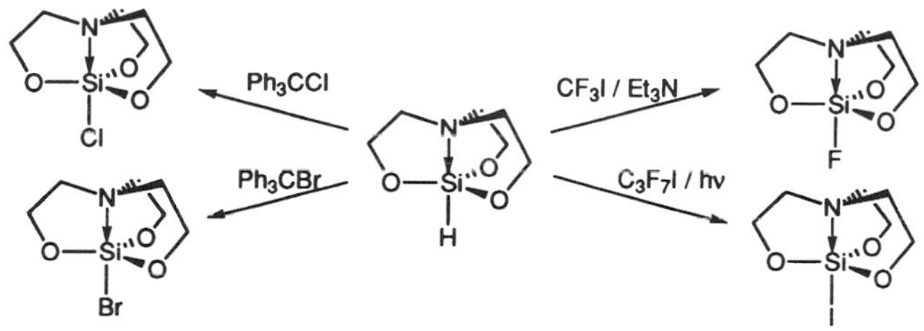

Scheme 4.18

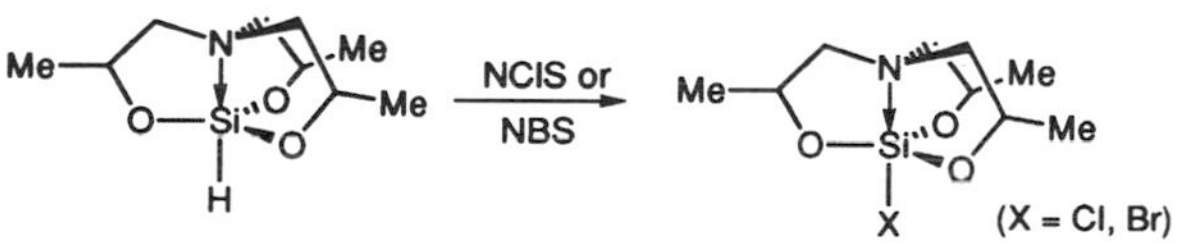

Me, Me, Me (H) — NCIS or NBS → Me, Me, Me (X) (X = Cl, Br) (4.19)

4.2.7.1.2 Structure The first X-ray structure determination of a silatrane was performed in 1968 by Turley and Boer[221] for phenylsilatrane **157** (R = Ph). They found that the silicon atom is pentacoordinated with a geometry at the silicon atom corresponding to a distorted TBP, the N atom and the phenyl group occupying axial positions. The N—Si distance was found to be 2.19 Å. Subsequently, two other distinct crystallographic forms of phenylsilatrane, β and γ, were described with Si—N lengths of 2.15 Å[222] and 2.13 Å,[223] respectively. In almost all the silatranes studied, the intramolecular Si—N distance lies between 2.0 and 2.2 Å.[216] The presence of an electron-withdrawing substituent at the silicon atom shortens the interatomic Si—N distance (2.02 Å in 1-chlorosilatrane[216]). Until now, the shortest Si—N length (1.96 Å) recorded is that found in the O-methylated silatrane **158**.[224]

Introduction of phenyl groups in the cage as in **159** or replacement of an oxygen atom by a CH_2 group as in **160** lengthens the Si—N distances to about 2.335 Å in both **159**[225] and **160**.[226] The Si—N distances in silatranes calculated from both *ab initio* and semiempirical methods are longer than those found in the crystal. However, it was found that the energy cost for constraining the Si—N distance in hydrosilatrane to a value similar to that observed in the crystal is less than 25 kJ mol^{-1}. This suggests that crystal forces may be responsible for the much shorter Si—N distance in the solid.[227]

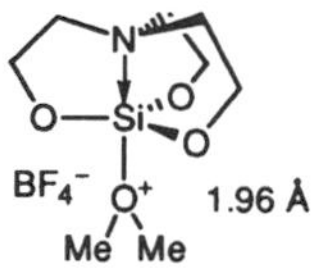

4.2.7.1.3 Pseudosilatranes Organosilicon ethers of diethylamines $R_2Si(OCH_2CH_2)_2NR'$, called "pseudosilatranes,"[228] exhibit bicyclic structures similar to silatranes, with nitrogen bonded to silicon. The Si—N distance depends on the substituents on the nitrogen atom. When this substituent is H as in **161a**[229] and **162**[230] or Me as in **161b**,[231] **163**,[232] and **164**,[233] the Si—N distances are relatively short, varying from 2.00 Å for **162** to 2.30 Å for **161a**, **163**, and **164** and 2.68 Å for **161b**. In these compounds, the geometry around the silicon atom is that of a more or less distorted trigonal bipyramid, the dioxoazasilacyclooctane ring being in a boat–boat conformation. With bulky R groups (Ph, ^{t}Bu), the Si—N distances are longer (around 3.1 Å)[231] and the geometry around the silicon atoms corresponds to a monocapped tetrahedron. The eight-membered ring is in a chair–chair conformation in **165** and in a boat–chair conformation in **166**. Such a boat–chair conformation of the eight-membered ring is also found in compound **167**[228], in which the geometry around the silicon atom is a monocapped tetrahedron with an Si—N distance of 2.968 Å. A related structure is also found in compound **168**[234] with a 10-membered ring, the N—Si distance being 2.727 Å.

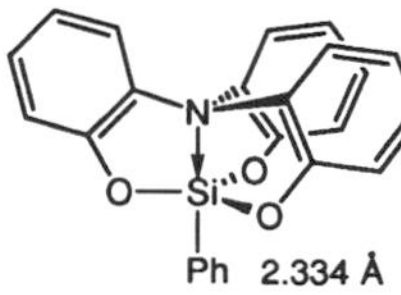

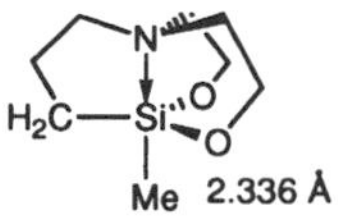

4.2.7.1.4 Reactivity Although the physicochemical properties of silatranes and of their derivatives have been studied in detail,[216,235,236] their reactivity has not been extensively explored because it is generally poor. Thus Frye[220] reported the unexpected stability toward solvolysis of 1-chlorosilatrane. However, 1-iodosilatrane displays electrophilic reactivities with various classes of organic and organometallic compounds.[219] Some of these reactions

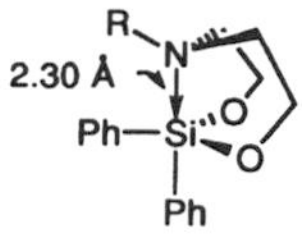

161 a : R = H
b : R = Me

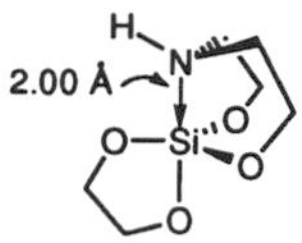

162

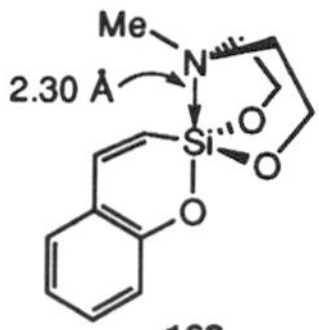

163

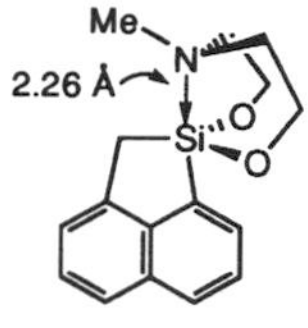

164

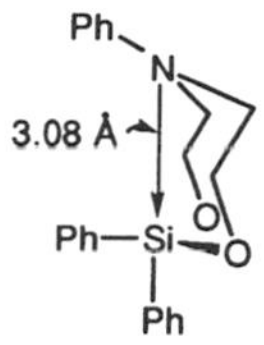

165

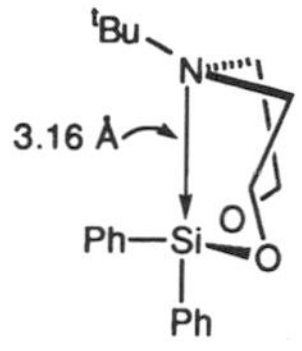

166

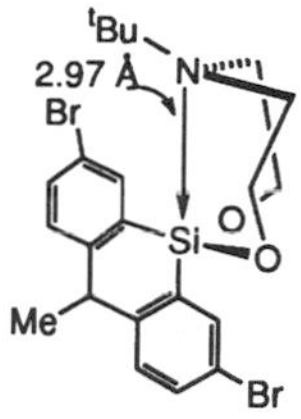

167

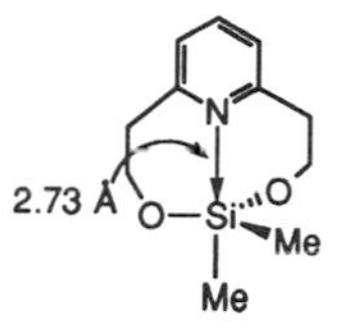

168

are indicated in Scheme 4.19. Silatrane itself is a weak reducing agent, but is able to reduce alkyl bromides, ketones, or acyl chlorides[237] (Scheme 4.20). However, its reactivity toward these compounds is lower than that of the pentacoordinate spirosilane **65**. It reacts also with *n*-butyllithium and Grignard reagents with cleavage of the Si—O bond.[238] 3,7,10-Trimethylsilatrane reacted immediately with HCl or HBr with evolution of dihydrogen[220] (Eq. 4.20). This redox reaction does not occur with Et_3SiH or $(EtO)_3SiH$. 1-Alkyl and 1-arylsilatranes reacted with *N*-bromosuccinimide,[239] *m*-chloroperbenzoic acid,[239] and $HgCl_2$[240] with cleavage of the Si—C bond (Scheme 4.21). 1-Arylsilatranes, but not 1-alkylsilatranes, were reduced by $LiAlH_4$ to the corresponding silane;[238] *n*-butyllithium converted both 1-aryl and 1-alkylsilatranes to the corresponding tetraorganosilanes[238] (Scheme 4.21). Allylsilatrane reacted with carbonyl compounds in the presence of a catalytic amount of a Lewis acid (Scheme 4.22) as does allylsilane, but in contrast with the pentacoordinate spirosilane **67c**, did not react with carbonyl compounds under conditions of nucleophilic activation (KF, NaOMe).[92]

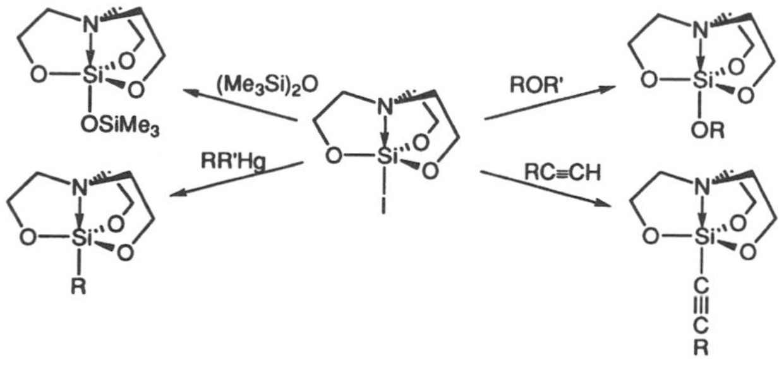

Scheme 4.19

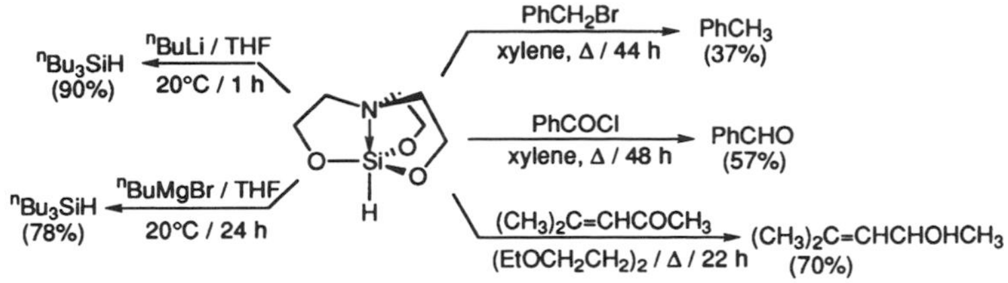

$^nBu_3SiH$ (90%) ← ^nBuLi / THF, 20°C / 1 h

$^nBu_3SiH$ (78%) ← ^nBuMgBr / THF, 20°C / 24 h

$PhCH_2Br$, xylene, Δ / 44 h → $PhCH_3$ (37%)

PhCOCl, xylene, Δ / 48 h → PhCHO (57%)

$(CH_3)_2C{=}CHCOCH_3$, $(EtOCH_2CH_2)_2$ / Δ / 22 h → $(CH_3)_2C{=}CHCHOHCH_3$ (70%)

Scheme 4.20

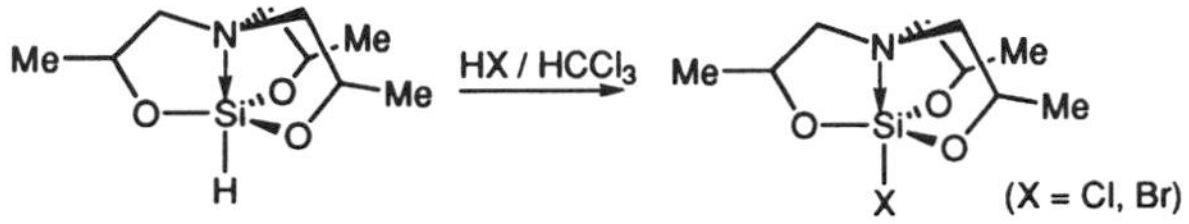

Me, Me, Me, H; HX / $HCCl_3$ → Me, Me, Me, X (X = Cl, Br) (4.20)

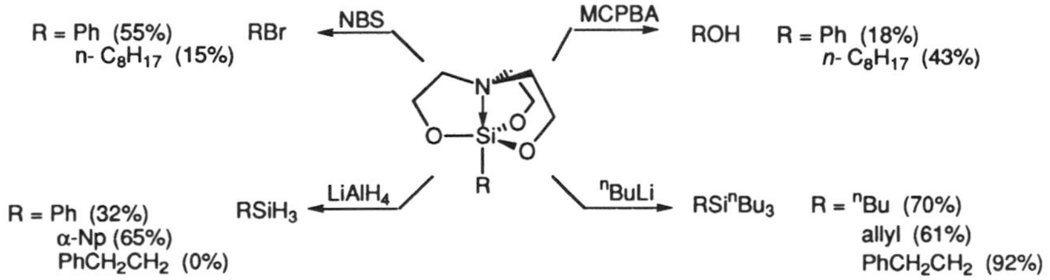

R = Ph (55%), n- C_8H_{17} (15%) RBr ← NBS

MCPBA → ROH R = Ph (18%), *n*- C_8H_{17} (43%)

R = Ph (32%), α-Np (65%), $PhCH_2CH_2$ (0%) $RSiH_3$ ← $LiAlH_4$

^nBuLi → RSi^nBu_3 R = ^nBu (70%), allyl (61%), $PhCH_2CH_2$ (92%)

Scheme 4.21

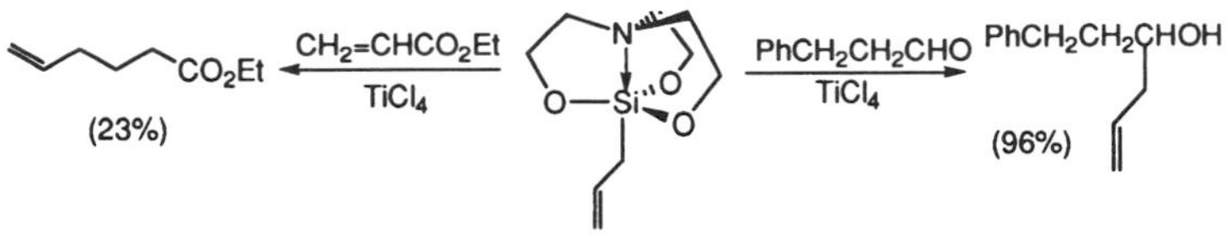

CO_2Et (23%) ← $CH_2{=}CHCO_2Et$, $TiCl_4$

$PhCH_2CH_2CHO$, $TiCl_4$ → $PhCH_2CH_2CHOH$ (96%)

Scheme 4.22

Acid-catalyzed hydrolysis of phenylsilatrane gave triethanolamine and probably polyphenylsilsequioxane.[241] It was suggested that the rate-determining step involves protonation of the nitrogen atom along with the breaking of the N→Si coordinative bond. The disilatranyl ferrocene **169** was hydrolyzed to the cross-linked solid silsesquioxane **170**[242] (Eq. 4.21).

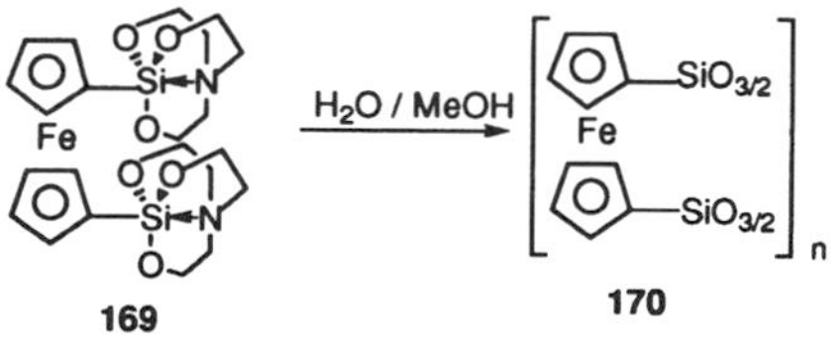

H_2O / MeOH

169 → **170** ($SiO_{3/2}$, $SiO_{3/2}$, Fe)$_n$ (4.21)

4.2.7.2 Azasilatranes Triazasilatranes **171** were synthesized more recently[243] by heating tris(dimethylamino)silanes with tris(2-aminoethyl)-amine[243,244] (Eq. 4.22). A multinuclear NMR spectroscopic study of these compounds was published in 1987.[245] X-ray analysis of 1-phenyl-2,8,9-triazasilatrane[246] showed that the silicon atom is pentacoordinated with a somewhat distorted TBP geometry, similar to that of 1-phenylsilatrane, with an N–Si bond length of 2.13 Å.

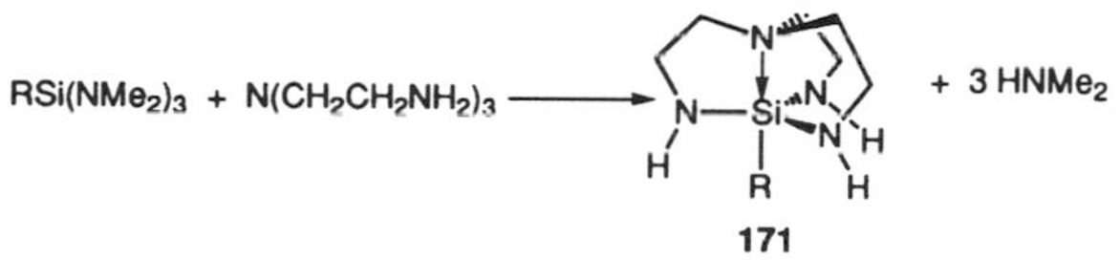

$RSi(NMe_2)_3 + N(CH_2CH_2NH_2)_3 \longrightarrow$ **171** $+ 3\ HNMe_2$ (4.22)

Silylation of the NH functional groups of azasilatranes by chlorosilanes has been studied.[244] 1-Hydroazasilatrane reacted with an excess of RMe_2SiCl (R = H, Me, Ph) to give silylation of the equatorial NH groups. In the case of 1-ethoxyazasilatrane, silylation of all three amino groups occurred on treatment with Me_2HSiCl, disilylation with Me_3SiCl and $PhMe_2SiCl$, and no silylation with $^tBuMe_2SiCl$. Compound **172** reacted with CF_3SO_3Me to give **173** in which the silicon atom is tetracoordinated[247] (Eq. 4.23). Azasilatranes solvolyze in methanol at room temperature to give the corresponding trimethoxysilanes.[248]

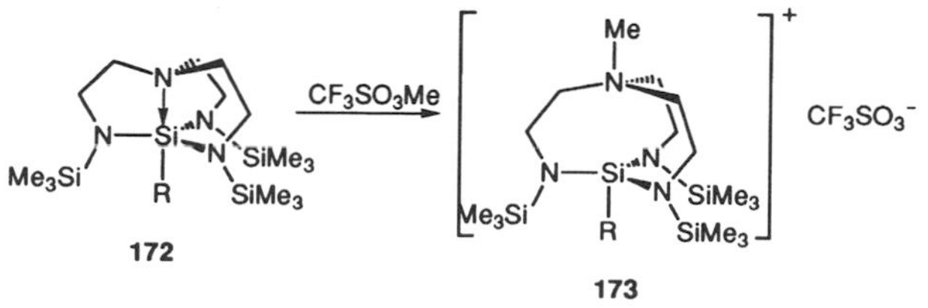

172 $\xrightarrow{CF_3SO_3Me}$ [**173**]$^+$ $CF_3SO_3^-$ (4.23)

4.3 HEXACOORDINATE SILICON COMPOUNDS

4.3.1 Dianionic Fluorosilicates

4.3.1.1 Preparation Organopentafluorosilicate compounds, $[RSiF_5]^{2-}$, were prepared first by Tansjö,[249] and their chemistry was subsequently developed by Müller and co-workers,[52,250,251] and by Kumada and co-workers.[252] In 1961, Tansjö[249] obtained organofluorosilicates by reaction of $RSi(NHR')_3$ with an excess of HF (Eq. 4.24) and subsequently by reaction of organotrifluorosilanes with primary amines[253,254] (Eq. 4.25). Müller and Dathe[255,256] found an easier way to prepare these pentafluorosilicates by addition of fluoride anions to $RSiF_3$. Kumada, and co-worker[257] extended this method by reacting $RSiCl_3$ with an aqueous solution of potassium fluoride and developed the use of organopentafluorosilicates in organic synthesis.

$$RSi(NHR')_3 \xrightarrow{6\ HF} [RSiF_5]^{2-}\ 2\,[R'NH_3]^+ + [R'NH_3]^+\ F^- \tag{4.24}$$

$$4\ R'NH_2 + 3\ RSiF_3 \longrightarrow [RSiF_5]^{2-}\ 2\,[R'NH_3]^+ + 2\ RSi(NHR')F_2 \tag{4.25}$$

Unlike organopentafluorosilicates, diorganotetrafluorosilicates are not very stable in solution. For example, $[Ph_2SiF_4]^{2-}$ decomposed during its formation into phenylpentafluorosilicate by cleavage of one Si—Ph bond[254] (Eq. 4.26). $[Cl_2CH(CH_3)SiF_4]^{2-}$ was prepared in aqueous solution, but it gradually decomposed in water.[258] However, dimethyldifluorosilane and even trimethylfluorosilane reacted with solid tetramethylammonium fluoride, under pressure, to form the corresponding tetra- and trifluorosilicates.[259]

$$Ph_2SiF_2 + 2\ RNH_3^+\ F^- \longrightarrow [Ph_2SiF_4]^{2-}\ 2\ RNH_3^+ \xrightarrow{RNH_3^+\ F^-} [PhSiF_5]^{2-}\ 2\ RNH_3^+ + PhH + RNH_2 \tag{4.26}$$

The sole X-ray structure determination of an organopentafluorosilicate was performed on the zwitterion **174**.[260] The geometry around the silicon atom corresponds to a slightly distorted octahedron with Si—F bond lengths of 1.685–1.740 Å.

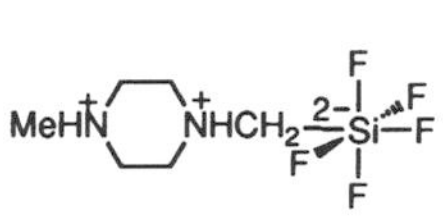

$$MeH\overset{+}{N}(CH_2CH_2)_2\overset{+}{N}HCH_2\text{—}\overset{2-}{Si}F_5$$

174

4.3.1.2 Reactivity Some reactions of organopentafluorosilicates $[RSiF_5]^{2-}$ (with R = Me or Ph) and $[R—CH=CH—SiF_5]^{2-}$ are summarized in Schemes 4.23 and 4.24, respectively.

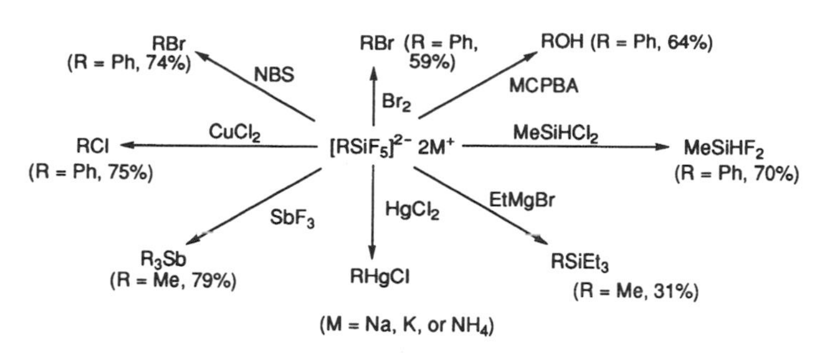

Scheme 4.23

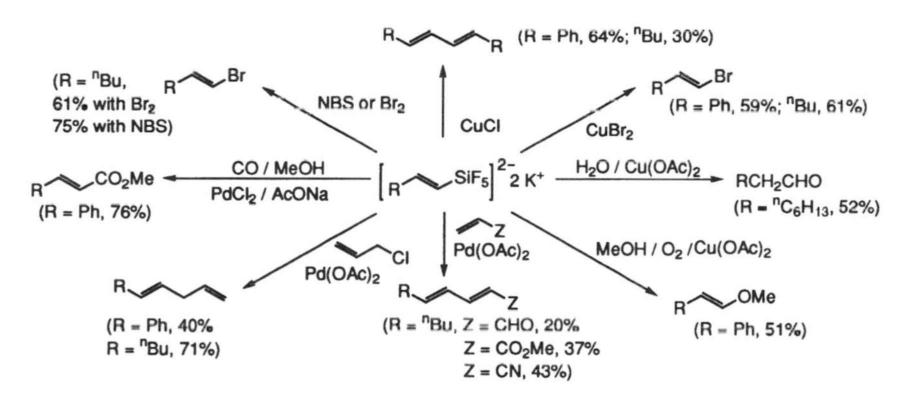

Scheme 4.24

Organopentafluorosilicates $[RSiF_5]^{2-}$ react with chlorosilanes to give chlorine–fluorine exchange[261] and with Grignard reagents to give tetraorganosilanes.[262] The hexafluorosilicate anion $[SiF_6]^{2-}$ also reacts with Grignard reagents,[263] but heating and long reaction times are necessary and the yields are poor. Cleavage of the Si—C bond in $[RSiF_5]^{2-}$ occurs on reaction with $HgCl_2$[264] or SbF_3,[265] $CuCl_2$,[266] NBS or Br_2,[257,267] and m-chloroperbenzoic acid[268] (Scheme 4.23). Cleavage of the Si—C bond in $[R—CH=CHSiF_5]^{2-}$ occurs on treatment with $Cu(OAc)_2$,[266] palladium salts,[269] NBS, Br_2,[257,267] $CuBr_2$,[266] CuCl[270] or AgF[271] (Scheme 4.24).

4.3.2 Spirosilicates

4.3.2.1 Dianionic Spirosilicates

4.3.2.1.1 Preparation and Structure Tris(1,2-benzenediolato)silicates **175** were first prepared by Rosenheim and co-workers[272] by reaction of cathecol with SiO_2 (Scheme 4.25). This reaction was subsequently studied by Weiss and co-workers[273] and Barnum.[274] The compounds may also be easily prepared by reaction of catechol with $Si(OMe)_4$[275] or $Si(OEt)_4$[86] under basic conditions. Moreover, catechol is such an efficient ligand that complexes **175** can be obtained from $[SiF_6]^{2-}$ $2M^+$ (M = Na, K)[276] (Scheme 4.25). The X-ray crystal structure determination of **175c**[277] has shown that the anion forms a nearly perfect octahedron centered at silicon.

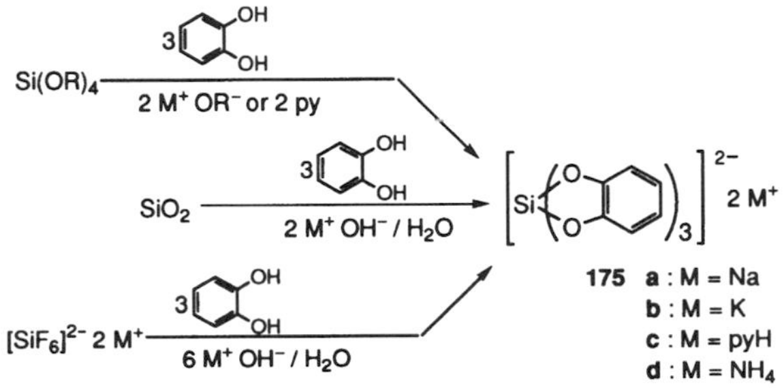

Scheme 4.25

SiO_2 also reacts with ethylene glycol in the presence of BaO to give the hexacoordinate silicate **176**.[278] The X-ray crystal structure analysis revealed two crystallographically independent molecules with an asymmetric unit in which each silicon atom has a slightly distorted octahedral geometry. It is noteworthy that the reaction of SiO_2 with ethylene glycol in the presence of alkali metal hydroxides afforded pentacoordinate complexes[98] (Scheme 4.7). Recently the hexacoordinate diorganospirosilicates **177**[279] were prepared by reaction of $(EtO)_2Si(NCS)_2$ with catechol in the presence of a base.

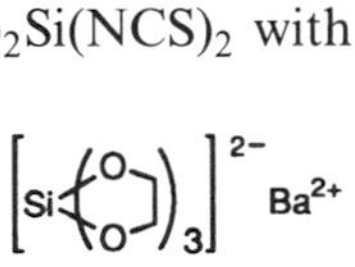

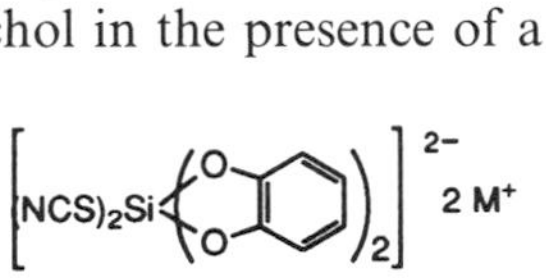

4.3.2.1.2 Reactivity Complex **175a** is very reactive towards organometallic derivatives (Scheme 4.26), allowing the preparation of various silicon compounds from silica or hexafluorosilicates. Thus, it reacts with Grignard reagents[275] giving tri- or tetraorganosilanes, depending on the R group of the Grignard reagent. The intermediate **178** can be transformed *in situ* into monofunctional organosilanes. Reaction occurs with Grignard reagents

RMgBr bearing a hydrogen atom in the β position, activated by Cp_2TiCl_2. Monofunctional organosilanes R_3SiH are obtained with primary Grignard reagents and dihydrogenosilanes R_2SiH_2 with secondary Grignard reagents.[117] **175a** can also be quantitatively reduced to SiH_4 by reaction with $LiAlH_4$ (Scheme 4.26). Barnum[274] observed that **175d** is stable in water and in dilute aqueous ammonia solution, but an immediate precipitate of hydrated silica is obtained in aqueous acid solution. Recently, it was shown that **175a** is also transformed into SiO_2 on treatment with the stoichiometric amount of water in methanolic solution.[280] Electrophilic reagents such as HCl react with **175b** to give catechol and the spirosiloxane **179**.[281]

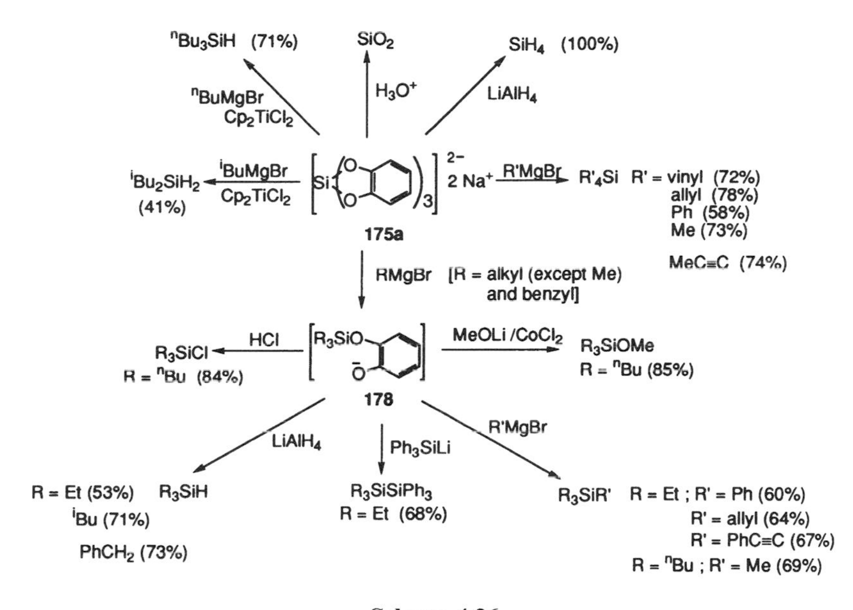

Scheme 4.26

4.3.2.2 Cationic Spirosilicon Complexes 2-Hydroxypyridine N-oxide smoothly transforms SiO_2 into the cationic complex **180**[282] under acidic conditions. Confirmation of the hexacoordination of the silicon atom came from partial resolution of **180** into optical enantioners on treatment with disodium(-)dibenzoyl-L-tartrate.[283]

4.3.3 Intermolecular Donation by Neutral Donors to an Organosilane

4.3.3.1 Neutral Complexes In addition to SiF_4-$2NH_3$, many complexes of analogous composition have been described.[8] The stability of the adducts SiX_4—$2NMe_3$ and SiX_4—$2PMe_3$ decreases in the order $SiF_4 > SiCl_4 > SiBr_4$

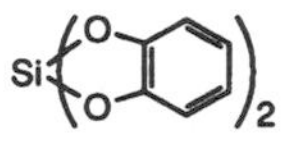

179

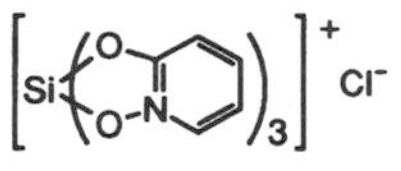

180

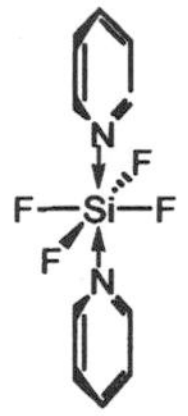

181

for the adducts with NMe_3, whereas the reverse order is observed for the adducts with PMe_3.[284] Pyridine behaves as a good ligand and a series of hexacoordinate complexes Me_2SiX_2—2py, $MeSiX_3$—2py, and SiX_4—2py (X = Cl, Br) has been prepared.[285] Crystal structure analysis of $SiCl_4$—$2PMe_3$[286] showed that the silicon atom has the expected octahedral geometry, with the donor ligands in the *trans* arrangement. Such a geometry was also found in complex **181**[287] with Si—F distances of 1.64 Å and N–Si distances of 1.93 Å.

Chelating diamines such as 2,2′-bipyridine and 1,10-phenanthroline are also good ligands for SiF_4[288] and for the chlorosilanes $SiCl_4$,[289] $RSiCl_3$,[290] and $(MeCl_2Si)_2$.[291] Complexation of SiF_4 by 2,2′-bipyridine gave rise to complex **182**, which adopts a slightly distorted octahedral geometry.[288] The N—Si distances (1.972 and 1.982 Å) are longer than in **181**. In complexes **183**,[292] **184**,[290] and **185**,[293] the N—Si interactions (1.95–2.02 Å) are in all cases opposite to Si—Cl bonds. The Si—Cl distances around the hexacoordinate silicon atom in **183** and **185** are longer (2.20–2.39 Å) than those for tetrahedral silicon atoms (2.03–2.08 Å).

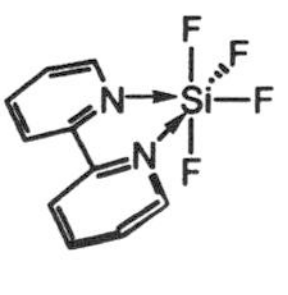

182

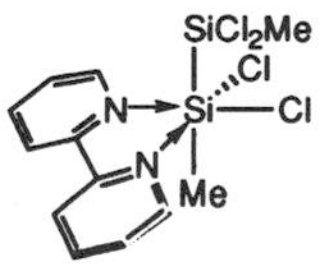

183

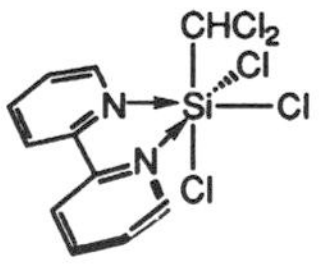

184

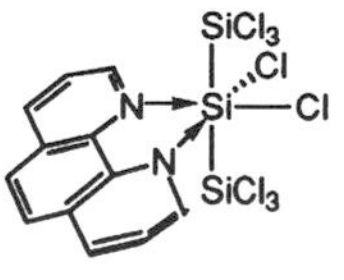

185

1,10-Phenantroline also adds to the spirosilane **36** (Eq. 4.27), giving the hexacoordinate complex **186**;[294] its X-ray structure determination has shown a slightly distorted octahedral framework with N—Si distances of 2.051 and 2.089 Å. In solution, minor amounts of two other disastereoisomers are also present. From ^{19}F NMR studies it was concluded that the interconversion of the three isomers and the eniantomerization of each of them takes place by dissociation of the phenanthroline ligand.

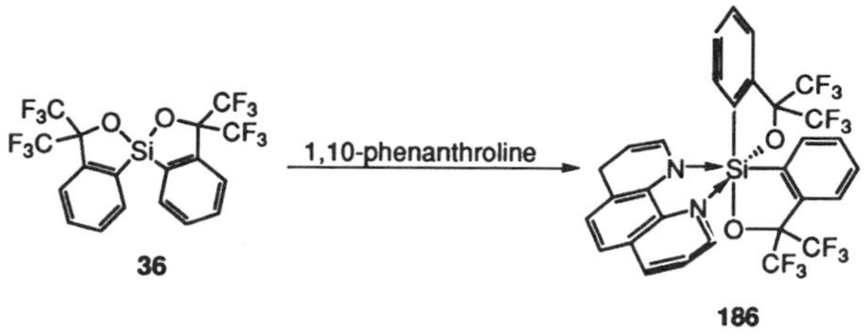

F_3C, F_3C, O, O, Si, CF_3, CF_3 —1,10-phenanthroline→ N, N, Si, O, O, CF_3, CF_3, CF_3, CF_3 (4.27)

36

186

4.3.3.2 Cationic Complexes

Diiodosilanes react with 2,2′-bipyridine[295] or 1,10-phenantroline[296] to give the cationic complexes **187** and **188**, respectively (Scheme 4.27). Complexes of this type could also be prepared by reaction of alcohols[297] or chlorine[298] with compound **189**[299] (Scheme 4.28). Tetracationic complexes **190** have been prepared both by reaction of 2,2′-bipyridine with Si_2Br_6[300] and by reaction of iodine[301] with compound **191**[302] (Scheme 4.29). Partial hydrolysis of **190b** gave the dicationic complex **192**, the X-ray structure determination of which showed the octahedral geometry of the molecule with the OH groups *cis* to each other.[303] The N—Si bond lengths *trans* to the OH groups are longer (2.005 Å) than the others (1.953 Å).

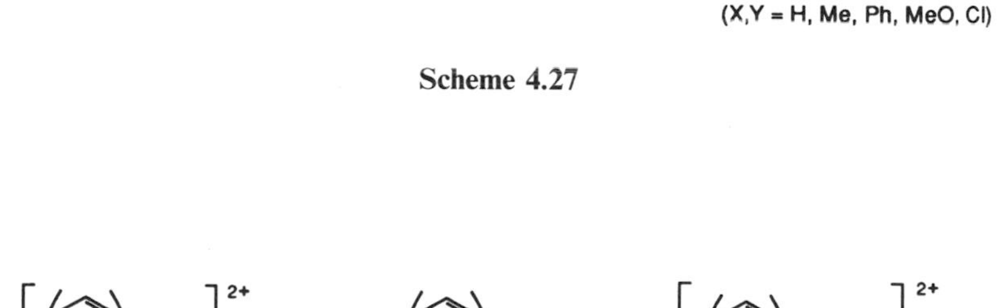

$[(\text{phen})_2SiXY]^{2+}$ $2\,I^-$ ←1,10-phenanthroline— $SiXYI_2$ —2,2′-bipyridine→ $[(\text{bipy})_2SiXY]^{2+}$ $2\,I^-$

188

187

(X,Y = H, Me, Ph, MeO, Cl)

Scheme 4.27

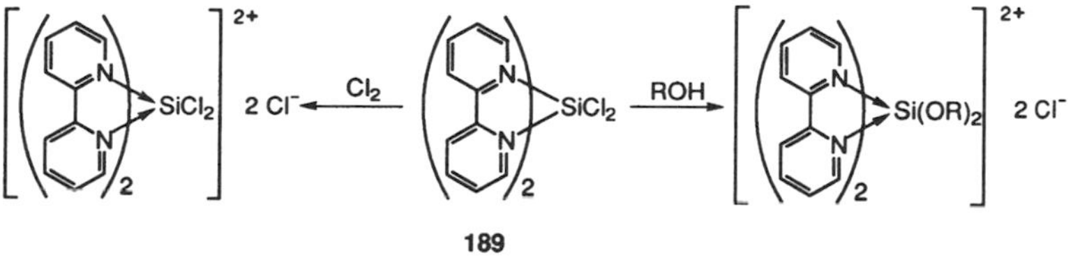

$[(\text{bipy})_2SiCl_2]^{2+}$ $2\,Cl^-$ ←Cl_2— $(\text{bipy})_2SiCl_2$ —ROH→ $[(\text{bipy})_2Si(OR)_2]^{2+}$ $2\,Cl^-$

189

Scheme 4.28

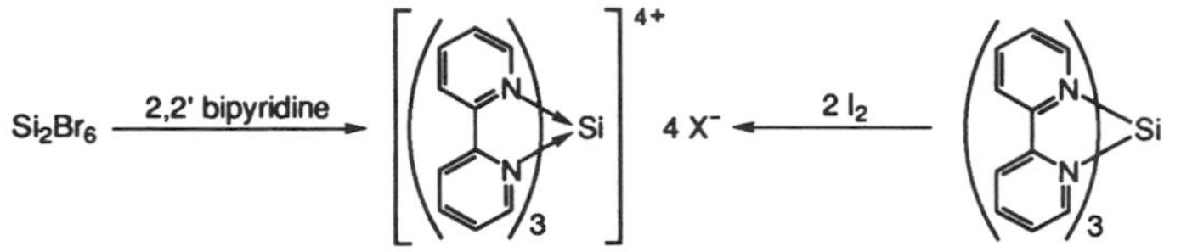

Si_2Br_6 —2,2' bipyridine→ [(bipy)$_3$Si]$^{4+}$ 4 X^- ←$2\ I_2$— (bipy)$_3$Si

190 **a** : X = Br
b : X = I

191

Scheme 4.29

4.3.4 N→Si Intramolecular Coordination to an Organosilane

Hexacoordination of the silicon atom can be achieved by two intramolecular chelations to a tetracoordinate silicon atom or by one intramolecular chelation to a pentacoordinate silicon atom.

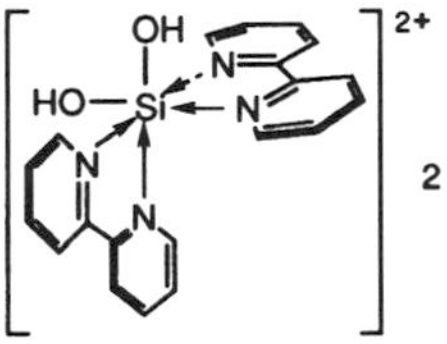

192 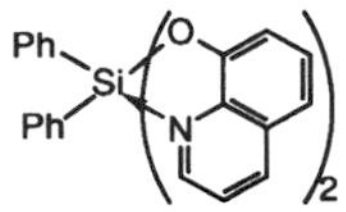 **193**

4.3.4.1 Preparation, Structure, and Fluxional Behavior The reaction of phenylsilazane with 8-hydroxyquinoline gave compound **193**[304,305] for which the hexacoordination of the silicon atom was inferred from the UV spectra. The X-ray structure determination of a similar complex **194**[306] confirmed the hexacoordination of the silicon atom with N—Si distances of 2.014 and 2.016 Å, the methyl group and the chlorine atom being in a *cis* arrangement like the N atoms, but the O atoms being *trans* to each other.

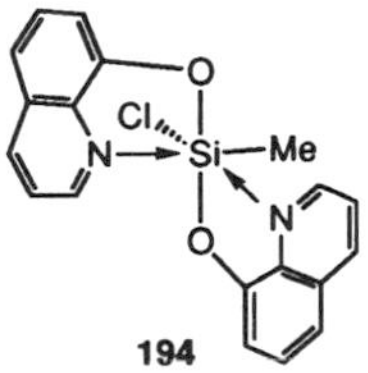

194 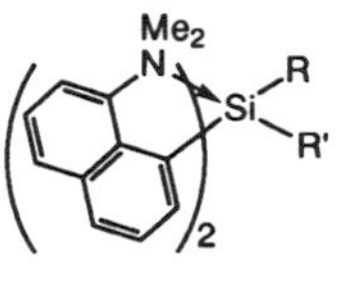 **195**

a : R = R' = H
b : R = H, R' = F
c : R = R' = C≡CH
d : R = R' = catecholato
e : R = R' = Cl
f : R = R' = F

The X-ray structure analyses of compounds **195a**,[307] **195b**,[307] **195c**,[308] **195d**[309] and **196**[307] showed that the silicon atoms are hexacoordinated, the geometries of these compounds being that of bicapped tetrahedrons with the Me_2N groups *cis* to each other. In **195a** and **195b**, one Me_2N group is *cis* with respect to a hydrogen atom, whereas the other is *trans* with respect to the other substituent (H or F). The values of the $C_{(1)}$—Si—$C_{(2)}$ angles are of 109.7° for **195a** and 108.2° for **195b**. In compounds **195c**, **195d**, and **196** the NMe_2 groups are also *cis*. They are *trans* to the monodentate ligands in **195c** and **196** and *trans* to the oxygen atoms of the cathecolate ligand in **195d**. In these compounds the bicapped tetrahedrons are distorted, with $C_{(1)}$—Si—$C_{(2)}$ angles of 134° in **195c**, 144.5° in **195d**, and 135.5° in **196**. Compounds **195** exhibit fluxional behavior in solution.[310] Dynamic NMR studies have shown unambiguously that a nondissociative mechanism operates. The free energies of activation were found to be 61.4 kJ mol^{-1} for **195b** and 63.5 kJ mol^{-1} for **195a**. Bicapped tetrahedral geometries were also found around the two silicon atoms of the silane **197**.[311] The N—Si distances *trans* to the Si—H bonds are longer (3.008 Å) than those *trans* to the Si—C bonds (2.681 Å).

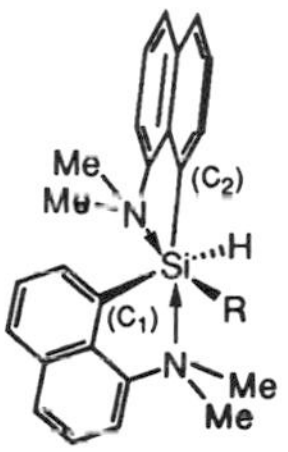

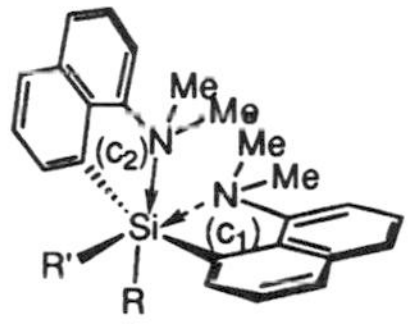

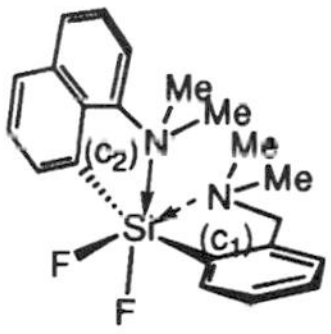

195 a : R = H
b : R = F

195 c : R = R' = C≡CH
d : R = R' =

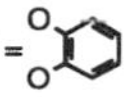

196

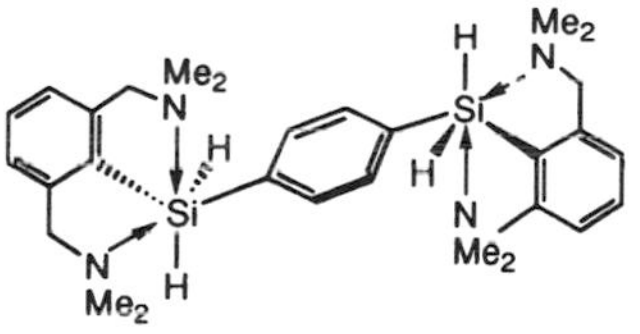

197

The fluxional behavior of complexes **198**[312] prepared according to Eq. 4.28 has been thoroughly studied. ^{29}Si NMR studies indicated that these compounds are hexacoordinated in solution; this was confirmed by the X-ray structure determination for **198** (X = Me, R = Ph).[313] The chlorine atom and the methyl group are *cis*, as are the two oxygen atoms, the two Me_2N groups being *trans*, with N—Si distances of 2.015 and 2.036 Å. The ^{1}H NMR spectra of these complexes are temperature dependent; two rate processes take place, one with a $\Delta G^{\ddagger}$ of 44.3–68.4 kJ mol^{-1} and the other with a $\Delta G^{\ddagger}$ of 62.7–77.3 kJ mol^{-1}. This behavior was explained by a mechanism involving 1,2 shift of adjacent ligands via a bicapped tetrahedral intermediate.

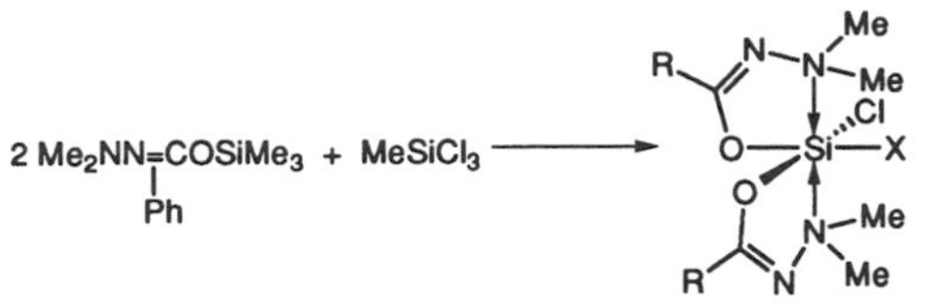

$$2\,Me_2NN{=}C(Ph)OSiMe_3 + MeSiCl_3 \longrightarrow \mathbf{198} \qquad (4.28)$$

Hexacoordinate fluorosilicates with one intramolecular coordination can be obtained by addition of a fluoride anion to a pentacoordinate di or trifluorosilane. However, there is no reaction in the case of pentacoordinate monofluorosilanes. The X-ray crystal structure analysis of **199**[314] showed the hexacoordination of the silicon atom with a slightly distorted octahedral geometry. The N—Si distance (2.213 Å) is shorter than the N—Si distance in the pentacoordinate trifluorosilane **83**[138] (2.318 Å). This short N—Si distance could be explained by a more extensive delocalization of the negative charge onto the fluorine atoms. The values of $\Delta G^{\ddagger}$ for fluorine exchange in solution were found to be 46 kJ mol^{-1} for **200** (R = H or Me) and 62.7 kJ mol^{-1} for **199**.

[K, 18-c-6]$^+$ **199** [K, 18-c-6]$^+$ **200**

A nearly octahedral geometry was also found for the silicate **201a**[315] with an N—Si distance of 2.157 Å. It was shown by dynamic NMR studies that the $\Delta G^{\ddagger}$ for N—Si bond-breaking is 43.9 kJ mol^{-1} for **201b** and that $\Delta G^{\ddagger}$ for the nondissociative stereoisomerization about silicon atom is lower than 29 kJ mol^{-1}.

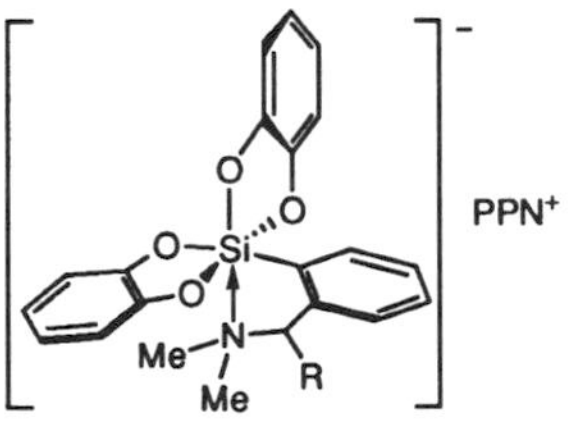

PPN$^+$ **201 a** : R = H; **b** : R = Me

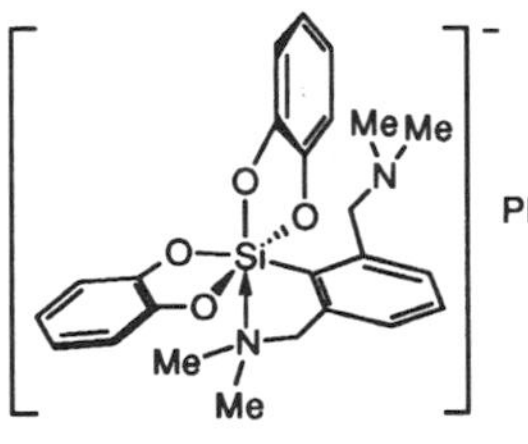

PPN$^+$ **202**

Compound **202**[316] was not obtained by the classical reaction; only the zwitterionic complex **203** was afforded (Eq. 4.29). A strong base (KH) was required to abstract the proton of **203**. X-ray crystal analyses showed that **202**[316b] and **203**[316a] have very similar structures to that of **201a**, with N—Si distances of

2.173 Å for **202** and 2.085 Å for **203**. In **203** the proton of the Me_2NH^+ group is involved in hydrogen bonding with one oxygen atom of a catecholate ligand. This induces a more perfect octahedral geometry in **203** than in **202**. The difficulty in preparing **202** is due to the unexpectedly high basicity of this compound, the pK_a of which was found to be 16.7[317] (much higher than that of "proton sponge," 12.3[318]). The gain in stability on going from **202** to **203** as well as the formation of hydrogen bonding in **203** may explain, in part, the high basicity of the silicate **202**. In solution, a dynamic coordination–decoordination of the Me_2N groups to the silicon center in **202** and in **204**[316b] was shown by dynamic NMR studies. The $\Delta G^{\ddagger}$ for these phenomena were estimated to be 46.4 kJ mol^{-1} for **202** and 38.5 kJ mol^{-1} for **204**.[317]

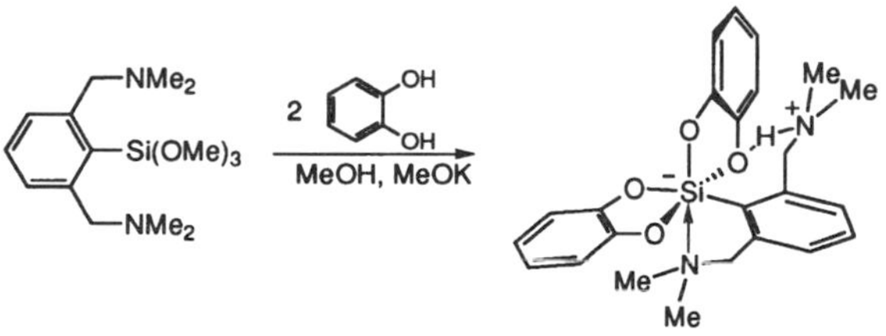
NMe2
Si(OMe)3
NMe2
2
OH
OH
MeOH, MeOK
Me
Me
H
N
O
Si
N
Me
Me

203

(4.29)

Inducement of intramolecular N→Si interaction at the silicon atom of silatranes has proved to be difficult.[319] Thus the X-ray structure analysis of **205** showed no coordination of the Me_2N group, whereas in **206** the silicon atom is hexacoordinated but with a highly distorted octahedral geometry. The N—Si distance (2.95 Å) indicates a weak N→Si interaction imposed by the geometry of the ligand.

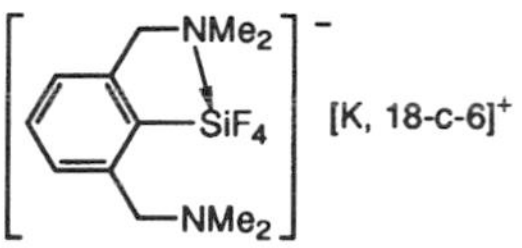
NMe2
SiF4
NMe2
[K, 18-c-6]+

204

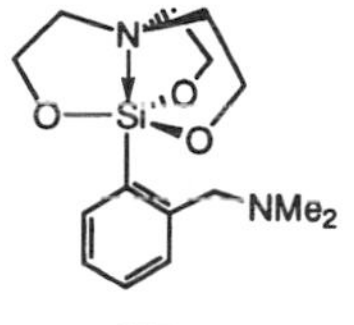
N
O
Si
O
O
NMe2

205

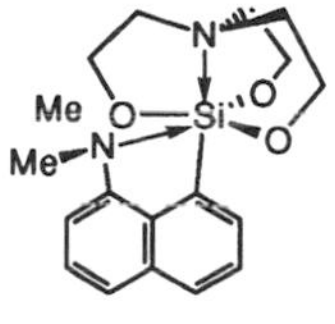
N
Me
O
Si
O
O
Me
N

206

The preparation of phthalocyanine[320] and naphthocyanine[321] derivates of silicon has been described. Naphthocyanine complexes are interesting because they display activity against cancerous cells.[322] The X-ray crystal stucture determination of **207**[320] as well as that of the porphyrin complex **208**[323] showed that, in both cases, the geometries around the silicon atoms correspond to octahedrons, slightly more distorted in **208** than in **207**. The four nitrogen atoms are coplanar, with the N—Si distances slightly longer in **207** (1.915–1.924 Å) than in **208**(1.848–1.887 Å).

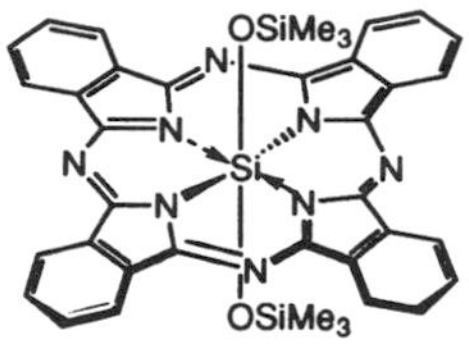

207

208

4.3.4.2 ***Reactivity*** The neutral hexacoordinate silicon complexes have their own, sometimes unexpected reactivity. Thus the dihydrosilane **195a** reacts with hydroxyl compounds (Scheme 4.30) but not with alkoxides.[324] The dichloro compound **195e** reacts readily with nucleophiles (Scheme 4.31), but the corresponding difluoro compound **195f** is completely inert.[324]

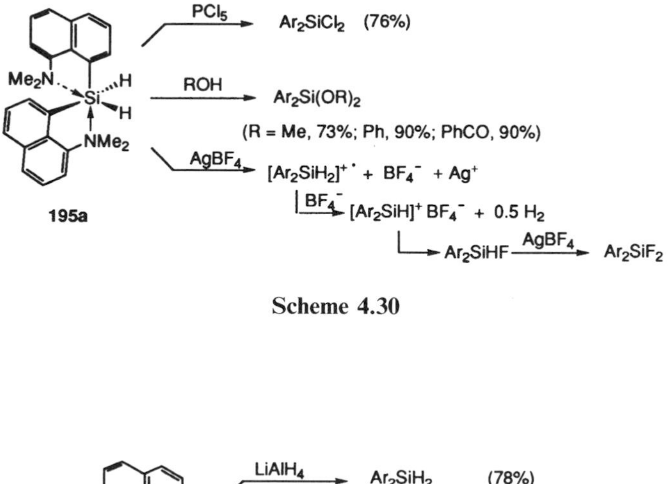

Scheme 4.30

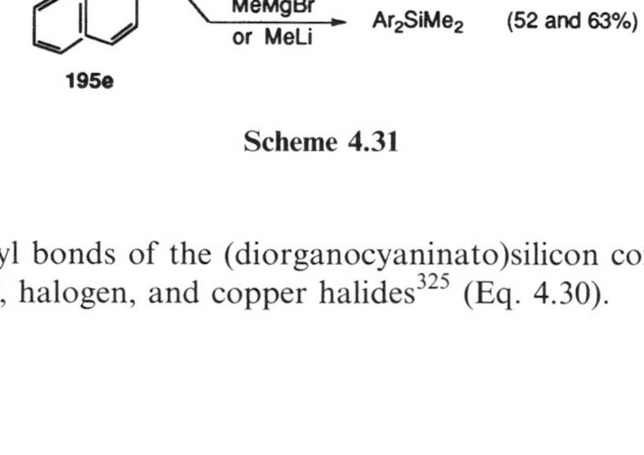

Scheme 4.31

Alkyl and aryl bonds of the (diorganocyaninato)silicon complexes **209** are cleaved by NBS, halogen, and copper halides[325] (Eq. 4.30).

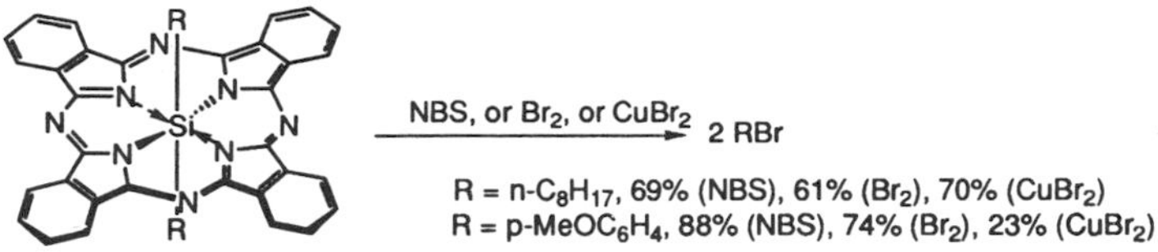
R
N
Si
N
R
NBS, or Br_2, or $CuBr_2$
2 RBr
R = n-C_8H_{17}, 69% (NBS), 61% (Br_2), 70% ($CuBr_2$)
R = p-$MeOC_6H_4$, 88% (NBS), 74% (Br_2), 23% ($CuBr_2$)

(4.30)

209

4.3.5 O→Si Intramolecular Coordinations to an Organosilane

4.3.5.1 Preparation, Structure, and Fluxional Behavior The first hexacoordinate complex with two O→Si intramolecular coordinations was the cationic complex **210** prepared by Dilthey[326] in 1903 (Scheme 4.32). Further work on the preparation of these complexes showed that the reaction product depends on the starting silane. From $Si(OAc)_4$ the neutral complex **211** is obtained,[327] whereas from $SiCl_4$ the reaction product is the cationic complex **210**.[326,328] Nevertheless, the neutral complex **212** could be obtained in poor yield[329] from $SiCl_4$ in excess. Neutral complexes **213**[328] and **214**[330] have been prepared from $RSiCl_3$ and Ph_2SiCl_2. However, from Me_2SiCl_2 only the tetravalent silane **215** was formed.[328]

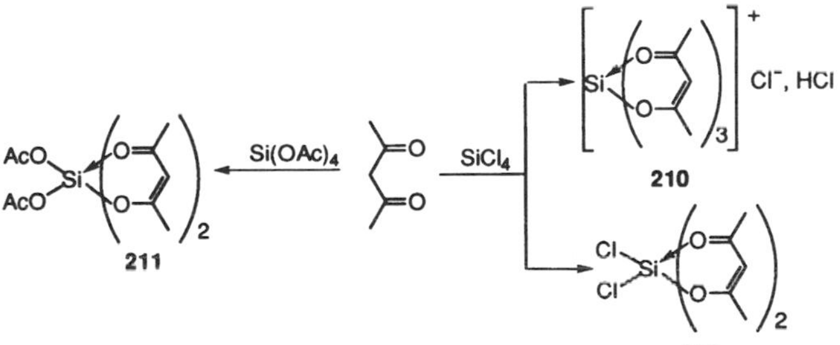
AcO
AcO
Si
2
211
Si(OAc)4
SiCl4
Si
3
+
Cl⁻, HCl
210
Cl
Cl
Si
2
212

Scheme 4.32

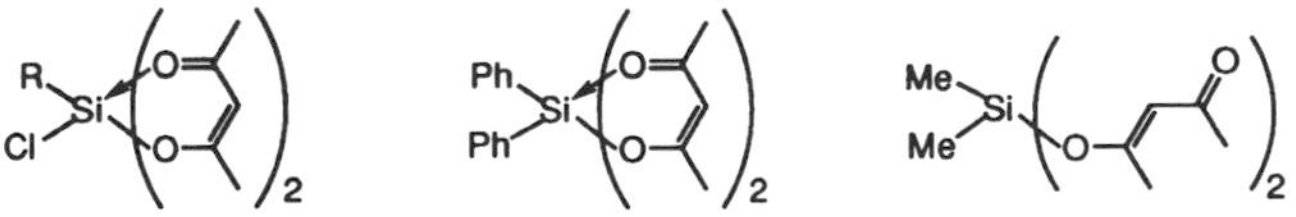
R
Cl
Si
2
Ph
Ph
Si
2
Me
Me
Si
2

213 (R = Me, Ph) **214** **215**

Tropolone can also give the cationic complex **216**[331] or neutral complexes **217**[207,332] depending on the starting silane (Scheme 4.33). The X-ray structure determination of **217** ($R = R' = Ph$) showed that the silicon atom is located at the center of a distorted octahedron with the two phenyl groups *cis*. The O—Si distances *trans* to the phenyl groups are longer (1.923 and 1.879 Å) than the others (1.793 and 1.805 Å). Dynamic ^{13}C NMR studies indicated that compounds **217** (R, R′ = Me, Ph, OMe) undergo ligand permutational isomerization.[332]

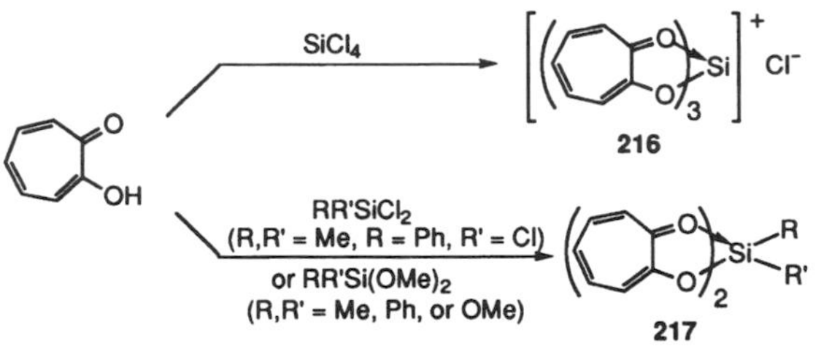

SiCl₄

RR'SiCl₂ (R,R' = Me, R = Ph, R' = Cl) or RR'Si(OMe)₂ (R,R' = Me, Ph, or OMe)

216

217

Scheme 4.33

The neutral complex **218**,[333] in which the hexacoordination about the silicon atom was inferred from IR spectroscopy, was prepared by reaction of $Cl_2Si(CH_2Cl)_2$ with *N*-trimethylsilylbutyrolactam.

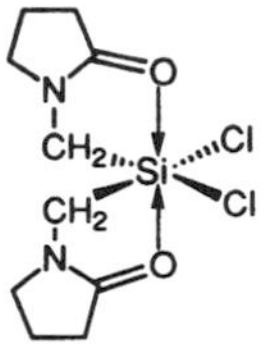

218

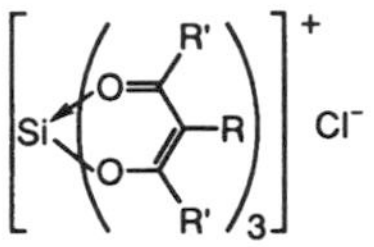

219

4.3.5.2 ***Reactivity*** It was shown that the hydrolysis of complexes **219** (R = H, R′ = Me, Ph) is more rapid in alkaline medium than under aqueous acidic conditions.[334,335] Similarly, hydrolysis of **216** is catalyzed by hydroxide ions.[336] Substituents at position 3 of 2,4-pentanedionates **219** (R′ = Me, R = benzyl, allyl, Me, Et, ^iPr, OAc, or Cl) also increase the rate of hydrolysis.[337] The rate of the racemization of some chiral chelates was found to be the same as that of hydrolysis.[335,338]

4.4 HEPTA AND OCTACOORDINATE SILICON COMPOUNDS

Very few compounds with coordination numbers higher than six are known. The first derivative that was believed to be heptacoordinated is $[SiF_7]^{3-}$ $3K^+$, obtained by heating $[SiF_6]^{2-}$ in dry air at 750°C (Eq. 4.31). However, it was shown by X-ray diffraction that it consists of an array of K^+, $[SiF_6]^{2-}$ and F^- ions.[339] The reaction of Me_2SiCl_2 with Schiff bases gave the cationic silicon compounds **220** (Eq. 4.32), for which heptacoordination of the silicon atom was suggested from IR and NMR data.[340] Unfortunately X-ray crystal structure analysis was not performed on these compounds.

$$2\,[SiF_6]^{2-}\,2\,K^+ \xrightarrow{750^\circ C} [SiF_7]^{3-}\,3\,K^+ + SiF_4 + KF \qquad (4.31)$$

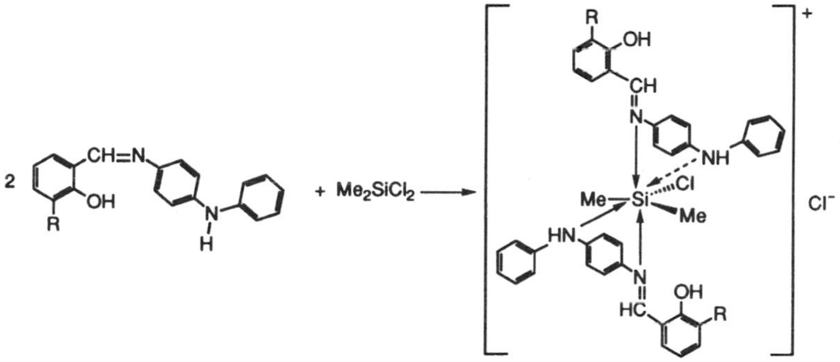

(4.32)

220

Recently, X-ray analysis showed that compounds **221**[341–343] and **222**[343] are heptacoordinated with tricapped tetrahedral geometries. **221a**[342] has a propeller shape with an approximate C_3 geometry, the Si—H bond coinciding with the axis of symmetry. The tetrahedral geometry of the silicon atom is retained, nitrogen coordination occurring *trans* to an Si—C bond with N—Si—C angles values between 172.8 and 178.7°. The lone pairs of the nitrogen atoms are directed toward the silicon atom with N—Si distances of 3.00, 3.00, and 3.04 Å, these being below the sum of the N and Si van der Waals radii (3.5 Å). The same geometry at the silicon atom was found in compound **222**,[343] but with slightly shorter Si—N distances (2.88, 2.88, and 2.93 Å) owing to the rigidity of the ligand. A similar geometry was also found in **221b**.[343] However, the molecule deviates from C_3 geometry with an N_1—Si—C_1 angle of 152.8°, the mean value of the two other N—Si—Ar angles being 175.3°. In this molecule, the three N→Si interactions are unequal, with N—Si distances of 3.00, 3.31, and 3.49 Å. That suggests that steric hindrance plays an important part in the formation of intramolecular chelation.

The X-ray crystal structure analysis of compound **223**[344] (Eq. 4.33) showed that the molecule has a C_2 axis and that the four NMe_2 groups are directed toward the silicon atom opposite to the Si—H and Si—Ar bonds. The N→Si coordinative bonds *trans* to the Si—H bonds are longer (3.12 Å) than the N→Si coordinative bonds *trans* to the phenyl groups (2.895 Å). The basic tetrahedral structure around the silicon atom is maintained, the four N→Si

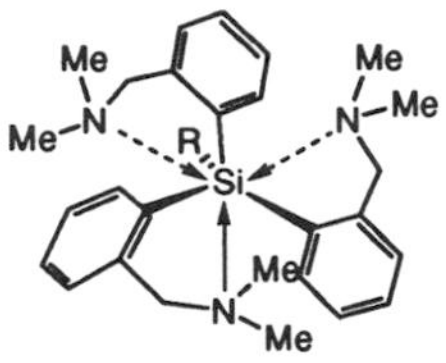

221 a : R = H
b : R = F

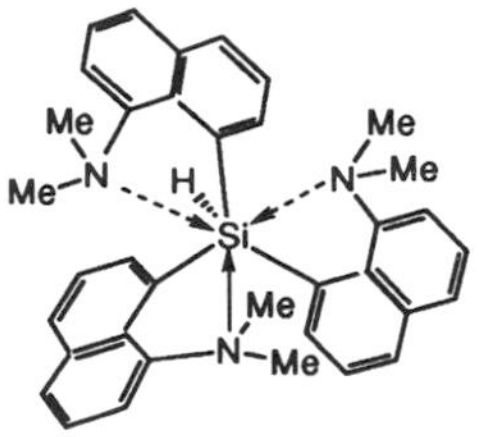

222

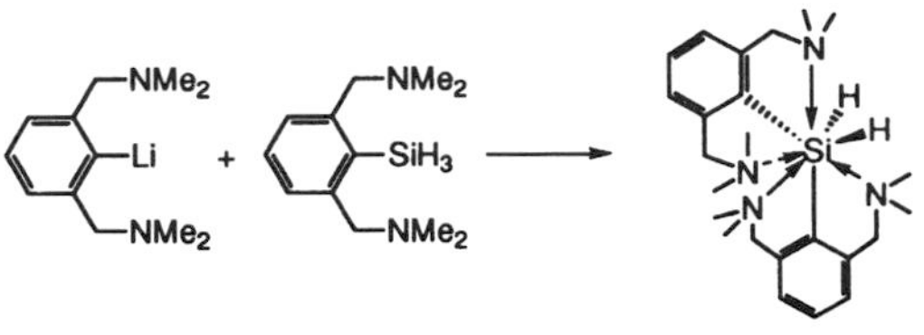

(4.33)

223

coordinative bonds forming another tetrahedron. In spite of the steric hindrance around the silicon atom, **223** reacted with HCl or HBr to give the silyl dication **224**[344] (Eq. 4.34). The X-ray crystal structures of compounds **225**[345] and **226**[346] in which the silicon atom is also [4 + 4] coordinated showed the tetracapped tetrahedral geometry of the silicon atom, with four weak intramolecular F→Si interactions (2.71 to 3.07 Å) in **225** and four weak intramolecular N→Si interactions (2.50 Å) in **226**.

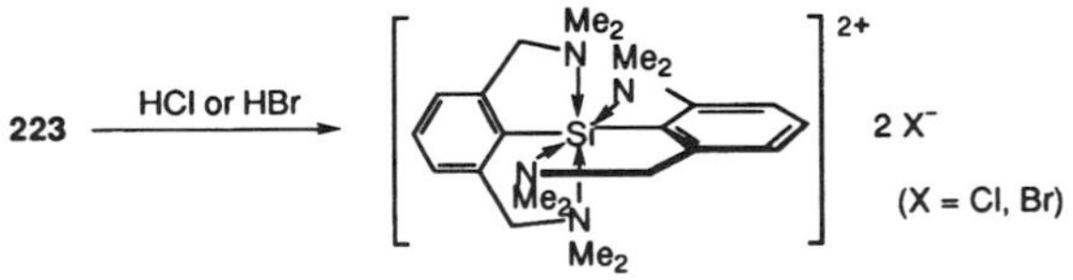

223 —HCl or HBr→ [224]$^{2+}$ 2 X^- (X = Cl, Br)

(4.34)

224

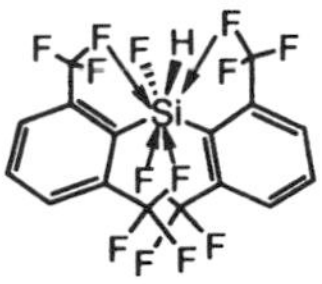

225

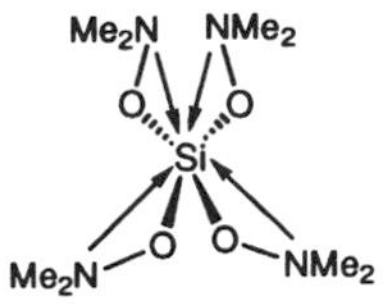

226

4.5 CONCLUSION

Since the discovery of the $[SiF_6]^{2-}$ anion, the chemistry of hypercoordinate silicon compounds has grown very rapidly, especially over the last 20 years. X-ray structure determinations have shown that pentacoordinate silicon compounds with electronegative ligands have trigonal bipyramidal or square pyramidal geometries (or lie between them), whereas hexacoordinate silicon compounds have octahedral geometries. They all show fluxional behavior in solution. Moreover, the pentacoordinate compounds are usually more reactive than the corresponding tetravalent silicon compounds.

New types of hypercoordinate silicon compounds, in which the extra coordination was achieved by one, two, three, or four intramolecular (X→Si) interactions between the silicon atom and electron-rich heteroatoms, have been described. In this series, pentacoordinate compounds have either trigonal bipyramidal geometries (with relatively short X—Si bonds) or monocapped tetrahedral geometries (with long X—Si bonds). Similarly, hexacoordinate compounds have octahedral or bicapped tetrahedral geometries. All the hepta- or octacoordinate silicon compounds obtained until now have tri- or tetracapped tetrahedral geometries. All these compounds show fluxional behavior in solution and have reactivities different from those of the tetravalent silicon analogs.

REFERENCES

1. J. L. Gay-Lussac, L. J. Thenard, *Mémoires de Physique et de Chimie de la Société d'Arcueil* **2**, 317 (1809).
2. J. Davy, *Phil. Trans.* **102**, 352 (1812).
3. J. J. Berzelius, *Ann. Chim. Phys.* **27**, 287 (1824).
4. J. A. A. Ketelaar, *Z. Kristallog.* **92**, 155 (1935).
5. (a) R. J. P. Corriu, C. Guérin, J. J. E. Moreau, *Top. Stereochem.* **15**, 43 (1984). (b) R. J. P. Corriu, C. Guérin, J. J. E. Moreau, *The Chemistry of Organic Silicon Compounds*. S. Patai and Z. Rappoport, Eds. (John Wiley, Chichester, 1989), chapter 4.
6. V. O. Gel'mbol'dt, A. A. Ennan, *Russ. Chem. Rev.* **58**, 371 (1989).
7. R. R. Holmes, *Chem. Rev.* **90**, 17 (1990).
8. I. R. Beattie, *Quart. Rev. Chem. Soc.* **17**, 382 (1963).
9. B. J. Aylett, *Progr. Stereochem.* **4**, 213 (1969).
10. S. N. Tandura, M. G. Voronkov, N. V. Alekseev, *Top. Curr. Chem.* **131**, 99 (1986).
11. M. G. Voronkov, L. I. Gubanova, *Main Group Met. Chem.* **10**, 209 (1987).
12. V. E. Shklover, Yu. T. Struchkov, M. G. Voronkov, *Russ. Chem. Rev.* **58**, 211 (1989).
13. E. Lukevics, O. Pudova, R. Sturkovich, *Molecular Structure of Organosilicon Compounds* (Ellis Horwood, Chichester, 1989).

14. R. J. P. Corriu, J. C. Young, *The Chemistry of Organic Silicon Compounds*, S. Patai, and Z. Rappoport, Eds (John Wiley, Chichester, 1989), chapter 20.
15. C. Chuit, R. J. P. Corriu, C. Reyé, J. C. Young, *Chem. Rev.* **93**, 1371 (1993).
16. (a) R. J. P. Corriu, G. Dabosi, M. Martineau, *J. Organomet. Chem.* **150**, 27 (1978); (b) R. J. P. Corriu, G. Dabosi, M. Martineau, *J. Organomet. Chem.* **154**, 33 (1978); (c) R. J. P. Corriu, G. Dabosi, M. Martineau, *J. Organomet. Chem.* **186**, 25 (1980).
17. R. J. P. Corriu, *J. Organomet. Chem.* **400**, 81 (1990).
18. J. A. Deiters, R. R. Holmes, *J. Am. Chem. Soc.* **112**, 7197 (1990).
19. M. S. Gordon, M. T. Carroll, L. P. Davis, L. W. Burggraf, *J. Phys. Chem.* **94**, 8125 (1990).
20. R. Müller, C. Dathe, *Z. Anorg. Allg. Chem.* **343**, 150 (1966).
21. H. C. Clark, K. R. Dixon, *J. Chem. Soc. Chem. Comm.* 717 (1967).
22. J. J. Harris, B. Rudner, *J. Am. Chem. Soc.* **90**, 515 (1968).
23. F. Klanberg, E. L. Muetterties, *Inorg. Chem.* **7**, 155 (1968).
24. R. K. Marat, A. F. Janzen, *Can. J. Chem.* **55**, 1167 (1977).
25. H. C. Clark, K. R. Dixon, J. G. Nicolson, *Inorg. Chem.* **8**, 450 (1969).
26. B. S. Ault, *Inorg. Chem.* **18**, 3339 (1979).
27. (a) A. S. Pilcher, H. L. Ammon, P. DeShong, *J. Am. Chem. Soc.* **117**, 5166 (1995); (b) D. Albanese, D. Landini, M. Penso, *Tetrahedron Lett.* **36**, 8865 (1995).
28. C. H. DePuy, V. M. Bierbaum, L. A. Flippin, J. J. Grabowski, G. K. King, R. J. Schmitt, S. A. Sullivan, *J. Am. Chem. Soc.* **102**, 5012 (1980).
29. (a) S. A. Sullivan, C. H. DePuy, R. Damrauer, *J. Am. Chem. Soc.* **103**, 480 (1981); (b) R. Damrauer, J. A. Hankin, *J. Organomet. Chem.* **521**, 93 (1996).
30. R. Damrauer, L. W. Burggraf, L. P. Davis, M. S. Gordon, *J. Am. Chem. Soc.* **110**, 6601 (1988).
31. C. H. DePuy, R. Damrauer, J. H. Bowie, J. C. Sheldon, *Acc. Chem. Res.* **20**, 127 (1987).
32. R. Damrauer, J. A. Hankin, *Chem. Rev.* **95**, 1137 (1995).
33. D. Schomburg, *J. Organomet. Chem.* **221**, 137 (1981).
34. D. Schomburg, R. Krebs, *Inorg. Chem.* **23**, 1378 (1984).
35. J. J. Harland, J. S. Payne, R. O. Day, R. R. Holmes, *Inorg. Chem.* **26**, 760 (1987).
36. R. K. Marat, A. F. Janzen, *Can. J. Chem.* **55**, 3845 (1977).
37 R. Damrauer, S. E. Danahey, *Organometallics* **5**, 1490 (1986).
38. S. E. Johnson, R. O. Day, R. R. Holmes, *Inorg. Chem.* **28**, 3182 (1989).
39. S. E. Johnson, J. S. Payne, R. O. Day, J. M. Holmes, R. R. Holmes, *Inorg. Chem.* **28**, 3190 (1989).
40. S. E. Johnson, J. A. Deiters, R. O. Day, R. R. Holmes, *J. Am. Chem. Soc.* **111**, 3250 (1989).
41. F. H. Carré, R. J. P. Corriu, C. Guérin, B. J. L. Henner, W. W. C. Wong Chi Man, *J. Organomet. Chem.* **347**, Cl (1988).
42. K. Tamao, T. Hayashi, Y. Ito, M. Shiro, *Organometallics* **11**, 182 (1992).
43. R. O. Day, C. Sreelatha, J. A. Deiters, S. E. Johnson, J. M. Holmes, L. Howe, R. R. Holmes, *Organometallics* **10**, 1758 (1991).

44. R. Damrauer, B. O'Connell, S. E. Danahey, R. Simon, *Organometallics* **8**, 1167 (1989).

45. M. S. Gordon, T. L. Windus, L. W. Burggraf, L. P. Davis, *J. Am. Chem. Soc.* **112**, 7167 (1990).

46. T. L. Windus, M. S. Gordon, L. P. Davis, L. W. Burggraf, *J. Am. Chem. Soc.* **116**, 3568 (1994).

47. (a) M. Kira, T. Hoshi, C. Kabuto, H. Sakurai, *Chem. Lett.* 1859 (1993); (b) M. Kira, T. Hoshi, H. Sakurai, *Chem. Lett.* 807 (1995).

48. R. O'Dell, *Tetrahedron Lett.* **36**, 5723 (1995).

49. (a) K. Tamao, T. Hayashi, Y. Ito, M. Shiro, *J. Am. Chem. Soc.* **112**, 2422 (1990); (b) K. Tamao, T. Hayashi, Y. Ito, M. Shiro, *Organometallics* **11**, 2099 (1992).

50. K. Ebata, T. Inada, C. Kabuto, H. Sakurai, *J. Am. Chem. Soc.* **116**, 3595 (1994).

51. D. Brondani, F. H. Carré, R. J. P. Corriu, J. J. E. Moreau, M. Wong Chi Man, *Angew. Chem. Int. Ed. Engl.* **35**, 324 (1996).

52. R. Müller, *Organometal. Chem. Rev.* **1**, 359 (1966).

53. R. Tacke, J. Becht, G. Mattern, W. F. Kuhs, *Chem. Ber.* **125**, 2015 (1992).

54. (a) R. Tacke, J. Becht, A. Lopez-Mras, W. S. Sheldrick, A. Sebald, *Inorg. Chem.* **32**, 2761 (1993); (b) R. Tacke, J. Becht, O. Dannappel, R. Ahlrichs, U. Schneider, W. S. Sheldrick, J. Hahn, F. Kiesgen, *Organometallics* **15**, 2060 (1996).

55. (a) R. J. P. Corriu, C. Guérin, B. J. L. Henner, W. W. C. Wong Chi Man, *Organometallics* **7**, 237 (1988); (b) J. L. Bréfort, R. J. P. Corriu, C. Guérin, B. J. L. Henner, W. W. C. Wong Chi Man, *Organometallics* **9**, 2080 (1990).

56. K. C. Kumara Swamy, V. Chandrasekhar, J. J. Harland, J. M. Holmes, R. O. Day, R. R. Holmes, *J. Am. Chem. Soc.* **112**, 2341 (1990).

57. B. Becker, R. Corriu, C. Guérin, B. Henner, Q. Wang, *J. Organomet. Chem.* **359**, C33 (1989).

58. (a) R. Corriu, C. Guérin, B. Henner, Q. Wang, *J. Organomet. Chem.* **365**, C7 (1989); (b) R. J. P. Corriu, C. Guérin, B. Henner, Q. Wang, *Organometallics* **10**, 2297 (1991).

59. B. Becker, R. J. P. Corriu, C. Guérin, B. Henner, Q. Wang, *J. Organomet. Chem.* **368**, C25 (1989).

60. R. J. P. Corriu, C. Guérin, B. J. L. Henner, Q. Wang, *Organometallics* **10**, 3574 (1991).

61. R. J. P. Corriu, C. Guérin, B. J. L. Henner, Q. Wang, *J. Organomet. Chem.* **439**, C1 (1992).

62. J. L. Bréfort, R. Corriu, C. Guérin, B. Henner, *J. Organomet. Chem.* **370**, 9 (1989).

63. B. Becker, R. J. P. Corriu, C. Guérin, B. J. L. Henner, *J. Organomet. Chem.* **369**, 147 (1989).

64. D. J. Hajdasz, R. R. Squires, *J. Am. Chem. Soc.* **108**, 3139 (1986).

65. D. J. Hajdasz, Y. Ho, R. R. Squires, *J. Am. Chem. Soc.* **116**, 10751 (1994).

66. J. D. Payzant, K. Tanaka, L. D. Betowski, D. K. Bohme, *J. Am. Chem. Soc.* **98**, 894 (1976).

67. D. A. Dixon, W. R. Hertler, D. B. Chase, W. B. Farnham, F. Davidson, *Inorg. Chem.* **27**, 4012 (1988).

68. K. Tamao, T. Hayashi, Y. Ito, *Bull. Soc. Chim. Fr.* **132**, 556 (1995).
69. H. J. Frohn, V. V. Bardin, *J. Organomet. Chem.* **501**, 155 (1995).
70. R. J. P. Corriu, C. Guérin, B. J. L. Henner, Q. Wang, *Organometallics* **10**, 3200 (1991).
71. L. L. Hench, J. K. West, *Chem. Rev.* **90**, 33 (1990).
72. C. J. Brinker, G. W. Scherer, *Sol Gel Science. The Physics and Chemistry of Sol Gel Processing* (Academic Press, New York, 1990).
73. D. A. Loy, K. J. Shea, *Chem. Rev.* **95**, 1431 (1995).
74. R. J. P. Corriu, D. Leclercq, *Angew. Chem. Int. Ed. Engl.* **35**, 1421 (1996).
75. R. Müller, L. Heinrich, *Chem. Ber.* **94**, 1943 (1961).
76. C. L. Frye, *J. Am. Chem. Soc.* **92**, 1205 (1970).
77. J. J. Harland, R. O. Day, J. F. Vollano, A. C. Sau, R. R. Holmes, *J. Am. Chem. Soc.* **103**, 5269 (1981).
78. D. Schomburg, *Z. Naturforsch.* **38B**, 938 (1983).
79. K. C. Kumara Swamy, C. Sreelatha, R. O. Day, J. M. Holmes, R. R. Holmes, *Inorg. Chem.* **30**, 3126 (1991).
80. R. R. Holmes, R. O. Day, J. S. Payne, *Phosphorous, Sulfur and Silicon* **42**, 1 (1989).
81. E. F. Perozzi, R. S. Michalak, G. D. Figuly, W. H. Stevenson, D. B. Dess, M. R. Ross, J. C. Martin, *J. Org. Chem.* **46**, 1049 (1981).
82. W. B. Farnham, R. L. Harlow, *J. Am. Chem. Soc.* **103**, 4608 (1981).
83. S. K. Chopra, J. C. Martin, *J. Am. Chem. Soc.* **112**, 5342 (1990).
84. W. H. Stevenson, J. C. Martin, *J. Am. Chem. Soc.* **104**, 309 (1982).
85. M. Kira, K. Sato, C. Kabuto, H. Sakurai, *J. Am. Chem. Soc.* **111**, 3747 (1989).
86. C. L. Frye, *J. Am. Chem. Soc.* **86**, 3170 (1964).
87. G. Cerveau, C. Chuit, R. J. P. Corriu, L. Gerbier, C. Reyé, J. L. Aubagnac, B. El Amrani, *Int. J. Mass Spectrom. Ion Processes* **82**, 259 (1988).
88. A. Boudin, G. Cerveau, C. Chuit, R. J. P. Corriu, C. Reyé, *Angew. Chem. Int. Ed. Engl.* **25**, 473 (1986).
89. A. Boudin, G. Cerveau, C. Chuit, R. J. P. Corriu, C. Reyé, *Bull. Chem. Soc. Jpn.* **61**, 101 (1988).
90. M. Kira, K. Sato, H. Sakurai, *J. Am. Chem. Soc.* **110**, 4599 (1988).
91. M. Kira, K. Sato, H. Sakurai, *J. Org. Chem.* **52**, 948 (1987).
92. G. Cerveau, C. Chuit, R. J. P. Corriu, C. Reyé, *J. Organomet. Chem.* **328**, C17 (1987).
93. A. Hosomi, S. Kohra, K. Ogata, T. Yanagi, Y. Tominaga, *J. Org. Chem.* **55**, 2415 (1990).
94. E. F. Perozzi, J. C. Martin, *J. Am. Chem. Soc.* **101**, 1591 (1979).
95. G. Cerveau, C. Chuit, E. Colomer, R. J. P. Corriu, C. Reyé, *Organometallics* **9**, 2415 (1990).
96. D. A. Loy, J. H. Small, K. J. Shea, *Organometallics* **12**, 1484 (1993).
97. K. J. Shea, D. A. Loy, J. H. Small, *Chem. Mater.* **4**, 255 (1992).

98. K. Y. Blohowiak, D. R. Treadwell, B. L. Mueller, M. L. Hoppe, S. Jouppi, P. Kansal, K. W. Chew, C. L. S. Scotto, F. Babonneau, J. Kampf, R. M. Laine, *Chem. Mater.* **6**, 2177 (1994).
99. B. Herreros, T. L. Barr, P. J. Barrie, J. Klinowski, *J. Phys. Chem.* **98**, 4570 (1994).
100. P. Kansal, R. M. Laine, *J. Am. Ceram. Soc.* **77**, 875 (1994).
101. R. Tacke, F. Wiesenberger, A. Lopez-Mras, J. Sperlich, G. Mattern, *Z. Naturforsch.* **47b**, 1370 (1992).
102. R. Tacke, A. Lopez-Mras, J. Sperlich, C. Strohmann, W. F. Kuhs, G. Mattern, A. Sebald, *Chem. Ber.* **126**, 851 (1993).
103. R. Tacke, A. Lopez-Mras, P. G. Jones, *Organometallics* **13**, 1617 (1994).
104. C. Strohmann, R. Tacke, G. Mattern, W. F. Kuhs, *J. Organomet. Chem.* **403**, 63 (1991).
105. R. Tacke, J. Sperlich, C. Strohmann, G. Mattern, *Chem. Ber.* **124**, 1491 (1991).
106. M. Mühleisen, R. Tacke, *Organometallics* **13**, 3740 (1994).
107. F. P. Boer, J. J. Flynn, J. W. Turley, *J. Am. Chem. Soc.* **90**, 6973 (1968).
108. D. Schomburg, *Z. Naturforsch.* **37B**, 195 (1982).
109. R. R. Holmes, R. O. Day, J. J. Harland, A. C. Sau, J. M. Holmes, *Organometallics* **3**, 41 (1984).
110. R. R. Holmes, R. O. Day, V. Chandrasekhar, J. J. Harland, J. M. Holmes, *Inorg. Chem.* **24**, 2016 (1985).
111. R. R. Holmes, R. O. Day, V. Chandrasekhar, J. M. Holmes, *Inorg. Chem.* **24**, 2009 (1985).
112. R. R. Holmes, R. O. Day, J. J. Harland, J. M. Holmes, *Organometallics* **3**, 347 (1984).
113. R. S. Berry, *J. Chem. Phys.* **32**, 933 (1960).
114. (a) T. Kawashima, N. Iwama, R. Okazaki, *J. Am. Chem. Soc.* **114**, 7598 (1992); (b) T. Kawashima, R. Okazaki, *Synlett*, 600 (1996).
115. W. H. Stevenson, S. Wilson, J. C. Martin, W. B. Farnham, *J. Am. Chem. Soc.* **107**, 6340 (1985).
116. W. H. Stevenson, J. C. Martin, *J. Am. Chem. Soc.* **107**, 6352 (1985).
117. A. Boudin, G. Cerveau, C. Chuit, R. J. P. Corriu, C. Reyé, *J. Organomet. Chem.* **362**, 265 (1989).
118. G. Cerveau, C. Chuit, R. J. P. Corriu, L. Gerbier, C. Reyé, *C. R. Acad. Sci. Paris Ser. II* **312**, 1311 (1991).
119. O. W. Webster, W. R. Hertler, D. Y. Sogah, W. B. Farnham, T. V. RajanBabu, *J. Am. Chem. Soc.* **105**, 5706 (1983).
120. D. Y. Sogah, W. B. Farnham, *Organosilicon and Bioorganosilicon Chemistry*, H. Sakurai, Ed (Ellis Horwood, Chichester, 1985), p. 219.
121. R. Rudman, W. C. Hamilton, S. Novick, T. D. Goldfarb, *J. Am. Chem. Soc.* **89**, 5157 (1967).
122. A. J. Blake, E. A. V. Ebsworth, A. J. Welch, *Acta Cryst. C* **40**, 895 (1984).
123. W. S. Sheldrick, W. Wolfsberger, *Z. Naturforsch.* **32B**, 22 (1977).
124. D. G. Anderson, A. J. Blake, S. Cradock, E. A. V. Ebsworth, D. W. H. Rankin, A. J. Welch, *Angew. Chem. Int. Ed. Engl.* **25**, 107 (1986).

125. A. R. Bassindale, J. Jiang *J. Organomet. Chem.* **446**, C3 (1993).
126. N. Kuhn, T. Kratz, D. Bläser, R. Boese, *Chem. Ber.* **128**, 245 (1995).
127. C. Zybill, G. Müller, *Organometallics* **7**, 1368 (1988).
128. C. Zybill, D. L. Wilkinson, C. Leis, G. Müller, *Angew. Chem. Int. Ed. Engl.* **28**, 203 (1989).
129. C. Leis, D. L. Wilkinson, H. Handwerker, C. Zybill, G. Müller, *Organometallics* **11**, 514 (1992).
130. H. J. Campbell-Ferguson, E. A. V. Ebsworth, *J. Chem. Soc. (A)*, 705 (1967).
131. J. Y. Corey, R. West, *J. Am. Chem. Soc.* **85**, 4034 (1963).
132. A. R. Bassindale, T. Stout, *J. Chem. Soc. Chem. Commun.* 1387 (1984).
133. K. Hensen, T. Zengerly, T. Müller, P. Z. Pickel, *Anorg. Allg. Chem.* **558**, 21 (1988).
134. G. van Koten, *Pure Appl. Chem.* **61**, 1681 (1989); G. van Koten, *Pure Appl. Chem.* **62**, 1155 (1990); J. T. B. H. Jastrzebski, G. van Koten, *Adv. Organom. Chem.* **35**, 241 (1993).
135. G. Klebe, M. Nix, H. Fuess, *Chem. Ber.* **117**, 797 (1984).
136. G. Klebe, K. Hensen, H. Fuess, *Chem. Ber.* **116**, 3125 (1983).
137. Yu. E. Ovchinnikov, Yu. T. Struchkov, N. F. Chernov, O. M. Trofimova, M. G. Voronkov, *J. Organomet. Chem.* **461**, 27 (1993).
138. F. Carré, R. J. P. Corriu, A. Kpoton, M. Poirier, G. Royo, J. C. Young, *J. Organomet. Chem.* **470**, 43 (1994).
139. G. Klebe, *J. Organomet. Chem.* **332**, 35 (1987).
140. G. Klebe, J. W. Bats, K. Hensen, *Z. Naturforsch.* **B38**, 825 (1977).
141. G. Klebe, J. W. Bats, K. Hensen, *J. Chem. Soc. Dalton Trans.* 1 (1985).
142. D. Kummer, S. C. Chaudhry, J. Seifert, B. Deppisch, G. Mattern, *J. Organomet. Chem.* **382**, 345 (1990).
143. D. Kummer, S. H. A. Halim, W. Kuhs, G. Mattern, *J. Organomet. Chem.* **446**, 51 (1993).
144. T. van den Ancker, B. S. Jolly, M. F. Lappert, C. L. Raston, B. W. Skelton, A. H. White, *J. Chem. Soc. Chem. Commun.*, 1990 (1006).
145. H. H. Karsch, F. Bienlein, A. Sladek, M. Heckel, K. Burger, *J. Am. Chem. Soc.* **117**, 5160 (1995).
146. R. Probst, C. Leis, S. Gamper, E. Herdtweck, C. Zybill, N. Auner, *Angew. Chem. Int. Ed. Engl.* **30**, 1132 (1991).
147. N. Auner, R. Probst, C. R. Heikenwälder, E. Herdtweck, S. Gamper, G. Müller, *Z. Naturforsch.* **B48**, 1625 (1993).
148. H. Handwerker, C. Leis, R. Probst, P. Bissinger, A. Grohmann, P. Kiprof, E. Herdtweck, J. Blümel, N. Auner, C. Zybill, *Organometallics* **12**, 2162 (1993).
149. J. Belzner, H. Ihmels, M. Noltemeyer, *Tetrahedron Lett.* **36**, 8187 (1995).
150. J. Boyer, C. Brelière, F. Carré, R. J. P. Corriu, A. Kpoton, M. Poirier, G. Royo, J. C. Young, *J. Chem. Soc. Dalton Trans.* 43 (1989).
151. C. Brelière, F. Carré, R. J. P. Corriu, M. Poirier, G. Royo, *Organometallics* **5**, 388 (1986).
152. (a) K. Tamao, M. Asahara, A. Kawachi, *J. Organomet. Chem.* **521**, 325 (1996); (b) K. Tamao, Y. Tarao, Y. Nakagawa, K. Nagata, Y. Ito, *Organometallics* **12**, 1113

(1993); (c) K. Tamao, K. Nagata, M. Asahara, A. Kawachi, Y. Ito, M. Shiro, *J. Am. Chem. Soc.* **117**, 11592 (1995).

153. R. Köster, G. Seidel, B. Wrackmeyer, K. Horchler, D. Schlosser, *Angew. Chem. Int. Ed. Engl.* **28**, 918 (1989).
154. L. M. Englehardt, P. C. Junk, W. C. Patalinghug, R. E. Sue, C. L. Raston, B. W. Skelton, A. H. White, *J. Chem. Soc. Chem. Commun.* 930 (1991).
155. R. J. P. Corriu, G. Royo, A. de Saxcé, *J. Chem. Soc. Chem. Commun.* 892 (1980).
156. R. J. P. Corriu, M. Poirer, G. Royo, *J. Organomet. Chem.* **233**, 165 (1982).
157. C. Brelière, F. Carré, R. J. P. Corriu, A. de Saxcé, M. Poirier, G. Royo, *J. Organomet. Chem.* **205**, C1 (1981).
158. G. Klebe, K. Hensen, *J. Chem. Soc. Dalton Trans.* 5 (1985).
159. R. J. P. Corriu, A. Kpoton, M. Poirier, G. Royo, A. de Saxcé, J. C. Young, *J. Organomet. Chem.* **395**, 1 (1990).
160. R. J. P. Corriu, A. Kpoton, M. Poirier, G. Royo, J. Y. Corey, *J. Organomet. Chem.*, **277**, C25 (1984).
161. J. Boyer, R. J. P. Corriu, A. Kpoton, M. Mazhar, M. Poirier, G. Royo, *J. Organomet. Chem.* **301**, 131 (1986).
162. R. J. P. Corriu, M. Mazhar, M. Poirier, G. Royo, *J. Organomet. Chem.* **306**, C5 (1986).
163. F. H. Carré, R. J. P. Corriu, G. F. Lanneau, Z. Yu, *Organometallics* **10**, 1236 (1991).
164. J. Boyer, C. Brelière, R. J. P. Corriu, A. Kpoton, M. Poirier, G. Royo, *J. Organomet. Chem.* **311**, C39 (1986).
165. P. Arya, R. J. P. Corriu, K. Gupta, G. F. Lanneau, Z. Yu, *J. Organomet. Chem.* **399**, 11 (1990).
166. R. J. P. Corriu, G. F. Lanneau, M. Perrot, *Tetrahedron Lett.* **28**, 3941 (1987).
167. R. J. P. Corriu, G. F. Lanneau, M. Perrot, *Tetrahedron Lett.* **29**, 1271 (1988).
168. R. J. P. Corriu, G. F. Lanneau, M. Perrot-Petta, V. D. Mehta, *Tetrahedron Lett.* **31**, 2585 (1990).
169. R. J. P. Corriu, G. F. Lanneau, M. Perrot-Petta, *Synthesis* 954 (1991).
170. R. J. P. Corriu, G. F. Lanneau, Z. Yu, *Tetrahedron* **49**, 9019 (1993).
171. R. J. P. Corriu, M. Poirier, G. Royo, *C. R. Acad. Sci. Paris Ser. II* **310**, 1337 (1990).
172. P. Arya, J. Boyer, R. J. P. Corriu, G. F. Lanneau, M. Perrot, *J. Organomet. Chem.* **346**, C11 (1988).
173. P. Arya, J. Boyer, F. Carré, R. Corriu, G. Lanneau, J. Lapasset, M. Perrot, C. Priou, *Angew. Chem. Int. Ed. Engl.* **28**, 1016 (1989).
174. R. J. P. Corriu, G. F. Lanneau, V. D. Mehta, *J. Organomet. Chem.* **419**, 9 (1991).
175. R. Corriu, G. Lanneau, C. Priou, *Angew. Chem. Int. Ed. Engl.* **30**, 1130 (1991).
176. (a) R. J. P. Corriu, G. F. Lanneau, B. P. S. Chauhan, *Organometallics* **12**, 2001 (1993); (b) R. Corriu, B. P. S. Chauhan, G. Lanneau, *Organometallics* **14**, 1646 (1995).
177. B. P. S. Chauhan, R. Corriu, G. Lanneau, C. Priou, N. Auner, H. Handwerker, E. Herdtweck, *Organometallics* **14**, 1657 (1995).

178. H. Kobayashi, K. Ueno, H. Ogino, *Organometallics* **14**, 5490 (1995).
179. H. Handwerker, M. Paul, J. Blümel, C. Zybill, *Angew. Chem. Int. Ed. Engl.* **32**, 1313 (1993).
180. R. Corriu, B. P. S. Chauhan, G. Lanneau, *Organometallics* **14**, 4014 (1995).
181. C. Chuit, R. J. P. Corriu, A. Mehdi, C. Reyé, *Angew. Chem. Int. Ed. Engl.* **32**, 1311 (1993).
182. (a) M. Chauhan, C. Chuit, R. J. P. Corriu, C. Reyé, *Tetrahedron Lett.* **37**, 845 (1996); (b) M. Chauhan, C. Chuit, R. J. P. Corriu, A. Mehdi, C. Reyé, *Organometallics* **15**, 4326 (1996).
183. V. A. Benin, J. C. Martin, M. R. Willcott, *Tetrahedron Lett.* **35**, 2133 (1994).
184. C. Brelière, F. Carré, R. Corriu, M. Wong Chi Man, *J. Chem. Soc. Chem. Commun.* 2333 (1994).
185. J. Belzner, D. Schär, B. O. Kneisel, R. Herbst-Irmer, *Organometallics* **14**, 1840 (1995).
186. F. P. Boer, F. P. van Remoortere, *J. Am. Chem. Soc.* **92**, 801 (1970).
187. M. Veith, A. Rammo, *J. Organomet. Chem.* **521**, 429 (1996).
188. K. D. Onan, A. T. McPhail, C. H. Yoder, R. W. Hillyard, *J. Chem. Soc. Chem. Commun.* 209 (1978).
189. M. G. Voronkov, Yu. L. Frolov, V. M. D'yakov, N. N. Chipanina, L. I. Gubanova, G. A. Gavrilova, L. V. Klyba, T. N. Aksamentova, *J. Organomet. Chem.* **201**, 165 (1980).
190. A. I. Albanov, L. I. Gubanova, M. F. Larin, V. A. Pestunovich, M. G. Voronkov, *J. Organomet. Chem.* **244**, 5 (1983).
191. V. A. Pestunovich, M. F. Larin, M. S. Sorokin, A. I. Albanov, M. G. Voronkov, *J. Organomet. Chem.* **280**, C17 (1985).
192. V. P. Baryshok, D. Monkhoobor, G. A. Gravilova, N. N. Chipanina, A. I. Albanov, M. G. Voronkov, *Metallorg. Khim.* **5**, 1306 (1992). [English translation, *Organomet. Chem. USSR* **5**, 639 (1992).]
193. A. A. Macharashvili, V. E. Shklover, Yu. T. Struchkov, G. I. Oleneva, E. P. Kramarova, A. G. Shipov, Yu. I. Baukov, *J. Chem. Soc. Chem. Commun.* 683 (1988).
194. A. A. Macharashvili, V. E. Shklover, N. Yu. Chernikova, M. Yu. Antipin, Yu. T. Struchkov, Yu. I. Baukov, G. I. Oleneva, E. P. Kramarova, A. G. Shipov, *J. Organomet. Chem.* **359**, 13 (1989).
195. O. B. Artamkina, E. P. Kramarova, A. G. Shipov, Yu. I. Baukov, A. A. Macharashivili, Yu. E. Ovchinnikov, Yu. T. Struchkov, *Zh. Obshch. Khim.* **63**, 2289 (1993). [English translation, *Russian J. Gen. Chem.* **63**, 1590 (1993).]
196. O. B. Artamkina, E. P. Kramarova, A. G. Shipov, Yu. I. Baukov, A. A. Macharashvili, Yu. E. Ovchinnikov, Yu. T. Struchkov, *Zh. Obshch. Khim.* **64**, 263 (1994). [English translation, *Russian J. Gen. Chem.* **64**, 237 (1994).]
197. A. A. Macharashvili, V. E. Shklover, Yu. T. Struchkov, Yu. I. Baukov, E. P. Kramarova, G. I. Oleneva, *J. Organomet. Chem.* **327**, 167 (1987).
198. M. G. Voronkov, V. A. Pestunovich, Yu. I. Baukov, *Metallorg. Khim.* **4**, 1210 (1991). [English translation, *Organomet. Chem. USSR* **4**, 593 (1991).]

199. A. O. Mozzhukhin, M. Yu. Antipin, Yu. T. Struchkov, A. G. Shipov, E. P. Kramarova, Yu. I. Baukov, *Metallorg. Khim.* **5**, 906 (1992). [English translation, *Organomet. Chem. USSR* **5**, 439 (1992).]

200. A. A. Macharashvili, V. E. Shklover, Yu. T. Struchkov, B. A. Gostevskii, I. D. Kalikhman, O. B. Bannikova, M. G. Voronkov, V. A. Pestunovich, *J. Organomet. Chem.* **356**, 23 (1988).

201. A. A. Macharashvili, V. E., Shklover, Yu. T. Struchkov, M. G. Voronkov, B. A. Gostevsky, I. D. Kalikhman, O. B. Bannikova, V. A. Pestunovich, *J. Organomet. Chem.* **340**, 23 (1988).

202. V. V. Negrebetsky, V. V. Negrebetsky, A. G. Shipov, E. P. Kramarova, Yu. I. Baukov, *J. Organomet. Chem.* **496**, 103 (1995).

203. A. Mix, U. H. Berlekamp, H. G. Stammler, B. Neumann, P. Jutzi, *J. Organomet. Chem.* **521**, 177 (1996).

204. (a) A. G. Shipov, E. P. Kramarova, O. B. Artamkina, Yu. I. Baukov, N. I. Pirogov, *Metallorg. Khim.* **4**, 1101 (1991). [English translation, *Organomet. Chem. USSR,*, **4**, 541 (1991).] (b) O. B. Artamkina, A. G. Shipov, M. Nasim, D. P. Krut'ko, G. S. Zaitseva, Yu. I. Baukov, *Metallorg Khim.* **4**, 694 (1991). [English translation, *Organomet. Chem. USSR*, **4**, 340 (1991).]

205. A. G. Shipov, E. P. Kramarova, O. B. Artamkina, G. I. Oleneva, N. A. Nepomnyashchaya, Yu. I. Baukov, *Zh. Obshch. Khim.* **65**, 272 (1995). [English translation, *Russian J. Gen. Chem.* **65**, 233 (1995).]

206. A. R. Bassindale, M. Borbaruah, *J. Chem. Soc. Chem. Commun.* 352 (1993).

207. E.L. Muetterties, H. Roesky, C. M. Wright, *J. Am. Chem. Soc.* **88**, 4856 (1966).

208. G. Schott, K. Golz, *Z. Anorg. Allg. Chem.* **383**, 314 (1971).

209. G. Schott, K. Golz, *Z. Anorg. Allg. Chem.* **399**, 7 (1973).

210. V. E. Shklover, Yu. T. Struchkov, O. G. Rodin, V. F. Traven, B. I. Stepanov, *J. Organomet. Chem.* **266**, 117 (1984).

211. Yu. E. Ovchinnikov, V. E. Shklover, Yu. T. Struchkov, V. F. Traven, O. G. Rodin, V. I. Rokitskaya, *J. Organomet. Chem.* **347**, 33 (1988).

212. V. E. Shklover, Yu. Ovchinnikov, Yu. T. Struchkov, V. I. Rokitskaya, O. G. Rodin, V. F. Traven, B. I. Stepanov, M. Yu. Aismont, *J. Organomet. Chem.* **301**, 273 (1986).

213. (a) R. O. Day, T. K. Prakasha, R. R. Holmes, H. Eckert, *Organometallics* **13**, 1285 (1994); (b) T. K. Prakasha, S. Srinivasan, A. Chandrasekaran, R. O. Day, R. R. Holmes, *J. Am. Chem. Soc.* **117**, 10003 (1995).

214. (a) H. H. Karsch, R. Richter, A. Schier, *Z. Naturforsch.* **B48**, 1533 (1993); (b) H. H. Karsch, R. Richter, E. Witt, *J. Organomet. Chem.* **521**, 185 (1996).

215. M. G. Voronkov, *Pure Appl. Chem.* **13**, 35 (1966).

216. M. G. Voronkov, V. M. D'yakov, S. V. Kirpichenko, *J. Organomet. Chem.* **233**, 1 (1982).

217. A. B. Fineston, *U.S. Pat.* 2953545 (1960).

218. C. L. Frye, G. E. Vogel, J. A. Hall, *J. Am. Chem. Soc.* **83**, 996 (1961).

219. M. G. Voronkov, V. P. Baryshok, L. P. Petukhov, V. I. Rakhlin, R. G. Mirskov, V. A. Pestunovich, *J. Organomet. Chem.* **358**, 39 (1988).

220. C. L. Frye, G. A. Vincent, W. A. Finzel, *J. Am. Chem. Soc.* **93**, 6805 (1971).

221. J. W. Turley, F. P. Boer, *J. Am. Chem. Soc.* **90**, 4026 (1968).

222. L. Parkanyi, K. Simon, J. Nagy, *Acta Cryst.* **B30**, 2328 (1974).

223. L. Parkanyi, J. Nagy, K. Simon, *J. Organomet. Chem.* **101**, 11 (1975).

224. R. J. Garant, L. M. Daniels, S. K. Das, M. N. Janakiraman, R. A. Jacobson, J. G. Verkade, *J. Am. Chem. Soc.* **113**, 5728 (1991).

225. F. P. Boer, J. W. Turley, J. J. Flynn, *J. Am. Chem. Soc.* **90**, 5102 (1968).

226. F. P. Boer, J. W. Turley, *J. Am. Chem. Soc.* **91**, 4134 (1969).

227. M. S. Gordon, M. T. Carroll, J. H. Jensen, L. P. Davis, L. W. Burggraf, R. M. Guidry, *Organometallics* **10**, 2657 (1991).

228. J. Y. Corey, N. P. Rath, C. S. John, E. R. Corey, *J. Organomet. Chem.* **399**, 221 (1990).

229. J. J. Daly, F. Sanz, *J. Chem. Soc. Dalton Trans.* 2051 (1974).

230. D. Schomburg, *Z. Anorg. Allg. Chem.* **493**, 53 (1982).

231. A. Kemme, J. Bleidelis, I. Urtane, G. Zelchan, E. Lukevics, *J. Organomet. Chem.* **202**, 115 (1980).

232. O. A. D'yachenko, L. O. Atovmyan, S. M. Aldoshin, T. L. Krasnova, V. V. Stepanov, E. A. Chernyshev, A. G. Popov, V. V. Antipova, *Izvest. Akad. Nauk. SSSR Ser. Khim.* 2648 (1974). [English Translation 2561 (1974).]

233. O. A. D'yachenko, L. O. Atovmyan, S. M. Aldoshin, N. G. Nomalenkova, A. G. Popov, V. V. Antipova, E. A. Chernyshev, *Izvest. Akad. Nauk. SSR Ser. Khim.* 1081 (1975). [Engl. Trans. 990 (1975).]

234. J. J. H. Edema, R. Libbers, A. Ridder, R. M. Kellogg, A. L. Spek, *J. Organomet. Chem.* **464**, 127 (1994).

235. M. G. Voronkov, *Topics Curr. Chem.* **84**, 77 (1979).

236. M. H. P. van Genderen, H. M. Buck. *Recl. Trav. Chim. Pays-Bas* **106**, 449 (1987).

237. M. T. Attar-Bashi, C. Eaborn, J. Vencl, D. R. M. Walton, *J. Organomet. Chem.* **117**, C87 (1976).

238. G. Cerveau, C. Chuit, R. J. P. Corriu, N. K. Nayyar, C. Reyé, *J. Organomet. Chem.* **389**, 159 (1990).

239. A. Hosomi, S. Iijima, H. Sakurai, *Chem. Lett.* 243 (1981).

240. J. D. Nies, J. M. Bellama, N. Ben-Zvi, *J. Organomet. Chem.* **296**, 315 (1985).

241. A. Daneshrad, C. Eaborn, R. Eidenschink, D. R. M. Walton, *J. Organomet. Chem.* **90**, 139 (1975).

242. G. Cerveau, R. J. P. Corriu, N. Costa, *J. Non-Cryst. Solids* **163**, 226 (1993).

243. E. Lukevics, G. Zelchans, I. I. Solomennikova, E. E. Liepins, J. I. Ankovska, I. Mazeika, *Zh. Obshch. Khim.* **47**, 109 (1977); *Chem. Abstr.* **86**, 171536j (1977).

244. D. Gudat, J. G. Verkade, *Organometallics* **8**, 2772 (1989).

245. E. Kupce, E. Liepins, A. Lapsina, G. Zelchans, E. Lukevics, *J. Organomet. Chem.* **333**, 1 (1987).

246. A. A. Macharashvili, V. E. Shklover, Yu. T. Struckhov, A. Lapsina, G. Zelchans, E. Lukevics, *J. Organomet. Chem.* **349**, 23 (1988).

247. D. Gudat, L. M. Daniels, J. G. Verkade, *J. Am. Chem. Soc.* **111**, 8520 (1989).

248. D. Gudat, J. G. Verkade, *Organometallics* **9**, 2172 (1990).

249. L. Tansjö, *Acta Chem. Scand.* **15**, 1583 (1961).

250. R. Müller, C. Dathe, D. Mross, *Chem. Ber.* **98**, 241 (1965).
251. R. Müller, *Z. Chem.* **24**, 41 (1984).
252. M. Kumada, K. Tamao, J. Yoshida, *J. Organomet. Chem.* **239**, 115 (1982).
253. L. Tansjö, *Acta Chem. Scand.* **18**, 456 (1964).
254. L. Tansjö, *Acta Chem. Scand.* **18**, 465 (1964).
255. R. Müller, C. Dathe, *J. Prakt. Chem.* **22**, 232 (1963).
256. R. Müller, C. Dathe, Z. *Anorg. Allg. Chem.* **341**, 41 (1965).
257. K. Tamao, J. Yoshida, H. Yamamoto, T. Kakui, H. Matsumoto, M. Takahashi, A. Kurita, M. Murata, M. Kumada, *Organometallics* **1**, 355 (1982).
258. R. Müller, R. Köhne, H. J. Model, C. Dathe, Z. *Anorg. Allg. Chem.* **368**, 243 (1969).
259. J. J. Moscony, A. G. MacDiarmid, *Chem. Commun.* 307 (1965).
260. R. Tacke, M. Mühleisen, *Angew. Chem. Int. Ed. Engl.* **33**, 1359 (1994).
261. R. Müller, C. Dathe, H. J. Frey, *Chem. Ber.* **99**, 1614 (1966).
262. R. Müller, *Z. Chem.* **5**, 220 (1965).
263. E. M. Soshestvenskaya, *Zh. Obshch. Khim.* **22**, 1122 (1952); *Chem. Abstr.* **47**, 8030d (1953).
264. R. Müller, C. Dathe, *Chem. Ber.* **98**, 235 (1965).
265. R. Müller, C. Dathe, *Chem. Ber.* **99**, 1609 (1966).
266. J. Yoshida, K. Tamao, T. Kakui, A. Kurita, M. Murata, K. Yamada, M. Kumada, *Organometallics* **1**, 369 (1982).
267. R. Müller, C. Dathe, *Z. Anorg. Allg. Chem.* **341**, 49 (1965).
268. K. Tamao, T. Kakui, M. Akita, T. Iwahara, R. Kanatani, J. Yoshida, M. Kumada, *Tetrahedron* **39**, 983 (1983).
269. J. Yoshida, K. Tamao, H. Yamamoto, T. Kakui, T. Uchida, M. Kumada, *Organometallics* **1**, 542 (1982).
270. J. Yoshida, K. Tamao, T. Kakui, M. Kumada, *Tetrahedron Lett.* 1141 (1979).
271. R. Müller, M. Dressler, C. Dathe, *J. Prak. Chem.* **312**, 150 (1970).
272. A. Rosenheim, B. Raibmann, G. Schendel, *Z. Anorg. Allg. Chem.* **196**, 160 (1931).
273. A. Weiss, G. Reiff, A. Weiss, *Z. Anorg. Allg. Chem.* **311**, 151 (1961).
274. (a) D. W. Barnum, *Inorg. Chem.* **9**, 1942 (1970); (b) D. W. Barnum, *Inorg. Chem.* **11**, 1424 (1972).
275. (a) A. Boudin, G. Cerveau, C. Chuit, R. J. P. Corriu, C. Reyé, *Angew. Chem. Int. Ed. Engl.* **25**, 474 (1986); (b) A. Boudin, G. Cerveau, C. Chuit, R. J. P. Corriu, C. Reyé, *Organometallics* **7**, 1165 (1988).
276. G. Cerveau, C. Chuit, R. J. P. Corriu, L. Gerbier, C. Reyé, *C. R. Acad. Sci. Paris Ser. II* **312**, 1311 (1991).
277. J. J. Flynn, F. P. Boer, *J. Am. Chem. Soc.* **91**, 5756 (1969).
278. M. L. Hoppe, R. M. Laine, J. Kampf, M. S. Gordon, L. W. Burggraf, *Angew. Chem. Int. Ed. Engl.* **32**, 287 (1993).
279. S. P. Narula, R. Shankar, M. Kumar, Meenu, *Inorg. Chem.* **33**, 2716 (1994).
280. G. Cerveau, R. J. P. Corriu, *C. R. Acad. Sci. Paris Ser. II* **316**, 35 (1993).

281. D. J. Ray, R. M. Laine, C. Viney, T. R. Robinson, *Polym. Prepr. (Am. Chem. Soc. Div. Polym. Chem.)* **32**, 550 (1991).

282. A. Weiss, D. R. Harvey, *Angew. Chem. Int. Ed. Engl.* **3**, 698 (1964).

283. D. R. Harvey, A. Weiss, *Angew. Chem. Int. Ed. Engl.* **4**, 357 (1965).

284. I. R. Beattie, G. A. Ozin, *J. Chem. Soc. (A)* 2267 (1969).

285. K. Hensen, Z. Busch, *Naturforsch.* **37B**, 1174 (1982).

286. H. E. Blayden, M. Webster, *Inorg. Nucl. Chem. Lett.* **6**, 703 (1970).

287. V. A. Bain, R. C. G. Killean, M. Webster, *Acta Crystallogr. Sect. B* **25**, 156 (1969).

288. A. D. Adley, P. H. Bird, A. R. Fraser, M. Onyszchuk, *Inorg. Chem.* **11**, 1402 (1972).

289. (a) U. Wannagat, *Angew. Chem.* **69**, 516 (1957); (b) U. Wannagat, K. Hensen, P. Petesch, F. Vielberg, *Monatsh. Chem.* **98**, 1415 (1967).

290. D. Kummer, S. C. Chaudhry, T. Debaerdemaeker, U. Thewalt, *Chem. Ber.* **123**, 945 (1990).

291. (a) D. Kummer, H. Köster, M. Speck, *Angew. Chem. Int. Ed. Engl.* **8**, 599 (1969); (b) D. Kummer, A. Balkir, H. Köster, *J. Organomet. Chem.* **178**, 29 (1979).

292. G. Sawitzki, H. G. von Schnering, *Chem. Ber.* **109**, 3728 (1976).

293. D. Kummer, S. C. Chaudhry, W. Depmeier, G. Mattern, *Chem. Ber.* **123**, 2241 (1990).

294. W. B. Farnham, J. F. Whitney, *J. Am. Chem. Soc.* **106**, 3992 (1984).

295. D. Kummer, K. E. Gaisser, T. Seshadri, *Chem. Ber.* **110**, 1950 (1977).

296. D. Kummer, T. Seshadri, *Chem. Ber.* **110**, 2355 (1977).

297. D. Kummer, H. Köster, *Z. Anorg. Allg. Chem.* **398**, 279 (1973).

298. D. Kummer, T. Seshadri, *Z. Anorg. Allg. Chem.* **428**, 129 (1977).

299. S. Herzog, F. Krebs, *Z. Chem.* **8**, 149 (1968).

300. D. Kummer, H. Köster, *Z. Anorg. Allg. Chem.* **402**, 297 (1973).

301. D. Kummer, K. E. Gaisser, J. Seifert, R. Wagner, *Z. Anorg. Allg. Chem.* **459**, 145 (1979).

302. S. Herzog, F. Krebs, *Naturwissenschaften* **50**, 330 (1963).

303. G. Sawitzki, H. G. von Schnering, D. Kummer, T. Seshadri, *Chem. Ber.* **111**, 3705 (1978).

304. M. Wada, R. Okawara, *Inorg. Nucl. Chem. Lett.* **5**, 355 (1969).

305. M. Wada, T. Suda, R. Okawara, *J. Organomet. Chem.* **65**, 335 (1974).

306. G. Klebe, D. Tran Qui, *Acta Cryst.* **C40**, 476 (1984).

307. C. Brelière, F. Carré, R. J. P. Corriu, M. Poirier, G. Royo, J. Zwecker, *Organometallics* **8**, 1831 (1989).

308. K. Boyer-Elma, F. H. Carré, R. J. P. Corriu, W. E. Douglas, *J. Chem. Soc. Chem. Commun.* 725 (1995).

309. F. Carré, G. Cerveau, C. Chuit, R. J. P. Corriu, C. Reyé, *New J. Chem.* **16**, 63 (1992).

310. C. Brelière, R. J. P. Corriu, G. Royo, J. Zwecker, *Organometallics* **8**, 1834 (1989).

311. F. Carré, C. Chuit, R. J. P. Corriu, A. Mehdi, C. Reyé, *Organometallics* **14**, 2754 (1995).

312. (a) I. Kalikhman, D. Kost, M. Raban, *J. Chem. Soc. Chem. Commun.* 1253 (1995); (b) D. Kost, I. Kalikhman, M. Raban, *J. Am. Chem. Soc.* **117**, 11512 (1995).
313. A. O. Mozzhukhin, M. Yu. Antipin, Yu. T. Struchkov, B. A. Gostevskii, I. D. Kalikhman, V. A. Pestunovich, M. G. Voronkov, *Metalloorg. Khim.* **5**, 658 (1992). [*Chem. Abst.* **117**, 234095w (1992).]
314. C. Brelière, F. Carré, R. J. P. Corriu, W. E. Douglas, M. Poirier, G. Royo, M. Wong Chi Man, *Organometallics* **11**, 1586 (1992).
315. F. Carré, G. Cerveau, C. Chuit, R. J. P. Corriu, C. Reyé, *Angew. Chem. Int. Ed. Engl.* **28**, 489 (1989).
316. (a) F. Carré, C. Chuit, R. J. P. Corriu, A. Mehdi, C. Reyé, *J. Organomet. Chem.* **446**, C6 (1993); (b) F. Carré, C. Chuit, R. J. P. Corriu, A. Fanta, A. Mehdi, C. Reyé, *Organometallics* **14**, 194 (1995).
317. C. Chuit, R. J. P. Corriu, A. Mehdi, C. Reyé, *Chem. Eur. J.* **2**, 342 (1996).
318. R. W. Alder, P. S. Bowman, W. R. S. Steele, D. R. Wintermann, *J. Chem. Soc. Chem. Commun.* 723 (1968).
319. F. Carré, G. Cerveau, C. Chuit, R. J. P. Corriu, N. K. Nayyar, C. Reyé, *Organometallics* **9**, 1989 (1990).
320. J. R. Mooney, C. K. Choy, K. Knox, M. E. Kenney, *J. Am. Chem. Soc.* **97**, 3033 (1975).
321. T. C. Wen, M. E. Liu, S. Y. Young, Y. S. Lee, K. M. Chen, C. N. Kuo, *J. Chin. Chem. Soc.* **40**, 141 (1993).
322. N. Brasseur, T. L. Nguyen, R. Langlois, R. Ouellet, S. Marengo, D. Houde, J. E. van Lier, *J. Med. Chem.* **37**, 415 (1994).
323. K. M. Kane, F. R. Lemke, J. L. Petersen, *Inorg. Chem.* **34**, 4085 (1995).
324. C. Brelière, R. J. P. Corriu, G. Royo, W. W. C. Wong Chi Man, J. Zwecker, *Organometallics* **9**, 2633 (1990).
325. K. Tamao, M. Akita, H. Kato, M. Kumada, *J. Organomet. Chem.* **341**, 165 (1988).
326. W. Dilthey, *Chem. Ber.* **36**, 923 (1903).
327. R. M. Pike, R. R. Luongo, *J. Am. Chem. Soc.* **88**, 2972 (1966).
328. R. West, *J. Am. Chem. Soc.* **80**, 3246 (1958).
329. D. W. Thompson, *Inorg. Chem.* **8**, 2015 (1969).
330. K. M. Taba, W. V. Dahlhoff, *J. Organomet. Chem.* **280**, 27 (1985).
331. E. L. Muetterties, C. M. Wright, *J. Am. Chem. Soc.* **86**, 5132 (1964).
332. M. Kira, L. C. Zhang, C. Kabuto, H. Sakurai, *Chem. Lett.* 659 (1995).
333. E. P. Kramarova, L. S. Smirnova, O. B. Artamkina, A. G. Shipov, Yu. I. Baukov, Yu. E. Ovchinnikov, A. O. Mozzhukhin, Yu. T. Struchkov, *Zh. Obshch. Khim.* **63**, 2275 (1993). [English translation, *Russian J. Gen. Chem.* **63**, 1580 (1993).]
334. R. G. Pearson, D. N. Edgington, F. Basolo, *J. Am. Chem. Soc.* **84**, 3233 (1962).
335. E. L. Muetterties, C. M. Wright, *J. Am. Chem. Soc.* **87**, 21 (1965).
336. S. K. Dhar, V. Doron, S. Kirschner, *J. Am. Chem. Soc.* **81**, 6372 (1959).
337. H. Kelling, H. U. Kibbel. *Z. Anorg. Allg. Chem.* **386**, 59 (1971).
338. T. Inoue, J. Fujita, K. Saito, *Bull. Chem. Soc. Jpn.* **48**, 1228 (1975).
339. D. L. Deadmore, W. F. Bradley, *Acta Cryst.* **15**, 186 (1962).

340. N. S. Biradar, V. L. Roddabasanagoudar, T. M. Aminabhavi, *Indian J. Chem.* **24A**, 873 (1985).
341. C. Brelière, F. Carré, R. J. P. Corriu, G. Royo, *Organometallics* **7**, 1006 (1988).
342. N. Auner, R. Probst, F. Hahn, E. Herdtweck, *J. Organomet. Chem.* **459**, 25 (1993).
343. C. Brelière, F. Carré, R. J. P. Corriu, G. Royo, M. Wong Chi Man, J. Lapasset, *Organometallics* **13**, 307 (1994).
344. F. Carré, C. Chuit, R. J. P. Corriu, A. Mehdi, C. Reyé, *Angew. Chem. Int. Ed. Engl.* **33**, 1097 (1994).
345. J. Braddock-Wilking, M. Schieser, L. Brammer, J. Huhmann, R. Shaltout, *J. Organomet. Chem.* **499**, 89 (1995).
346. N. W. Mitzel, D. W. H. Rankin, Abstracts of *XIth International Symposium on Organosilicon Chemistry*, Montpellier, France, September 1–6, 1996, p. PA 8.

5 Hypercoordinate Silicon Species in Organic Syntheses

MITSUO KIRA
Department of Chemistry, Graduate School of Science, Tohoku University, Sendai, Japan
Photodynamics Research Center, The Institute of Physical and Chemical Research (RIKEN), Sendai, Japan

LUO CHENG ZHANG
Photodynamics Research Center, The Institute of Physical and Chemical Research (RIKEN), Sendai, Japan

5.1 INTRODUCTION

Pentacoordinate silicon species exist as reactive intermediates or even as isolable compounds, in contrast to the corresponding pentacoordinate carbon species, which are usually proposed only as transition states of S_N2-type substitution reactions. The existence of pentacoordinate and hexacoordinate species is one of the distinctive properties of Group 14 elements heavier than carbon. A pentacoordinate silicon species should be regarded not only as an intermediate in the nucleophilic substitution of a tetracoordinate silicon species, but also as an intermediate in the transformation from tetracoordinate to hexacoordinate silicon species. The pentacoordinate silicon center may thus offer a unique reaction site for organic syntheses.

$$SiX_4 \underset{-Y^-}{\overset{+Y^-}{\rightleftharpoons}} \left[Y{-}SiX_4 \right]^- \underset{-Z^-}{\overset{+Z^-}{\rightleftharpoons}} \left[YZSiX_4 \right]^{2-} \quad (5.1)$$

Extensive studies have focused on the synthetic utility of hypercoordinate silicon species, since the pioneering work by Kumada and Tamao on a series of hexacoordinate alkyl- and arylpentafluorosilicates.[1] There are already a number of review articles on the structure and reactivity of hypercoordinate

Chemistry of Hypervalent Compounds, edited by Kin-ya Akiba.
ISBN 0-471-24019-2

organosilicon species,[2–4] and therefore, we will deal only with relatively recent studies of synthetically useful reactions of hypercoordinate organosilicon compounds and their mechanistic aspects in this chapter. In particular, C—C bond formation using pentacoordinate allylsilicon reagents and silyl enolates, catalytic cross-coupling reactions with pentacoordinate silicon species, and reduction of carbonyl compounds using pentacoordinate hydridosilicon species will be discussed in detail.

5.2 ALLYLATION OF CARBONYL COMPOUNDS WITH PENTACOORDINATE ALLYLIC SILICON SPECIES

Among a number of C—C bond formation reactions, the reactions of allylic metals have attracted special attention because of their synthetic analogy to the aldol addition of metal enolates. Thus, the methodology and reagents suitable for the regio- and stereocontrolled allylation of carbonyl compounds are currently developing as an important objective in this area.

Two synthetically important reaction modes of allylic trimethylsilanes with carbonyl compounds have been reported. First, the allylation of ketones and acetals by allylic trimethylsilanes in the presence of Lewis acids such as $TiCl_4$ or BF_3 gives the corresponding homoallylic alcohols in a regiospecific manner; the synthetic usefulness of these allylations has been well established and reviewed.[4-5] The driving force of the reaction is attributed to the enhancement of electrophilicity of the allylic γ-carbon, due to the effective hyperconjugation between an Si—C σ bond and a π system, as well as to the activation of carbonyl carbon by a Lewis acid.

Me(Me)C=CH–CH$_2$SiMe$_3$ + R^{1}R^{2}C=O —1) $TiCl_4$, CH_2Cl_2; 2) Hydrolysis→ CH$_2$=CH–C(Me)(Me)–C(R^1)(R^2)OH (5.2)

Second, the allylation of aldehydes and ketones is performed using allylic trimethylsilanes in the presence of a catalytic amount of fluoride ions.[6] Although the reaction also gives the corresponding homoallyl alcohols in high yields, the exclusive regiosepecificity, which is characteristic of the Lewis-acid-catalyzed allylation using allylic trimethylsilanes, is lost.

Me(Me)C=CH–CH$_2$SiMe$_3$ + R^{1}R^{2}C=O —1) F^-; 2) Hydrolysis→ CH$_2$=CH–C(Me)(Me)–C(R^1)(R^2)OH + Me$_2$C=CH–CH$_2$–C(R^1)(R^2)OH (5.3)

Two controversial mechanisms have been proposed for the fluoride-ion catalyzed allylation using allylic trimethylsilanes, as shown in Scheme 5.1: (1) an allylic anion is involved as an intermediate formed via the nucleophilic substitution of allylic trimethylsilanes by fluoride ions,[6] and (2) a fluoride ion adds to an allylic trimethylsilane to give the pentacoordinate allylic silicon compound, which reacts with a carbonyl compound directly to give allylation products.[7]

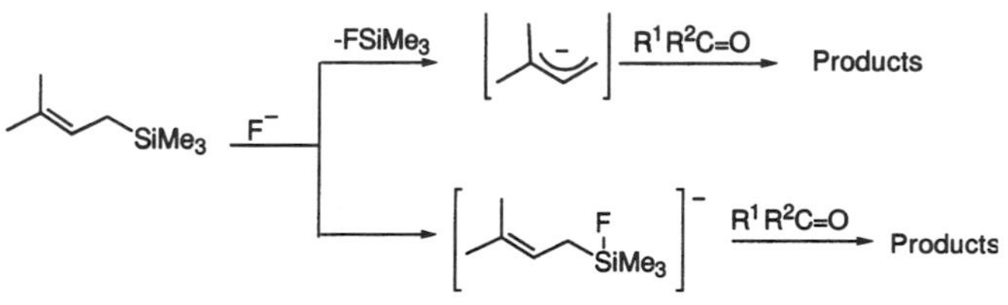

Scheme 5.1

5.2.1 Substituent Effects on Fluoride-Ion Catalyzed Allylation Using Allylic Silanes

Since it is well known that hypercoordinate silicon compounds are stabilized by electronegative and small-size substituents,[2–4,8] the mechanisms for the fluoride-ion catalyzed allylation of carbonyl compounds using allylic silanes may depend on the substituents at silicon. As shown in Table 5.1, the reactivity and regioselectivity of the reactions of prenylsilanes fluorinated at the silicon atoms with benzaldehyde in the presence of TBAF are quite different, depending on the number of fluorines at one silicon.[9] In the presence of a catalytic amount of TBAF, both prenyltrimethylsilane and prenyldimethylfluorosilane give the corresponding homoallylic alcohols as a mixture of regioisomers during the reactions with benzaldehydes. However, prenyldifluoromethylsilane does not react with benzaldehyde even in the presence of the stoichiometric amount of TBAF.

TABLE 5.1 Reactions of PrenylSiMe$_x$F$_{3-x}$ with Benzaldehyde in the Presence of Fluoride Ions

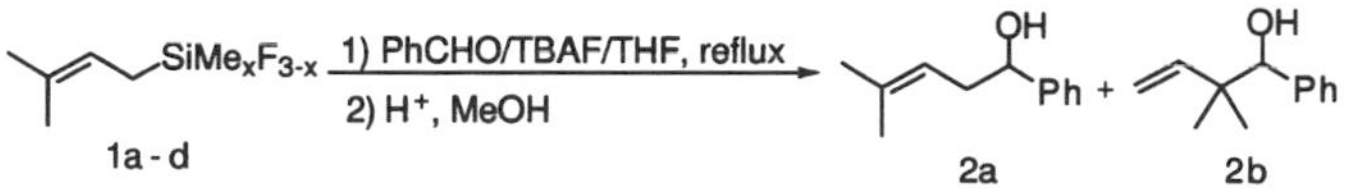

(5.4)

Prenylsilane (x=)	TBAF/equiv.	Reaction Time (h)	Products, Yield (%)		Relative Yield (2a/2b)
			2a	**2b**	
1a (3)	0.05	40	23	17	58/42
1b (2)	0.05	40	23	15	61/39
1c (1)	2.0	48	—	< 1	—
1d (0)	2.0	1	0	91	0/100

Quite interestingly, prenyltrifluorosilane reacts smoothly with benzaldehyde in the presence of the stoichiometric amount of TBAF to give 2,2-dimethyl-1-phenylbut-3-ene-1-ol in a regiospecific manner. These results suggest not only that the two mechanisms operate and cross over, depending on the number of fluorine substituents, but also that a pentacoordinate allylic silicon species, once formed, works as a facile regiospecific allylation reagent of aldehydes.

5.2.2 Allylation with Pentacoordinate Allylic Tetrafluorosilicon Species

In a pentacoordinate allylic silicon species, the pentacoordinate silicon moiety can act as a strong electron-donating substituent for an allylic π system to enhance the electrophilicity of the allylic γ-carbon. In addition, the pentacoordinate silicon can interact as a Lewis acid with a carbonyl oxygen to form a hexacoordinate silicon complex. If silicon in a pentacoordinate species exhibits both functions, the species may serve as a useful reagent for regiospecific and highly stereoselective allylation of carbonyl compounds via a cyclic six-membered transition state. A number of reactions based on the concepts mentioned above have been explored in recent years.

Typically, highly diastereoselective crotylation of various aldehydes has been achieved using stereohomogeneous (*E*)- and (*Z*)-crotyltrifluorosilanes in the presence of cesium fluoride (Eqs 5.5 and 5.6).[10–11] The crotyltrifluorosilanes are prepared stereohomogeneously and are very stable configurationally and storable for a couple of months at 0°C without any change in the purity, as expected from the weak tendency to undergo the metallotropic rearrangement. Thus the crotyltrifluorosilanes would be taken as ideal reagents for diastereoselective crotylation of aldehydes.

3E —CsF/RCHO→ [transition state with SiF_4, H, O, R]‡ → → 4A (5.5)

3Z —CsF/RCHO→ [transition state with SiF_4, H, O, R]‡ → → 4S (5.6)

Typically, benzaldehyde reacts with (*E*)-crotyltrifluorosilane (*E*/*Z* = 99/1) in the presence of an equimolar amount of cesium fluoride to afford *anti*-1-phenyl-2-methyl-3-buten-1-ol in 92% yield with the *anti*/*syn* ratio of 99/1, while a similar reaction using (*Z*)-ctrotyltrifluorosilane (*E*/*Z* = 1/99) gave the *syn* isomer of the homoallyl alcohol in 96% yield with *anti*/*syn* = 1/99.[11]

5.2.3 Allylation with Other Pentacoordinate Allylic Silicon Species

The allylation of aldehydes with allyltrifluorosilanes is feasible in the presence of triethylamine and a wide variety of hydroxy compounds under mild conditions. Results show a similar stereochemical outcome to the allylation using cesium fluoride as a promoter.[12] The reaction is promoted even by a trace amount of water as a hydroxy compound. The allyltrifluorosilane/ROH/Et_3N reagent system is less reactive than the allyltrifluorosilane/CsF system to discriminate between linear and α-branched alkanals; prenylation of primary aldehydes proceeds very smoothly, whereas no reaction occurs with secondary and tertiary aldehydes. Allyltrialkoxysilanes have been revealed to react also with aldehydes in the presence of catechol via the pentacoordinate allylic bis(catecholato)silicon species.[13]

R^1, R^2 allyl-SiF_3 + RCHO —[1) $R^3OH/Et_3N/CH_2Cl_2$, rt; 2) Hydrolysis]→ R–CH(OH)–C(R^2)(R^1)–CH=CH_2 (5.7)

R^3OH = Catechol, 2,2'-Dihydroxybiphenyl, PhCH(OH)COOH, $NH_2(CH_2)_2OH$, MeOH, H_2O

R^1, R^2 allyl-$Si(OEt)_3$ + RCHO —[1) catechol /Et_3N; 2) Hydrolysis]→ R–CH(OH)–C(R^2)(R^1)–CH=CH_2 (5.8)

A number of anionic and neutral pentacoordinate allylic silicon compounds have been isolated and characterized as stable molecules.[14–15] Expectedly, these allylic pentacoordinate silicon species react with various aldehydes without any catalyst, indicative of the intrinsic high reactivity of pentacoordinate allylsilicon species towards aldehydes, whereas the reactivity depends on the ligands on silicon. The isolated anionic pentacoordinate allylic silicon compounds **5** and **6** are less reactive than *in situ* generated allyltetrafluorosilicon and the related pentacoordinate allylic silicon compounds described above. Compounds **5** and **6** react with aromatic aldehydes but do not react with aliphatic aldehydes. The pentacoordinate allylic silicon species having rigid TBP (trigonal bipyamidal) structures like **8** do not react even with aromatic aldehydes. In contrast, neutral allylic silicon compounds **7** react not only with aromatic aldehydes but with aliphatic aldehydes, smoothly.[16]

5.2.4 Allylation Using Allyltrichlorosilanes in the Presence of a Base

Kobayashi and co-workers have recently reported a good modification of the method for allylation of aldehydes via pentacoordinate silicon routes.[17–19] Thus, allylic trichlorosilanes have been found to react with various aldehydes

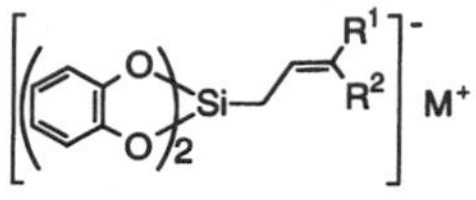

5

M = Li, PPN; R^1, R^2 = H or Me

$PPN^+ = [Ph_3P{=}N{=}PPh_3]^+$

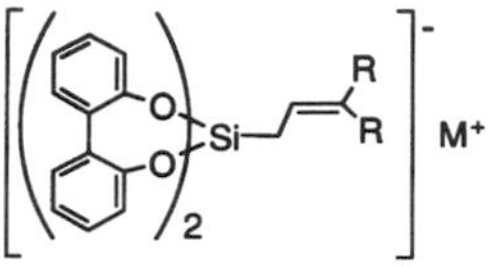

6

M = Li; R = H or Me

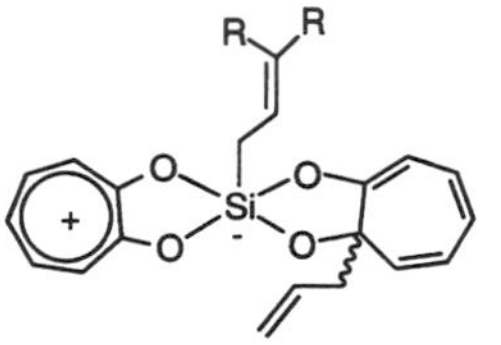

7

R = H or Me

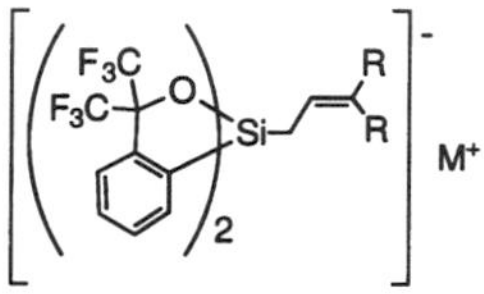

8

M = Li, PPN; R = H or Me

in DMF at 0°C to give the corresponding homoallyl alcohols in quite high yields in a regiospecific and highly diastereoselective manner. In these reactions, the use of DMF is crucial. In dichloromethane, acetonitrile, benzene, ether, and THF, no reaction occurs, while in HMPA, the corresponding homoallylic alcohols are given in moderate yields. Since the diastereoselectivity of the reactions using (*E*)- and (*Z*)-crotyltrichlorosilanes is excellent, similar to the reactions using crotyltrifluorosilane/CsF, the former reactions would proceed via cyclic transition states having hexacoordinate silicon with DMF as a ligand.

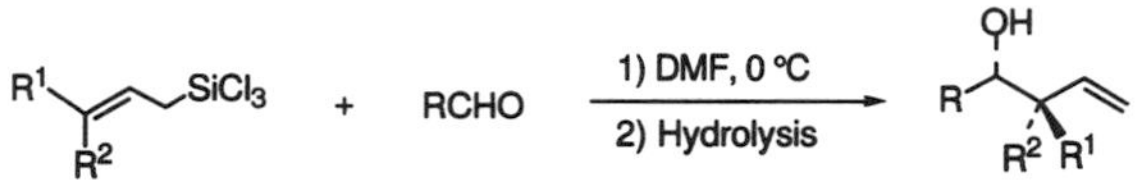

$$R^1R^2C{=}CHCH_2SiCl_3 + RCHO \xrightarrow[\text{2) Hydrolysis}]{\text{1) DMF, 0 °C}} RCH(OH)CR^1R^2CH{=}CH_2 \quad (5.9)$$

It is a synthetic advantage that these reactions can be carried out in one pot without isolation of the intermediate allylic trichlorosilane obtained from an allylic halide/trichlorosilane or 1,3-diene/trichlorosilane. Instead of DMF, several chiral bases have been used to achieve enantioselective allylation of aldehydes (*vide infra*).[20]

5.2.5 Allylation of Aldehydes Using Allylsilacyclobutanes

1-Allylsilacyclobutanes have been found to react with aldehydes without any catalysts to give the corresponding homoallylic alcohols with high regio- and stereoselectivity.[21]

R = Ph, n-C_6H_{13}, c-C_6H_{11}; R^1, R^2 = H, Ph, Alkyl

(5.10)

Ab initio MO calculations showed that the reaction of an allylsilacyclobutane model with formaldehyde proceeds via a transition state involving pentacoordinate silicon species with a lower activation barrier than the reaction of allylsilane or methyl-substituted allylsilanes.[22] The high reactivity of the allylsilacyclobutanes in comparison to common allyltrialkylsilanes is attributed to a strong Lewis acidic nature of the silicon center. The mechanistic aspects will be discussed more in detail in Section 5.2.6.

5.2.6 Mechanistic Aspects

Stereochemical control in the addition of crotylmetals to aldehydes has been extensively investigated from a mechanistic viewpoint. Reactions of crotyltrimethylsilanes with aldehydes in the presence of Lewis acids have been found to proceed via acyclic transition states to give the corresponding homoallyl alcohols with the *syn* stereochemistry as the major product, regardless of the stereochemistry of the starting crotylsilanes.[23] In contrast, pentacoordinate crotylsilicates exhibit the quite different diastereoselectivity in the reactions with aldehydes, where the *anti*- and *syn*-homoallyl alcohols are given exclusively from (E)- and (Z)-crotylsilicates, respectively; the diastereoselectivity is similar to that demonstrated by the crotylation using crotylboronates.[24] Reaction of a triethylammonium bis(catecholato)allylic silicon species with benzaldehyde gave the corresponding optically active homoallyl alcohols (Eq. 5.11).[25] The stereoselectivity of the allylation is thus characterized by the six-membered cyclic transition state having a chair conformation, as shown in the previous section.

PhCHO/Catechol/Et_3N, 80 °C, 16 h; 50% ee → 51% ee + 50% ee (90:10)

(5.11)

The high reactivity of pentacoordinate allylic silicon species compared to the tetracoordinate silicon analogs may be caused by the significant Lewis acidity giving hexacoordinate silicates and by enhanced nucleophilicity of the allylic γ-carbon due to σ–π conjugation (Figure 5.1). Whereas a tetracoordinate silyl group with electronegative substituents is usually less electron-donating than a trialkylsilyl group, the pentacoordinate silicon group having an additional ligand may be even more electron-donating to the allyl π systems than the trialkylsilyl group. The cyclic transition state may be favored, since the

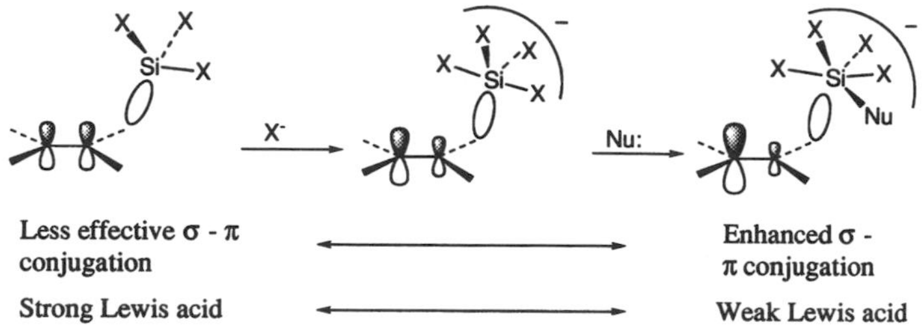

Figure 5.1 Schematic presentation of the effects of hypercoordination on the HOMO of allylsilanes and the Lewis acidity.

hexacoordination at silicon may further increase the electron-donating ability of the silyl group and, in turn, the nucleophilicity of the allylic γ-carbon due to σ–π conjugation. The importance of the nucleophillic attack of the allylic γ-carbon was also exhibited by substituent effects on the relative reactivity towards substituted benzaldehydes (ρ value for the Hammett plot was 1.18).[15]

The higher migratory aptitude of an allyl group on a *hexacoordinate* silicon atom towards carbonyl carbon compared with that on a *pentacoordinate* silicon atom has been clearly shown by the facile thermal isomerization of an isolated hexacoordinate diallylic bis(tropolonato)silicon compound; the migration occurs at 0°C, whereas the produced pentacoordinate allylic silicon does not react under these conditions.[16]

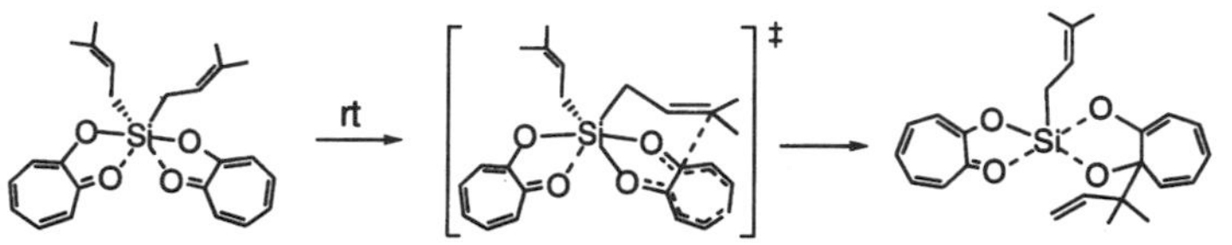

(5.12)

Transition structures and energies for the reaction of allyltetrafluorosilicate with formaldehyde were investigated using *ab initio* MO calculations with the MIDI-4(*) basis set.[26]

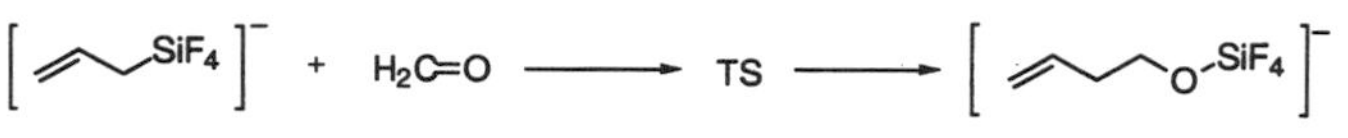

(5.13)

Among various transition structures, a cyclic chair, three cyclic boats, and a linear form were searched for. Although no boat transition structure has been found, the hexacoordinate chair form was found to be 50.2 kcal mol^{-1} lower in energy than a linear pentacoordinate one. The cyclic chair and linear transition structures are shown in Figure 5.2. A C—C bond forming distance of 2.22 Å in the chair form is comparable to that for the allylation of formaldehyde with allylborane (2.28 Å) and allylboronic acid (2.18 Å) calculated with the 3-21G

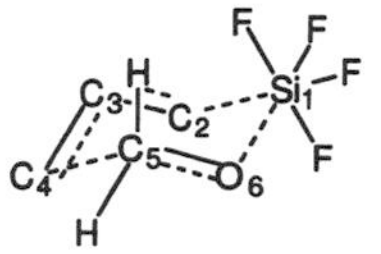

R (C2-C3)	= 1.422 Å	∠2-1-6	= 83.9 °		
R (C3- C4)	= 1.376	∠1-6-5	= 127.2		
R (C4---C5)	= 2.224	∠6-5-4	= 106.0		
R (C5-O)	= 1.267	∠5-4-3	= 92.5		
R (O---Si)	= 1.902	∠4-3-2	= 125.3		
R (Si-C)	= 2.134	∠3-2-1	= 109.7		

Chair TS

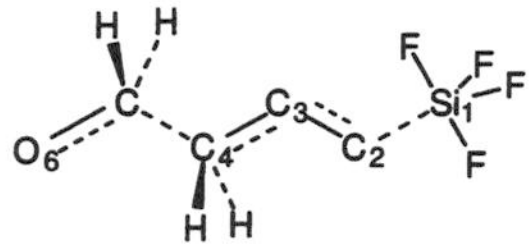

R (C2-C3)	= 1.351 Å	∠3-4-5-6 = -175.6 °	
R (C3- C4)	= 1.481		
R (C4---C5)	= 1.616		
R (C5-O)	= 1.340		
R (Si-C)	= 2.783		

Linear TS

Figure 5.2 Structures of transition states for the reaction of allyltetrafluorosilicate with formaldehyde determined using MIDI-4(*) basis set.

basis set.[27] One can derive from the structural parameters that the chair transition structure is an early (reactant-like) one, while the linear transition state structure is a late (product-like) one. Given the chair transition structure, the relative energies of axial and equatorial conformations of alkyl groups were assessed theoretically by replacing an appropriate hydrogen with a standard methyl group. Single point calculations with MIDI-4(*) level indicate that a methyl group attached to the carbonyl carbon prefers the equatorial position over the axial position by 18.7 kcal mol^{-1}, which is compared to the value for the similar reaction of allylborane (5.5 kcal mol^{-1} with the 3-21G basis set).[27] The larger relative energy of the allylsilicate suggests its higher stereoselectivity compared with the allylborane.

The origin of the high reactivity of 1-allylsilacyclobutane towards aldehydes has recently been investigated using *ab initio* MO theory.[22] The reaction of *tetracoordinate* allylsilanes with formaldehyde has been found to proceed *via* a pentacoordinate silicon transition state. The oxygen of aldehyde attacks the apical site of the silicon center, whereas the allyl group departs from the silicon center in the equatorial plane without causing pseudorotation. The calculations have revealed that the barrier height is related to the arrangement of bonds at the silicon center. The electron-accepting ability and acidic hardness of the silicon center correlate well with the calculated activation energies of the reaction; the high reactivity of the allylsilacyclobutane is explained by the large electron-accepting ability and the small acidic hardness of silicon in the small ring.

5.2.7 Synthetic Application of Pentacoordinate Allylic Silicon Species

5.2.7.1 Allylation of Various Carbonyl Compounds α-Hydroxyketones and α-ketocarboxylic acids react with allylic trifluorosilanes and trialkoxysilanes without protection of the hydroxy group in the presence of triethylamine, yielding the corresponding *tert*-homoallyl alcohols in an extremely highly regio- and diastereospecific manner.[28–29]

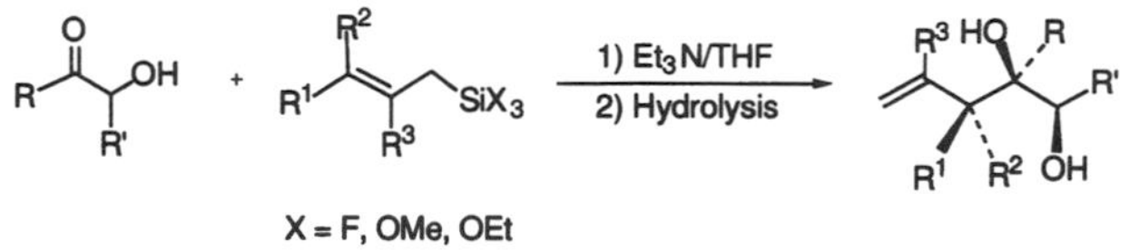

X = F, OMe, OEt

(5.14)

X = F, OMe, OEt

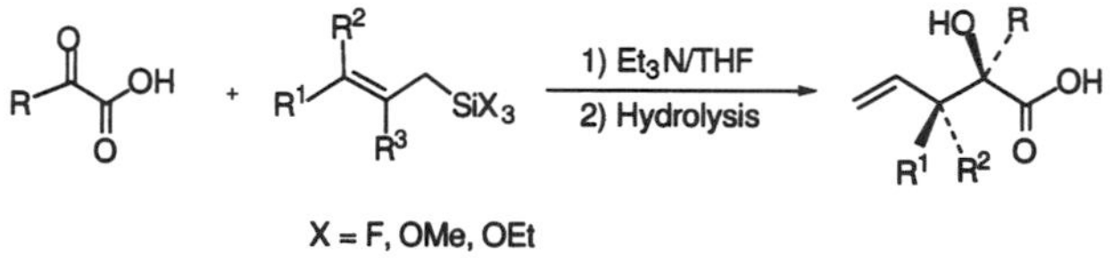

(5.15)

Typically, reactions of benzoin (R = R′ = Ph in Eq. 5.14) with (*E*)- and (*Z*)-crotyltrifluorosilanes gave the corresponding 3-methyl-1,2-diphenylpent-4-ene-1,2-diols with 100% 1,2- and 2,3-diastereoselectivities. These regio- and diastereospecificities, as well as the enhanced reactivity of the α-hydroxy ketones, suggest strongly that the reaction proceeds via the 1,3-bridged cyclohexane-like transition state, as shown below.

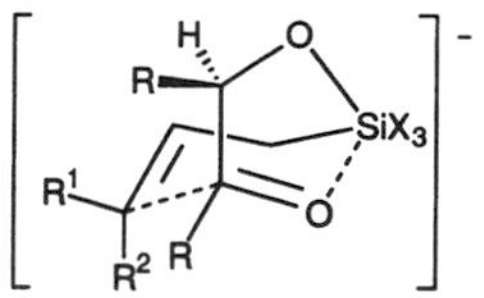

Allyltrifluorosilane/triethylamine systems are also useful for allylation of various α- and β-functional ketones, such as several β-hydroxy- and β-amino-α, β-eneones, and enolizable 1,3-[29] and 1,2-diketones.[30]

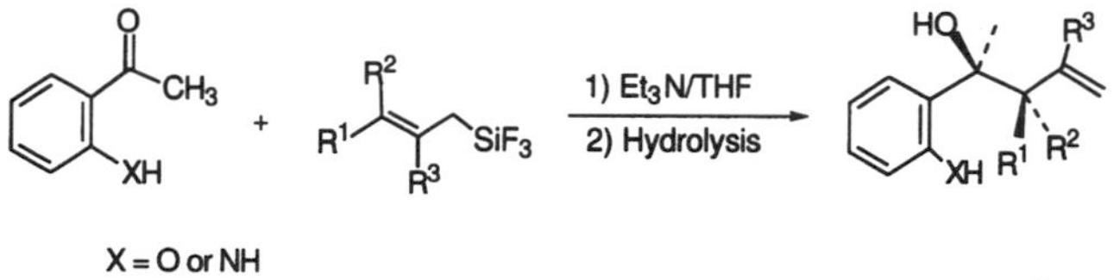

X = O or NH

(5.16)

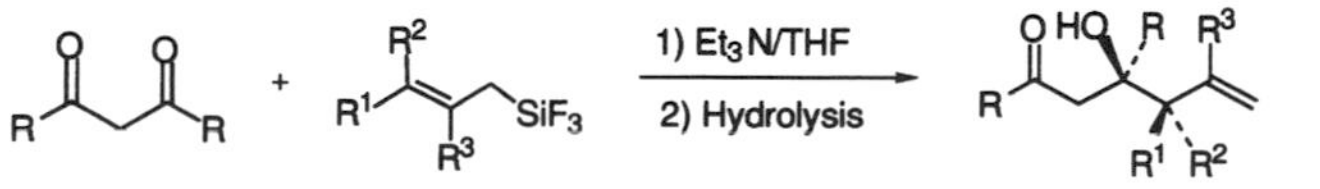

(5.17)

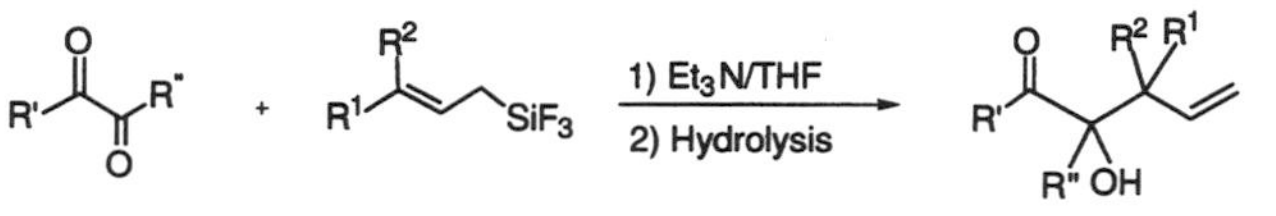

(5.18)

5.2.7.2 Enantioselective Allylation Using Pentacoordinate Allylic Silicon Species Enantioselective synthesis of homoallyl alcohols by reactions of allyl- and crotylsilanes with carbonyl compounds has received much attention in recent years. One of the most common strategies to accomplish asymmetric allylation is the use of chiral organosilicon compounds, where the silicon atom is connected by chiral modifiers. However, the stereoselectivity obtained thus far by this route has been modest.[31] One reason would be the extended antiperiplanar transition states in the Lewis-acid mediated allylation, where the chiral auxiliary would exert little effect on the stereochemical outcome of the reaction. Quite recently, a significant advance has been achieved by Denmark and coworkers in the asymmetric allylation with allyltrichlorosilanes in the presence of chiral Lewis bases **9–11**, but the enantioselectivity of the products is still unsatisfactory.[32] Typically, a reaction of benzaldehyde with allyltrichlorosilane in CH_2Cl_2 in the presence of a chiral phosporamide **9** (**a**; R^1 = Me, R^2 = $-(CH_2)_5-$; 1 equiv.) at −78°C for 6 h gave the corresponding homoallyl alcohol in 85% yield with 63% ee. Increasing the bulk of substituents reduced the rate and/or selectivity of the allylation. The phosphoramide-promoted process is mechanistically related to the reactions of allyltrichlorosilanes in DMF, which were developed by Kobayashi and co-workers. Reactions of (*E*)- and (*Z*)-crotyltrichlorosilanes with benzaldehyde using **9a** as the promoter proceeded smoothly to afford the known homoallylic alcohols with high diastereoselectivity and levels of enantioselectivity similar to those in the allylation.

ArCHO + R^1/R^2-crotyl-$SiCl_3$ —(additive, −78°C; solvent)→ homoallyl alcohol (OH, Ar, R^2, R^1) (5.19)

Optical active additives:

9 **10** **11**

We have found recently that 2-allyl-1,3,2-dioxasilolane-4,5-dicarboxylic esters (**12**) react with carbonyl compounds without any catalyst to give the homoallyl alcohol enantioselectively.[33]

Although chemical yields of the allylation of benzaldehyde with a methylallylsilane (**12a**) did not depend significantly on the solvent, the ee value of the product decreased with increasing polarity of the solvent; toluene (24% ee) > CH_2Cl_2 (18% ee) > THF (16% ee) > DMF (3% ee). These results suggest that

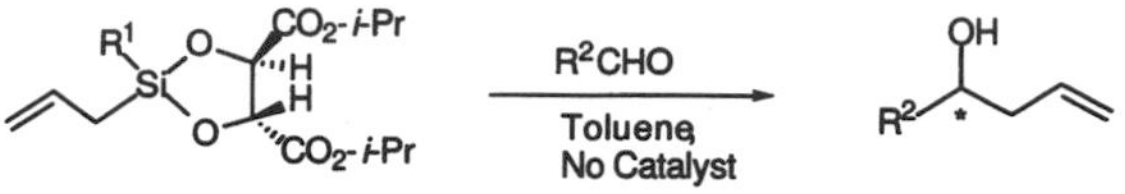

$$\text{12} \xrightarrow[\text{Toluene, No Catalyst}]{R^2CHO} R^2CH(OH)CH_2CH{=}CH_2 \qquad (5.20)$$

12a: R^1 = Me, 12b: R^1 = Ph,
12c: R^1 = *i*-Pr, 12d: R^1 = Allyl
12e: R^1 = Cl, 12f: R^1 = O-*t*-Bu.

the intramolecular coordination of a carbonyl oxygen of the tartrate ester to the silicon atom at the transition state is indispensable for the chiral induction; polar solvents would make the intramolecular coordination less favorable. Further evidence for the intramolecular carbonyl coordination was given by the following observations: (1) neither allyltrimethoxysilane (**13**) nor allylmethyl(1,2-benzenediolato)silane (**14**) react with aldehydes under the same reaction conditions as described above. (2) Allylation of aldehydes with a pentacoordinate allyldifluorosilicate (**15**), which was prepared by the reaction of allyltrifluorosilane and a tartrate ester, occurred at −20°C, but with poor enantioselectivity (< 1% ee).

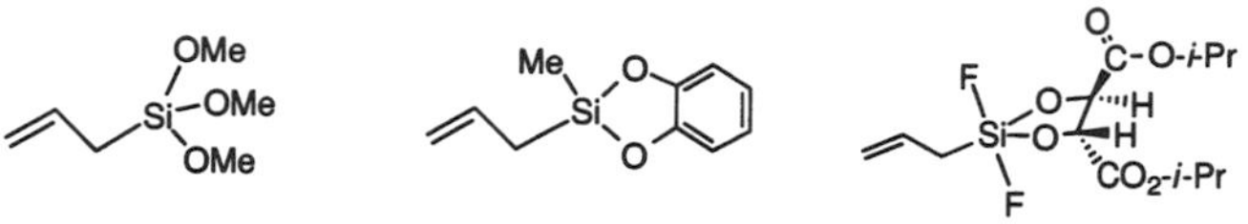

13 **14** **15**

The remarkable improvement of the enantioselectivity of allylation using tetracoordinate silicon species having a tartrate ligand could be due to the following reason. Thus, the pentacoordination of the silicon atom not only enhances the nucleophilicity of the γ-carbon of the allylsilane due to σ–π conjugation, but also brings the chiral centers of the tartrate ester to the proximity of the reaction site to result in the increased enantioselectivity. The asymmetric induction achieved by using **12a–d** is consistent with the formation of the major product via a transition state T, as shown in Scheme 5.2. During allylation of an aldehyde with a pentacoordinate allylsilicon compound, there are two possible directions for the attack of the aldehyde to the trigonal bipyramidal silicon P: paths A and B, as shown in Scheme 5.2. It can be reasonably assumed that path A is favored over path B, because in the latter, the repulsive interaction exists between the aldehydic oxygen atom and the ester group on the β-face, which is free from coordination to silicon. Since in **12e** and **12f**, pentacoordinate silicon species having the electronegative chlorine and alkoxy groups at the axial position (P′) are formed preferably, the opposite chirality is induced in the products via path A′.

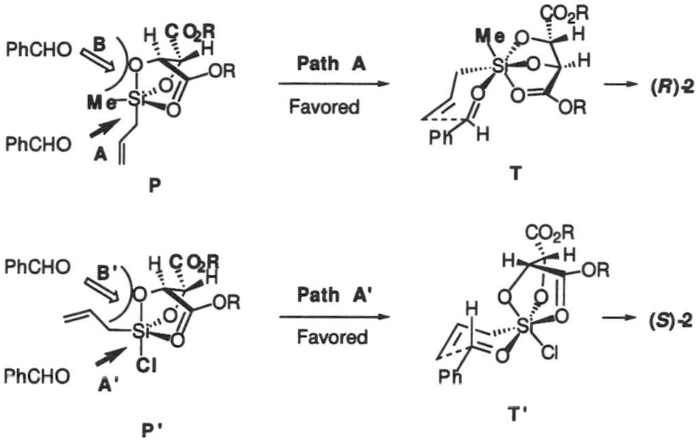

Scheme 5.2

At the almost same time as our report, Wang and co-workers[34] reported a similar study. Whereas they proposed that there was the coordination of an amine to the silicon atom in the transition state, such coordination is not required, since the present allylation takes place in the absence of amines.

5.3 UNCATALYZED ALDOL REACTIONS WITH SILYL ENOL ETHERS

Aldol addition reaction of trialkylsilyl enol ethers with aldehydes is well known as a powerful and selective C—C bond-forming reaction.[35] However, reactions of enoxysilanes derived from ketones, acids or esters with carbonyl compounds proceed usually in the presence of the stoichiometric or a catalytic amount of a Lewis acid, fluoride ion, high pressure, elevated temperatures, etc., as an activator. The fluoride-ion-catalyzed aldol reaction of enol silyl ethers has been considered usually to proceed via a nonchelated extended transition state.[36] As an analogy with the allylation of aldehydes with pentacoordinate allylic silicon species, modification of the substituents at silicon may change the transition state of aldol reaction with silyl enol ethers from the extended to the cyclic transition state, which would regulate the stereochemistry of the aldol addition.

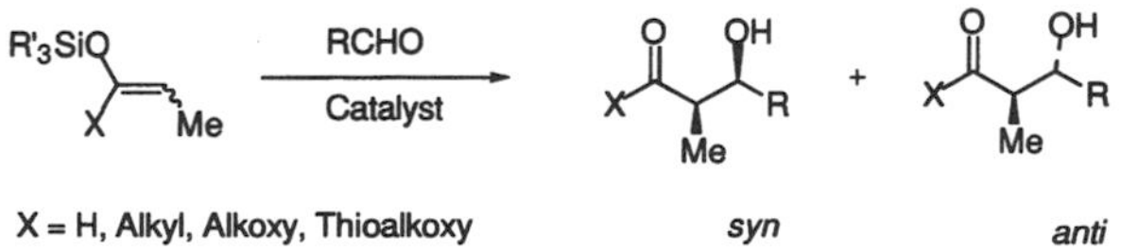

(5.21)

Recently, enoxysilacyclobutanes derived from esters and thiol esters have been found by the groups of Denmark[37] and Myers,[38] to react with various aldehydes without catalysts. Thus, whereas the reaction of enoxysilane **16** with benzaldehyde at 100°C showed absolutely no sign of reaction after 66 h, enoxysilane **17** reacted quite slowly under the same conditions ($t_{1/2} = 800$ min). In contrast to ketone-derived enoxysilacyclobutanes, the silacyclobutyl *O,O*-, *O,S*- and *O,N*-ketene acetals **18**–**20** were extremely reactive towards aldehydes.

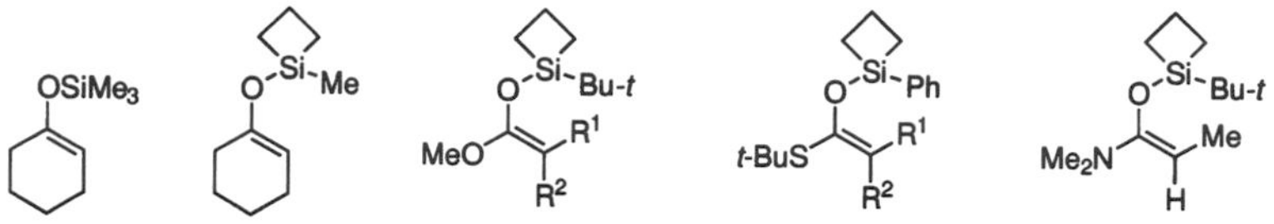

16 17 18 19 20

18: R^1 = H, R^2 = Me; R^1 = Me, R^2 = H

19: R^1 = H, R^2 = Me; R^1 = Me, R^2 = H

The aldol reaction of silacyclobutyl *O,O*-ketene acetal **18** (*E*/*Z* = 95/5) with benzaldehyde in $CDCl_3$ gave the corresponding alcohols after desilylation with HF/THF with *syn*/*anti* ratio of 95/5, while the pure *Z*-isomer of **18** reacted sluggishly with opposite weak *anti* selectivity (*syn*/*anti* = 42/58). The results indicate that the reaction is highly diastereoselective, but the *E*-isomer reacts nearly 600 times faster than the *Z*-isomer.

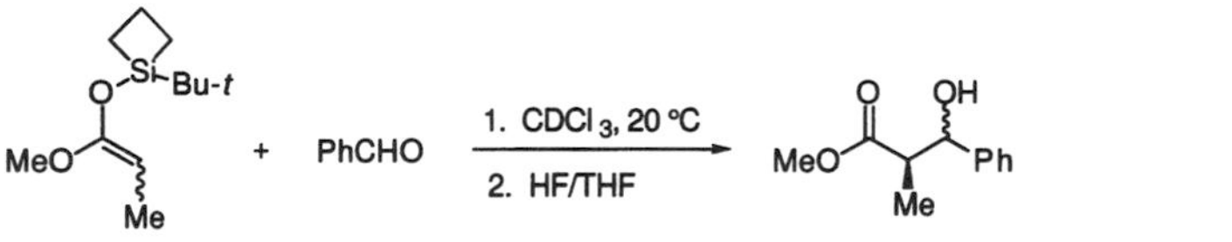

+ PhCHO → (1. $CDCl_3$, 20 °C; 2. HF/THF)   (5.22)

E/Z = 95/5   94% yield (syn/anti = 95/5)

From double-label crossover experiments, this thermal aldol reaction has been concluded to proceed via a cyclic transition state involving pentacoordinate silicon; it is suggested by modeling studies that the boat transition state is preferred over the chair form, which explains the diastereoselectivity. Whereas no spectroscopic evidence for complexation of the silacyclobutanes with aldehydes was obtained, the high reactivity of enol silacyclobutanes is ascribed to the strain release at the pentacoordinate silicon transition state.

Kobayashi and co-workers have recently reported that novel silyl enolates, prepared *in situ* from ketones and dimethylsilyl ditriflate in the presence of a tertiary amine, react smoothly with not only aldehydes, but acetals and α, β-unsaturated ketones without catalysts under mild conditions, to give the corresponding adducts with high diastereselectivities.[39] Typically, a reaction of benzaldehyde with the silyl enol ether prepared from diethyl ketone and triethylamine gave the corresponding adduct in 78% yield with the *syn*/*anti* ratio of 91/9.

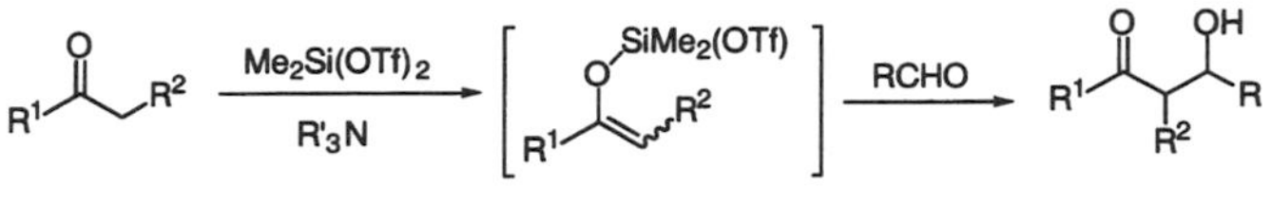

(5.23)

In these silyl enol ethers, the Lewis acidity on silicon is enhanced, due to the strong electron-withdrawing substitutent. This makes the aldol reactions via cyclic transition states involving pentacoordinate silicon more likely.

Michael addition promoted by CsF in the presence of tetraalkoxysilane may also proceed via a pentacoordinate silicon transition state, since the silyl enol ethers having highly electron-accepting trialkoxysilyl groups are expected to be produced during the reaction.[40]

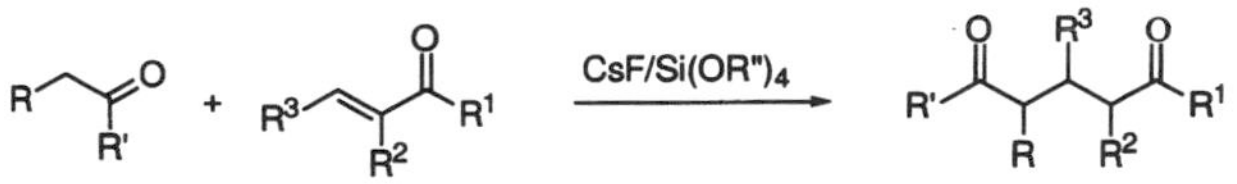

(5.24)

Very recently, Denmark and co-workers have reported interesting aldol reactions with trichlorosilyl enolates, which are isolated as pure substances by distillation.[41]

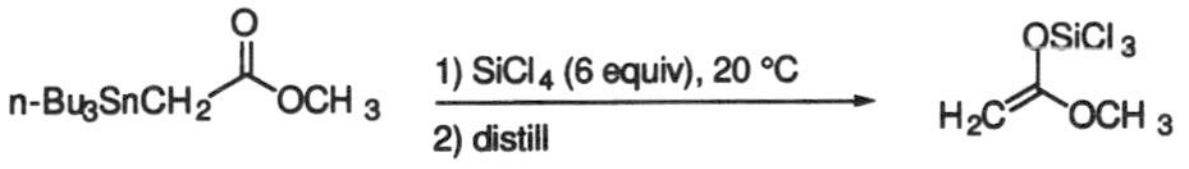

(5.25)

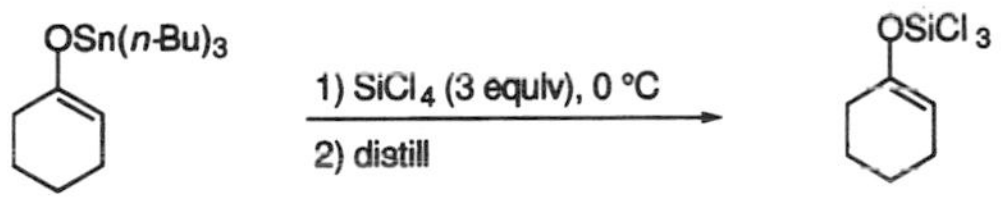

(5.26)

The ketene acetal **21** reacts spontaneously with a number of aldehydes at −80°C without a promoter to give the corresponding aldol addition products in good yields. The reactions of trichlorosilyl enolates with aldehydes are greatly promoted by 0.1 equivalent HMPA. When a chiral phosphoramide such as **10** is used as a promoter, the reaction takes place enantioselectively, but the enantioselectivity is usually low because of the competitive background reaction. However, the reactions of **22** with benzaldehyde and (*E*)-cinnamaldehyde occur with quite high *anti* selectivity (65–99/1) and enantioselectivity (88–93% ee). A chairlike transition state having a hexacoordinate silicate is suggested for the aldol reaction with trichlorosilyl enolates.

$$H_2C{=}C(OSiCl_3)OCH_3\ (\mathbf{21}) + PhCHO \xrightarrow[\text{2) aq. NaHCO}_3]{\text{1) CH}_2\text{Cl}_2,\ 0\,^\circ\text{C}} PhCH(OH)CH_2CO_2CH_3 \quad \text{98 \% yield} \tag{5.27}$$

$$\mathbf{22} + PhCHO \xrightarrow[\text{2) aq.NaHCO}_3]{\text{1) CH}_2\text{Cl}_2,\ (S,S)\text{-10, (0.1 mol), -78}\,^\circ\text{C}} \text{2-(hydroxy(phenyl)methyl)cyclohexanone} \quad \text{95 \% yield, 93 \%ee} \tag{5.28}$$

5.4 REDUCTION OF CARBONYL COMPOUNDS WITH HYPERCOORDINATE HYDRIDOSILICATES

Since the literature prior to 1992 for this topic has been thoroughly reviewed by Corriu's group,[2–4] our discussion will concentrate on the mechanistic aspects and more recent studies.

A number of studies have been reported on the reduction of carbonyl compounds using tetracoordinate hydridosilanes under nucleophilic catalysis. The following are typical examples:[42–47]

$$PhMe_2SiH + R'R''C{=}O \xrightarrow[\text{HMPA}]{(Et_2N)_3S^+Me_3SiF_2^- \text{ or TBAF}} R'R''CH(OSiMe_2Ph) \tag{5.29}$$

$$(RO)_3SiH + R'R''C{=}O \xrightarrow[\text{2) H}_3\text{O}^+]{\text{1) cat. ROM/Et}_2\text{O}} R'R''CH(OH) \quad M = Li, Na, K \tag{5.30}$$

$$R'R''C{=}O \xrightarrow[\text{2) H}_3\text{O}^+]{\text{1) Cl}_3\text{SiH/o-C}_6\text{H}_4\text{(OLi)}_2\text{/0}\,^\circ\text{C}} R'R''CH(OH) \tag{5.31}$$

Whereas pentacoordinate hydridosilicates are usually postulated as reactive species in the reduction of carbonyl compounds, in some cases pentacoordinate hydridosilicates were isolated and used as reducing reagents without catalysts.[48–50]

In these reductions, Si—H bonds in pentacoordinate hydridosilicates are taken to be more reactive than those in tetracoordinate silanes. In addition, it is anticipated that the pentacoordinate silicon serves as a Lewis acid and can accommodate a carbonyl compound to give a hexacoordinate silicon inter-

$$(RO)_3SiH + ROK \longrightarrow HSi(OR)_4^- K^+ \xrightarrow[2)\ H_3O^+]{1)\ R'R''C=O/THF} R'R''CHOH \tag{5.32}$$

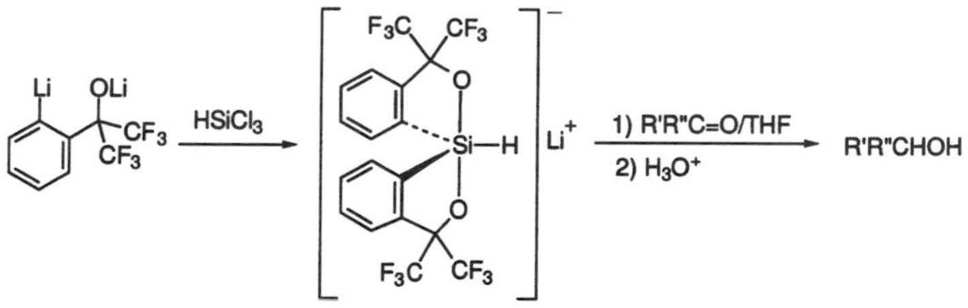

(5.33)

mediate, as suggested in the allylation of aldehydes using pentacoordinate allylic silicon species; however, there is no direct evidence for the carbonyl coordination prior to the hydride transfer.[44,51] A highly nucleophilic hydride intermediate can attack the carbonyl carbon to give the corresponding alcohols after hydrolysis (Scheme 5.3). An excellent linear Hammett plot for the reduction of substituted benzaldehydes with bis(biphenyl-2,2′-diolato)hydridosilicate has been observed with a large positive ρ value (2.3).[44] A significant deuterium isotope effect has been shown in the reduction of acetophenone with $PhMe_2SiD$ in the presence of fluoride ions in HMPA ($k_H/k_D = 1.50$).[51] These results suggest that the hydride transfer occurs at the rate-determining step, whereas the detailed mechanisms may depend on the reagents, solvents, and other conditions. Alternatively, the single-electron-transfer–hydrogen-atom abstraction mechanism has been proposed for the fluoride-ion-catalyzed reduction of carbonyl compounds using phenyldimethylsilane.[52]

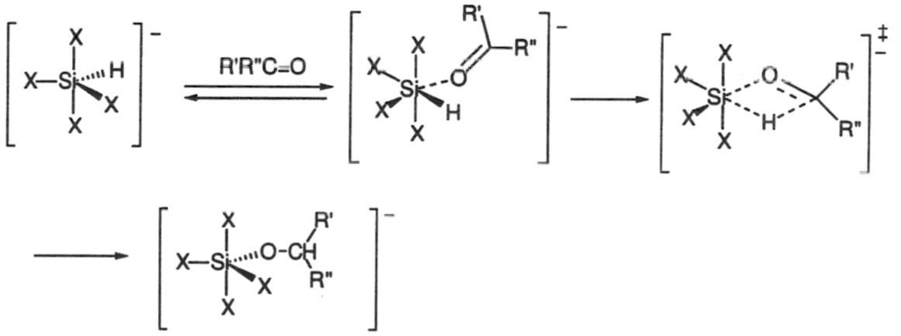

Scheme 5.3

Reduction of α, β-epoxy ketones by trimethoxysilane in the presence of a catalytic amount of lithium methoxide gives the corresponding alcohols.[53] The *anti*/*syn* selectivity has been found to depend significantly on the solvent: *anti* selectivity in ether, whereas *syn* selectivity in HMPA. It is suggested that in the less polar diethyl ether the reaction proceeds via a chelation model as mentioned above, whereas in polar HMPA, the chelation of the ketone is prevented by coordination of HMPA to the silicon, and the reaction proceeds via a Felkin–Ahn-type transition state.

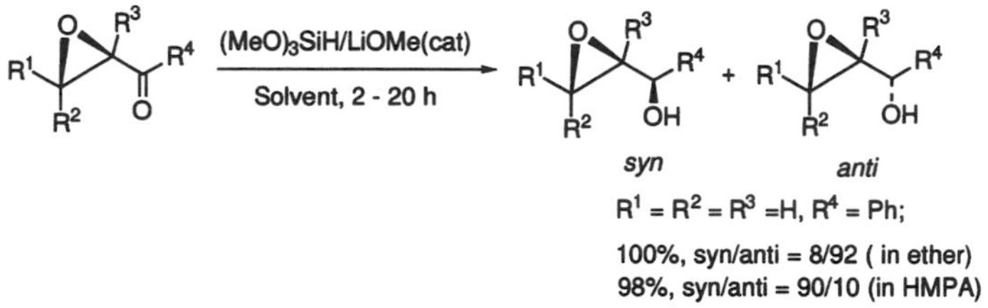

$$\text{(5.34)}$$

(MeO)$_3$SiH/LiOMe(cat)
Solvent, 2 - 20 h

syn *anti*

$R^1 = R^2 = R^3 = H$, $R^4 = Ph$;
100%, syn/anti = 8/92 (in ether)
98%, syn/anti = 90/10 (in HMPA)

Enantioselective reductions using the hydridosilicate strategy have been achieved. The groups of Hosomi,[54] Burk,[55] and Salvadori[56] have reported enantioselective reduction of aromatic ketones using trialkoxysilanes in the presence of chiral alkali metal alkoxides with enantioselectivities from 20 to 80% ee.

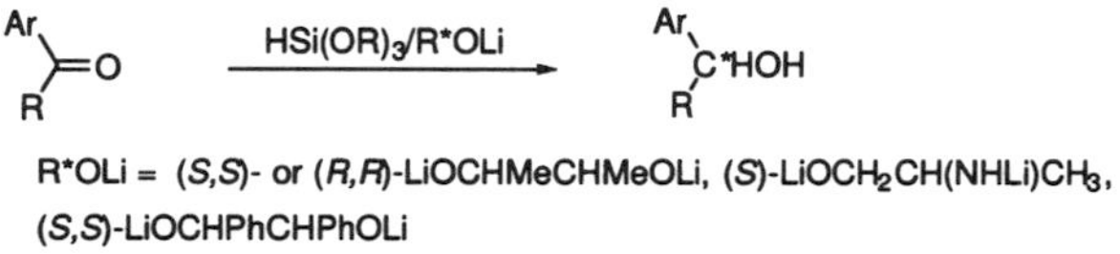

$$\text{Ar(R)C=O} \xrightarrow{\text{HSi(OR)}_3/\text{R*OLi}} \text{Ar(R)C*HOH} \quad (5.35)$$

R*OLi = (*S,S*)- or (*R,R*)-LiOCHMeCHMeOLi, (*S*)-LiOCH$_2$CH(NHLi)CH$_3$, (*S,S*)-LiOCHPhCHPhOLi

Neutral pentacoordinate hydridosilicates have been found to serve also as mild reducing reagents of carboxylic acids,[57] esters,[58] acid chlorides,[59] and CO_2.[60]

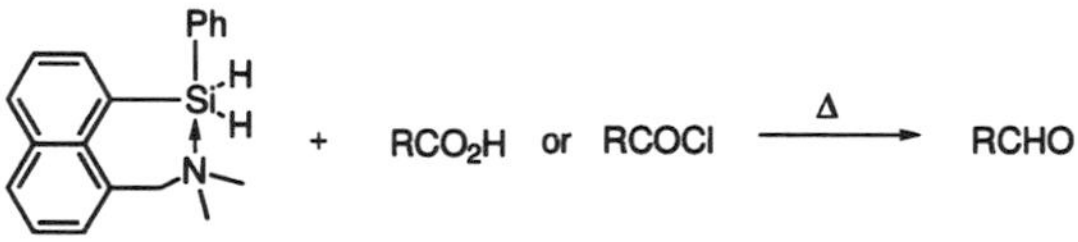

$$\text{+ RCO}_2\text{H or RCOCl} \xrightarrow{\Delta} \text{RCHO} \quad (5.36)$$

Erythro diols are obtained in mild and neutral conditions, through the intramolecular nucleophilic catalyzed hydrosilylation of diketones and hydroxy-ketones with the aminoaryltrihydrosilanes.[61]

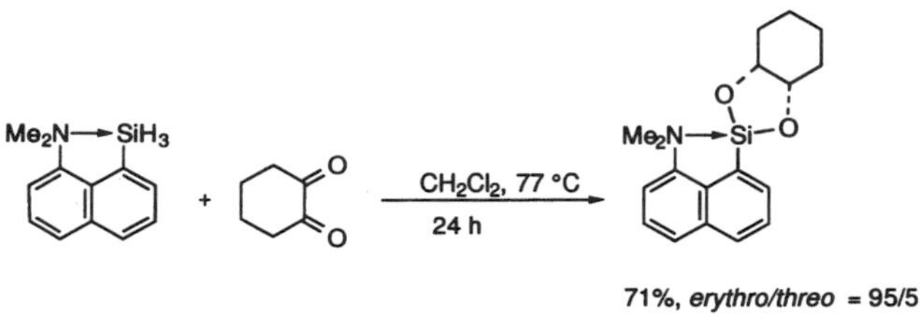

$$\xrightarrow[\text{24 h}]{\text{CH}_2\text{Cl}_2\text{, 77 °C}} \quad (5.37)$$

Me$_2$N→SiH$_3$ + ... → Me$_2$N→Si–O

71%, *erythro/threo* = 95/5

5.5 TRANSITION-METAL-CATALYZED CROSS-COUPLING REACTIONS OF HYPERCOORDINATE SILICATES WITH ORGANIC HALIDES

Homocoupling reactions of various hexacoordinate alkenylpentafluorosilicates in the presence of an equimolar amount of silver(I) and copper(II) salts were first reported by Tamao and Kumada in 1979.[62–63] This field has seen great development since the report of palladium-catalyzed cross-coupling of various organosilanes with alkenyl-, aryl-, and allylhalides promoted by a stoichiometric amount of fluoride ions by Hatanaka and Hiyama in 1988.[64] The following are typical examples of synthetically useful cross-coupling reactions.[65–69]

SiMe3 + 1-iodonaphthalene —TASF, (η^3-AllylPdCl)$_2$ (cat.), HMPA, 50 °C→ 1-vinylnaphthalene 98% (5.38)

n-C_6H_{11}CH=CHSiMe$_2$F + ICH=CHC_6H_{11}-n —TASF, (η^3-AllylPdCl)$_2$ (cat.), THF→ n-C_6H_{11}CH=CHCH=CHC_6H_{11}-n 83% (5.39)

SiMe(OMe)$_2$ + 1-iodonaphthalene —(η^3-AllylPdCl)$_2$/P(OEt)$_3$ (cat.), TBAF, THF, 50 °C→ 1-Np 96% (5.40)

O_2N–C$_6$H$_4$–I + [vinyl-Si(catecholate)$_2$]$^-$ [Et$_3$NH]$^+$ —PdCl$_2$(PhCN)$_2$, P(OEt)$_3$(cat.), Dioxane, reflux→ O_2N–C$_6$H$_4$–CH=CH$_2$ 84% (5.41)

Although the details of the reaction mechanism have not been elucidated, the reactions are believed to involve pentacoordinate organosilicon species as reactive intermediates generated by the attack of a fluoride ion to the tetracoordinate silicon. The electrophilicity of the organic group is thus enhanced, allowing for the transfer of the organic group from silicon to palladium catalyst in the transmetallation step. Among the reactions of a series of (E)-n-$C_6H_{11}CH{=}CHSiMe_{3-n}F_n$ ($RSiMe_{3-n}F_n$) with 1-iodonaphthalene, $RSiMe_2F$ showed the highest reactivity and $RSiMeF_2$ was slightly less reactive.[66] Contrary to expectations, $RSiF_3$ in addition to $RSiMe_3$ was unreactive under the cross-coupling reaction conditions.

The stereochemistry of the cross-coupling reaction of chiral alkylsilanes with aryl triflates was found to be influenced by the reaction temperatures and the solvent.[70]

Me Ph SiF$_3$ + TfO COMe → Me Ph COMe (5.42)

50 °C, THF: nearly complete retention

60 °C, HMPA-THF (1/10): predominantly inversion

Highly γ-selective cross-coupling reactions of allylic trifluorosilanes with aryl, alkenyl, and allyl halides have been achieved using palladium catalyst in the presence of fluoride ions in THF.[71]

SiF$_3$ + I COMe —TBAF, $Pd(PPh_3)_4$, THF / 80 °C, 19 h→ COMe (5.43)

95%

REFERENCES

1. M. Kumada, K. Tamao, J. Yoshida, *J. Organomet. Chem.* **239**, 115 (1982).
2. R. J. P. Corriu, C. Guérin, J. J. E. Moreau, *Top Stereochem.* **15**, 43 (1984).
3. R. J. P. Corriu, J. C. Young, in *The Chemistry of Organic Silicon Compounds*, S. Patai and Z. Rappoport, Eds (John Wiley, Chichester, 1989), Part 2, Chapter 20.
4. C. Chuit, R. J. P. Corriu, C. Reye, J. C. Young, *Chem. Rev.* **93**, 1371 (1993).
5. H. Sakurai, *Pure Appl. Chem.* **54**, 1 (1982); H. Sakurai, *Pure Appl. Chem.* **57**, 1759 (1985); H. Sakurai, *Synlett*, 1 (1989); A. Hosomi, *Acc. Chem. Res.*, **21**, 200 (1988); D. Schinzer, *Synthesis*, 263 (1988).
6. A. Hosomi, A. Shirahata, H. Sakurai, *Tetrahedron Lett.* **33**, 3043 (1978); J.-P. Pillot, J. Dunogues, R. Calas, *Tetrahedron Lett.* **22**, 1871 (1976).
7. G. Majetich, A. M. Casares, D. Chapman, M. Behnke, *Tetrahedron Lett.* **29**, 1909 (1983); G. Majetich, A. M. Casares, D. Chapman, M. Behnke, *J. Org. Chem.* **51**, 1745 (1986); G. Majetich, R. W. Desmond, J. J. Soria, *J. Org. Chem.* **51**, 1753 (1986).
8. R. R. Holmes, *Chem. Rev.* **90**, 17 (1990); R. Tacke, J. Becht, A. Lopez-Mras, J. Sperlich, *J. Organomet. Chem.* **446**, 1 (1993); E. C. Lingafelter, *Coord. Chem. Rev.* **1**, 151 (1966); G. A. Miller and E. D. Schlemper, *Inorg. Chim. Acta.* **30**, 131 (1978); U. Teutsch and H. Schmidtke, *J. Chem. Phys.* **84**, 6034 (1986).
9. M. Kira, M. Kobayashi, H. Sakurai, unpublished work.

10. M. Kira, M. Kobayashi, H. Sakurai, *Tetrahedron Lett.* **28**, 4081 (1987).
11. M. Kira, T. Hino, H. Sakurai, *Tetrahedron Lett.* **30**, 1099 (1989).
12. M. Kira, K. Sato, H. Sakurai, *J. Am. Chem. Soc.* **112**, 257 (1990).
13. A. Hosomi, S. Kohra, Y. Tominaga, *J. Chem. Soc., Chem. Commun.* 1517 (1987).
14. M. Kira, K. Sato, H. Sakurai, *J. Am. Chem. Soc.* **110**, 4599 (1988).
15. A. Hosomi, S. Kohra, Y. Tominaga, *J. Org. Chem.* **55**, 2415 (1990).
16. M. Kira, L-C. Zhang, C. Kabuto, H. Sakurai, *Organometallics* **15**, 5335 (1996).
17. S. Kobayashi, K. Nishio, *Tetrahedron Lett.* **34**, 3453 (1993).
18. S. Kobayashi, K. Nishio, *J. Org. Chem.* **59**, 6620 (1994).
19. S. Kobayashi, K. Nishio, M. Yasuda, *Synlett* 153 (1996).
20. S. E. Denmark, D. M. Coe, N. E. Pratt, and B. D. Griedel, *J. Org. Chem.* **59**, 6161 (1994).
21. K. Matsumoto, K. Oshima, K. Utimoto, *J. Org. Chem.* **59**, 7152 (1994).
22. K. Omoto, Y. Sawada, H. Fujimoto, *J. Am. Chem. Soc.* **118**, 1750 (1996).
23. T. Hayashi, Y. Kabeta, I. Hamachi, M. Kumada, *Tetrahedron Lett.* **24**, 2865 (1983).
24. R. W. Hoffman, H. J. Zeiss, *J. Org. Chem.* **46**, 1309 (1981).
25. T. Hayashi, Y. Matsuoka, T. Kiyoi, Y. Ito, S. Kohra, Y. Tominaga, A. Hosomi, *Tetrahedron Lett.* **29**, 5667 (1988).
26. M. Kira, K. Sato, H. Sakurai, M. Hada, M. Izawa, J. Ushio, *Chem. Lett.* 387 (1991).
27. Y. Li, K. N. Houk, *J. Am. Chem. Soc.* **111**, 1236 (1989).
28. K. Sato, M. Kira, H. Sakurai, *J. Am. Chem. Soc.* **111**, 6429 (1989).
29. M. Kira, K. Sato, K. Sekimoto, R. Gewald, H. Sakurai, *Chem. Lett.* 281 (1995).
30. R. Gewald, M. Kira, H. Sakurai, *Synthesis* 111 (1996).
31. E. Torres, G. L. Larson, G. J. Mcgarvey, *Tetrahedron Lett.* **29**, 1355 (1988); S. J. Hathaway, L. A. Paquette, *J. Org. Chem.* **48**, 3351 (1983); L. Cappi, A. Mordini, M. Taddei, *Tetrahedron Lett.* **28**, 969 (1987); T. H. Chan, D. Wang, *Tetrahedron Lett.* **30**, 3041 (1989).
32. S. E. Denmark, D. M. Coe, N. E. Pratt, B. D. Griedel, *J. Org. Chem.* **59**, 6161 (1994).
33. L-C. Zhang, M. Kira, H. Sakurai, *Chem. Lett.* 129 (1997).
34. Z. Wang, D. Wang, X. Sui, *J. Chem. Soc. Chem., Commun.* **2261** (1996).
35. C. Gennari, in *Comprehensive Organic Synthesis, Vol. 2, Additions to C-Xπ Bonds Part 2*, C. H. Heathcock, Ed. (Pergamon Press, Oxford, 1991), p. 629; D.A. Evans, J. V. Nelson, T. R. Taber, in *Topics in Stereochemistry*, E. L. Eliel, S. H. Wilen, Eds (Wiley Interscience, New York, 1983), Vol. 13, p. 1; C. H. Heathcock, in *Asymmetric Synthesis*, J. D. Morrison, Ed. (Academic Press, New York, 1984), Vol. 3, Chapter 2; T. Mukaiyama, *Org. React.* **28**, 203 (1982).
36. E. Nakamura, M. Shimizu, I. Kuwajima, J. Sakata, K. Yokoyama, R. Noyori, *J. Org. Chem.* **48**, 932 (1983); I. Kuwajima, E. Nakamura, *Acc. Chem. Res.* **18**, 181 (1985); R. Noyori, I. Nishida, J. Sakata, *Tetrahedron Lett.* **21**, 2085 (1980); R. Noyori, I. Nishida, J. Sakata, *J. Am. Chem. Soc.* **105**, 1598 (1983); E. Nakamura, S. Yamago, D. Machii, I. Kuwajima, *Tetrahedron Lett.* **29**, 2207 (1988); C. Chuit,

R. J. P. Corriu, C. Reye, *J. Organometal. Chem.* **358**, 57 (1988); J. Boyer, R. J. P. Corriu, R. Perz, C. Reye, *J. Organometal. Chem.* **184**, 157 (1980).

37. S. E. Denmark, B. D. Griedel, D. M. Coe, *J. Org. Chem.* **58**, 988 (1993); S. E. Denmark, B. D. Griedel, D. M. Coe, M. E. Schnute, *J. Am. Chem. Soc.* **116**, 7026 (1994).
38. A. G. Myers, S. E. Kephart, H. Chen, *J. Am. Chem. Soc.* **114**, 7922 (1992).
39. S. Kobayashi, K. Nishio, *J. Org. Chem.* **58**, 2647 (1993).
40. R. J. P. Corriu, R. Perz, *Tetrahedron Lett.* **26**, 1311 (1985); C. Chuit, R. J. P. Corriu, C. Reye, *Tetrahedron Lett.* **23**, 5531 (1982); J. Boyer, R. J. P. Corriu, R. Perz, C. Reye, *Tetrahedron* **39**,117 (1983); T. V. RajanBabu, *J. Org. Chem.* **49**, 2083 (1984); T. V. Rajanbabu, D. W. Ovenall, G. S. Reddy, D. Y. Sogah, *J. Am. Chem. Soc.* **110**, 5841 (1988). W. J. Brittain, *J. Am. Chem. Soc.* **110**, 7440 (1988).
41. S. E. Denmark, S. B. D. Winter, X. Su, K-T. Wong, *J. Am. Chem. Soc.* **118**, 7404 (1996).
42. M. Fujita, T. Hiyama, *J. Am. Chem. Soc.* **106**, 4629 (1984); M. Fujita, T. Hiyama, *J. Am. Chem. Soc.* **107**, 8294 (1985).
43. J. L. Brefort, R. Corriu, C. Guérin, and B. Henner, *J. Organomet. Chem.* **370**, 9 (1989); R. J. P. Corriu, C. Guérin, B. Henner, Q. Wang, *J. Organomet. Chem.* **365**, C7 (1989); B. Becker, R. J. P. Corriu, C. Guérin, B. Henner, Q. Wang, *J. Organomet. Chem.* **368**, C25 (1989).
44. M. Kira, K. Sato, H. Sakurai, *J. Org. Chem.* **52**, 948 (1987).
45. A. Hosomi, H. Hayashida, S. Kohra, Y. Tominaga, *J. Chem. Soc., Chem. Commun.* **18**, 1411 (1986); *idem, Tetrahedron Lett.* **29**, 959 (1988).
46. Yu. Goldberg, E. Abele, M. Shymanska, E. Lukevics, *J. Organomet. Chem.* **372**, C9 (1989).
47. S. Kobayashi, M. Yasuda, I. Hachiya, *Chem. Lett.* 407 (1996).
48. M. Kira, K. Sato, H. Sakurai, *Chem. Lett.* 2243 (1987).
49. S. K. Chopra, J. C. Martin, *J. Am. Chem. Soc.* **112**, 5342 (1990).
50. B. Becker, R. J. P. Corriu, C. Guérin, B. Henner, Q. Wang, *J. Organomet. Chem.* **359**, C33 (1989).
51. M. Fujita, T. Hiyama, *Tetrahedron Lett.* **28**, 2263 (1987).
52. D. Yang, D. D. Tanner, *J. Org. Chem.* **51**, 2267 (1986).
53. M. Hojo, A. Fujii, C. Murakami, H. Aihara, A. Hosomi, *Tetrahedron Lett.* **36**, 571 (1995).
54. S. Kohra, Y. Tominaga, Y. Matsuoka, H. Hayashida, A. Hosomi, *Tetrahedron Lett.* **29**, 89 (1988); Y. Tominaga, Y. Matsuoka, H. Hayashida, S. Kohra, A. Hosomi, *Tetrahedron Lett.* **29**, 5771 (1988).
55. M. J. Burk, J. E. Feaster, *Tetrahedron Lett.* **33**, 2099 (1987).
56. D. Pini, A. Iuliano, P. Salvadori, *Tetrahedron, Asymmetry* **3**, 693 (1992).
57. R. J. P. Corriu, G. F. Lanneau, M. Perrot, *Tetrahedron Lett.* **28**, 3941 (1987); R. J. P. Corriu, G. F. Lanneau, M. Perrot, *Tetrahedron Lett.* **29**, 1271 (1988).
58. C. Chuit, R. J. P. Corriu, R. Perz, C. Reye, *Synthesis* 981 (1982).
59. R. J. P. Corriu, G. F. Lanneau, M. Perrot-Petta, V. D. Mehta, *Tetrahedron Lett.* **31**, 2585 (1990).

60. P. Arya, J. Boyer, R. J. P. Corriu, G. F. Lanneau, M. Perrot, *J. Organomet. Chem.* **346**, C11 (1988).
61. R. J. P. Corriu, G. F. Lanneau, Z. Yu, *Tetrahedron* **49**, 9019 (1993).
62. J. Yoshida, K. Tamao, T. Kakui, M. Kumada, *Tetrahedron Lett.* 1141 (1979).
63. K. Tamao, H. Matsumoto, T. Kakui, M. Kumada, *Tetrahedron Lett.* 1137 (1979).
64. Y. Hatanaka, T. Hiyama, *Tetrahedron Lett.* **29**, 97 (1988).
65. Y. Hatanaka, T. Hiyama, *J. Org. Chem.* **53**, 918 (1988).
66. Y. Hatanaka, T. Hiyama, *J. Org. Chem.* **54**, 268 (1989).
67. Y. Hatanaka, K. Mabui, T. Hiyama, *Tetrahedron Lett.* **30**, 2403 (1989).
68. A. Hosomi, S. Kohra, Y. Tominaga, *Chem. Pharm. Bull.* **36**, 4622 (1988).
69. K. Tamao, K. Kobayashi, Y. Ito, *Tetrahedron Lett.* **30**, 6051 (1989).
70. Y. Hatanaka, T. Hiyama, *J. Am. Chem. Soc.* **112**, 7793 (1990).
71. Y. Hatanaka, Y. Ebina, T. Hiyama, *Tetrahedron Lett.* **31**, 2719 (1990); Y. Hatanaka, Y. Ebina, T. Hiyama, *J. Am. Chem. Soc.* **113**, 7075 (1991).

6 Synthesis of Hypervalent Organophosphorus Compounds and Their Reactions

TAKAYUKI KAWASHIMA
Department of Chemistry, Graduate School of Science, The University of Tokyo, Tokyo, Japan

6.1 INTRODUCTION

Recently, much attention has been paid to the chemistry of hypervalent compounds, which have electronic structures with formal valence shell electrons over octet, because of interest in their unique structures and reactivity. The organophosphorus compounds usually have a trivalent or pentavalent phosphorus, because phosphorus is classified as a Group 15 element of the periodic table. Therefore, pentacoordinate (hypervalent) organophosphorus compounds are neutral chemical species, and hence pentacovalent inorganic phosphorus compounds such as PF_5 has been known as stable compounds since the 19th century.

In 1949, Wittig and co-workers succeeded in the synthesis of pentaphenylphosphorane as the first example of a pentacovalent organophosphorus compound having five carbon ligands and could realize the long-standing dream of Staudinger.[1]

It is well-known that the so-called Wittig reaction was discovered during an investigation to attempt the synthesis of alkyl derivatives of phosphorane (Scheme 6.1). However, the original form of the Wittig reaction was reported by Staudinger in 1920 (Scheme 6.2).[2]

At that time, nobody noticed the direct relationship between hypervalent chemistry and the Wittig reaction, because a betaine was considered as an intermediate of the reaction.

However, hypervalent organophosphorus chemistry has been independently developed by a new finding that the Ramirez reaction affords oxyphosphoranes by oxidative addition of α-diketones with trivalent phosphines.[3]

Chemistry of Hypervalent Compounds, edited by Kin-ya Akiba.
ISBN 0-471-24019-2

$$Ph_4P^+ I^- + PhLi \longrightarrow Ph_5P$$
$$Ph_3P^+CH_3 I^- + PhLi \longrightarrow Ph_3P{=}CH_2 + PhH$$
$$Ph_3P{=}CH_2 + Ph_2C{=}O \longrightarrow Ph_2C{=}CH_2 + Ph_3P{=}O$$

Scheme 6.1

$$Ph_3P{=}CPh_2 + PhN{=}C{=}O \longrightarrow Ph_2C{=}C{=}NPh + Ph_3P{=}O$$

Scheme 6.2

Westheimer's proposal that a pentacoordinate species is involved as an intermediate in the hydrolysis of phosphoric esters,[4] and confirmation of the existence of stereomutation (pseudorotation) came by the investigation of the stereochemistry of the reaction and dynamic studies by nuclear magnetic resonance (NMR) spectroscopy.[4,5]

Thanks to the large contributions of Wittig and many other synthetic chemists, the Wittig reaction has been widely utilized as an excellent method for the synthesis of olefins and as a key step of the synthesis of natural products.[6] Recently, many mechanistic studies of the Wittig reaction have been carried out for the purpose of elucidating the origin of its stereochemistry. Vedejs and then Maryanoff found by low-temperature ^{31}P NMR spectroscopy that pentacoordinate 1,2-oxaphosphetanes are only detectable intermediates and decompose to give stereospecifically the corresponding olefins in the Wittig reaction of nonstabilized phosphorus ylides with aldehydes (Scheme 6.3), elucidating that the intermediates of the Wittig reaction are hypervalent species.[7]

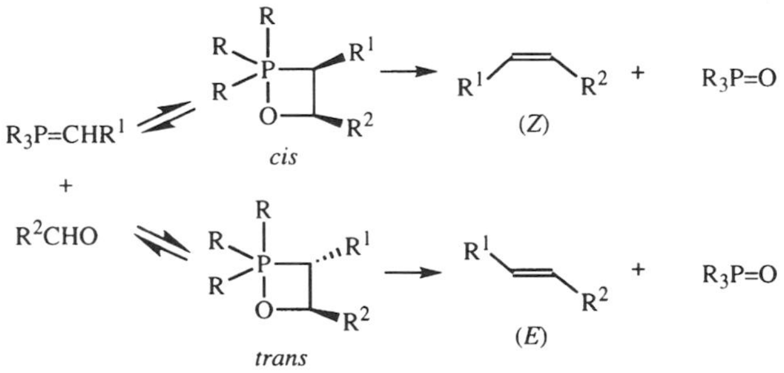

$R_3P{=}CHR^1$ + R^2CHO ⇌ *cis* oxaphosphetane → $R^1CH{=}CHR^2$ (*Z*) + $R_3P{=}O$

$R_3P{=}CHR^1$ + R^2CHO ⇌ *trans* oxaphosphetane → $R^1CH{=}CHR^2$ (*E*) + $R_3P{=}O$

Scheme 6.3

In sharp contrast to the chemistry of hypervalent pentacoordinate compounds, that of hexacoordinate and heptacoordinate hypervalent phosphorus compounds has not been sufficiently investigated so far and the reports are quite limited. However, very recently, two reviews on hexacoordinate phosphorus compounds were reported, in which some compounds described here are involved.[8]

Most of the reviews on hypervalent organophosphorus compounds have been reported before 1979,[9] so the author wishes to describe mainly the papers published after 1980, including the author's work on hypervalent 1,2-oxaphosphetanes and 1,2-azaphosphetidines.

6.2 SYNTHESIS OF HYPERVALENT ORGANOPHOSPHORUS COMPOUNDS

6.2.1 Pentacoordinate Organophosphorus Compounds

6.2.1.1 Monophosphorus Compounds

6.2.1.1.1 By Nucleophilic Addition to Phosphonium Salts As described above, in 1949, Wittig was able to synthesize pentaphenylphosphorane by the reaction of tetraphenylphosphonium iodide with phenyllithium. Attempted synthesis of pentamethylphosphorane by this method was unsuccessful and resulted in the formation of methylenetrimethylphosphorane. The reaction using methyltriphenylphosphonium iodide also undergoes deprotonation to give methylenetriphenylphosphorane, which was reacted with benzophenone to afford 1,1-diphenylethylene and triphenylphosphine oxide (Scheme 6.1).

This method using nucleophiles and phosphonium salts can be utilized to synthesize various kinds of pentacovalent compounds. As shown in Scheme 6.4, spirophosphoranes **1a** and **1b** with two 2,2′-biphenylenes were synthesized as stable compounds by the corresponding spirophosphonium salts **2** with nucleophiles. Even the reaction with $LiAlH_4$ and $NaBH_4$ leads to a stable P–H compounds, for example, **1c**.[9a,10]

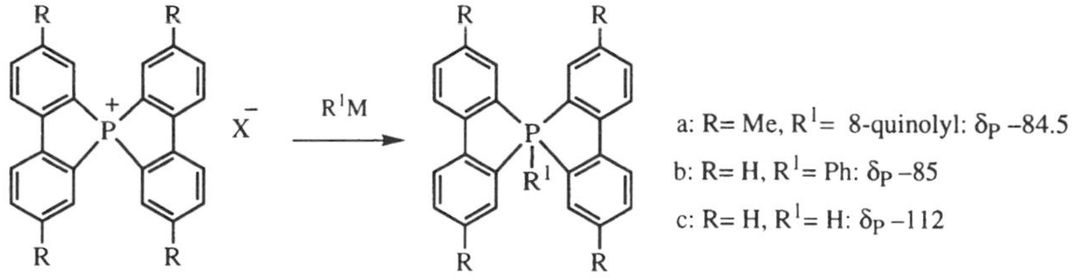

R R R R P+ X− R^1M → P R^1 R R R R

2 **1**

a: R= Me, R^1= 8-quinolyl: δ_P −84.5

b: R= H, R^1= Ph: δ_P −85

c: R= H, R^1= H: δ_P −112

Scheme 6.4

Similarly, Granoth and Martin reported the synthesis of hydridophosphorane **3** from spirophosphonium salt **4**, which was prepared by the reaction of hydroxyphoshorane **5** with trifluoromethanesulfonic acid, as shown in Scheme 6.5.[11] The reaction of **4** with Grignard reagents nicely gave the corresponding phosphoranes **7**.[11b] Bidentate ligands having trifluoromethyl groups instead of methyl groups are the so-called Martin ligands, which have been utilized to stabilize various kinds of hypervalent species. Some of them are summarized in his reviews.[11d,e]

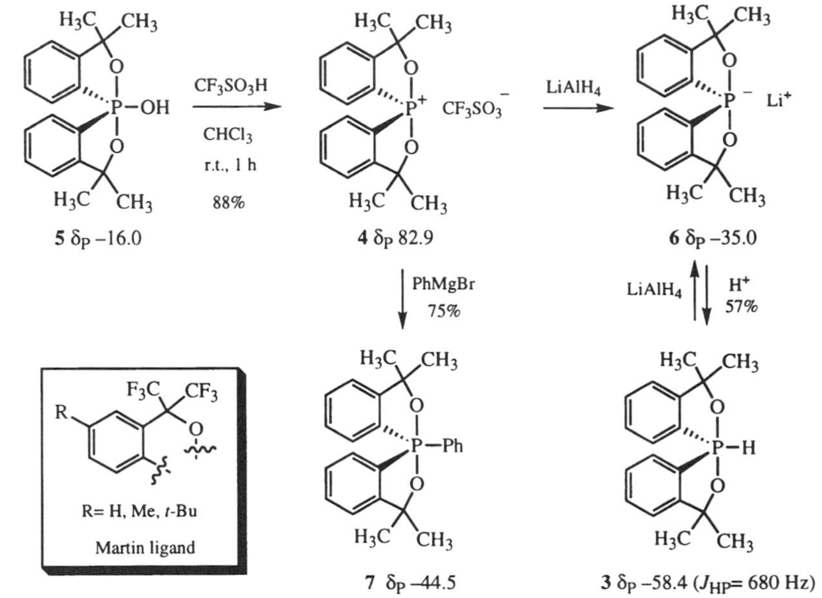

Scheme 6.5

6.2.1.1.2 By Ligand Exchange Although the preparation of pentaoxyphosphorane $(PhO)_5P$ using PCl_5 and phenol was described for the first time in 1927 and was confirmed by Russian workers in 1959,[12] in 1968, a reinvestigation by Ramirez and co-workers showed conclusively that the earlier method did not give $(PhO)_5P$. The compound was prepared by the reaction of phenol with PCl_5 in the presence of collidine at 0°C as shown in Scheme 6.6.[13]

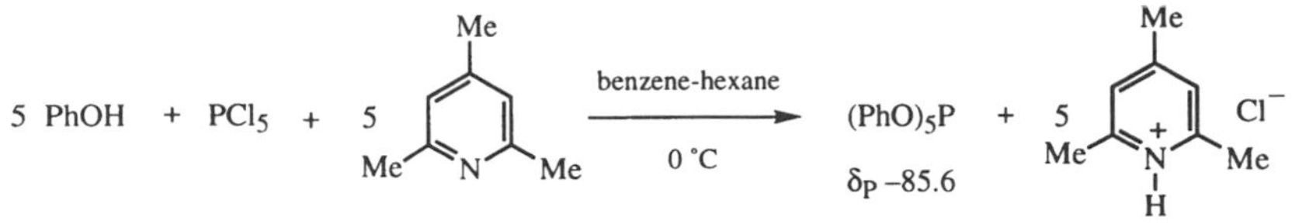

Scheme 6.6

The reaction of $(PhO)_5P$ with one mole of catechol in CH_2Cl_2 at 25°C gave catecholtriphenoxyphosphorane **8** and the reaction with second catechol afforded the corresponding spirophosphorane **9** (Scheme 6.7).[5]

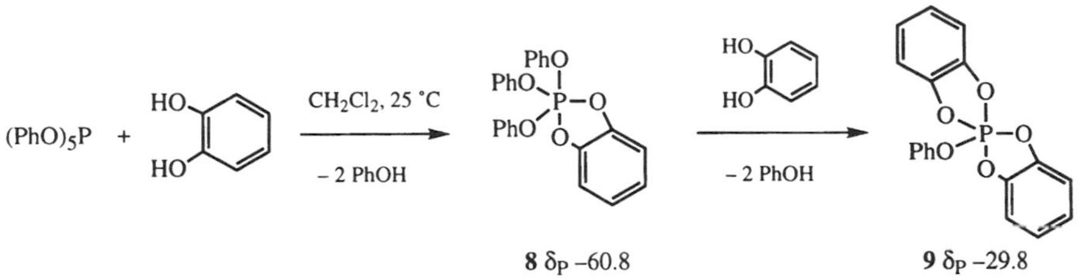

8 δ_P −60.8 **9** δ_P −29.8

Scheme 6.7

6.2.1.1.3 By Oxidative Addition In the course of extensive research on oxyphosphoranes, Ramirez found the so-called Ramirez reaction involving oxidative addition of the lone pair of the phosphorus to α-diketones or *o*-quinones and elucidated that pentacoordinate phosphorus compounds are more stable, as the number of electron-withdrawing groups such as alkoxyl group increases on the central phosphorus.[3]

As shown in Scheme 6.8, Schmutzler and co-workers synthesized **10** as an example of the compounds having two pentacoordinate phosphorus centers in the molecule by the reaction of the corresponding bis(chlorophosphino)alkanes **11** with tetrachloro-*o*-benzoquinone (*o*-choranil).[14]

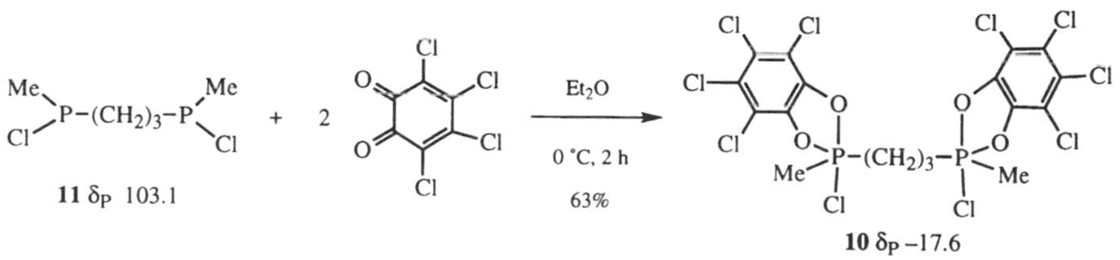

11 δ_P 103.1 **10** δ_P −17.6

Scheme 6.8

An intramolecular Ramirez-type reaction was reported by Schmidpeter and co-workers for the synthesis of ring-fused spiro compound **12**, the crystal structure of which was determined by X-ray crystallographic analysis (Scheme 6.9).[15]

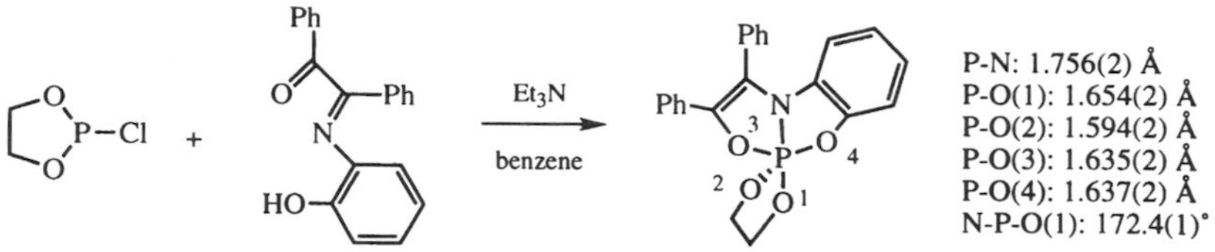

P-N: 1.756(2) Å
P-O(1): 1.654(2) Å
P-O(2): 1.594(2) Å
P-O(3): 1.635(2) Å
P-O(4): 1.637(2) Å
N-P-O(1): 172.4(1)°

12 δ_P (CH_2Cl_2) −16.4

Scheme 6.9

As shown in Scheme 6.10, Arduengo and co-workers reported that low-coordinate hypervalent phosphorus compound **13a** undergoes the Ramirez reaction with *o*-chloranil to give **14**.[16]

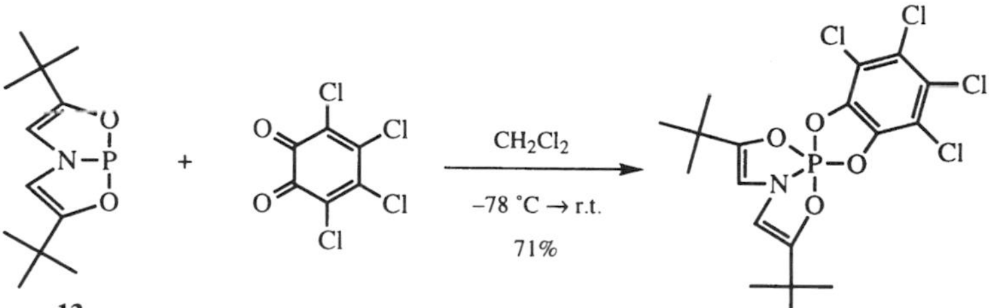

13a δ_P (CD_2Cl_2) 187

14 δ_P (CD_2Cl_2) −10.2

Scheme 6.10

For a recent example, phosphorane **15** with three different heterocycles was synthesized by the reaction of aminophosphine **16** with hexafluorobiacetyl, along with an unusual elimination of MeCl. Use of two mol equivalents of HFA instead of α-diketone gave the corresponding 1,3,2-dioxaphospholane **17** (Scheme 6.11).[17]

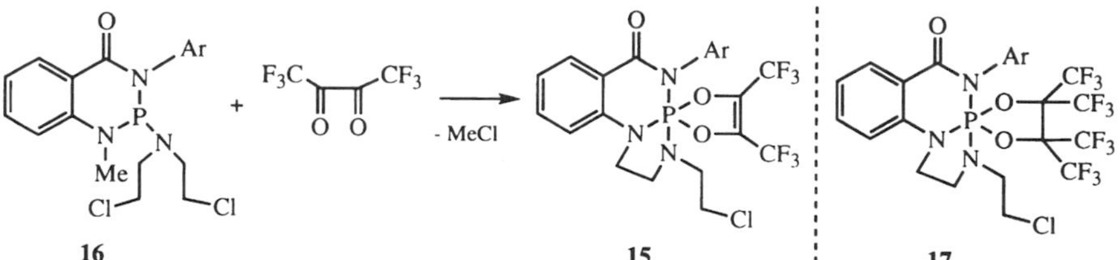

Scheme 6.11

Hydridophosphoranes can be usually synthesized by intramolecular insertion of the lone pair of the phosphorus to a hydroxyl group in the phosphorus ester or amidite intermediates. For example, **18** was synthesized as shown in

Scheme 6.12.[18] The photoreaction of **18** with alkyl disulfides provides a new synthetic method for pentacoordinate alkylthiophosphoranes **19a–d**.[19]

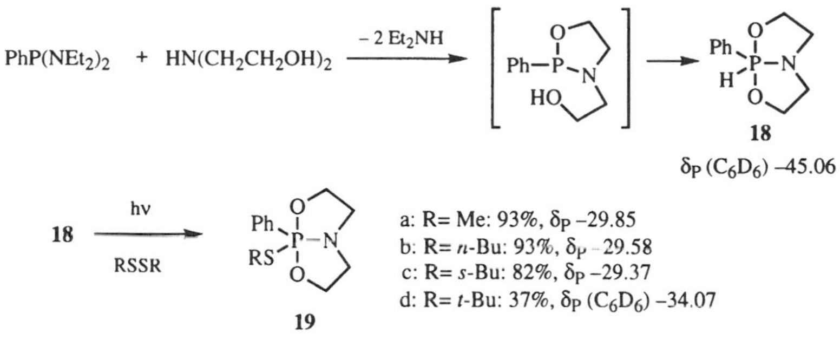

$PhP(NEt_2)_2$ + $HN(CH_2CH_2OH)_2$ —($-2\ Et_2NH$)→ [intermediate] → **18**

δ_P (C_6D_6) −45.06

18 —(hν, RSSR)→ **19**

a: R= Me: 93%, δ_P −29.85
b: R= *n*-Bu: 93%, δ_P −29.58
c: R= *s*-Bu: 82%, δ_P −29.37
d: R= *t*-Bu: 37%, δ_P (C_6D_6) −34.07

Scheme 6.12

Houalla and co-workers reported that the reaction of bicyclophosphine **20** with diethyleneglycohol, dihydroxyethyl sulfide or diethanolamine gives bis(hydridobicyclophosphorane)s **21**, **22**, and **23**, respectively, which were reacted with diols in the presence of CCl_4 and Et_3N (the Todd reaction) to give macrocycles containing two bicyclophosphorane moieties (**24–26**) along with compounds containing three and four bicyclophosphorane moieties (Scheme 6.13).[20] These compounds could be separated by column chromatography.

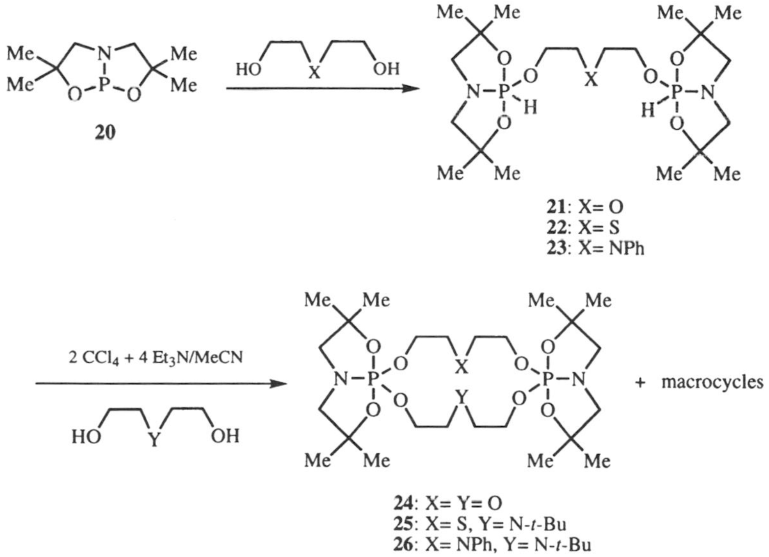

20 —(HO–X–OH)→ **21**: X= O; **22**: X= S; **23**: X= NPh

—($2\ CCl_4 + 4\ Et_3N$/MeCN, HO–Y–OH)→ + macrocycles

24: X= Y= O
25: X= S, Y= N-*t*-Bu
26: X= NPh, Y= N-*t*-Bu

Scheme 6.13

Triaryldifluorophosphoranes **27–29** were synthesized by the reaction of triarylphosphines with dimethylaminosulfur trifluoride, whose X-ray analysis showed that they have trigonal bipyramidal (TBP) structures with two fluorine atoms at the apical positions (Scheme 6.14).[21]

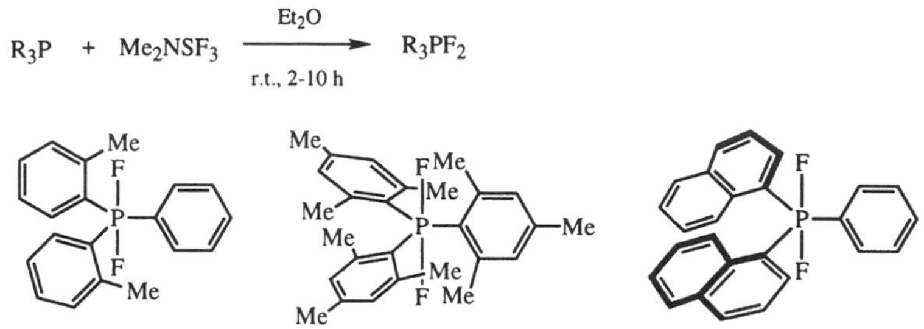

$$R_3P + Me_2NSF_3 \xrightarrow[\text{r.t., 2-10 h}]{Et_2O} R_3PF_2$$

27 (60%)	**28** (68%)	**29** (54%)
δ_P −41.2 (J_{PF}= 630 Hz)	δ_P −40.4 (J_{PF}= 654 Hz)	δ_P −21.3 (J_{PF}= 644 Hz)
P-F: 1.683(2) Å	P-F: 1.673(2) Å	P-F: 1.675(1) Å
F-P-F: 178.5(2)°	F-P-F: 179.6(2)°	F-P-F: 178.99(8)°

Scheme 6.14

6.2.1.1.4 By 1,3-Dipolar Cycloaddition Phosphorus ylides can react with 1,3-dipolar reagents such as nitrile oxides and nitrones to give five-membered heterocycles with a pentacoordinate phosphorus **30**[22] and **31**,[23] respectively (Scheme 6.15).

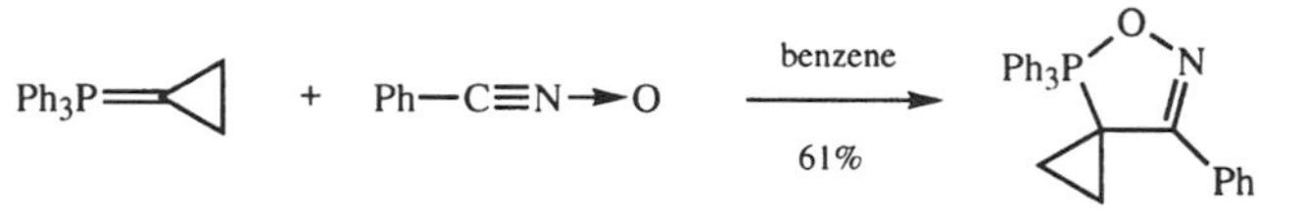

Ph_3P=cyclopropylidene + Ph—C≡N→O $\xrightarrow[61\%]{\text{benzene}}$ **30** δ_P −34.2

$Ph_3P{=}CH_2$ + PhN(O)=CHPh $\xrightarrow[93\%]{Et_2O}$ **31** δ_P −58.6

Scheme 6.15

6.2.1.1.5 By Transannular Interaction Verkade and co-workers reported the synthesis of bicyclic phosphite **32** by the reaction of triethanolamine with triaminophosphine or trialkyl phosphites. Attempted alkylation of **32** with Meerwein reagent resulted in the formation of phosphatranes **33**, whose nitrogen coordinates to the phosphorus by transannular interaction (Scheme 6.16).[24a,b]

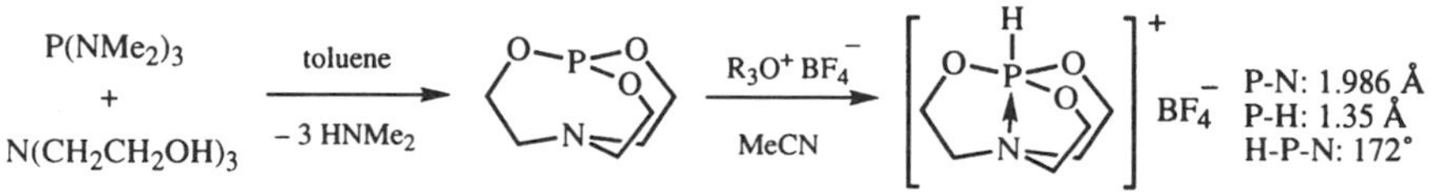

$P(NMe_2)_3$ + $N(CH_2CH_2OH)_3$ —toluene, $-3\ HNMe_2$→ **32** δ_P (C_6H_6) 115.2 —$R_3O^+ BF_4^-$, MeCN→ **33** BF_4^- δ_P −20.9 $^1J_{PH}$= 779.6 Hz

P-N: 1.986 Å
P-H: 1.35 Å
H-P-N: 172°

Scheme 6.16

The structure of **33** was determined to be distorted TBP by X-ray crystallographic analysis.[24a] The bond length between nitrogen and phosphorus atoms is 1.986 Å. The reaction of tris(2-aminoethyl)amine instead of triethanolamine with diaminochlorophosphine in the presence of triethylamine gave the P-protonated phosphatranes **34a**,[24c] **34b**,[24d] and **34c**[24d] (Scheme 6.17). It is very interesting that the protonation did not occur on the nitrogen, but the phosphorus, and that the protonation occurred even under basic conditions, indicating that the phosphorus center is more basic than the nitrogen of triethylamine. The distances between transannular nitrogen and the phosphorus in **35** are in the region from 3.33 to 2.19 Å, depending on the substituents (R^1) on the phosphorus. Further details are presented in a review.[24e]

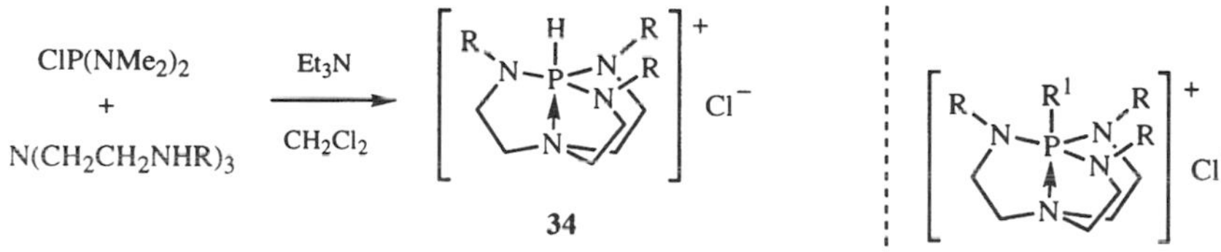

$ClP(NMe_2)_2$ + $N(CH_2CH_2NHR)_3$ —Et_3N, CH_2Cl_2→ **34** Cl^- | **35** Cl^-

a: R= Me (δ_P −10.6, $^1J_{PH}$= 491 Hz) P-N^a: 1.967 Å
b: R= H (δ_P −42.9, $^1J_{PH}$= 453 Hz) P-N^a: 2.078(4) Å
c: R= CH_2Ph (δ_P −11.0, $^1J_{PH}$= 506 Hz)

Scheme 6.17

As shown in Scheme 6.18, Akiba and co-workers reported that the transannular interaction was also observed in the system of dibenzoazaphosphocins **36**.[25] Chlorination of **36** with $SOCl_2$ gave pentacoordinate bicyclophosphorane **37** with chlorine and nitrogen atoms at apical positions.

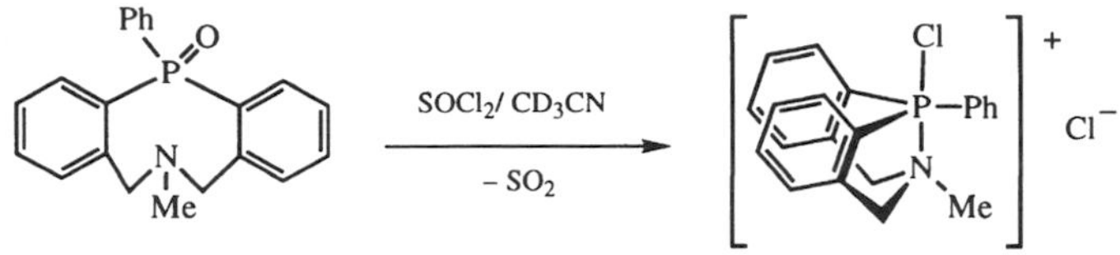

$SOCl_2$/ CD_3CN

- SO_2

36 δ_P 26.0

37 δ_P −8.3

Scheme 6.18

6.2.1.1.6 By Cyclization and Dehydration Hellwinkel reported the synthesis of pentacoordinate phosphorus compound **39** by intramolecular cyclization and dehydration of the corresponding phosphine oxide **38** under thermal or acidic conditions.[26] At almost the same time, Granoth found that treatment of bis(2-carboxyphenyl)phenylphosphine oxide **40** with an acid in water gave the corresponding spiroacylphosphorane **41**.[27] Also, (2,6-dicarboxyphenyl)diphenylphosphine oxide **42** could be converted to **43** by treatment with acetic acid.[26] Similarly, the reaction of bis(2-carboxyphenyl)phosphinic acid **44** with CF_3CO_2H afforded the corresponding hydroxyphosphorane **45** as the first example of a chemical species which has been considered as an important reactive intermediate of the hydrolysis of phosphoric esters (Scheme 6.19).[28]

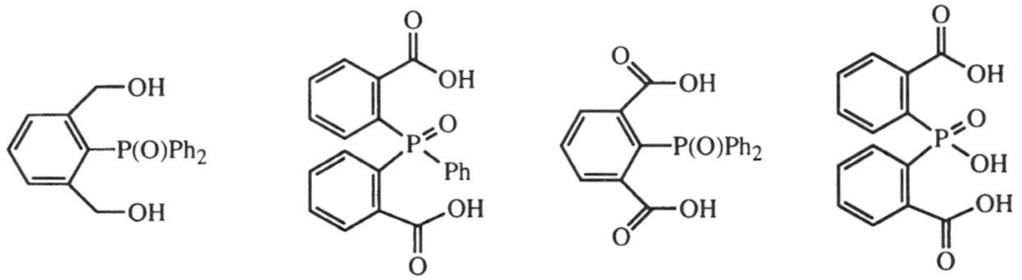

38 **40** **42** **44**

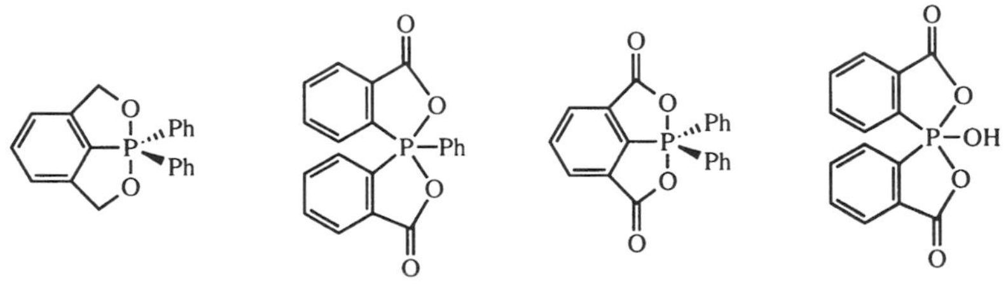

39 δ_P −22.9
98% (200 °C)
23% (H_3O^+)

41 δ_P −60.8
80% from dipotassium salt

43 δ_P −46.1

45 δ_P −38.5
75%

Scheme 6.19

6.2.1.2 Diphosphorus Compounds with a Direct P—P Bond The reaction of hydridophosphorane **46** with chlorodiphenylphosphine in the presence of Et_3N gave a $\sigma^3P—\sigma^5P$ type compound **47** with the coupling constant $^1J_{PP} = 322.2$ Hz.[29] A $\sigma^4P—\sigma^5P$ type compound **49** was obtained by treatment of the corresponding $\sigma^3P—\sigma^5P$ type compound **48** with one equivalent of sulfur (Scheme 6.20).[30]

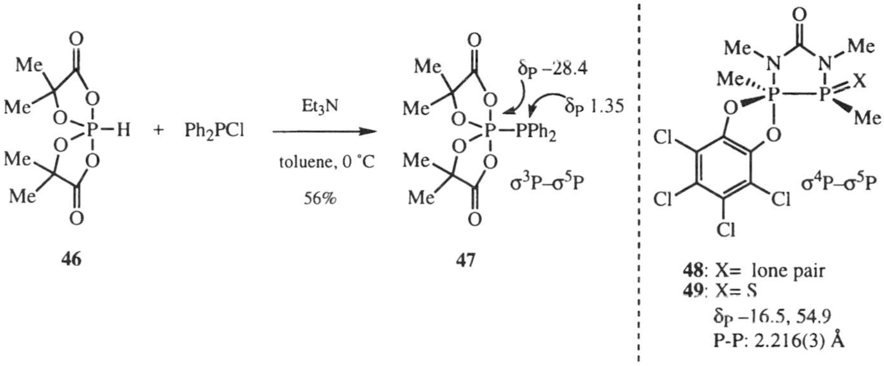

Scheme 6.20

Two interesting symmetric compounds **50** and **51** bearing a $\sigma^5P—\sigma^5P$ bond were synthesized by sequential treatment of hydridophosphorane **52** with *n*-BuLi and fluorophosphorane **53** and Ramirez reaction of cyclic diphosphine **54** with *o*-chloranil, respectively.

The P—P bonds of **50** and **51** are equatorial and apical bonds, respectively (Schemes 6.21[31] and 6.22[32]). It is not easily understandable why the apical P—P bond of **51** is shorter than the equatorial P—P bond of **50**.

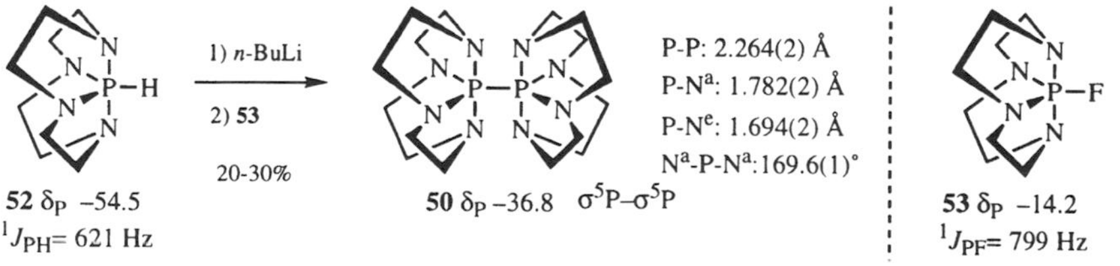

Scheme 6.21

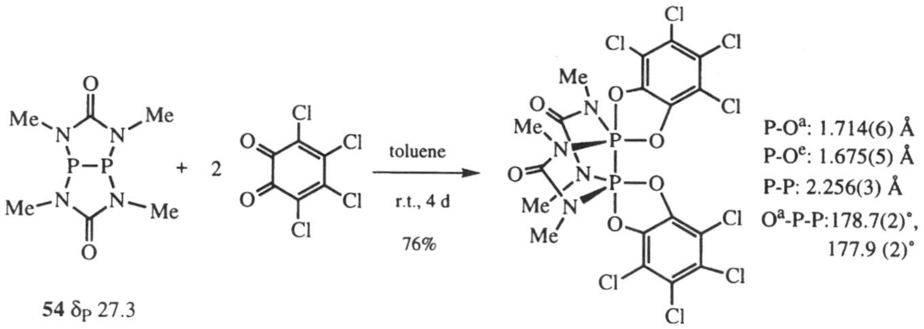

54 δ_P 27.3

toluene
r.t., 4 d
76%

51 δ_P −44.0 σ^5P–σ^5P

P-O^a: 1.714(6) Å
P-O^e: 1.675(5) Å
P-P: 2.256(3) Å
O^a-P-P: 178.7(2)°, 177.9 (2)°

Scheme 6.22

For an asymmetric example, Lamandé and co-workers synthesized **55** by the reaction of pentacoordinate P—H compound **56** with P—Cl compound **57** and showed that **55** has a P—P coupling constant of 750 Hz, which is one of the largest values reported for a single P—P bond (Scheme 6.23).[29]

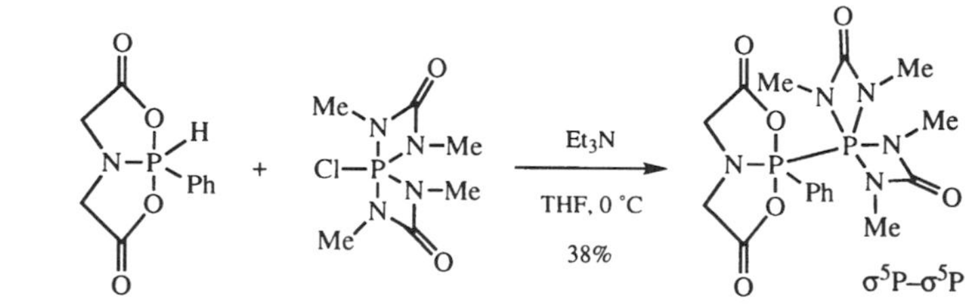

Et_3N
THF, 0 °C
38%

56 δ_P (CH_2Cl_2) −51.4
$^1J_{PH}$ = 780 Hz

57

55 δ_P −40.9, −68.1
$^1J_{PP}$ = 749.6 Hz
σ^5P–σ^5P

Scheme 6.23

6.2.1.3 Pseudorotamers with the Least Electronegative Group at the Apical Position Holmes and co-workers found that tetraoxyphoranes **58–60**, which were prepared by the Ramirez-type reaction of the corresponding phosphinites, sometimes have TBP structures with the carbon ligand, the least electronegative ligand, at the apical position (Scheme 6.24).[33] From their studies on the substituent effects, sterical repulsion of the *t*-butyl group of the benzene ring was claimed to disfavor equatorial and apical occupancy of the eight-membered ring.

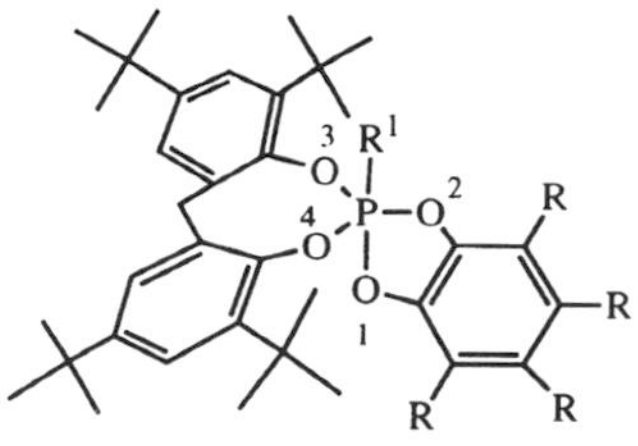

58: R= Cl, R^1= Ph (99%) δ_P –40.77
59: R= Cl, R^1= Et (67%) δ_P –23.94
60: R= R^1= Ph (71%) δ_P –47.4

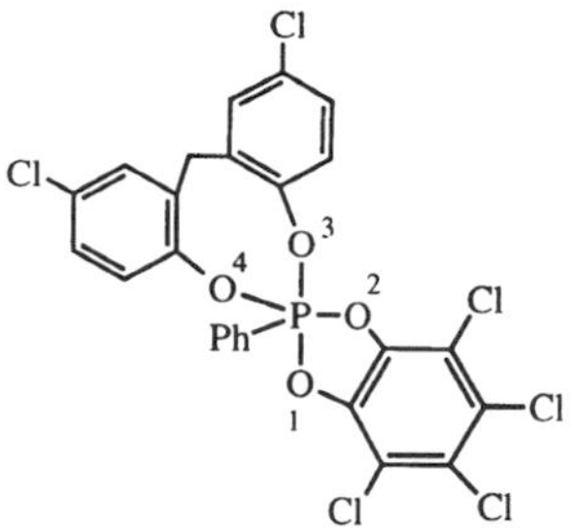

61 (40.3%) δ_P –32.29

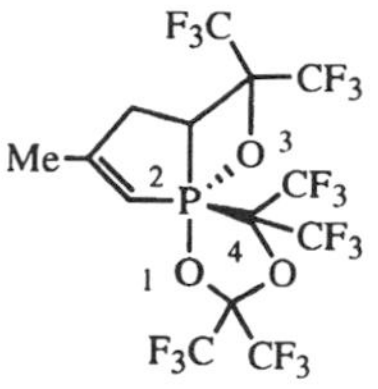

62 (15%) δ_P 8.6

Scheme 6.24

Therefore, in compound **61** having chlorine atoms at the *para*-positions without the *t*-butyl groups, two rings occupy apical and equatorial positions, while the carbon occupies the equatorial position, because of an increase in apicophilicity of the benzene ring. Selected bond lengths and angles are shown in Table 6.1. A ring-fused spirophosphorane **62** was synthesized by du Mont and co-workers and shown by X-ray crystallography to have a distorted TBP with the bridge head carbon at the apical position, as shown in Scheme 6.24.[34]

Very recently, as shown in Scheme 6.25, Akiba and co-workers found that the reaction of racemic hydridophosphorane **63** having two Martin ligands with three equivalents of *n*-BuLi, followed by the usual workup, gives hydridophosphorane **64**.[35] The structure was determined by X-ray crystallographic analysis. Hydrogen occupies an apical position; very interestingly, **64** undergoes cyclization with evolution of hydrogen to give the corresponding C-apical phosphorane **65** and O-apical pseudorotamer **66a**. C-Apical phosphorane **65** gradually isomerizes to **66a**. The C-apical isomer **65** was estimated to be around 12 kcal mol^{-1} less stable than O-apical isomer **66a** by the kinetic studies using **65** and **66b**.

6.2.1.4 Optically Active Phosphoranes For an example of optically active phosphoranes having a derivative of optically active aminoalcohol, McClure and co-workers succeeded in the synthesis of compound **67**, which does not undergo epimerization below 60°C.[36] Moriarty and co-workers synthesized compound **68**, having an optically active aminoalcohol as monodentate ligand, and could separate each diastereomer.[37] The X-ray analysis showed it has a TBP structure. Isomerization of **68** to **69** does not occur at room temperature, but on heating at 90°C a 1:1 mixture was obtained. The energy barrier of this process was estimated to be $\Delta G^{\ddagger} = 27$ kcal mol^{-1}, which is the highest barrier of pseudorotation of phosphoranes among those reported so far (Scheme 6.26).

TABLE 6.1 Selected Bond Lengths and Angles

	P—C (Å)	P—O(1) (Å)	P—O(2) (Å)	P—O(3) (Å)	P—O(4) (Å)	C—P—O(1) (deg.)	O(1)—P—O(3) (deg.)
58	1.838(8)	1.760(6)	1.662(5)	1.610(6)	1.602(6)	176.1(3)	89.6(3)
59	1.830(9)	1.760(6)	1.672(6)	1.598(4)	1.598(4)	174.2(4)	90.1(2)
60	1.860(7)	1.710(5)	1.630(5)	1.614(5)	1.610(5)	176.2(3)	93.4(2)
61	1.798(12)	1.747(8)	1.633(8)	1.661(8)	1.611(8)	94.5(5)	169.7(4)
62	1.899(2)	1.720(2)	—[a]	1.655(2)	—[b]	163.33(9)	85.35(8)

[a] P—C(2): 1.769(2) (Å).
[b] P—C(4): 1.932 (Å).

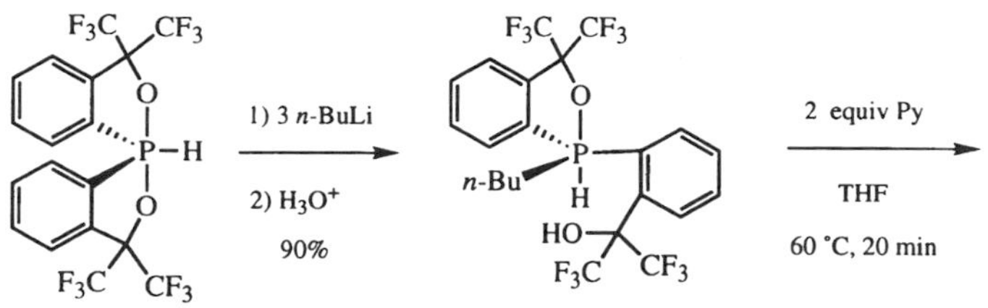

1) 3 *n*-BuLi
2) H_3O^+
90%

2 equiv Py
THF
60 °C, 20 min

63 δ_P −45.8
$^1J_{HP}$= 729 Hz

64 δ_P −34.4
$^1J_{HP}$= 273 Hz

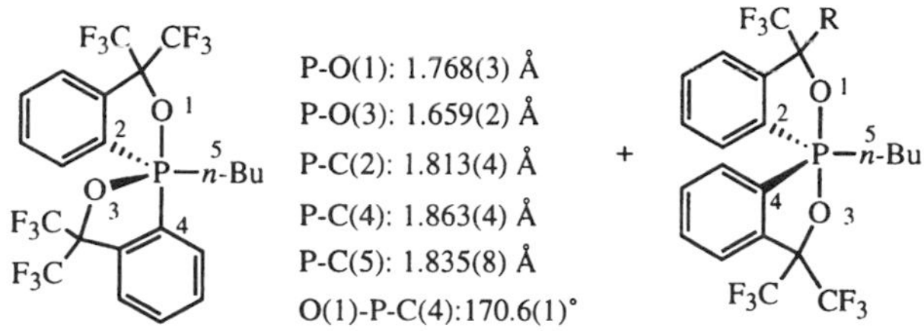

P-O(1): 1.768(3) Å
P-O(3): 1.659(2) Å
P-C(2): 1.813(4) Å
P-C(4): 1.863(4) Å
P-C(5): 1.835(8) Å
O(1)-P-C(4): 170.6(1)°

+

P-O(1): 1.765(2) Å
P-O(3): 1.753(2) Å
P-C(2): 1.818(2) Å
P-C(4): 1.820(2) Å
P-C(5): 1.835(8) Å
O(1)-P-C(3): 175.79(6)°

65 δ_P −3.5 (71%)
$^1J_{C4,P}$= 88 Hz

66a: R= CF_3 δ_P −18.8 (29%)
$^1J_{C4,P}$= 160 Hz
66b: R= CH_3

Scheme 6.25

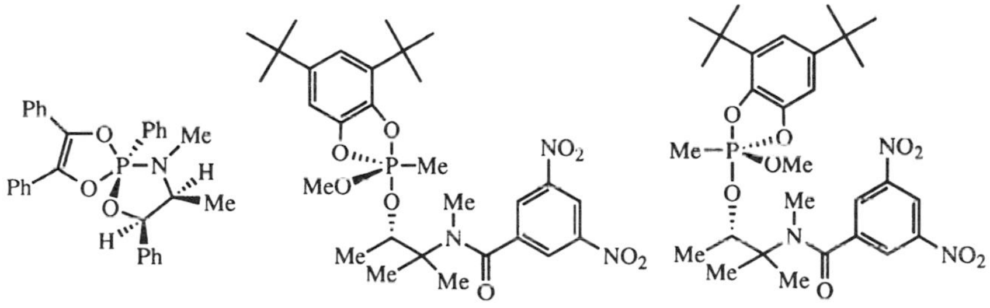

67 δ_p −35.32
$[\alpha]_D$ +12.0°

68 δ_p −20.70
$[\alpha]_D^{25}$ +9.40°

69 δ_p −20.17
$[\alpha]_D^{25}$ +2.94°

Scheme 6.26

Hellwinkel and co-workers intended to obtain optically active pentacoordinate phosphorane **70** by treatment of optically active hexacoordinate inonic compound **71** with dilute HCl, but the resulting **70** was racemic.[38] In order to avoid racemization, methyl-substituted compound **72** was used. Interestingly, **72** underwent regioselective bond cleavage to give optically active **73** as the first example of optically active phosphoranes which has chirality only at the phosphorus center (Scheme 6.27).

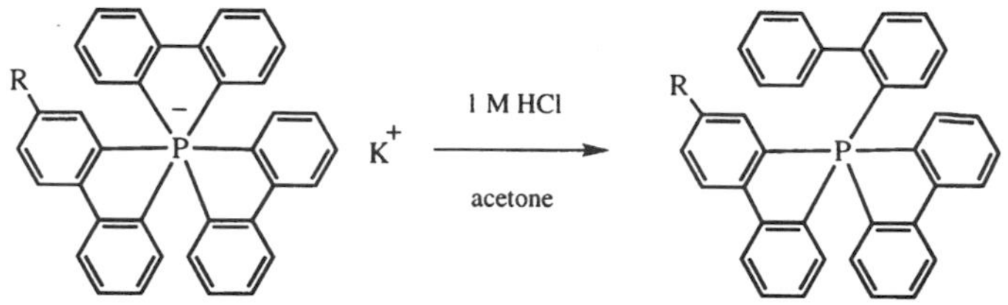

1 M HCl

acetone

71: R= H

72: R= Me $[\alpha]^{24}_{578}$ −1870°

70: R= H

73: R= Me $[\alpha]^{24}_{578}$ −94°

Scheme 6.27

As shown in Scheme 6.28,[39] Akiba and co-workers succeeded in the synthesis of optically active phosphorane **74** (81%) by separation of diastereomer **75**, followed by reduction with $LiAlH_4$ in refluxing Et_2O. Absolute configuration of **75** was determined by X-ray crystallographic analysis. Optically active hydridophosphorane **63** was also synthesized in 67% yield via unique C—P bond cleavage by the reaction of optically resoluted diastereomer **76** with MeLi in THF at −78°C, and was demonstrated to be a useful synthetic intermediate of optically active phosphoranes (Scheme 6.28).[40] Enantiomers of **74** and **63** were also synthesized from diastereomers of **75** and **76**, respectively. Thus an optically active pair of enatiomeric phosphoranes of rigid stereochemistry with asymmetry only upon the pentacoordinate phosphorus atom could be obtained.

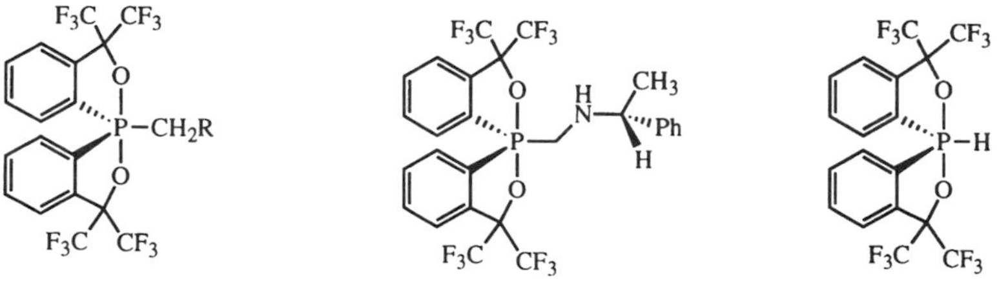

75: R= CO_2-(-)Men δ_P −25.4

$[\alpha]^{21}_{436}$ +11.1° (*c* 1.02, $CHCl_3$)

74: R= CH_2OH δ_P −21.5

$[\alpha]^{21}_{436}$ +108° (*c* 1.02, $CHCl_3$)

76 δ_P −24.6

$[\alpha]^{20}_{436}$ +73° (*c* 1.22, $CHCl_3$)

63 δ_P −45.9

$[\alpha]^{20}_{436}$ −16° (*c* 1.07, $CHCl_3$)

Scheme 6.28

6.2.2 Synthesis of Hexacoordinate and Heptacoordinate Organophosphorus Compounds

6.2.2.1 Synthesis of Hexacoordinate Compounds

6.2.2.1.1 By addition of Bidentate and Monodentate Ligands Lithium tris(2,2′-biphenylene)phosphate **77a** was obtained by the reaction of the corresponding phosphonium salt **2** with a dianion of the bidentate ligand (Scheme 6.29).[8a,38] Phosphazene derivative ($(NPCl_2)_3$) and PCl_5 can be used as the starting materials instead of the phosphonium salt. Similarly, (*o*-benzenedioxy)bis(2,2′-biphenylene)-, bis(*o*-benzenedioxy)(2,2′-biphenylene)-, and tris(*o*-benzenedioxy)-phosphate anions **77b–d** were synthesized by the reaction using catechol in the presence of a base. For these hexacoordinate compounds, a close relationship can be shown between the ^{31}P chemical shifts and the number of *o*-benzenedioxy groups. In going from **77a** (δ_P – 181) to **77b** (δ_P – 147), **77c** (δ_P – 106), and **77d** (δ_P – 82), each catechol substituent lowers the δ ^{31}P values by around 30 ppm.[9a] These types of compounds are stable to air and moisture, but reactive towards bromine and an acid to give the corresponding pentacoordinate phosphoranes, respectively. This provides a novel synthetic method for the pentacoordinate phosphorus compounds, by which optically active phosphorane **73** could be obtained, as shown in Section 6.2.1.4.

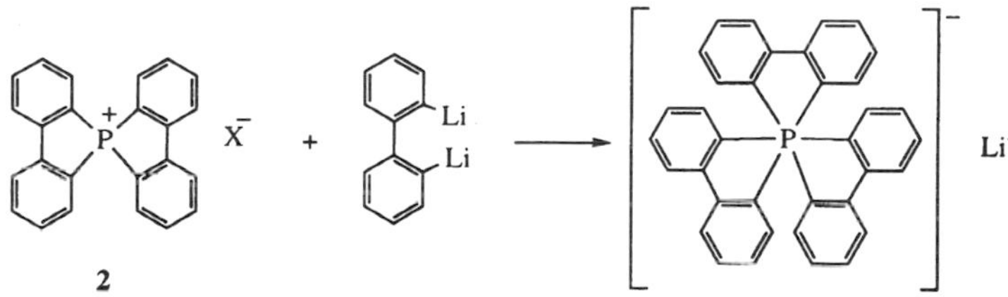

2

77a δ_P −181

Scheme 6.29

Trippett and co-workers synthesized hexacoordinate compounds **78** and **79** by the reaction of pentacoordinate phosphorane **80** with *p*-FC_6H_4ONa and CsF, respectively (Scheme 6.30).[41] The isomerization from first-formed *trans*-isomers **78** and **79** to *cis*-isomers **81** and **82** occurred, indicating that *cis* isomers are thermodynamically more stable than *trans* ones. This can be explained by taking into consideration a characteristic feature of a three-center–four-electron bond.

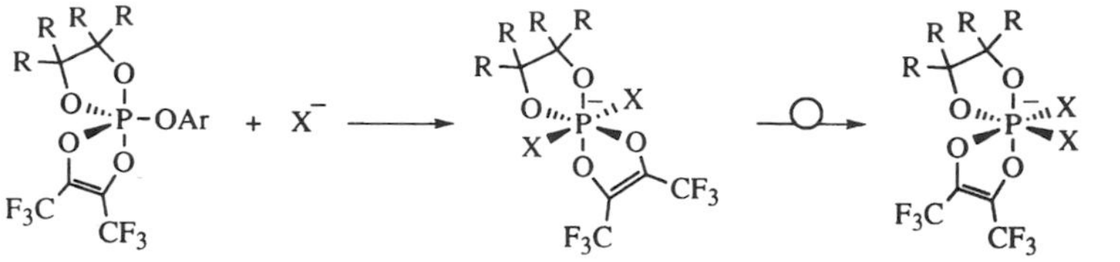

80

78: X= *p*-FC_6H_4O

79: X= F

81: X= *p*-FC_6H_4O

82: X= F

Scheme 6.30

6.2.2.1.2 By Ligand Exchange Followed by Intramolecular Nitrogen Coordination The reaction of dimethylamidinium **83** with PCl_5 gave a new class of neutral phosphorus compound **84** with a hexacoordinate phosphorus center.[42] The X-ray crystallographic analysis showed that **84** has a octahedral structure with two nitrogen and four chlorine atoms as the ligands (Scheme 6.31).[43]

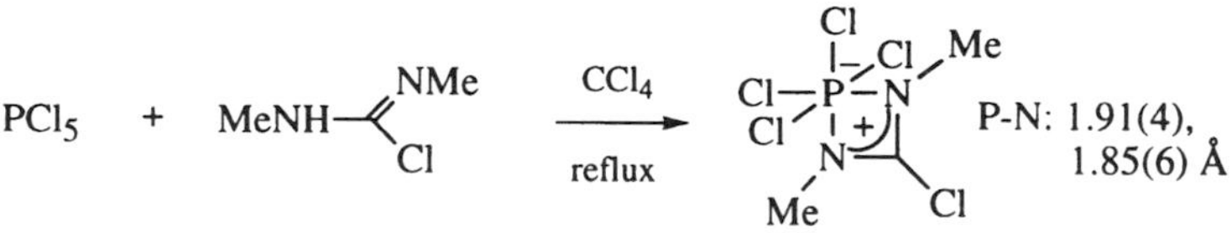

PCl_5 + MeNH–C(=NMe)Cl (**83**) $\xrightarrow[\text{reflux}]{CCl_4}$ **84** δ_P −202; P-N: 1.91(4), 1.85(6) Å

Scheme 6.31

Schmutzler and co-workers investigated the reaction of various pentacoordinate fluorophosphoranes **85** with 8-trimethylsiloxyquinoline (**86**) or with 2-(*N*,*N*-dimethylamino)acetoxytrimethylsilane (**87**).[44] Interestingly, the reaction of a five-membered fluorophosphorane nicely gave hexacoordinate **88**, whereas in the case of six-membered fluorophosphorane the product was not hexacoordinate compounds, but pentacoordinate **89** without intramolecular coordination. The structures of **88** and **90** were determined to be octahedral by X-ray crystallographic analysis. Structural data for compound **90** are shown as a typical example in Scheme 6.32.

Similarly, Kumara Swamy and co-workers reported syntheses of several tricyclic hexacoordinate phosphoranes with varying ring sizes.[45] The reaction of aminophosphorane **91** with 8-hydroxyquinoline, followed by intramolecular coordination, gave tricyclic compound **92** (Scheme 6.33).

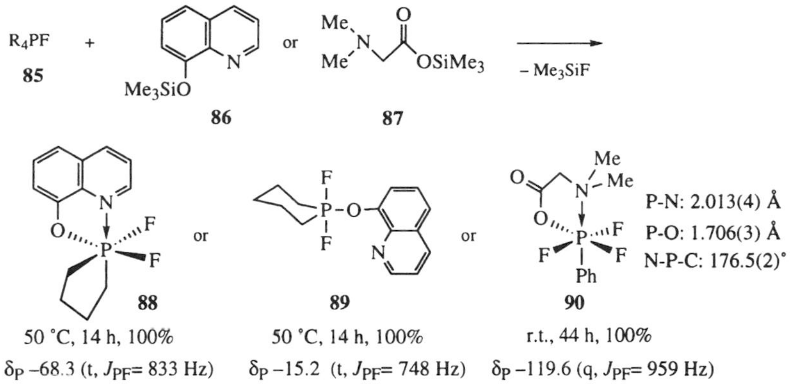

R_4PF **85** + **86** or **87** → $-Me_3SiF$

88 50 °C, 14 h, 100% δ_P −68.3 (t, J_{PF}= 833 Hz)

89 50 °C, 14 h, 100% δ_P −15.2 (t, J_{PF}= 748 Hz)

90 r.t., 44 h, 100% δ_P −119.6 (q, J_{PF}= 959 Hz)

P-N: 2.013(4) Å
P-O: 1.706(3) Å
N-P-C: 176.5(2)°

Scheme 6.32

91 δ_P −40.6

toluene, reflux, overnight, $-HNEt_2$

92 δ_P −100.5

P-N: 1.956(2) Å N-P-O(3): 172.58(8)°
P-O(1): 1.709(2) Å O(1)-P-O(4): 173.23(8)°

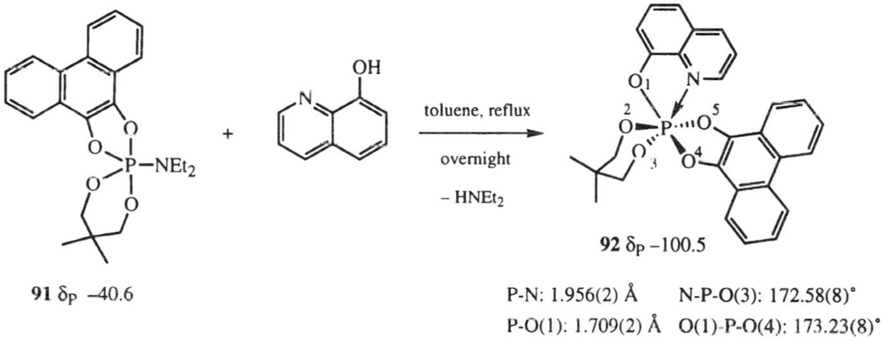

Scheme 6.33

6.2.2.1.3 By Transannular Interaction Although many oxyphosphoranes with various ring sizes have been synthesized so far, the compounds with eight-membered ring containing a sulfur atom were shown by X-ray analysis to have a hexacoordinate phosphorus based on the transannular interaction of the sulfur with the phosphorus. The distances between the sulfur and the phosphorus are scattered from 3.504 Å to 2.362 Å, reflecting the weak and strong interaction. Those of **93**[46] and **94**[47] are 2.504 Å and 2.362 Å, respectively, indicating that the latter has the strongest interaction (Scheme 6.34).

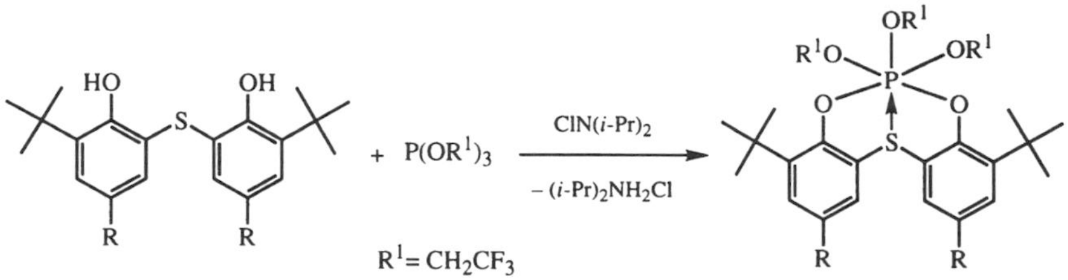

+ $P(OR^1)_3$ — $ClN(i\text{-}Pr)_2$, $-(i\text{-}Pr)_2NH_2Cl$ →

$R^1 = CH_2CF_3$

93: R= *t*-Bu (80%) δ_P −82.4 P-S: 2.504(3) Å

94: R= Me (76%) δ_P −82.3 P-S: 2.362(2) Å

Scheme 6.34

6.2.2.1.4 By Introduction of Tetradentate Ligands Hexacoordinate phosphorus(V)octaethylporphyrins were reported by Gouterman and co-workers for the first time.[48a] Dichlorophosphorus(V)tetraphenylporphyrin **95** was synthesized by the reaction of parent porphyrin **96** with $POCl_3$.[48b] Various kinds of hexacoordinate phosphorus compounds having a porphyrin ligand have been synthesized by using **95** as a key synthetic intermediate and some of them have been reported.[49] Shimidzu and co-workers[50] found that the reaction of **95** with various alcohols afforded axial dialkoxy phosphorus(V)-porphyrin derivatives **97** by efficient substitution of axial chloride ligands.

Similarly, bis(thienylalkoxy)- and bis(oligothienylalkoxy)-phosphorus(V)-porphyrins were prepared for the electrochemical synthesis of one-dimensional donor–acceptor polymers containing oligothiophenes and phosphorus porphyrins.[51] Interestingly, the reaction of **95** with various diols gave an oligomeric mixture of wheel-and-axle-shaped molecules. For example, trimer **98** of 1,2-ethanediol adduct is shown in Scheme 6.35.[52]

However, Akiba and co-workers[53] reported the preparation of dichlorophosphorus(V)octaethylporphyrin **99** by the reaction of octaethylporphyrin **100** with PCl_3 and the synthesis of the first dialkylphosphorus(V)porphyrin **101a,b** by treatment of **99** with trialkylaluminum in CH_2Cl_2, which was reported for antimony porphyrins.[54] The R—P=O compounds **102a,b** were synthesized nicely by the reaction of the parent porphyrin **100** with $RPCl_2$ in the presence of 2,6-dimethylpyridine in CH_2Cl_2 at room temperature, followed by rapid air oxidation during workup. Conjugated acids **103a,b** of **102a,b** were obtained by treatment of dilute HCl (Scheme 6.36).[53] From the X-ray crystallographic analysis it was found that **102a** and **103a-ClO$_4$** have very different structures: namely, the P—O bond lengths are 1.487(8) Å and 1.635(5) Å for **102a** and **103a-ClO$_4$**, respectively, and the porphyrin core of **102a** is almost planar, in contrast to **103a-ClO$_4$**, which is severely ruffled.

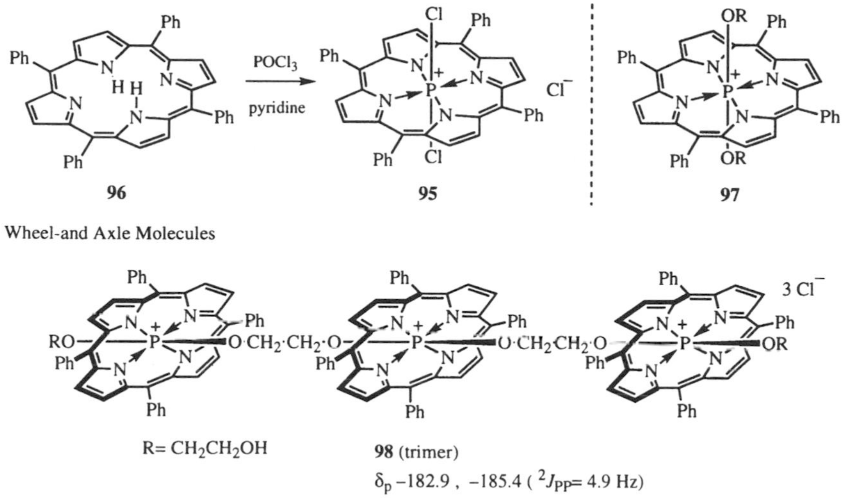

96 → 95 ($POCl_3$, pyridine); 97

Wheel-and Axle Molecules

R= CH_2CH_2OH

98 (trimer)

δ_P −182.9, −185.4 ($^2J_{PP}$= 4.9 Hz)

Scheme 6.35

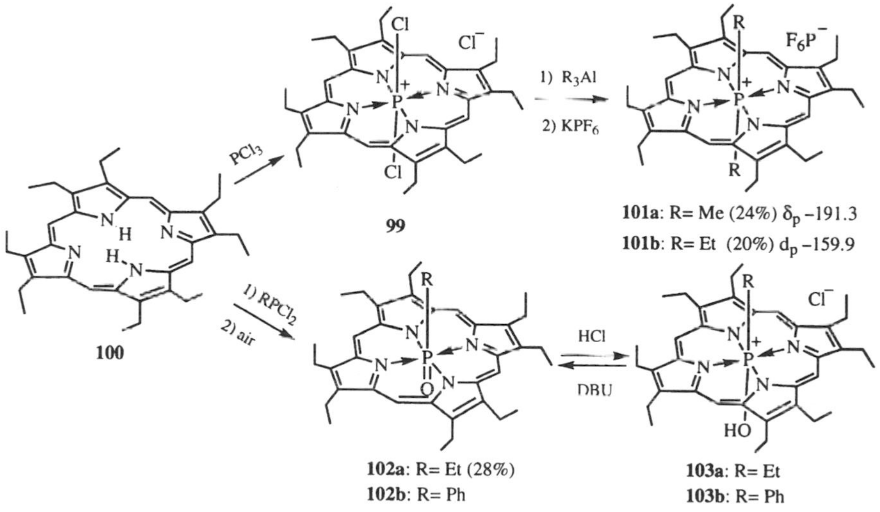

100 → **99** (PCl_3) → **101** (1) R_3Al; 2) KPF_6)

100 → **102** (1) $RPCl_2$; 2) air) ⇌ **103** (HCl / DBU)

101a: R= Me (24%) δ_P −191.3

101b: R= Et (20%) d_P −159.9

102a: R= Et (28%)

102b: R= Ph

103a: R= Et

103b: R= Ph

Scheme 6.36

For the purpose of putting a "bottom on the basket" of *p*-*t*-butylcalix[4]arene by connecting the oxygens with a single "hypervalent" phosphorus atom, the reaction of *p*-*t*-butylcalix[4]arene (**104**) with hexamethylphosphorous triamide was carried out to give the novel zwitterionic hexacoordinate phosphorus derivative **105**. The deprotonation of **105** with *n*-BuLi afforded **106** (Scheme 6.37).[55] The X-ray analysis of **106** showed that the skeleton of calix[4]arene takes a cone conformation, which was suggested by ^{1}H NMR spectroscopy,

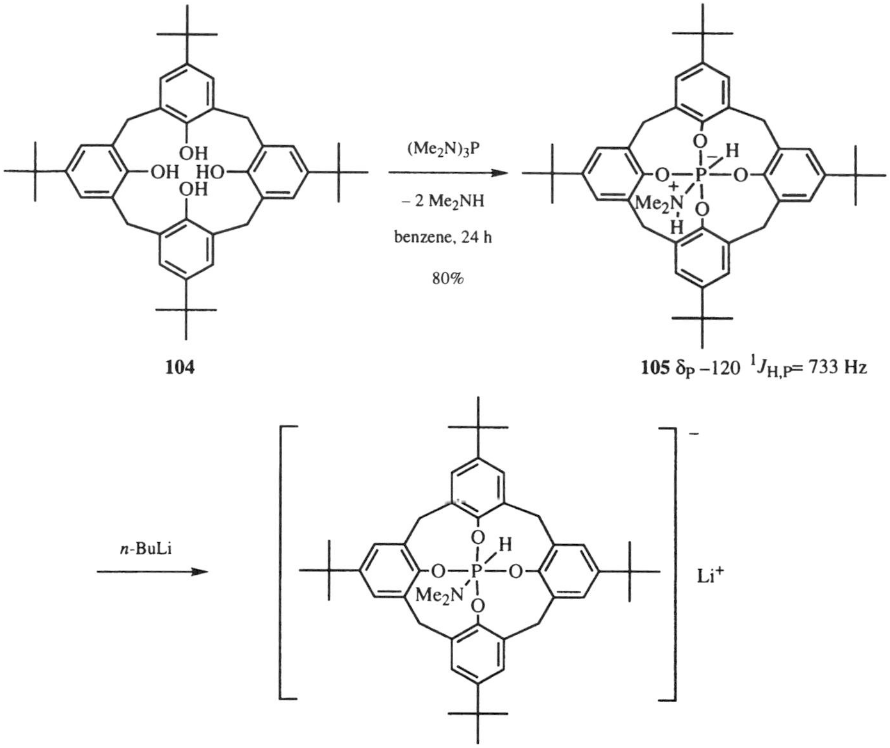

$(Me_2N)_3P$

$- 2\ Me_2NH$

benzene, 24 h

80%

104

105 δ_P −120 $^1J_{H,P}$= 733 Hz

n-BuLi

Li^+

106 δ_P −113 $^1J_{H,P}$= 836 Hz

Scheme 6.37

and that a hexacoordinate phosphorus atom with P—H group inside the basket and dimethylamino group outside makes the bottom of the molecule.[56] Attempted synthesis of *p*-*t*-butylcalix[4]areneP(H), pentacoordinate phosphorane with four oxygen and one hydrogen atoms was unsuccessful.

6.2.2.1.5 By Intramolecular Addition of Anionic Moiety It is well-known that trivalent phosphorus compounds are reacted with heterocumulenes such as carbon dioxide, isocyanate, isothiocyanate, and so on to give zwitterionic species. Shevchenko applied the reaction to compound **107** with tricoordinate and pentacoordinate phosphorus atoms in the molecule. The reaction of **107** with isocyanate gave a mixture of **108** and **109**, which were formed via nucleophilic attack of oxygen and nitrogen of an ambident anion, respectively (Scheme 6.38).[57]

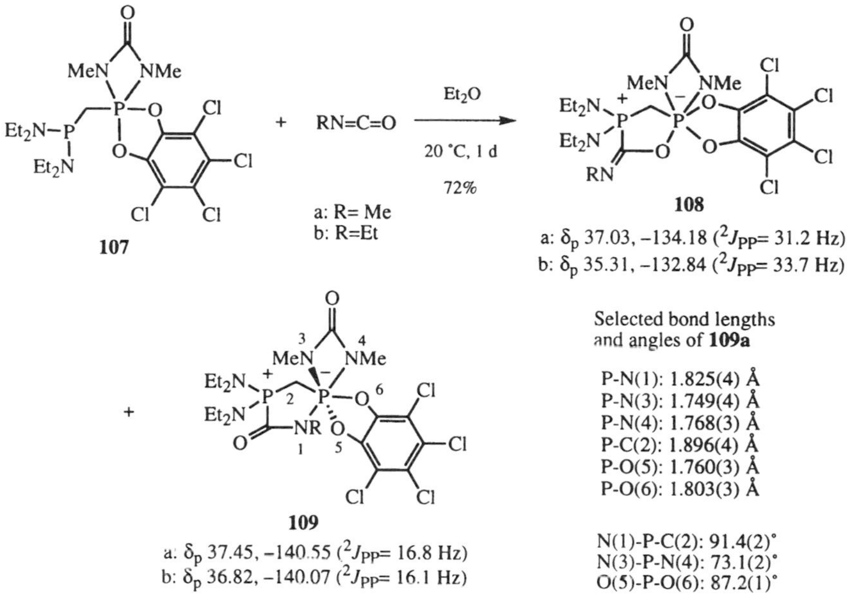

107 + RN=C=O (a: R= Me; b: R=Et) —Et_2O, 20 °C, 1 d, 72%→ **108** + **109**

108
a: δ_P 37.03, −134.18 ($^2J_{PP}$= 31.2 Hz)
b: δ_P 35.31, −132.84 ($^2J_{PP}$= 33.7 Hz)

109
a: δ_P 37.45, −140.55 ($^2J_{PP}$= 16.8 Hz)
b: δ_P 36.82, −140.07 ($^2J_{PP}$= 16.1 Hz)

Selected bond lengths and angles of **109a**

P-N(1): 1.825(4) Å
P-N(3): 1.749(4) Å
P-N(4): 1.768(3) Å
P-C(2): 1.896(4) Å
P-O(5): 1.760(3) Å
P-O(6): 1.803(3) Å

N(1)-P-C(2): 91.4(2)°
N(3)-P-N(4): 73.1(2)°
O(5)-P-O(6): 87.2(1)°

Scheme 6.38

6.2.2.2 Synthesis of Heptacoordinate Compounds The triarylphosphines **110** and **111** having amino groups at appropriate positions were synthesized. It was shown by X-ray analyses that they have a tetrahedral structure involving three carbons and a lone pair, in which three amino groups weakly interact with the phosphorus from the backside of the carbon ligand. The average of angles C—P—C is 107.2°, indicating that the tetrahedral structure still remains. The alkylation of **110** with iodomethane gave the corresponding *N*-methylated compound **112**, whereas that of **111** afforded *P*-methylated compound **113** (Scheme 6.39).[58] Although the X-ray analysis has not been carried out yet, it is expected that this compound has a heptacoordinate phosphorus.

6.3 REACTIONS OF PENTACOORDINATE PHOSPHORANES

6.3.1 Thermolysis

The thermolysis of pentaphenylphosphorane (Ph_5P) at 130°C gave benzene, biphenyl, triphenylphosphine, and so on.[59a] The reaction in the presence of styrene and the reaction in CCl_4 resulted in polymerization of styrene and the

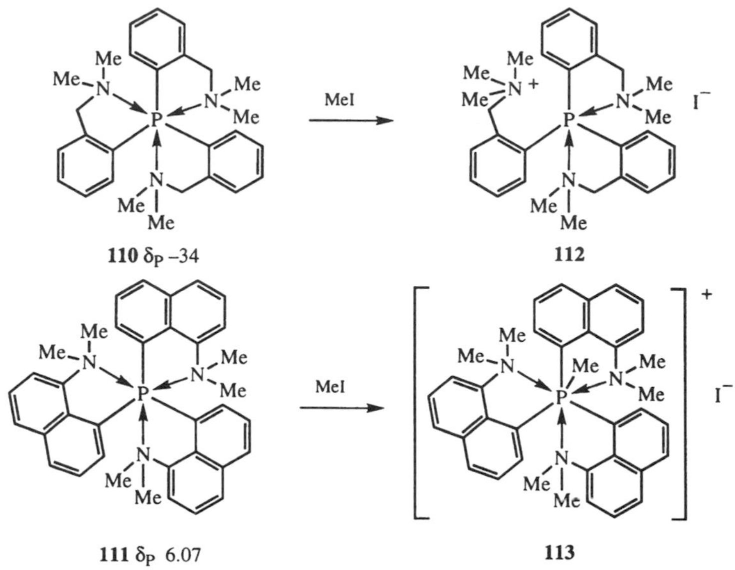

110 δ_P −34

112

111 δ_P 6.07

113

Scheme 6.39

formation of Ph_4PCl, respectively, suggesting that the reaction proceeds via a radical pathway (Scheme 6.40).[59b]

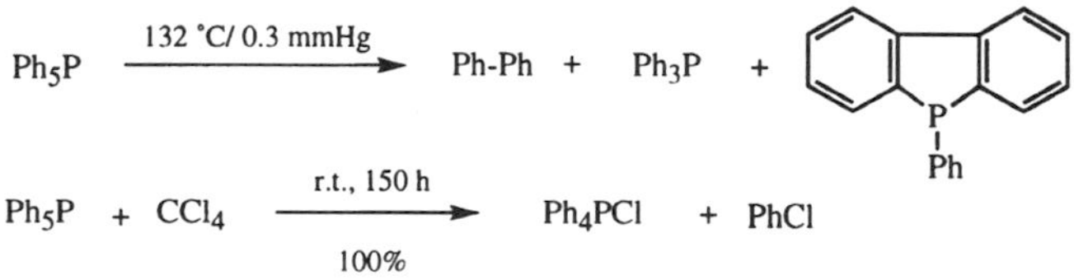

Ph_5P —(132 °C/ 0.3 mmHg)→ Ph-Ph + Ph_3P +

Ph_5P + CCl_4 —(r.t., 150 h; 100%)→ Ph_4PCl + PhCl

Scheme 6.40

Spirophosphorane **1b** was heated over its melting point to give a formal ligand-coupling product, ring-expanded nine-membered cyclophosphine **114**; however, no discussion on the mechanism has been reported (Scheme 6.41).[9a]

A typical reaction of oxyphosphoranes involves the formation of phosphonium oxide by the P—O bond cleavage. *Trans*-1,3,2-dioxaphospholane **115** gives the corresponding *cis*-oxirane, as shown in Scheme 6.42.[60]

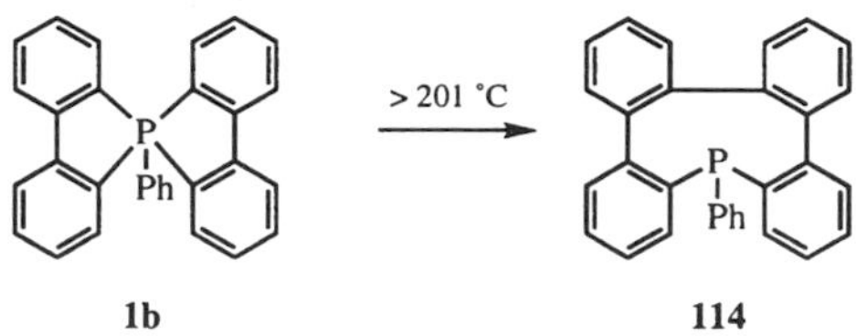

1b 114

Scheme 6.41

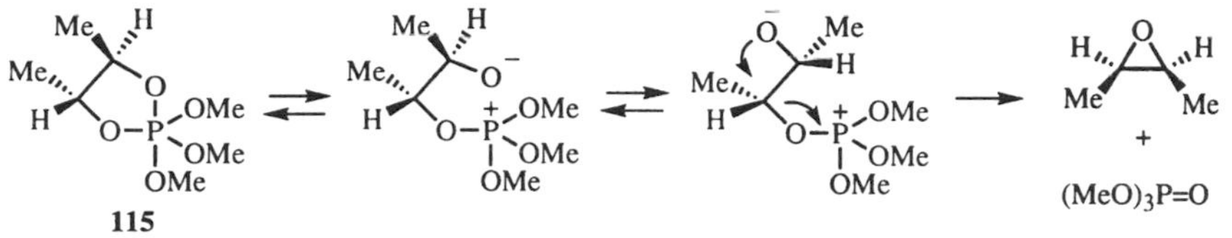

115

Scheme 6.42

6.3.2 Reactions of 1,3,2-Dioxaphospholenes and 1,2-Oxaphospholenes with Electrophiles

As described in Reference 3, there have been reported many reactions of 1,3,2-dioxaphospholenes **116**; most of them are reactions with electrophiles at the 4-carbon. The reaction with carbonyl compounds gives the corresponding oxyphosphoranes **117**, whereas the reaction with halogen, followed by Arbuzov reaction, affords acylhalomethyl phosphates **118** along with haloalkanes (Scheme 6.43). A driving force of these reactions seems to be the formation of P=O double bond, except for the reaction with carbonyl compounds.

116 117 118

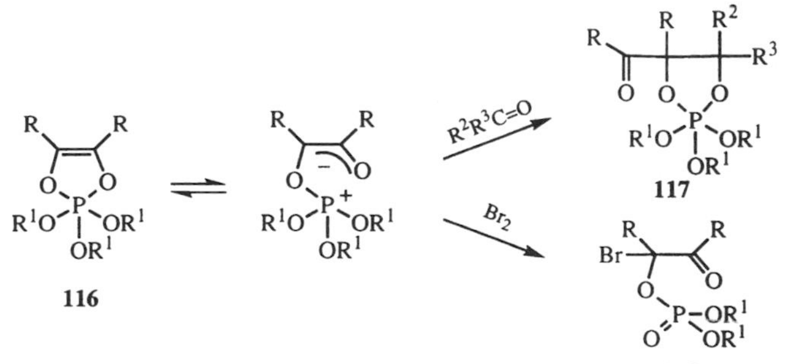

Scheme 6.43

McClure and co-workers extensively investigated applications of phosphoranes **119** obtained by the Ramirez reaction of α, β-unsaturated ketones with phosphites to the synthesis of phosphonate-functionalized compounds. For example, the synthesis of uracil phosphonate derivatives **120** was reported (Scheme 6.44).[61]

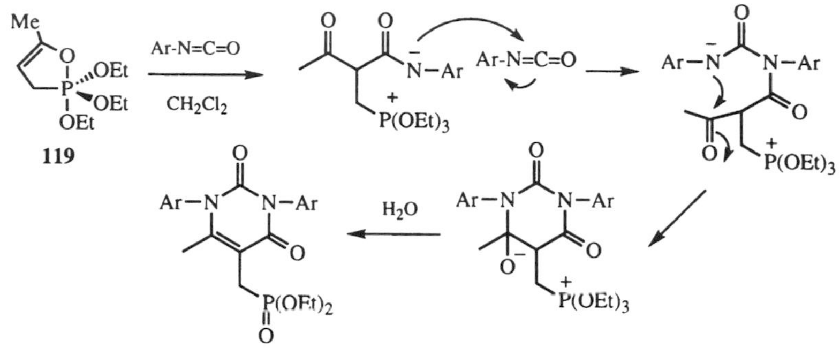

120 a: Ar= Ph (92%) δ_P 27.4
b: Ar= *p*-BrC_6H_4 (93%)

Scheme 6.44

6.4 SYNTHESIS AND REACTIONS OF HYPERVALENT 1,2-HETERAPHOSPHETANES

6.4.1 Pentacoordinate 1,2-Oxaphosphetanes

The first stable pentacoordinate 1,2-oxaphosphetane **121** was reported by Birum and Matthews.[62a] Later, several pentacoordinate 1,2-oxaphosphetanes **112**,[62b] **123**,[62c] **124**,[62d] **125**,[62e] **126**,[62f] **127**,[62g] and **128**,[62g] which are stable at room temperature, have been synthesized by the reaction of phosphorus ylides or *in-situ*-generated phosphorus ylides with carbonyl compounds (Scheme 6.45). The structures of most of them were determined by X-ray crystallographic analysis.

In the course of the study on the trapping of an intermediate of the Horner–Emmons reaction, an oxidophosphorane, the author and co-workers have found a novel and general synthetic route to isolable pentacoordinate 1,2-oxaphosphetanes **130** via bis(hydroxyalkyl)phosphine oxides **131** from methylphosphine oxide **129** bearing the Martin ligand, as shown in Scheme 6.46.[63] This method was nicely applicable to various aliphatic or aromatic aldehydes and ketones. The key step of the present method was an intramolecular cyclization and dehydration, as mentioned in Section 6.2.1.1.6.

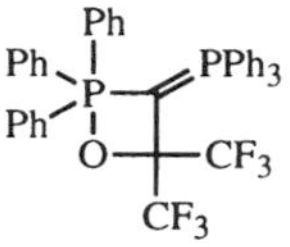

121 (76%)
δ_P 7.3, –54.0
$^2J_{PP}$= 47 Hz

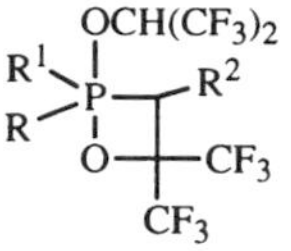

122
R= R^1= Ph; R^2= Me
δ_P (C_6H_6) –32.1

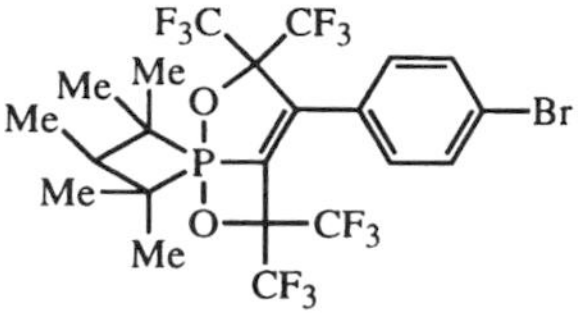

123
δ_P –31.6

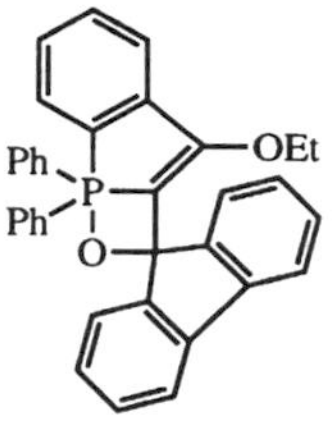

124
δ_P –57.6

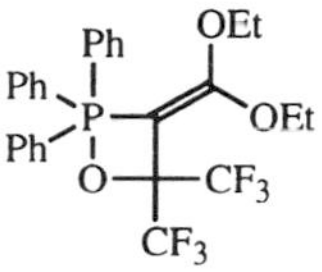

125 (92%)
δ_P –49.09

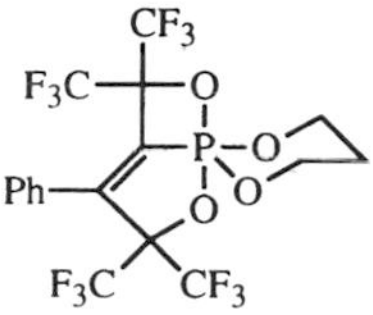

126 (91%)
δ_P –1.36

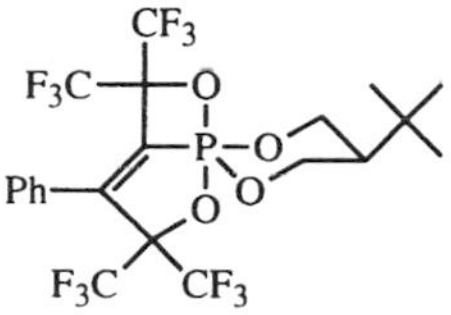

127 (30%)
δ_P (C_6D_6) –3.18

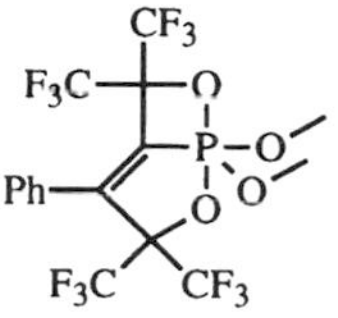

128 (43%)
δ_P (C_6D_6) –9.16

Scheme 6.45

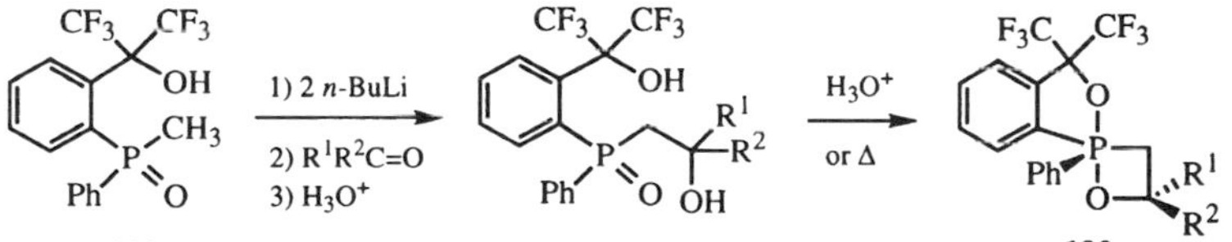

129

1) 2 *n*-BuLi
2) $R^1R^2C{=}O$
3) H_3O^+

131
a: R^1= R^2= *p*-FC_6H_4
b: R^1= R^2= Ph
c: R^1= R^2= CF_3

H_3O^+ or Δ

130
δ_P –30.3 ~ –39.1

Scheme 6.46

The direct introduction of a substituent to the 3-position of the oxaphosphetane ring was tried, because of the high stability and ready handling of **130**. Interestingly, the 3-lithio derivative **132a** resonates at δ – 28 in the ^{31}P NMR, indicating that this still maintains a pentacoordinate structure, in sharp contrast to the Schlosser reaction including 2-oxidoalkylidenephosphoranes, which have been considered to be formed by the lithiation of *anti* betaines.[64] The reaction of **132a** with electrophiles afforded stereoselectively 3-substituted 1,2-oxaphosphetanes with a *cis* configuration between electrophiles and the 2-phenyl group. The author and co-workers succeeded in the synthesis of 3-methoxycarbonyl-1,2-oxaphosphetane **133a** from **132b** as the first example of a pentacoordinate 1,2-oxaphosphetane bearing an electron-withdrawing group such as a carbonyl group at the 3-position, which has never been observed even spectroscopically (Scheme 6.47).[65]

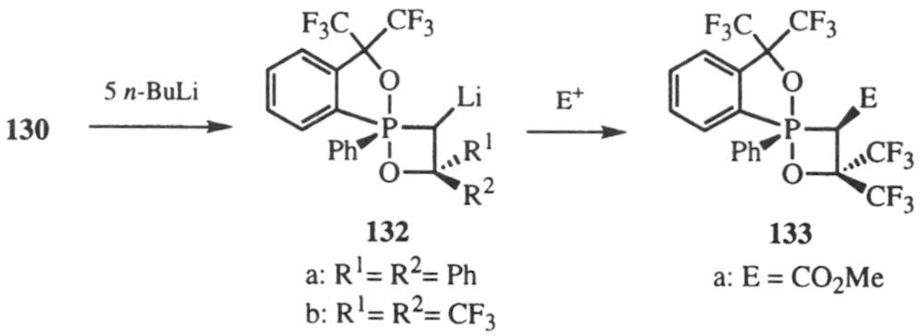

130 —5 *n*-BuLi→ 132 —E^+→ 133

132
a: $R^1 = R^2 =$ Ph
b: $R^1 = R^2 = CF_3$

133
a: E = CO_2Me

Scheme 6.47

The X-ray crystallographic analysis indicated that **130a** and **133a** have a distorted TBP structure with two oxygen atoms at the apical positions and three carbons at the equatorial positions. The four-membered rings are slightly puckered; the torsion angles are 9.7(2)° and 4.7(6)°, and hence the four-membered ring of **133a** is closer to a plane than that of **130a**. Since **130a** and **133a** are considered as formal intermediates of the Wittig reaction of nonstabilized and stabilized ylides, respectively, the results strongly support the following proposal of Vedejs concerning the origin of the stereochemistry of the Wittig reaction.[6e,66] That is, the Wittig reaction of nonstabilized phosphorus ylides proceeds via an early transition state emerging from the cross approach of ylides and carbonyl compounds, leading to a puckered four-membered ring with a favored formation of (*Z*)-oxaphosphetanes, whereas the Wittig reaction of stabilized ylides passes through a late transition state achieved by their parallel approach to give mainly (*E*)-oxaphosphetanes with a planar four-membered ring.

The thermolysis of **130** and **133a** around 100°C gave quantitatively the corresponding olefin and cyclic phosphinate. Monitoring of the reaction by ^{19}F NMR showed that the reaction is first order in the substrate. From kinetic studies involving substituent effects, solvent effects, and measurement of activation parameters, it was found that the olefin formation from these 1,2-oxaphosphetanes, the second step of the Wittig reaction, proceeds via a slightly polar transition state, as shown in Scheme 6.48. The relative rates at 111.3°C were calculated to be 1:11,400:478,000 (**130c:130b:133a**), indicating that the olefin formation was largely accelerated by the introduction of a methoxycarbonyl group at the 3-position.

From the author's recent studies on tetracoordinate 1,2-oxaphosphetanes, which were synthesized by the intramolecular dehydration of β-hydroxyalkylphosphonic acid monoesters and β-hydroxyalkylphosphinic acids with dicyclo-

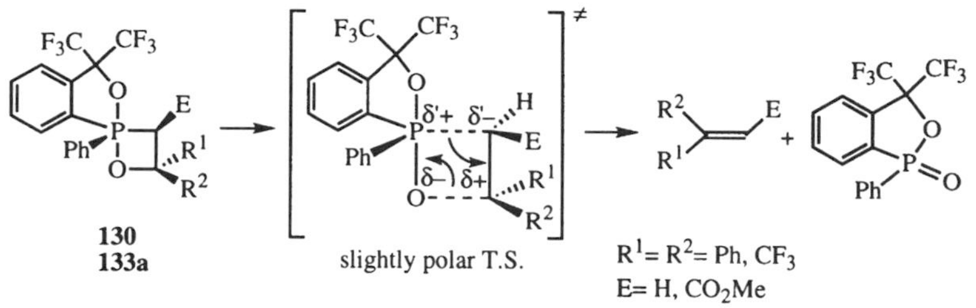

Scheme 6.48

hexyl carbodiimide,[67] a spirophosphorane with two oxaphosphetane rings in the molecule was expected to undergo a double-olefin extrusion. 1,5-Dioxa-4λ^5-phosphaspiro[3.3]heptane **134** as a novel pentacoordinate phosphorane was synthesized by the intramolecular cyclization and dehydration of bis(hydroxyalkyl)phosphine oxide **135** with Appel reagent ($Ph_3P—CCl_4$) in the presence of Et_3N. The X-ray crystallographic analysis showed a distorted TBP structure. Interestingly, upon heating at 190°C **134** did not undergo double-olefin extrusion, but afforded one equivalent of olefin and vinylphosphinic acid **136**, which was most likely formed by isomerization of intermediary tetracoordinate 1,2-oxaphosphetane emerging with the olefin under the reaction conditions (Scheme 6.49).[68] Very recently, double-olefin extrusion was achieved by using 4,4′,4′,4′-tetraarylspirobi[1,2-oxaphosphetane] and its derivatives.[69]

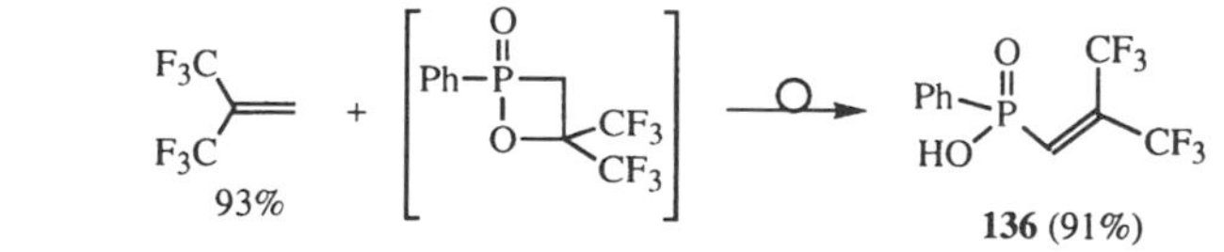

Scheme 6.49

The selected bond lengths and angles of the oxaphosphetane rings of **121–124**, **126–128**, **130a**, **133a**, and **134** are summarized in Table 6.2.

TABLE 6.2 Selected Bond Lengths and Angles of Oxaphosphetane Rings

	P—O (Å)	P—C (Å)	O—P—C (deg.)
121	2.012(14)	1.760(20)	71.3(7)
122	1.83(2)	1.79(1)	75.5(6)
123	1.85	1.81	71.6
124	1.835	1.784	—
126	1.799(5)	1.785(6)	73.4(2)
127	1.810(2)	1.772(3)	74.4(1)
128	1.783(2)	1.777(3)	75.0(1)
130a	1.728(2)	1.808(4)	77.4(1)
133a	1.781(6)	1.823(9)	75.5(3)
134	1.757(2)	1.818(3)	77.2(1)
	1.758(2)	1.814(4)	77.2(1)

6.4.2 Hexacoordinate 1,2-Oxaphosphetanides

Recently, Evans and co-workers reported that hexacoordinate 1,2-oxaphosphetanides **137** were observed as complicating signals by low-temperature ^{31}P NMR spectroscopy in the stereoselective condensation reaction of lithium enolates of alkoxycarbonylmethylspirooxyphosphoranes **138** with benzaldehyde (Scheme 6.50).[70]

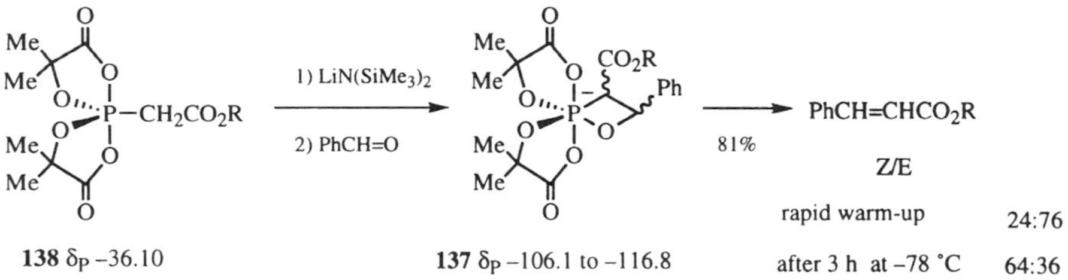

Scheme 6.50

Hexacoordinate 1,2-oxaphosphetanides **140** bearing the Martin ligand were synthesized by the author's and Akiba's laboratories from the corresponding hydroxy derivatives **139**; however, their thermolysis did not give the corresponding olefin, but benzophenone and methylphosphorane were given, except for the reaction of hexafluoroacetone adduct **140b**. The isomerization of the first formed hexacoordinate isomer to the most stable isomer **141** was observed by NMR spectroscopy (Scheme 6.51).[71,72]

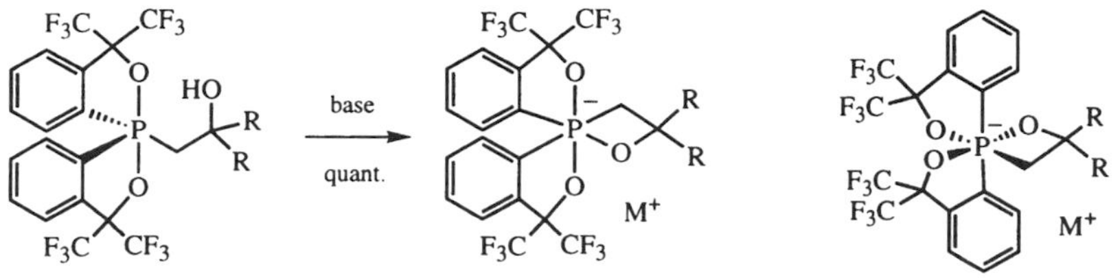

139

R= Ph δ_P −20.6
R= CF_3 δ_P −21.6

base, quant.

140

a: R= Ph; M= K, 18-c-6
b: R= CF_3; M= K, 18-c-6
c: R= Ph; M= K

141

δ_P ($(CD_3)_2CO$) −122.76
δ_P ($(CD_3)_2CO$) −118.37
δ_P (CD_3CN) −121.8

Scheme 6.51

6.4.3 Pentacoordinate 1,2-Azaphosphetidines

As it was shown that the intramolecular cyclization and dehydration is very useful for the synthesis of phosphoranes, as described above, the method was applied to β-aminoalkylphosphine oxide **142** bearing the Martin ligand.[73] Use of Mitsunobu reagent instead of Appel reagent, which was effective for the synthesis of **134**, gave pentacoordinate 1,2-azaphosphetidine **143**, a nitrogen analog of 1,2-oxaphosphetane **130** at the 1-position, namely, an intermediate of the Wittig reaction using Schiff's bases instead of carbonyl compounds (Scheme 6.52).[74]

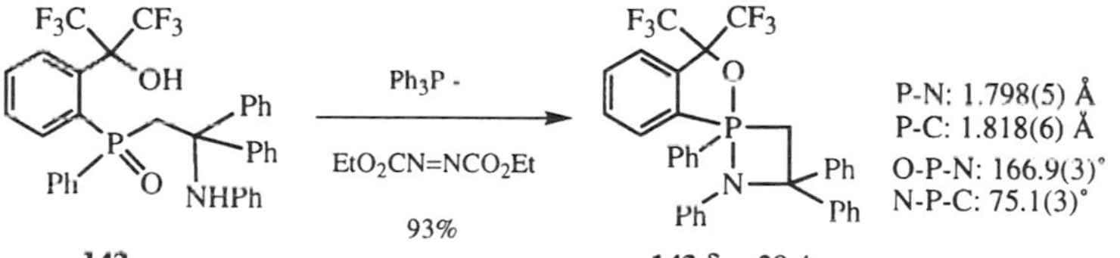

142

Ph_3P - $EtO_2CN{=}NCO_2Et$, 93%

143 δ_p −29.4

P-N: 1.798(5) Å
P-C: 1.818(6) Å
O-P-N: 166.9(3)°
N-P-C: 75.1(3)°

Scheme 6.52

When **143** was dissolved in d_8-toluene, one more signal appeared at higher field and the ratio was around 6:1 (Scheme 6.53). From NMR spectroscopic analysis this signal could be assigned to be an N-equatorial and C-apical pseudorotational isomer **144**. Very recently,[75] such a pseudorotamer **145** could be isolated as a pure state by the introduction of methoxycarbonyl group at the 3-position, followed by deprotonation–protonation at the 3-position, although X-ray analysis has not been done. It is very interesting that pseudorotation occurs before undergoing the Wittig-type reaction, because it suggests the calculated result that pseudorotation from an O-apical to a C-apical isomer takes place before the olefin formation, the second step of the Wittig reaction.

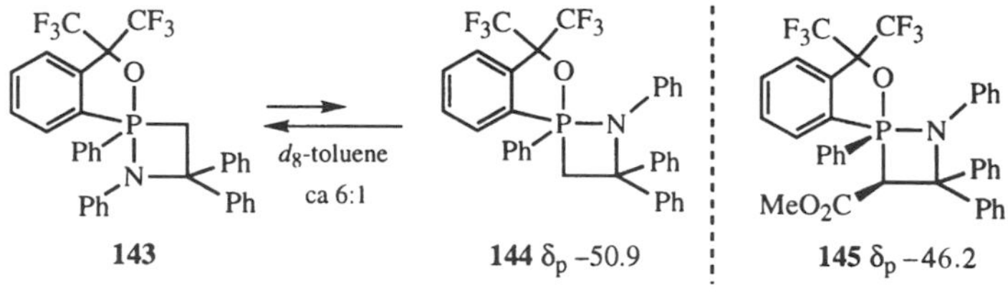

Scheme 6.53

With regard to pentacoordinate and hexacoordinate four-membered phosphorus compounds containing a nitrogen atom, 1,3,2-diazaphosphetidines such as **55**, **84**, **108**, and **109** have been already described in Sections 6.2.1.2, 6.2.2.1, 6.2.1.5 and 6.2.1.5, respectively. Some other examples: **146a**[76a] and **146b**,[76b] along with 1,3,2-oxazaphosphetidine **147**[77] are shown in Scheme 6.54.

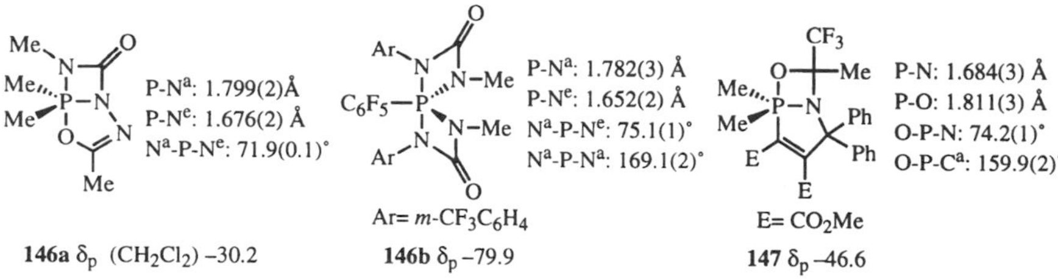

Scheme 6.54

6.4.4 Pentacoordinate 1,2-Thiaphosphetanes

Thioketones and -aldehydes also react with phosphorus ylides. Sometimes the isolated products are analogous with those of the carbonyl Wittig olefination, namely olefin plus phosphine sulfide, but often episulfide formation is observed,[78] in at least one case stereoselectively.[79] It has been assumed that these reactions proceed via a thiaphosphetane intermediate, but such a species has not previously been observed directly in thio-Wittig reactions under suitable reaction conditions. Erker and co-workers treated methylenetriphenylphosphorane with thiobenzophenone in d_8-toluene at −50°C to give pentacoordinate 1,2-thiaphosphetane **148**, which was stable ($t_{1/2} > 1$ h) up to −20°C (Scheme 6.55).[80] The reaction using the ^{13}C-enriched compounds $Ph_3P{=}^{13}CH_2$ and $Ph_2^{13}C{=}S$ revealed the phosphorus–carbon–carbon connectivity: $^{13}CH_2$ signal δ_C 64.4, dd, $^1J_{PC} = 93$ Hz, $^1J_{CC} = 37$ Hz; $^{13}CPh_2$ signal at δ 50.4, d, $^1J_{CC} = 37$ Hz. Above −20°C the thiaphosphetane slowly decomposes to give triphenylphosphine and 2,2-diphenylthiirane (> 95%). At higher temperatures, the secondary reaction products PPh_3 and $Ph_2C(—CH_2—S—)$ disappear, and the final products are Ph_3PS and $CH_2{=}CPh_2$. For the thiaphosphetane to thiirane transformation the path involving retroaddition, followed by thiophilic attack of the ylide to form betaine was concluded to be unlikely from the results of cross experiment.

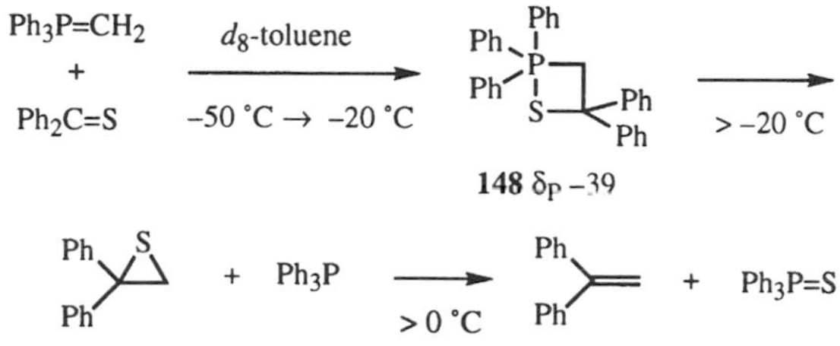

Scheme 6.55

6.4.5 Hypervalent 1,2-Dihydro-1,2-Diphosphetes

Very recently, Regitz and Bertrand and their co-workers found that the reaction of Bertrand's carbene with phosphaalkyne gave the unique four-membered ring, $1\sigma^4$, $2\sigma^2$-diphosphete **149**, which can be regarded as a cyclic phosphoranylidenephosphane. Alkylation of **149** with MeOTf followed by addition of *o*-chloranil afforded 1,2-dihydro-$1\sigma^4$, $2\sigma^5$-diphosphete **150**, whereas the reaction with two mol equivalents of *o*-chloranil gave the corresponding $1\sigma^4, 2\sigma^6$-diphosphete **151**, whose structure was determined to be octahedral by X-ray crystallographic analysis (Scheme 6.56).[81]

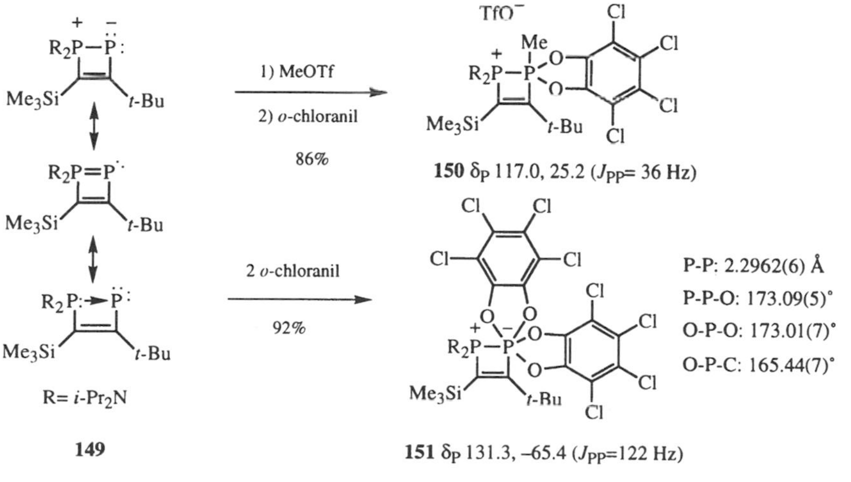

Scheme 6.56

6.5 SYNTHESIS AND REACTIONS OF LOW-COORDINATE HYPERVALENT PHOSPHORUS COMPOUNDS

6.5.1 10-P-3 Compounds

The first tricoordinate hypervalent phosphorus compound, 10-P-3, 3,7-di-*t*-butyl-5-aza-2,8-dioxa-1-phosphabicyclo[3.3.0]octa-2,4,6-triene (**13a**) was reported by Arduengo and co-workers in 1983. **13** was easily synthesized by condensation of a diketo amine ligand, **152**, with a phosphorus trihalide in the presence of three equivalents of a base, as shown in Scheme 6.57. The X-ray analysis showed it has a T-shape structure. The Ramirez-type reactions have been mentioned in Section 6.2.1.1.3. Recent reviews cover 10 years of research of this field.[82]

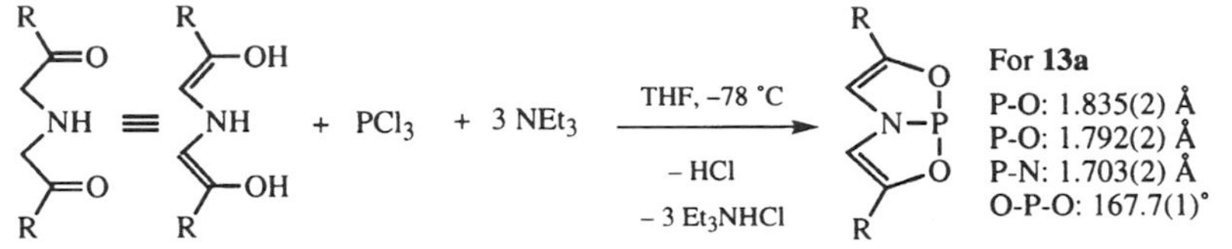

R R R
=O -OH O
NH ≡ NH + PCl_3 + 3 NEt_3 THF, −78 °C N−P
=O -OH − HCl O
R R − 3 Et_3NHCl R

152 13

For **13a**
P-O: 1.835(2) Å
P-O: 1.792(2) Å
P-N: 1.703(2) Å
O-P-O: 167.7(1)°

a: R= *t*-Bu (80%) δ_P 187
b: R= 1-Ad (63%) δ_P 185

Scheme 6.57

6.5.2 10-P-4 Anionic Compounds

Phosphoranides are hypervalent anionic phosphorus species, having more than four electron pairs in the phosphorus valence shell; they may also be carefully regarded as the conjugate bases of phosphoranes containing P—H bonds. The ions possess single bonds between phosphorus and its substituents, and usually have the structure with the lone pair of electrons equatorial in a pseudo-TBP arrangement (10-P-4) as shown below. For a typical stable example, Granoth and Martin prepared lithium derivative **6** either by deprotonation of the phosphorane **3** by $LiAlH_4$ or by $LiAlH_4$ reduction of triflate salt **4**, as shown in Scheme 6.5. Schomburg and co-workers isolated spirocyclic phosphoranide **153** from the reaction of Et_3N with phosphorane **154**, and determined its structure by X-ray crystallography (Scheme 6.58).[83] The chemical shift value of ^{31}P NMR in solution strongly suggests that **153** has a ring-opening phosphite structure. The structure of $Ph(CN)_2ClP^-$ Et_4N^+ (**155**) was also determined by X-ray crystallography to have one of the 10-P-4 structures, which have been interpreted as representing "frozen" intermediates for nucleophilic addition to a phosphorus (III) center.[84] The chemistry of this kind of hypervalent species was also described in a recent review.[85]

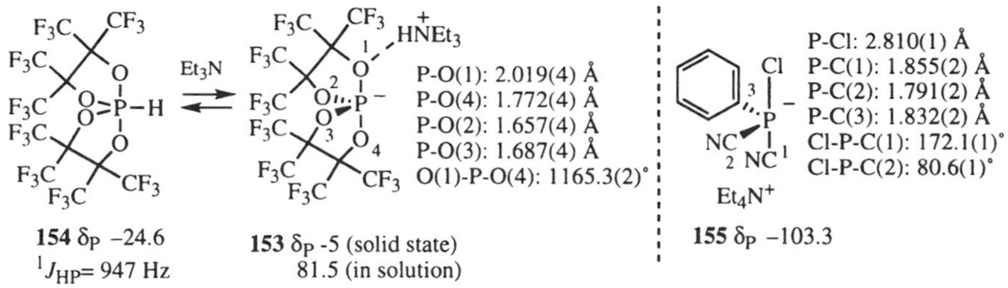

154 δ_P −24.6
$^1J_{HP}$= 947 Hz

153 δ_P -5 (solid state)
81.5 (in solution)

P-O(1): 2.019(4) Å
P-O(4): 1.772(4) Å
P-O(2): 1.657(4) Å
P-O(3): 1.687(4) Å
O(1)-P-O(4): 1165.3(2)°

155 δ_P −103.3

P-Cl: 2.810(1) Å
P-C(1): 1.855(2) Å
P-C(2): 1.791(2) Å
P-C(3): 1.832(2) Å
Cl-P-C(1): 172.1(1)°
Cl-P-C(2): 80.6(1)°

Scheme 6.58

Karsch and co-workers reported the synthesis of a spirocyclic cation **156** with a PP_4 skeleton and a 10 electron spiro P atom.[86] 1,2,3-Triphosphetene **158** was prepared by the reaction of PCl_3 with diphosphinomethanide **157** in the presence of TMEDA along with phosphorus ylide **159**. Cationic species **160** was obtained from **159** and $NaPPh_6$, which was reacted with **158** to give **156**. The X-ray analysis confirmed that the central atom is tetravalent; however, it does not have a distorted tetrahedral environment, but exhibits a distorted pseudo-TBP framework. The free equatorial position indicates the presence of a lone pair. The $^{31}P\{^1H\}$ NMR spectra of **156** in solution reveal a rapid scrambling of the equatorial and axial positions at room temperature. The signal of the central phosphorus atom is shifted to high field (δ − 76.26). The average coupling constant is $^1J_{PP}$ = 137.3 Hz. On cooling to −107°C, this scrambling process is essentially frozen, so an A_2B_2M spin system can be observed (Scheme 6.59).

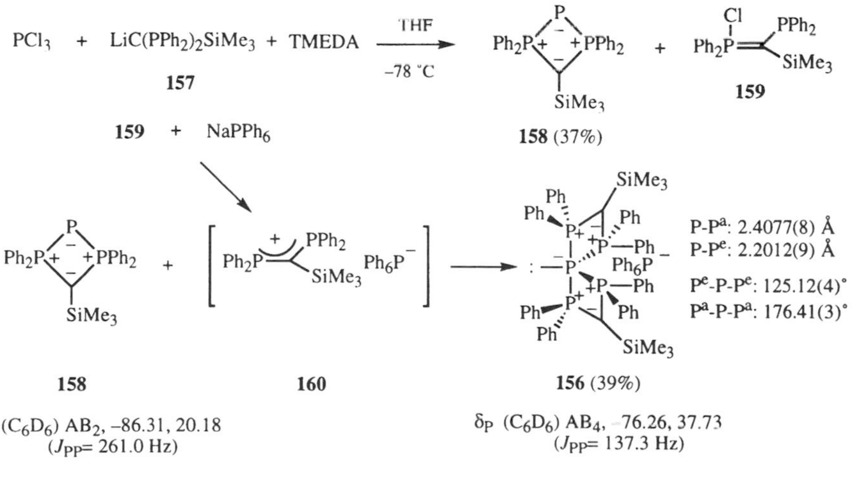

PCl_3 + $LiC(PPh_2)_2SiMe_3$ + TMEDA (157) —THF, −78 °C→ 158 (37%) + 159

159 + $NaPPh_6$

158 + 160 → 156 (39%)

P-P^a: 2.4077(8) Å
P-P^e: 2.2012(9) Å
P^e-P-P^e: 125.12(4)°
P^a-P-P^a: 176.41(3)°

158: δ_P (C_6D_6) AB_2, −86.31, 20.18 (J_{PP}= 261.0 Hz)

156: δ_P (C_6D_6) AB_4, −76.26, 37.73 (J_{PP}= 137.3 Hz)

Scheme 6.59

Finally, the author wishes to introduce some reviews on the ^{31}P NMR spectroscopy, which has become a routine and useful method to get information about the structures both in solution and in solid state of organophosphorus compounds and has strongly contributed to the development of this research field.[87]

REFERENCES

1. G. Wittig, M. Rieber, *Ann.* **562**, 187 (1949).
2. H. Staudinger, J. Meyer, *Chem. Ber.* **53**, 72 (1920).
3. F. Ramirez, *Pure Appl. Chem.* **9**, 337 (1964); F. Ramirez, *Bull. Soc. Chim. Fr.* 2443 (1966).
4. F. H. Westheimer, *Acc. Chem. Res.* **1**, 70 (1968).
5. F. Ramirez, *Acc. Chem. Res.* **1**, 168 (1968). F. Ramirez, *Bull. Soc. Chim. Fr.* 3491 (1968). K. Mislow, *Acc. Chem. Res.* **3**, 321 (1970). Berry pseudorotation (BPR) mechanism, see: R. S. Berry, *J. Chem. Phys.* **32**, 933 (1960). Turnstile rotation (TR) mechanism, see: P. Gillespie, P. Hoffman, H. Klusacek, D. Marquarding, S. Pfohl, F. Ramirez, E. A. Tsolis, I. Ugi, *Angew. Chem. Int. Ed. Engl.* **10**, 687 (1971); I. Ugi, D. Marquarding, H. Klusacek, P. Gillespie, F. Ramirez, *Acc. Chem. Res.* **4**, 288 (1971).
6. (a) A. Maercker, *Organic Reactions* **14**, 270 (1965); (b) D. J. H. Smith, in *Comprehensive Organic Chemistry*, D. H. R. Barton, W. D. Ollis, Eds (Pergamon, Oxford, 1979), Vol. 2, pp. 1316–1329; (c) I. Gosney, A. G. Rowley, in *Organophosphorus Reagents in Organic Synthesis*, J. I. G. Cadogan, Ed. (Academic Press, New York, 1979), pp. 17–153; (d) B. E. Maryanoff, A. B. Reitz, *Phosphorus, Sulfur and Silicon* **27**, 167 (1986); (e) B. E. Maryanoff, A. B. Reitz, *Chem. Rev.* **89**, 863 (1989).
7. E. Vedejs, K. A. J. Snoble, *J. Am. Chem. Soc.* **95**, 5778 (1973); E. Vedejs, G. P. Meier, K. A. J. Snoble, *ibid.* **103**, 2823 (1981); A. B. Reitz, S. O. Nortey, A. D. Jordan, Jr., M. S. Mutter, B. E. Maryanoff, *J. Org. Chem.* **51**, 3302 (1986); B. E. Maryanoff, A. B. Reitz, M. S. Mutter, R. R. Inners, H. R. Almond, Jr., R. R. Whittle, R. A. Olofson, *J. Am. Chem. Soc.* **108**, 7664 (1986); E. Vedejs, C. F. Marth, R. Ruggeri, *Ibid.* **110**, 3940 (1988).
8. R. R. Holmes, *Chem. Rev.* **96**, 927 (1996); C. Y. Wong, D. K. Kennepohl, R. G. Cavell, *ibid.* **96**, 1917 (1996).
9. (a) D. Hellwinkel, in *Organic Phosphorus Compounds*, G. M. Kosolapoff, L. Maier, Eds (John Wiley & Sons, New York, 1972), Vol. 3, Chapter 5B, pp. 185–339; (b) D. J. H. Smith, in *Comprehensive Organic Chemistry*, D. H. R. Barton, W. D. Ollis, Eds (Pergamon Press, Oxford, 1979), Vol. 2, Part 10.6, pp. 1233–1256; (c) K. Burger, in *Organophosphorus Reagents in Organic Synthesis*, J. I. G. Cadogan, Ed (Academic Press, New York, 1979), Chapter 11, pp. 467–510.
10. D. Hellwinkel, *Chem. Ber.* **98**, 576 (1968).
11. (a) I. Granoth, J. C. Martin, *J. Am. Chem. Soc.* **100**, 7434 (1978); (b) *Idem*, *ibid.***101**, 4618 (1978); (c) *Idem*, *ibid.* **101**, 4623 (1978); (d) J. C. Martin, E. F. Perozzi, *Science* **191**, 154 (1976); (e) J. C. Martin, *ibid.* **221**, 509 (1983).

12. I. N. Zhmurova, A. V. Kirsanov, *J. Gen. Chem. USSR* **29**, 1668 (1959).
13. F. Ramirez, A. J. Bigler, C. P. Smith, *J. Am. Chem. Soc.* **90**, 3507 (1968).
14. N. Weferling, R. Schmutzler, *Chem. Ber.* **122**, 1465 (1989).
15. A. Schmidpeter, D. Schomburg, W. S. Sheldrick, J. H. Weinmaier, *Angew. Chem. Int. Ed. Engl.* **15**, 781 (1976).
16. A. J. Arduengo, III, C. A. Stewart, F. Davidson, D. A. Dixon, J. Y. Becker, S. A. Culley, M. B. Mizen, *J. Am. Chem. Soc.* **109**, 627 (1987).
17. I. Neda, C. Melnicky, A. Vollbrecht, A. Fischer, P. G. Jones, A. Martens-Von Salzen, R. Schmutzler, U. Niemeyer, B. Kutscher, J. Engel, *Phosphorus, Sulfur, and Silicon* **109–110**, 629 (1996).
18. D. Houalla, T. Mouheich, M. Sanchez, R. Wolf, *Phosphorus* **5**, 229 (1975).
19. T. Kawashima, W. G. Bentrude, *J. Am. Chem. Soc.* **101**, 3981 (1979); W. G. Bentrude, T. Kawashima, B. A. Keys, M. Garroussian, W. Heide, D. A. Wedegaertner, *Ibid.* **109**, 1227 (1987).
20. D. Houalla, L. Moureau, S. Skouta, M.-R. Mazieres, *Phosphorus, Sulfur, and Silicon* **103**, 199 (1995).
21. R. R. Holmes, J. M. Holmes, R. O. Day, K. C. Kumara Swamy, V. Chandrasekhar, *Phosphorus, Sulfur, and Silicon* **103**, 153 (1995).
22. H. J. Bestmann, R. Kunstmann, *Chem. Ber.* **102**, 1816 (1969).
23. J. Wulff, R. Huisgen, *Angew. Chem.* **79**, 472 (1967).
24. (a) J. C. Clardy, D. S. Milbrath, J. P. Springer, J. G. Verkade, *J. Am. Chem. Soc.* **98**, 623 (1976); (b) D. S. Milbrath, J. G. Verkade, *ibid.* **99**, 6607 (1977); (c) C. Lensink, S. K. Xi, L. M. Daniels, J. G. Verkade, *ibid.* **111**, 3478 (1989); (d) M. A. H. Laramy, J. G. Verkade, *ibid.* **112**, 9421 (1990); (e) J. G. Verkade, *Acc. Chem. Res.* **26**, 483 (1993).
25. K.-y. Akiba, K. Okada, K. Ohkata, *Tetrahedron Lett.* **27**, 5221 (1986).
26. D. Hellwinkel, W. Krapp, *Chem. Ber.* **111**, 13 (1978).
27. Y. Segall, I. Granoth, A. Kalir, and E. D. Bergmann, *J. Chem. Soc. Chem. Commun.* 399 (1975).
28. Y. Segall, I. Granoth, *J. Am. Chem. Soc.* **100**, 5130 (1978).
29. L. Lamandé, A. Munoz, *Tetrahedron* **46**, 3527 (1990).
30. D. Schomburg, N. Weferling, R. Schmutzler, *J. Chem. Soc. Chem. Commun* 609 (1981).
31. J. E. Richman, R. O. Day, R. R. Holmes, *J. Am. Chem. Soc.* **102**, 3955 (1980).
32. H. W. Roesky, D. Amirzadeh-Asl, W. S. Sheldrick, *J. Am. Chem. Soc.* **104**, 2919 (1982).
33. N. V. Timosheva, A. Chandrasekaran, T. K. Prakasha, R. O. Day, R. R. Holmes, *Inorg. Chem.* **35**, 6552 (1996).
34. S. Vollbrecht, A. Vollbrecht, J. Jeske, P. G. Jones, R. Schmutzler, W.-W. du Mont, *Chem. Ber.* **130**, 819 (1997).
35. S. Kojima, K. Kajiyama, M. Nakamoto, K.-y. Akiba, *J. Am. Chem. Soc.* **118**, 12866 (1996).
36. C. K. McClure, C. W. Grote, B. A. Lockett, *J. Org. Chem.* **57**, 5195 (1990).

37. R. M. Moriarty, J. Hiratake, K. Liu, A. Wendler, A. K. Awasthi, R. Gilardi, *J. Am. Chem. Soc.* **113**, 9374 (1991).
38. D. Hellwinkel, *Chem. Ber.* **99**, 3642 (1966).
39. S. Kojima, K. Kajiyama, K.-y. Akiba, *Tetrahedron Lett.* **35**, 7037 (1994).
40. S. Kojima, K. Kajiyama, K.-y. Akiba, *Bull. Chem. Soc. Jpn* **68**, 1785 (1995).
41. J. J. H. M. Font Freide, S. Trippett, *J. Chem. Soc. Chem. Commun.* 157 (1980); *Idem, ibid.* 934 (1980).
42. H. P. Latscha, P. B. Hormuth, *Angew. Chem. Int. Ed. Engl.* **7**, 299 (1968).
43. M. L. Ziegler, J. Weiss, *Angew. Chem. Int. Ed. Engl.* **8**, 455 (1969).
44. R. Krebs, R. Schmutzler, D. Schomburg, *Polyhedron* **8**, 731 (1989).
45. M. A. Said, M. Pülm, R. Herbst-Irmer, K. C. Kumara Swamy, *J. Am. Chem. Soc.* **118**, 9841 (1996).
46. T. K. Prakasha, R. O. Day, R. R. Holmes, *Inorg. Chem.* **31**, 1913 (1992).
47. R. R. Holmes, T. K. Prakasha, R. O. Day, *Inorg. Chem.* **32**, 4360 (1993).
48. (a) P. Sayer, M. Gouterman, C. R. Connell, *J. Am. Chem. Soc.* **99**, 1082 (1977); (b) C. A. Marrese, C. J. Carrano, *Inorg. Chem.* **22**, 1858 (1983).
49. P. Sayer, M. Gouterman, C. R. Connell, *Acc. Chem. Res.* **15**, 73 (1982); M. Tsutsui, C. J. Carrano, *J. Coord. Chem.* **7**, 79 (1977); M. Gouterman, P. Sayer, E. Shankland, J. P. Smith, *Inorg. Chem.* **20**, 87 (1981); S. Mangani, E. F. Meyer, Jr., D. L. Cullen, M. Tsutsui, C. J. Carrano, *Ibid.* **22**, 400 (1983).
50. K. Kunimoto, H. Segawa, T. Shimidzu, *Tetrahedron Lett.* **33**, 6327 (1992); H. Segawa, K. Kunimoto, A. Nakamoto, T. Shimidzu, *J. Chem. Soc. Perkin Trans. 1*, 939 (1992).
51. H. Segawa, N. Nakayama, T. Shimidzu, *J. Chem. Soc. Chem. Commun.* 784 (1992).
52. H. Segawa, K. Kunimoto, K. Susumu, M. Taniguchi, T. Shimidzu, *J. Am. Chem. Soc.* **116**, 11193 (1994).
53. Y. Yamamoto, R. Nadano, M. Itagaki, K.-y. Akiba, *J. Am. Chem. Soc.* **117**, 8287 (1995).
54. K.-y. Akiba, Y. Onzuka, H. Hirota, M. Itagaki, Y. Yamamoto, *Organometallics* **13**, 2800 (1994).
55. D. V. Khasnis, M. Lattman, C. D. Gutsche, *J. Am. Chem. Soc.* **112**, 9422 (1990).
56. D. V. Khasnis, J. M. Burton, M. Lattman, H. Zhang, *J. Chem. Soc. Chem. Commun.* 562 (1991).
57. I. V. Shevchenko, *Tetrahedron Lett.* **36**, 2021 (1995).
58. C. Chuit, R. J. P. Corriu, P. Monforte, C. Reyé, J.-P. Declereq, A. Dubourg, *Angew. Chem. Int. Ed. Engl.* **32**, 1430 (1993).
59. (a) G. Wittig, G. Geissler, *Ann.* **580**, 44 (1953); (b) G. A. Razuvaev, N. A. Osanova, *Dokl. Akad. Nauk SSSR* **104**, 552 (1955); *Chem. Abstr.* **50**, 11268 (1956).
60. P. D. Bartlctt, M. E. Landis, M. J. Shapiro, *J. Org. Chem.* **42**, 1661 (1977).
61. C. K. McClure, C. W. Grote, A. L. Rheingold, *Tetrahedron Lett.* **34**, 983 (1993).
62. (a) G. H. Brium, C. N. Matthews, *J. Chem. Soc. Chem. Commun.* 137 (1967); G. Chioccola, J. J. Daly, *J. Chem. Soc. (A)* 568 (1968); (b) F. Ramirez, C. P. Smith, J. F. Pilot, *J. Am. Chem. Soc.* **90**, 6726 (1968); (c) H. A. E. Aly, J. H. Barlow, D. R. Russell, D. J. H. Smith, M. Swindles, S. Trippett, *J. Chem. Soc. Chem. Commun.*

449; (1976); (d) H. J. Bestmann, K. Roth, E. Wilhelm, R. Böhme, H. Burzlaff, *Angew. Chem. Int. Ed. Engl.* **18**, 876 (1979); (e) R. W. Saalfrank, W. Paul, H. Liebenow, *Ibid.* **19**, 713 (1980); (f) Y. Huang, A. M. Arif, W. G. Bentrude, *J. Am. Chem. Soc.* **113**, 7800 (1991); Y. Huang, A. E. Sopchik, A. M. Arif, W. G. Bentrude, *Ibid.* **115**, 4031 (1993); (g) Y. Huang, W. G. Bentrude, *Ibid.* **117**, 12390 (1995); Y. Huang, A. M. Arif, W. G. Bentrude, *Phosphorus, Sulfur, and Silicon* **103**, 225 (1995).

63. T. Kawashima, K. Kato, R. Okazaki, *J. Am. Chem. Soc.* **114**, 4008 (1992).
64. M. Schlosser, K. F. Chritmann, *Angew. Chem. Int. Ed. Engl.* **5**, 126 (1966). M. Schlosser, K. F. Chritmann, *Synthesis* 38 (1969).
65. T. Kawashima, K. Kato, R. Okazaki, *Angew. Chem. Int. Ed. Engl.* **32**, 869 (1993).
66. E. Vedejs, C. F. Marth, *J. Am. Chem. Soc.* **110**, 3948 (1988).
67. T. Kawashima, M. Nakamura, A. Nakajo, N. Inamoto, *Chem. Lett.* 1483 (1994); T. Kawashima, H. Takami, R. Okazaki, *ibid.* 1987 (1994).
68. T. Kawashima, H. Takami, R. Okazaki, *J. Am. Chem. Soc.* **116**, 4509 (1994).
69. T. Kawashima, R. Okazaki, R. Okazaki, *Angew. Chem. Int. Ed. Engl.*, **36**, 2500 (1997).
70. M. L. Bojin, S. Barkallah, S. A. Evans, Jr., *J. Am. Chem. Soc.* **118**, 1549 (1996).
71. T. Kawashima, K. Watanabe, R. Okazaki, *Tetrahedron Lett.* **38**, 551 (1997).
72. S. Kojima, K.-y. Akiba, *Tetrahedron Lett.* **38**, 547 (1997).
73. T. Kawashima, T. Soda, R. Okazaki, *Angew. Chem. Int. Ed. Engl.* **35**, 1096 (1996).
74. H. J. Bestmann, F. Seng, *Angew. Chem. Int. Ed. Engl.* **2**, 393 (1963).
75. T. Kawashima, T. Soda, R. Okazaki, unpublished results.
76. (a) A. Schmidpeter, J. Luber, D. Schomburg, W. S. Sheldrick, *Chem. Ber.* **109**, 3581 (1976); (b) H. W. Roesky, K. Ambrosius, M. Banek, W. S. Sheldrick, *Ibid.* **113**, 1847 (1980).
77. W. S. Sheldrick, D. Schomburg, A. Schmidpeter, T. von Criegern, *Chem. Ber.* **113**, 55 (1980).
78. For a review, see: E. Schaumann, in *The Chemistry of Double-bonded Functional Groups*, S. Patai, Ed. (John Wiley, New York, 1989), p. 1269.
79. E. Vedejs, D. A. Perry, R. G. Wilde, *J. Am. Chem. Soc.* **108**, 2985 (1986).
80. S. Wilker, C. Laurent, C. Sarter, C. Puke, G. Erker, *J. Am. Chem. Soc.* **117**, 7293 (1995).
81. M. Sanchez, R. Réau, F. Dahan, M. Regitz, G. Bertrand, *Angew. Chem. Int. Ed. Engl.* **35**, 2228 (1996).
82. For a review; see: A. J. Arduengo, III, C. A. Stewart, *Chem. Rev.* **94** 1215 (1994).
83. D. Schomburg, W. Storzer, R. Bohlen, W. Kuhn, G.-V. Röschenthaler, *Chem. Ber.* **116**, 3301 (1983).
84. R. M. K. Deng, K. B. Dillon, W. S. Sheldrick, *J. Chem. Soc. Dalton Trans.* 551 (1990).
85. K. B. Dillon, *Chem. Rev.* **94**, 1441 (1994).
86. H. H. Karsch, E. Witt, F. E. Hahn, *Angew. Chem. Int. Ed. Engl.* **35**, 2242 (1996).
87. G. Marvel, in *Progress in NMR Spectroscopy*, J. W. Emsley, J. Feeney, L. H. Sutcliffe, Eds. (Pergamon Press, Oxford, 1966), Vol. 1, p. 251; M. M.

Crutchfield, C. H. Dungan, J. H. Letcher, V. Mark, J. R. Van Wazer, ^{31}P Nuclear Magnetic Resonance, in *Topics in Phosphorus Chemistry*, M. Grayson, E. J. Griffith, Eds (Interscience Publishers, New York, 1967), Vol. 5; G. Mavel, in *Annual Reports on NMR Spectroscopy*, E. F. Mooney, Ed. (Academic Press, New York, 1973) Vol. 5b, p. 1; D. G. Gorenstein, *Progr. NMR Spectroscopy* **16**, 1 (1983); D. G. Gorenstein, *Phosphorus-31 NMR: Principles and Applications* (Academic Press, New York, 1984); *Phosphorus-31 NMR Spectral Properties in Compound Characterization and Structural Analysis*, L. D. Quin, J. G. Verkade, Eds. (VHC, New York, 1994).

7 Hypervalent Sulfuranes as Transient and Isolable Structures: Occurrence, Synthesis, and Reactivity

JÓZEF DRABOWICZ

Center of Molecular and Macromolecular Studies, Polish Academy of Sciences, Department of Organic Sulfur Compounds, Łódź, Poland

7.1 INTRODUCTION

The aim of this chapter is to give the reader an overview of the field of organic hypervalent sulfur chemistry. At present a large number of stable organosulfur compounds with ligand numbers from 1 to 6 can be prepared and, after isolation, handled under typical laboratory conditions.[1] Taking into account the scope and definition of hypervalent molecules (see Chapters 1 and 2) it became evident that all compounds with the number of ligands N equal to 5 or 6 fulfilled the formal conditions to be considered as members of this group of sulfur derivatives. Consequently, this chapter will be devoted to the presentation of the chemistry of these two classes of sulfur compounds in which the central sulfur has expanded its valence shell from eight to 10 or 12 electrons, respectively.

In the first class of hypervalent sulfur derivatives the most common group of compounds comprises δ-sulfuranes **1**, which, according to a general systematic scheme proposed by Martin and co-workers,[2] are considered as 10-S-4 species.

R1, R2, R3, R4 — S —:

1

Having accepted the definition of sulfuranes as a class of compounds in which sulfur has expanded its formal valence shell from 8 to 10 electrons, we

Chemistry of Hypervalent Compounds, edited by Kin-ya Akiba.
ISBN 0-471-24019-2

should consider another group of sulfuranes, the so-called II-sulfuranes.[3] Due to the fact that in these compounds, commonly encountered as sulfur ylides, the most important zwitterionic resonance structure, **2a**, is polarized from sulfur to carbon, they should be considered as 8-S-3 species. It means that this group of tricoordinated organosulfur derivatives, which are very useful synthetically,[4-6] do not, however, fit the conditions of hypervalency.

2a ⟷ **2b**

Another two groups of hypervalent sulfur derivatives in which the central sulfur atom has expanded its valence shell to 10 electrons are sulfuranide anions **3** (they should be considered as 10-S-3 species), and 10-S-5 species **4** (sulfurane oxide **4a**, sulfurane imines **4b**) and **5** (persulfonium cation).[7]

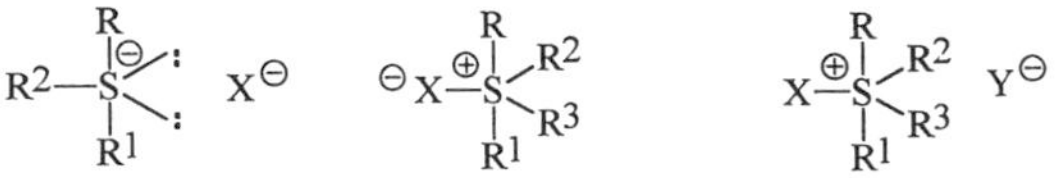

3 **4** a, X=O b, X=NR **5**

Hypervalent sulfur species with 12 electrons in their valence shells, commonly named persulfuranes, are designated as 12-S-6 species **6**.[7]

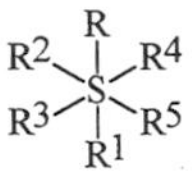

6

Short discussions on the reactivity and properties of hypervalent sulfur species may be found in many review articles[3,7–10] and books[1b,1c] devoted to sulfur chemistry. Recently, in a chapter on high-coordinated sulfur compounds,[11] various aspects of the chemistry of δ-sulfuranes and their oxides and persulfuranes have been discussed in a systematic and comprehensive way. For these reasons, and the limited space available here, this chapter will not be exhaustive, but will instead attempt to refer the interested reader to the appropriate literature. Although formally this chemistry started as early as 1873, with the preparation of a highly unstable SCl_4,[12] only the last 25 years have witnessed rapid development in this field. Therefore, this chapter is based mainly on results published after 1970, including very recent reports from 1995 and 1996.

7.2 BASIC STRUCTURAL RELATIONSHIP AND REACTIVITY PATTERN

Because the nature of bonding of hypervalent molecules is presented in detail in the first two chapters of this book, there is no need to repeat such a presentation in this chapter, devoted to the chemistry of a single group of hypervalent compounds. Considering the molecular geometry of hypervalent sulfur molecules, it should be noted that 10-S-4 and 10-S-5 species always form a trigonal bipyramidal structure (TBP geometry) and persulfuranes (12-S-6 species) have a square bipyramidal structure (SBP geometry) as shown in Scheme 7.1.[7,11]

Sulfuranes
(10-S-3; 10-S-4; 10-S-5)
TPB geometry

Persulfuranes
(12-S-6)
SBP geometry

Scheme 7.1

The molecular geometry of 10-S-3 species can also be considered formally as TBP geometry. For example, X-ray analysis shows that in **7** the collinear apical bromine ligands are perpendicular to the plane of the equatorial ligand.[13]

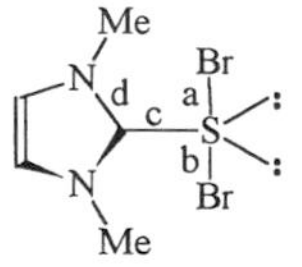

Angle (deg)	
ab	(186.0)
ac	(87.0)
acd	(96.0)

7

In hypervalent molecules which can be described as 10-S-4 and 10-S-5 species, the apical and equatorial ligands can interchange with each other to form a special type of stereoisomer, the so-called permutational isomers.[14] As early as 1960, Berry[15] proposed a mechanism called pseudorotation for nondissociative permutational isomerism of such TBP structures. This mechanism, in which a pairwise exchange of the two equatorial and two apical ligands results from the single pseudorotation step, via an SP structure, is shown in Scheme 7.2.

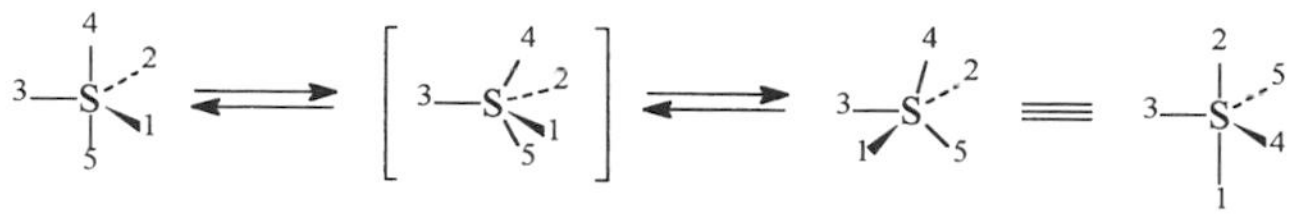

Scheme 7.2 The Berry-type mechanism of pseudorotation.

According to this mechanism, interconversion takes place with the substituent **3** stationary (pivot), whereas the angles between it and the two other equatorial substituents and two apical substituents decrease and increase respectively until an SP intermediate is formed. Further movement in this intermediate results in the formation of a new TBP in which the orginal equatorial substituents become apical and vice versa. A closely related mechanism called the "turnstile" rotation was proposed soon after by Ugi and Ramirez.[16] In this mechanism, pictured in Scheme 7.3, interchange of ligands results from the internal movement of one equatorial and one apical substituent against the remaining three substituents.

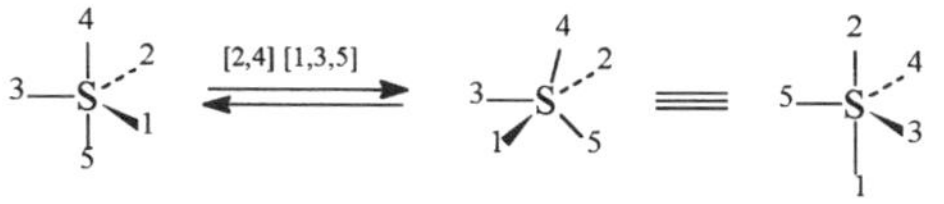

[2,4] [1,3,5]

Scheme 7.3 The turnstile rotation (TR) mechanism of pseudorotation.

It is interesting to note that the turnstile rotation was shown to be a higher-energy process in comparison with the Berry pseudorotation sequence. It should also be noted here that if a pseudorotation mechanism interconverting one 10-S-4 sulfurane structure to another requires passing through a very unstable sulfurane with an apical lone pair, its probability becomes very small.[7,11]

Very recently, Furukawa and co-workers[17] have reported, taking into account theoretical studies on the ligand-coupling reaction of tetramethyl sulfurane, that there is a possibility to consider a non-Berry pseudorotation (NBPR) process in which one apical and two equatorial ligands interchange their position in one step (Scheme 7.4).

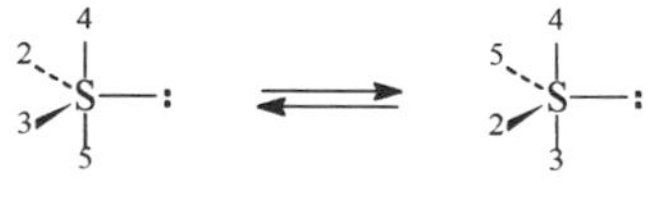

Scheme 7.4 The NBPR mechanism of pseudorotation.

Considering the topological properties of TBP structures, the NBPR process is equivalent to two successive BPR interchanges involving the lone electron pair as a pivot. The theoretical calculations [RHF/3-21G(*)] have shown that the barriers to NBPR are much higher than those of BPR, but they are lower than those of ligand coupling.[17]

Earlier an alternative mechanism through a planar transition state was proposed for interconverting two-wedge-shaped sulfuranes. This mechanism, pictured in Scheme 7.5, has been named the cuneal inversion.[18]

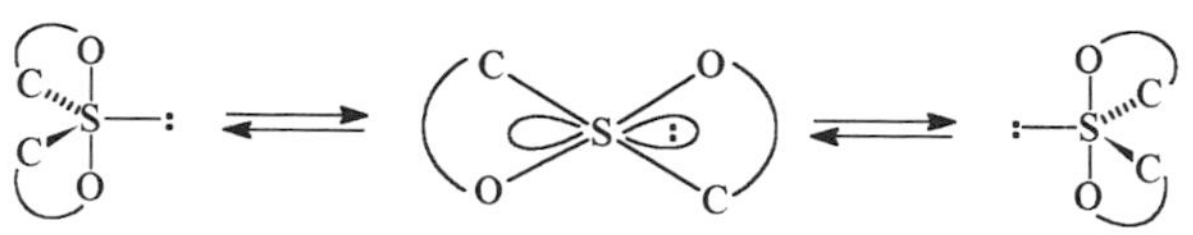

Scheme 7.5 The cuneal inversion of sulfuranes.

The interconversion of stereoisomers of octahedral persulfuranes by a dissociative mechanism involving a 10-S-5 cationic intermediate has been proposed.[19] According to this suggestion the Lewis acid-catalyzed isomerization of the all *trans* 12-S-6 difluoropersulfurane **8** to its *cis*, *cis*, *trans* isomer **9** involves the 10-S-5-persulfonium cation **10**, derived from **8** by heterolytic dissociation of a fluoride anion (Scheme 7.6).

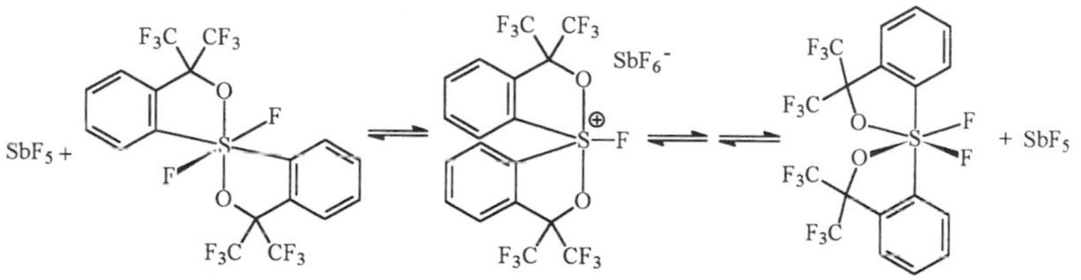

Scheme 7.6 The SbF_5-catalyzed isomerization of difluoropersulfurane.

The chemistry of hypervalent sulfuranes, like other hypervalent molecules, is dominated by the weak three-center–four-electron hypervalent bond. Such bonds are expected to be weak because two of the four electrons that participate in the bonding of the apical ligands are in a nonbonding molecular orbital. For example, their weakness is evident from structural data of the tetraalkoxysulfurane **11** in which the apical bonds are 0.12 Å longer than the equatorial bonds.[20]

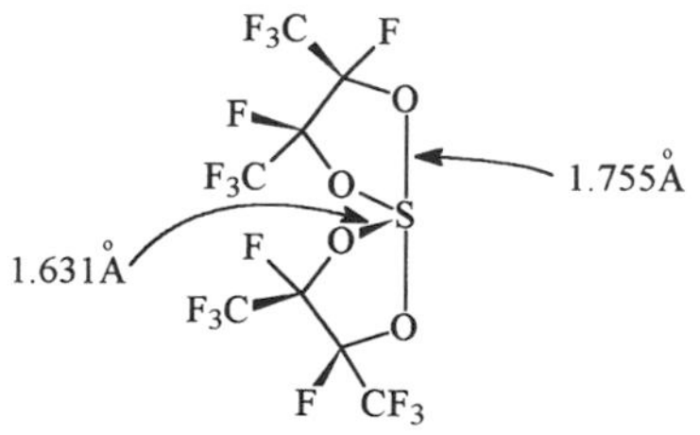

The calorimetrically determined[21,22] average bond dissociation energies (BDE) for the apical hypervalent S—O bonds in the acyclic sufurane **12**

(22.5 kcal mol^{-1}) and the spirosulfurane **13** (28 kcal mol^{-1}) are much lower than those found in their analogs containing the 2-center–2-electron bonds such as HO—S—OH (BDE = 81.4 kcal mol^{-1}) and ROS(O)OR (BDE = 65 kcal mol^{-1}).[23] The strength of the weak hypervalent bonds of **12** and **13** is comparable to the strength of covalent O—O bonds of organic peroxides (BDE values in range 30–38 kcal mol^{-1})[24] and N—N bonds of hydrazines (BDE = 22 kcal mol^{-1} for diphenyl hydrazine).[24]

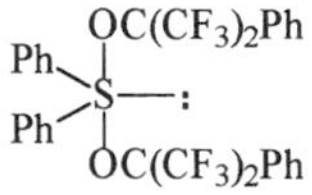

12

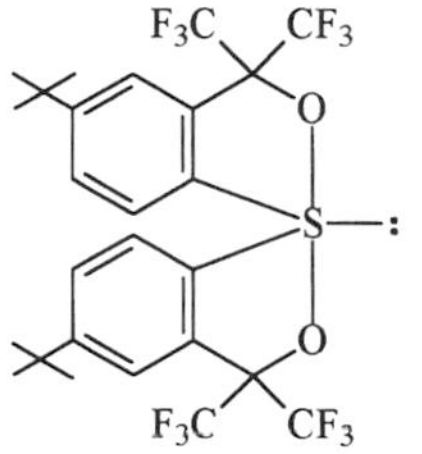

13

The most essential feature of sulfuranes as a species having an expanded valence shell is their relatively low stability caused by the tendency of the central sulfur atom to resume the normal valency by extruding either a ligand bearing a pair of electrons or a pair of ligands that afford stable compounds with an octet around the sulfur atom (Scheme 7.7).

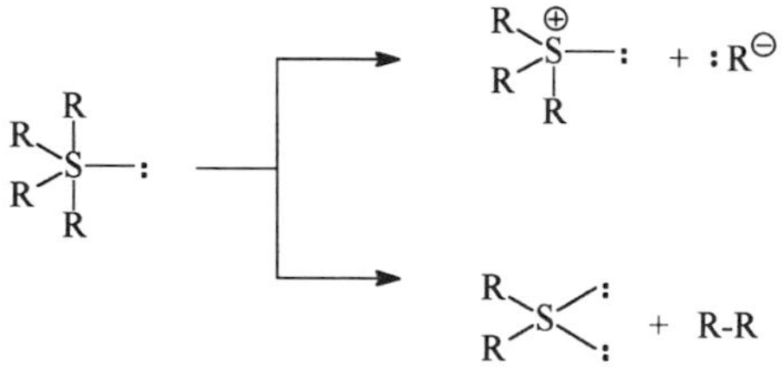

Scheme 7.7

Heterolysis of one of the O—S bonds in sulfurane **12** to form a 8-S-3 sulfonium ion **14** has been suggested to be a key step in the dehydration of tertiary and secondary alcohols to the corresponding olefins. A dissociative mechanism of this synthetically very useful reaction[25–27] is pictured in Scheme 7.8 with *t*-butyl alcohol as a substrate.[27]

A spectacular use of this sulfurane as a dehydration agent is illustrated by Eq. 7.1.[26] Other very complex alcohols were similarly converted to the corresponding unsaturated hydrocarbons.[26c–h]

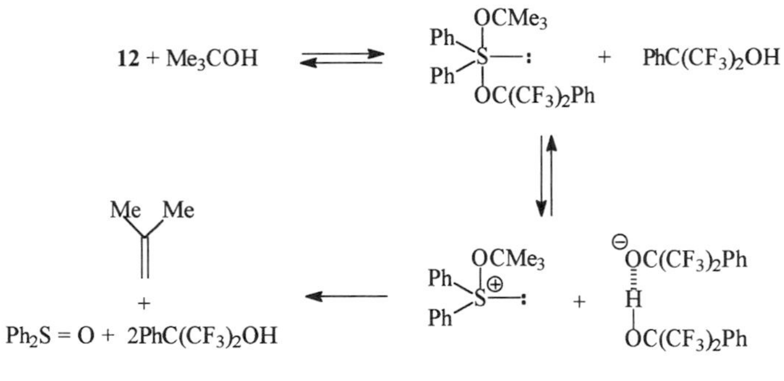

Scheme 7.8

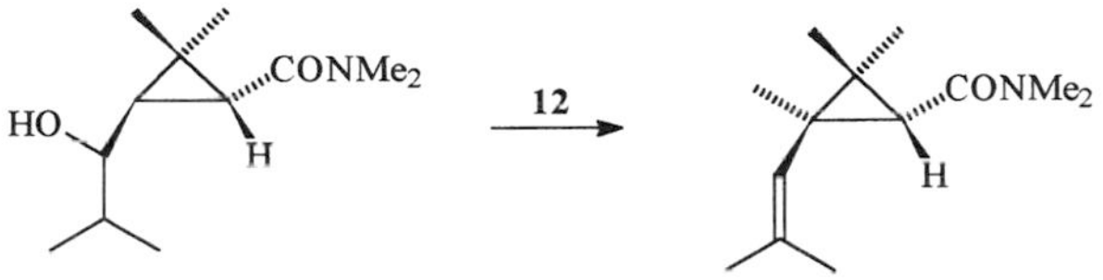

$$(7.1)$$

However, primary alcohols react with **12** to form unsymmetric ethers **15**, especially in the absence of structural features, increasing the acidity of the β protons (Eq. 7.2).[27]

$$\mathrm{ROH} + \mathbf{12} \longrightarrow \mathrm{RO{-}C(CF_3)_2{-}Ph} \quad (\mathbf{15}) \tag{7.2}$$

An one-step synthesis of simple epoxides and other more complex cyclic ethers from diols and **12** has also been reported.[28,29]

The tendency to resume the normal valency is a driving force for the formation of sultine **16** either by the room-temperature methylation of the tetraethylammonium salt of sulfuranide oxide **17** by methyl trifluoromethanesulfonate, or by heating (> 30 min at 210°C) of either methylsulfurane oxide **18** or methoxysulfurane **19** (Scheme 7.9).[30]

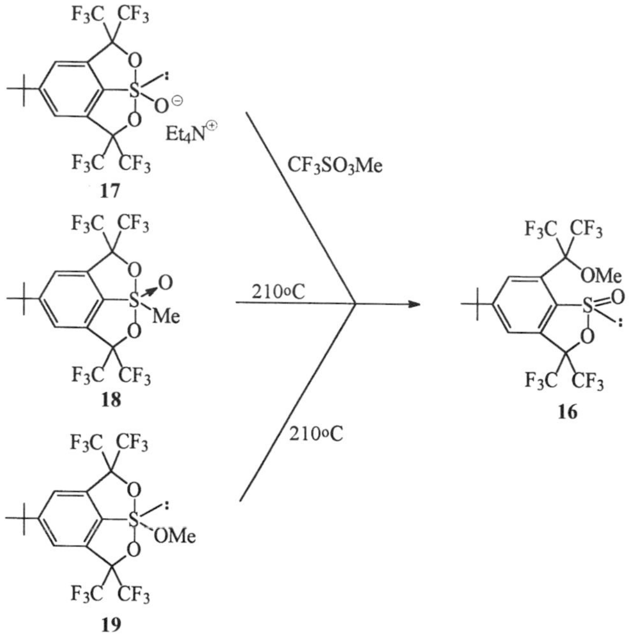

Scheme 7.9

In this context it is interesting to note that the hypervalent bonding is an important factor effecting the pyridine methylation reaction by a series of sulfurane structures **16** and **18–22**.

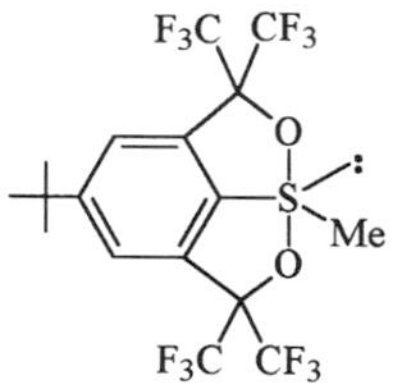

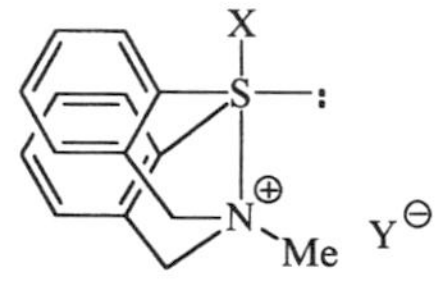

21 X=Me; Y=PF_6
22 X=OMe; Y=$SbCl_6$

Compounds **16** and **18–19** yielded only *N*-methylpyridinium sulfuranide oxide **23** in each case (Eq. 7.3).[30]

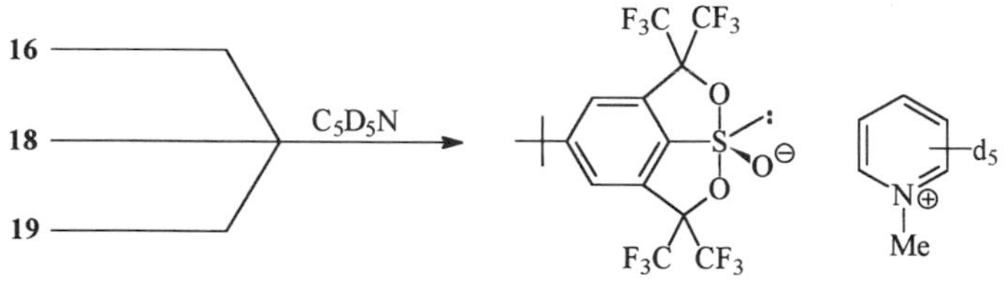

(7.3)

Demethylation of the sulfurane **20** by pyridine-d_5 was found to give *N*-methylpyridinium-d_5 sulfuranide **24** (Eq. 7.4).[30]

$$20 \xrightarrow[C_5D_5N]{5\%NaOD/D_2O} \mathbf{24} + \text{N-methylpyridinium-}d_5 \quad (7.4)$$

Similar demethylation of the cationic thiazocinium systems **21** and **22** afforded cyclic structures **25** and **26** as the final products (Eq. 7.5).[30]

$$\mathbf{21} \text{ or } \mathbf{22} \xrightarrow{C_5D_5N} \mathbf{25}/\mathbf{26} + \text{N-methylpyridinium-}d_5 \quad (7.5)$$

25; X = lone electron pair
26; X = O

The experimental data listed in Table 7.1, which includes also the pertinent values for diphenylmethylsulfonium tetrafluoroborate **27**, has led to the following conclusions.[30]

TABLE 7.1 Rate of Methylation of Pyridine-d_5 by Sulfuranes with OMe and SMe Ligands

Compound	ΔH^* (kcal mol^{-1})	ΔS^* (eu)	k (sec^{-1}) at 298.2K	Rates Relative to **16**	Rates Relative to **21**
16	18.5	−25.1	2.26×10^{-7}	1	0.15
18	21.1	−11.3	6.29×10^{-6}	27.8	4.25
19	14.3	−19.7	1.00×10^{-2}	4.42×10^{4}	6.76×10^{3}
20	18.5	−23.2	1.45×10^{6}	6.42	0.98
21	23.1	−7.82	1.38×10^{6}	6.11	1
22	17.8	−18.5	4.54×10^{-5}	201.0	30.7
27	18.1	−10.3	1.90×10^{-3}	8.41×10^{3}	1.28×10^{3}

Source: Reference 30.

The rate of demethylation of **21** by pyridine-d_5 is 1300 times slower than for diphenylmethylsulfonium tetrafluoroborate **27**. This decrease in rate results from the fact that during demethylation of **21**, both bonds which form the three-center–four-electron (3c–4e) hypervalent bond are cleaved, forming **25** with no hypervalent bond. Therefore, the transition-state barrier is a higher one than for **27**, where only the S—Me bond is broken in the attack by pyridine. Moreover, methoxysulfuranes **19** and **22** are demethylated faster by pyridine-d_5 than are S—Me sulfuranes **18** and **21** because the oxygens of **19** and **22** remaining on the sulfur atom after demethylation help to stabilize the final products. Sulfurane **19** methylates pyridine much faster than **22** does and is also faster than **18** or **16**, although **18**, **19**, and **16** all form the same sulfurane oxide **23**. The fastest demethylation of **19** results from the fact that transition state provides stabilization by having electrons donated to the antibonding O—S—O orbital; as the equatorial oxygen becomes more negatively charged the Me group moves to pyridine from **19**. The small methylation rate increase (4.3 times) for sulfurane oxide **18** relative to sulfurane **20** indicates that the oxygen atom of **18** does not become much more negative as the Me group leaves, although the sulfur becomes less positive as the S—O bond is broken.

Due to the presence of the hypervalent bond, protons of the equatorial S-methyl group of a 10-S-5 sulfurane oxide are much reduced in acidity when the four methyl groups adjacent to the apical oxygen in **27** are replaced by trifluoromethyl group affording **18**.[31] It was found that the deuterium exchange rate for the equatorial methyl group in **27** is 86,000 times faster than for the CF_3-substituted **18** (Eq. 7.6).

$$\mathbf{27},\ R{=}Me;\ \mathbf{18},\ R{=}CF_3 \xrightarrow[\text{B}]{CD_3OD} \mathbf{28},\ R{=}Me;\ \mathbf{29},\ R{=}CF_3 \quad (7.6)$$

This results from the fact that in the hypervalent-bond-containing structures the trifluoromethyl groups of the tridentate ligands of **18**, more than the methyl groups of **27**, strongly stabilized the symmetric species, relative to an unsymmetric species obtained by cleaving one of the apical bonds. Therefore, stabilization of anions **30** or **31**, which are formed as intermediates in the deuterium exchange processes, would be provided more by CH_3 groups than by CF_3 groups. In consequence the CF_3 substituent effects make **18** less acidic than **27**.[31]

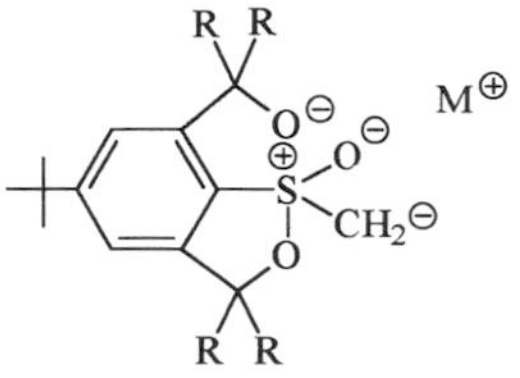

30, R=Me
31, $R=CF_3$

The second typical conversion of the δ-sulfuranes is the ligand-coupling reaction. During the 1960s, numerous reactions involving trivalent, tricoordinated sulfonium derivatives and nucleophiles were suggested to proceed through ligand coupling within δ-sulfuranes formed as intermediate since the finding of diaryls in the treatment of trialyl sulfonium salts with aryllithium.[32–38] In the middle of the 1980s the Oae and Furukawa[39] groups discovered that ligand coupling plays an important role in the chemistry of hypervalent sulfuranes formed in the reactions of the selected group of sulfoxides with organometallic reagents. Taking into account the structural features of the δ-sulfuranes, it is reasonable to consider that ligand coupling should take place between two apicophilic ligands, as was assumed[40] to occur during the thermal decomposition of the perfluoromethoxy sulfurane **32**, which afforded as the final products di-trifluoromethyl sulfide **33** and di-trifluoromethyl peroxide **34** (Eq. 7.7).

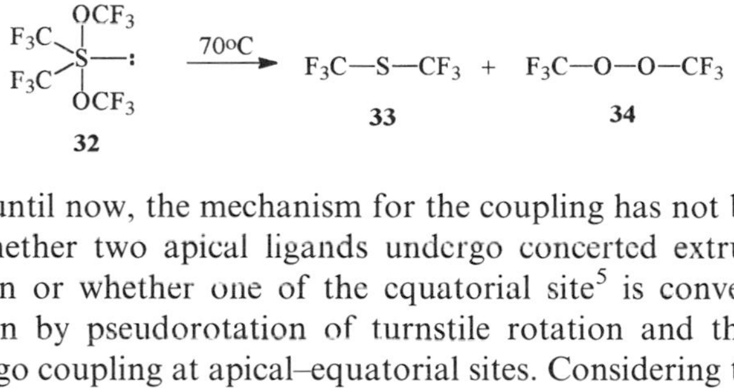

$$(F_3C)_2S(OCF_3)_2 \; (\mathbf{32}) \xrightarrow{70^{\circ}C} F_3C{-}S{-}CF_3 \; (\mathbf{33}) + F_3C{-}O{-}O{-}CF_3 \; (\mathbf{34}) \quad (7.7)$$

However, until now, the mechanism for the coupling has not been resolved, no matter whether two apical ligands undergo concerted extrusion at their apical position or whether one of the equatorial site[5] is converted into the apical position by pseudorotation of turnstile rotation and then these two ligands undergo coupling at apical–equatorial sites. Considering the stereoelectronic requirements for the ligand-coupling reactions, the second possibility seems more sound because the carbon atom attached to the coupling center retains its configuration. This observation strongly supports the view that the reaction should proceed via concerted front–side attack of one ligand on the other one via a three-membered transition state without bond breaking, as shown by the general equation 7.8.

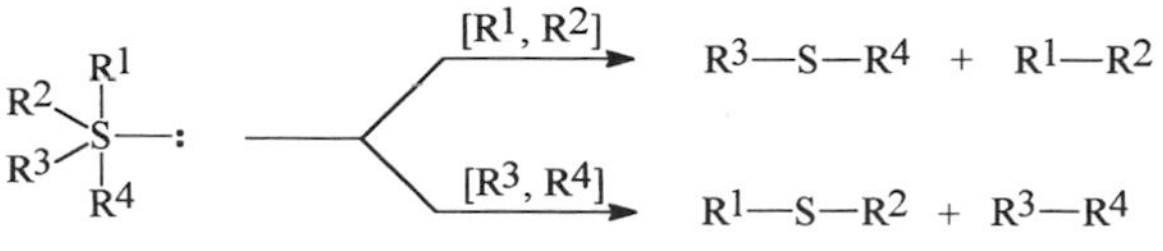

$$R^1R^2R^3R^4S \; (\mathbf{32}) \begin{cases} \xrightarrow{[R^1, R^2]} R^3{-}S{-}R^4 + R^1{-}R^2 \\ \xrightarrow{[R^3, R^4]} R^1{-}S{-}R^2 + R^3{-}R^4 \end{cases} \quad (7.8)$$

Very extensive and detailed studies carried out recently in the Oae and Furukawa groups,[39] which, due to the space limitation cannot be discussed in this chapter, fully supported this mechanistic proposal.

Another group of very important conversions in the chemistry of the δ-sulfuranes constitute ligand-exchange reactions which are represented by the general equation 7.9.

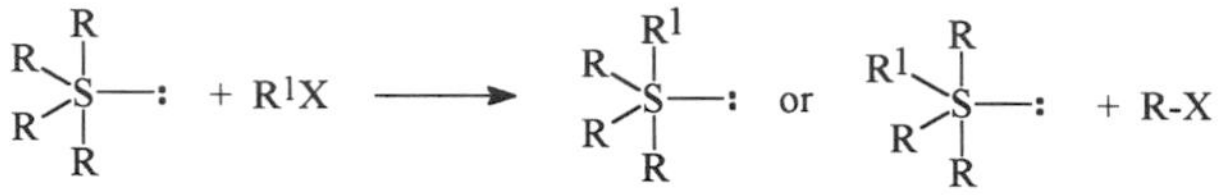

(7.9)

Their importance results mainly from the fact that most of the stable fluorine-containing sulfuranes have been prepared by such conversions using sulfur tetrafluoride or its mono- or di-substituted derivatives as substrates.[11] It is obvious that in these reactions associative mechanisms for nucleophilic displacements at the central sulfur atom should operate. However, mechanistic studies on this topic have been almost completly neglected. It has only been shown that the reaction of SF_4 with $F^{\ominus}$ gives a observable species $SF_5^{\ominus}$.[41] Analogous transition state **35** has also been suggested to explain retention of configuration in the hydrolysis of optically active chlorosulfurane **36** to the corresponding hydroxysulfoxide **37** (Scheme 7.10).[42]

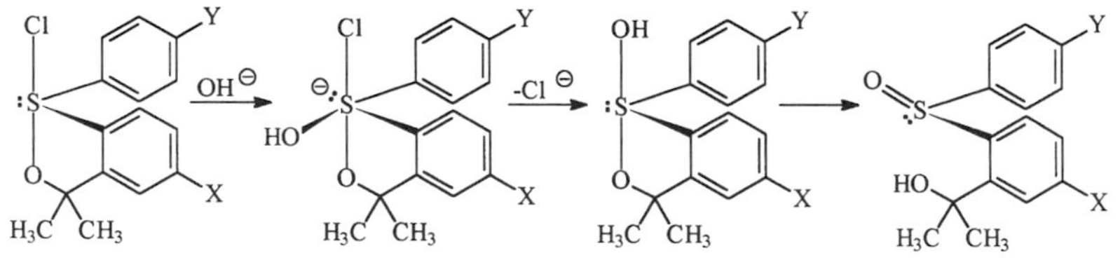

Scheme 7.10

7.3 OCCURRENCE AND SYNTHESIS OF PARTICULAR SULFURANES

7.3.1 10-S-3 Sulfuranes

Occurrence of 10-S-3 sulfurane structures **38** as transition states has been proved in some nucleophilic displacements on sulfenyl sulfur (Eq. 7.10). This subject has been reviewed in depth by Kice.[43]

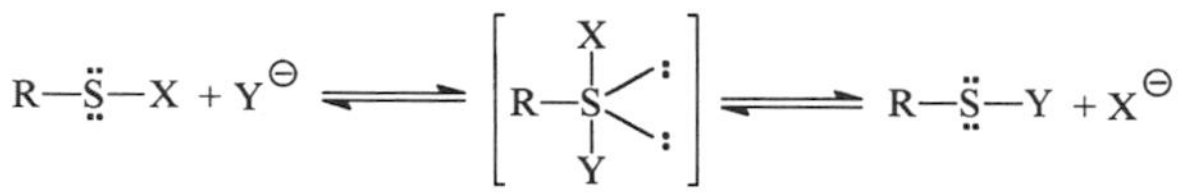

$$\mathrm{R{-}\ddot{\underset{..}{S}}{-}X + Y^{\ominus}} \rightleftharpoons \left[\mathrm{R{-}S(X)(Y)}\right] \rightleftharpoons \mathrm{R{-}\ddot{\underset{..}{S}}{-}Y + X^{\ominus}} \tag{7.10}$$

38

Many examples of isolable 10-S-3 species are presented by Hayes and Martin[7] in their review on hypervalent sulfuranes. They discuss the chemistry of thiathiophthenes **39**[44] and the first isolable anionic 10-S-3 species, sulfuranide anions **40** and **41**, which are stabilized by the presence of tridentate ligands.[45,46]

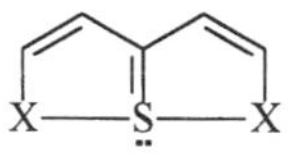

39 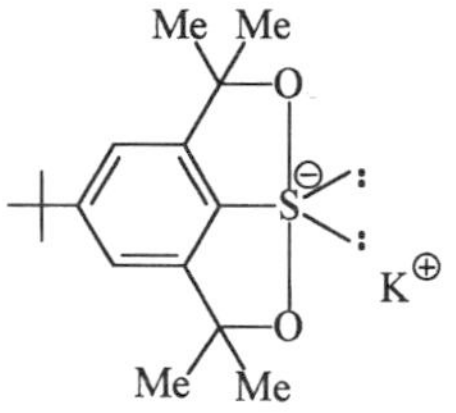 **40** 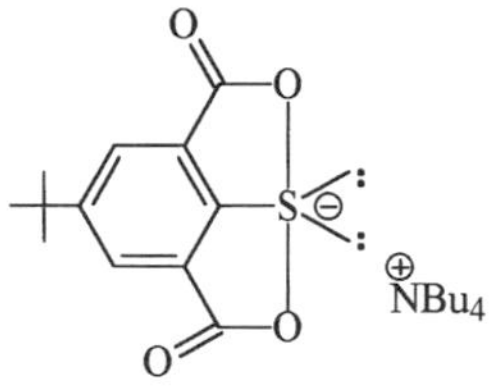 **41**

7.3.2 10-S-4 Sulfuranes

This group of hypervalent sulfur derivatives commonly named δ-sulfuranes possesses four σ bonds to sulfur in addition to the lone electron pair. Their involvement in many reactions, detection by a spectroscopic method, and isolation as stable compounds have been discussed in great detail by Drabowicz and colleagues.[11] δ-Sulfuranes have been proposed as reactive intermediates in the following reactions of divalent sulfur compounds:

1. photo-oxidation of sulfides[47–51]
2. chemical oxidation of sulfides[52–56]
3. chlorination of sulfenyl chlorides[57–60]
4. addition of sulfenyl chlorides to unsaturated C—C bond.[61–63]

Similarly, their formation has been proposed in the reactions of sulfonium salts with organometallic reagents and other nucleophiles[14,64–67] in pyrolysis of sulfonium salts,[68] and in alkylation of β-keto esters with sulfonium salts.[69] δ-Sulfuranes as reactive intermediates have been suggested in the following reactions of sulfinyl derivatives and other tetravalent organosulfur compounds:

1. racemization of sulfinyl derivatives[70–76]
2. oxidation of sulfinyl derivatives[77–80]

3. reduction of sulfinyl derivatives[81–88]
4. nucleophilic exchange reactions of sulfinyl derivatives[89–103]
5. selected reactions of sulfimines.[104–108]

Until now only a few reports describing spectroscopic evidence for the formation of δ-sulfurane intermediates have been published.[109–113] Among them there are very recent papers on the first detection of 2,2′-biphenylylene-diphenyl sulfurane **42** and tetraphenyl sulfurane **43** by ^{1}H, ^{13}C, and CH-COSY NMR experiments at low temperatures.[114–115]

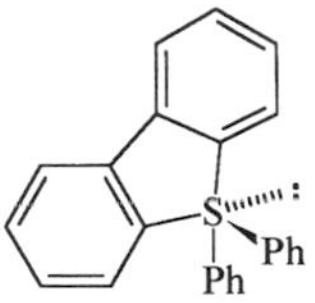

42 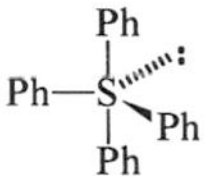 **43**

The last two decades have witnessed growing interest in the search for synthetic procedures that allow preparation and isolation of stable sulfuranes. As a consequence, this class of organosulfur compounds, which for almost 80 years had been limited to a few perhalogenated members, has been growing rapidly.

Among them, a very common and stable group of compounds comprises the organic sulfur trifluorides **44** and **45**, which can be prepared by fluorination of the corresponding disulfides (Eq. 7.11)[116–118] or by the exchange reaction of sulfur tetrafluoride with free or silylated alcohols or amines (Eq. 7.12).[119–128]

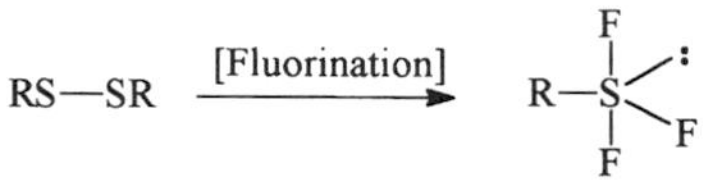

$$\text{RS—SR} \xrightarrow{\text{[Fluorination]}} \text{R—}SF_3 \quad \textbf{44} \tag{7.11}$$

$$SF_4 + \xrightarrow[\text{RXSiMe}_3]{\text{RXH or}} \text{RX—}SF_3 \quad \textbf{45}\text{, X=O or NR} \tag{7.12}$$

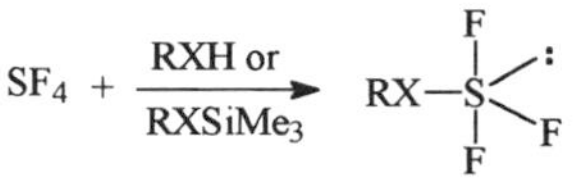

Similar exchange reactions were used for the preparation of bis-(dialkylamino)-sulfur difluorides **46**[129–131] and difluorosulfuranes **47** and **48**[132] from dialkylaminosulfur trifluorides and trifluoromethylsulfur trifluorides.

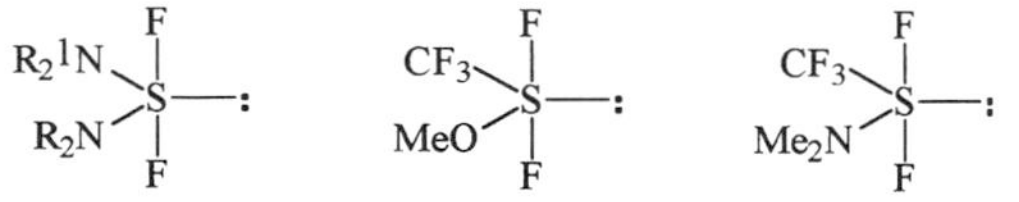

46 **47** **48**

Among a few synthetic procedures leading to monochlorosulfuranes **49**, the most general is the oxidation of the appropriate sulfide-alcohols **50** with one equivalent of *t*-butyl hypochlorite (Eq. 7.13).[42,133–134]

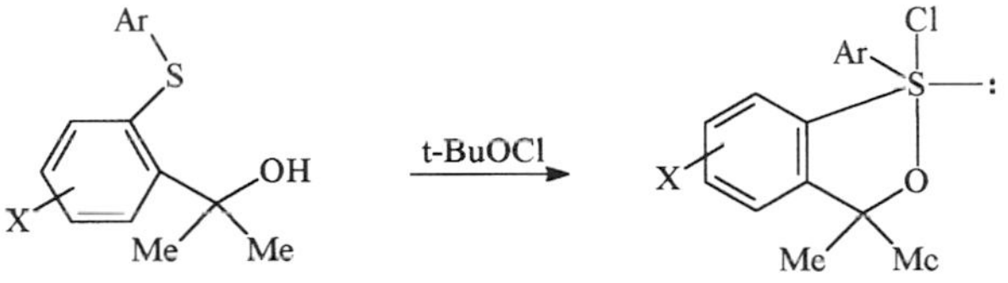

50 $\xrightarrow{\text{t-BuOCl}}$ **49** (7.13)

In similar way the azachlorosulfurane **51** can be prepared by reacting diamide sulfide **51a** with *t*-butyl hypochlorite (Eq. 7.14).[135]

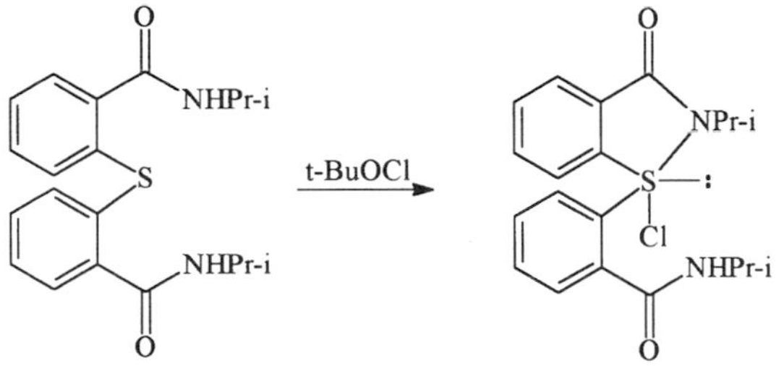

51a $\xrightarrow{\text{t-BuOCl}}$ **51** (7.14)

A series of stable bicyclosulfuranes **52**–**55** with alkoxy ligands were prepared by the *t*-butyl hypochlorite-induced oxidative ring closure of the corresponding hydroxysulfides (Eq. 7.15).[42,136]

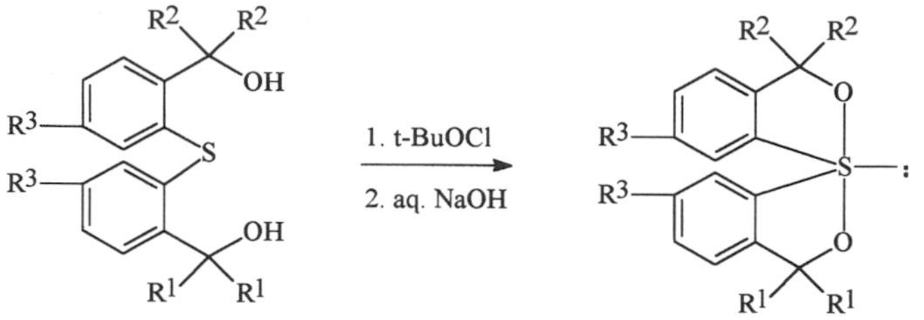

56-59 → (1. t-BuOCl; 2. aq. NaOH) → 52-55 (7.15)

56 and **52**, $R^1=R^2=R^3=H$
57 and **53**, $R^1=R^2=Me$, $R^3=H$
58 and **54**, $R^1=R^2=CF_3$, $R^3=H$
59 and **55**, $R^1=R^2=CF_3$, $R^3=t\text{-Bu}$

A similar procedure applied to the sulfidodiol **60** allowed isolation of a very stable dialkoxysulfurane **61** (Eq. 7.16).[137–138]

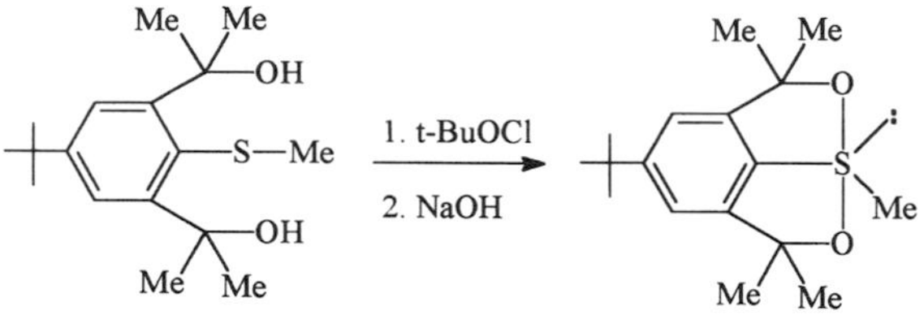

60 → (1. t-BuOCl; 2. NaOH) → 61 (7.16)

Due to the presence of the two trifluoromethyl groups and the five-membered ring, which stabilize sulfurane structures, it was possible to obtain monocyclic diaryloxysulfuranes **62**, **63**, and **64** in the reaction of difluorosulfurane **65** with silylated *o*-catechols **66** (Eq. 7.17).[139]

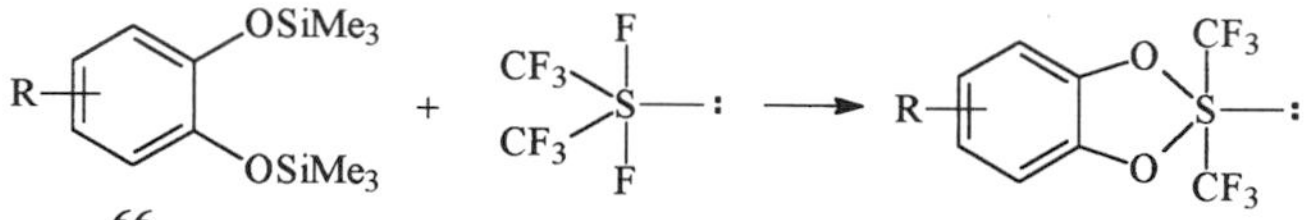

66 + 65 → (7.17)

62, R=H
63, R=3-Me
64, R=3,5-di-t-Bu

A series of acyclic and cyclic di-perfluoroalkoxysulfuranes was prepared by the reaction of perfluoromethyl hypochlorite with acyclic and cyclic sulfur (II) compounds[40] (Eqs 7.18 and 7.19).

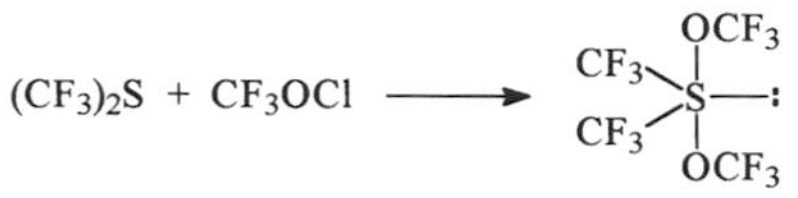

(CF3)2S + CF3OCl ⟶ CF3, CF3, OCF3, OCF3, S, : **67** (7.18)

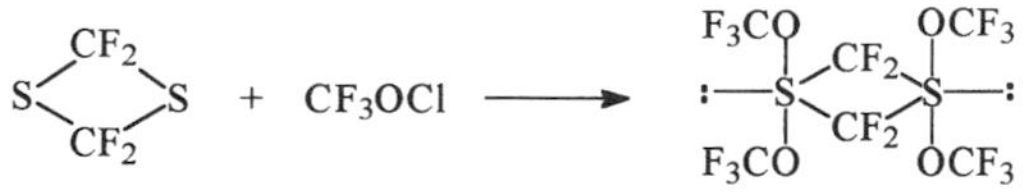

S(CF2)2S + CF3OCl ⟶ F3CO, OCF3, CF2, S, S, CF2, F3CO, OCF3 **68** (7.19)

Only a few alkoxyacyloxysulfuranes have been described. Among them are compounds **69** and **70**, in which the presence of a tridentate ligand stabilizes the structures, and which were prepared by oxidative conversion of the corresponding carboxylic acids **71** and **72** (Eq. 7.20).[134,140]

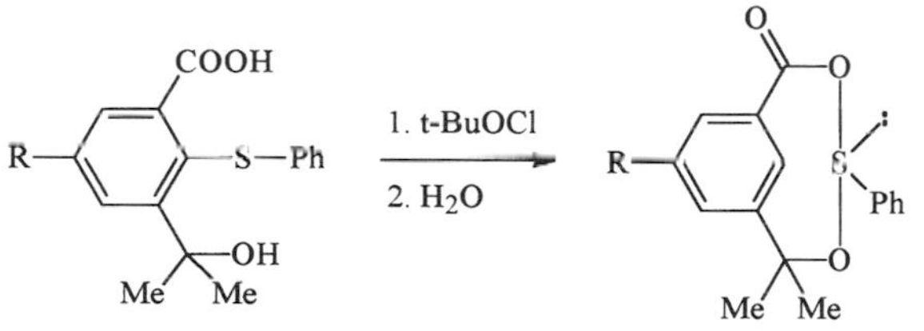

COOH, R, S—Ph, OH, Me, Me

1. t-BuOCl
2. H_2O

O, O, R, S, Ph, O, Me, Me (7.20)

71, R=H
72, R=t-Bu

69, R=H
70, R=t-Bu

The most general synthesis of diacyloxysulfuranes **73** is based on the oxidation of bis-(2-carboxyaryl) sulfides **74a–k** with different halogenating agents (Eq. 7.21).[141,142]

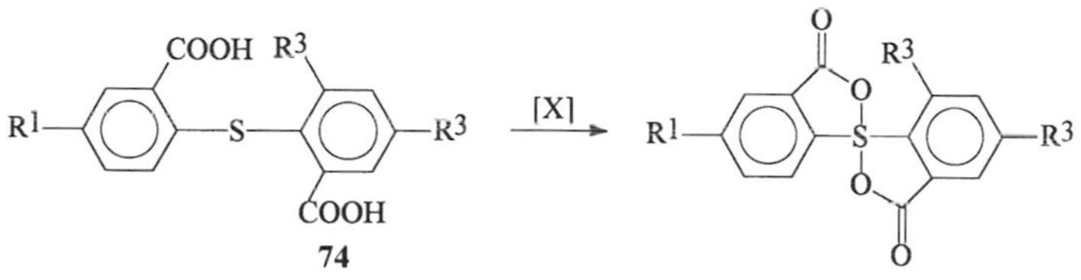

COOH R3, R1, S, R3, COOH **74** —[X]→ O, R3, O, R1, S, R3, O, O **73** (7.21)

X = halogenation agent

	a	b	c	d	e	f	g	h	i	j	k
R1=	H	NO_2	Cl	NHAc	OMe	NO_2	Cl	NHAc	OMe	H	H
R2=	H	H	H	H	H	NO_2	Cl	NHAc	OMe	Cl	N
R3=	H	H	H	H	H	H	H	H	H	Cl	N

By reacting lithium salts of *o*-bis-hydroxyphenols **75** with SF_4, several stable symmetric tetraoxysulfuranes **76** have been prepared (Eq. 7.22).[143]

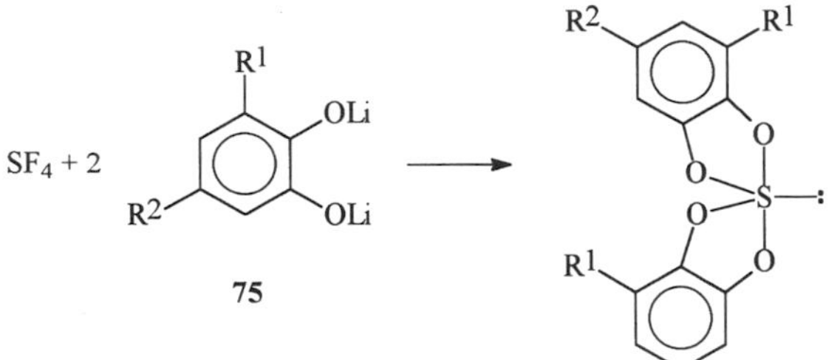

SF_4 + 2 [75: R¹, OLi, OLi, R²] ⟶ [76: R², R¹, O, O, S, O, O, R¹, R²] (7.22)

75

(**a**) R^1=R^2=H
(**b**) R^1=Me, R^2=H
(**c**) R^1=*t*-Bu

76

A few tetraalkoxysulfuranes **77** and **78** were synthetized via insertion reaction of the sulfoxylic esters **79** and **80** into the O—O bond in dioxetanes **81** and **82** (Eq. 7.23).[144]

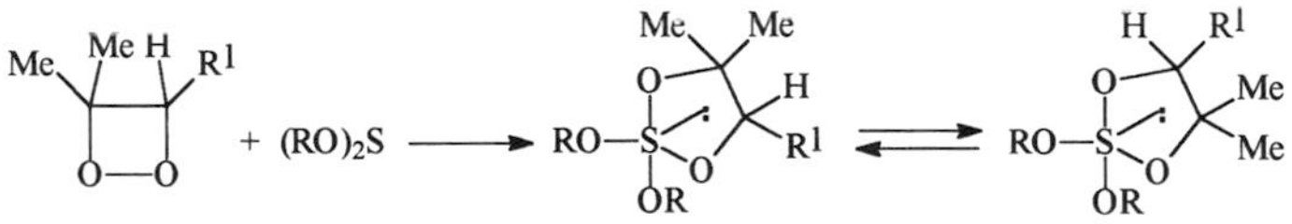

[Me, Me H, R¹, O—O] + $(RO)_2S$ ⟶ [Me, Me, O, H, RO—S, R¹, O, OR] ⇌ [H, R¹, O, Me, RO—S, Me, O, OR] (7.23)

81 R^1=H **79** R=Me **77** R=Me, R^1=H
82 R^1=Me **80** R=n-Pr **78** R=*n*-Pr, R^1=H

The first stable sulfurane with four C—S bonds, namely bis (2,2′-biphenylene) sulfurane **83**, was prepared only recently starting from dibenzothiophene S-oxide **84** via the corresponding sulfonium salts **85**, as shown in Eq. 7.24.[145]

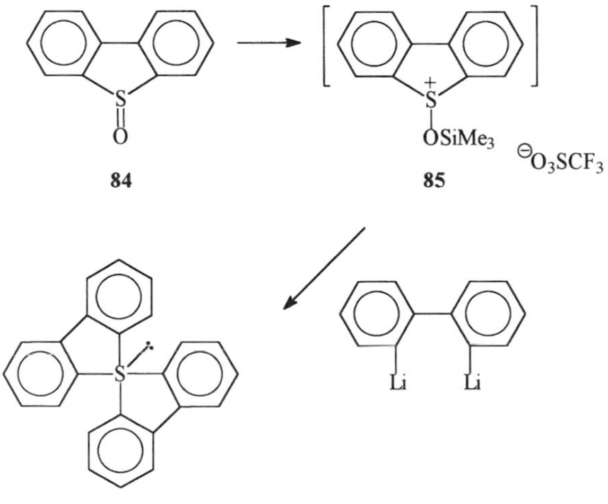

[84: S, O] ⟶ [85: $\overset{+}{S}$, $OSiMe_3$, ${}^{\ominus}O_3SCF_3$] ⟶ (Li, Li) [83: S] (7.24)

84

85

83

7.3.3 10-S-5 Sulfuranes

The most common group of this type of hypervalent sulfur derivative comprises the sulfurane oxides. Much less common are their analogs in which the S—O bond is replaced by an S—C or S–heteroatom bond.

Sulfurane oxides as reactive intermediates have been proposed in the following reactions of hexavalent pentacoordinated sulfur compounds:

1. nucleophilic substitution at sulfonyl sulfur[146–151]
2. decomposition of sulfones in the presence of strong inorganic base[152]
3. hydride reduction of diaryl sulfones occurring via arylsulfoxonium salts[153]
4. chlorine oxidation of sulfinyl derivatives such as sulfinyl chlorides and the selected group of sulfoxides[154–155]
5. transsulfonylation between aromatic sulfones and arenes[156]

The chemistry of stable sulfurane oxides began in 1937 with a patent[157] describing the preparation of sulfur oxytetrafluoride **86** from thionyl fluoride and elemental fluorine (Eq. 7.25).[158]

$$SOF_2 + F_2 \longrightarrow F_4S{=}O \quad (7.25)$$

86

Since that time a variety of isolable sulfurane oxides containing at least two fluorine atoms have been prepared using the oxide **86**, azasulfurane oxides **87**[159–160] and aryloxy derivatives of sulfur oxytetrafluoride **88**.[161]

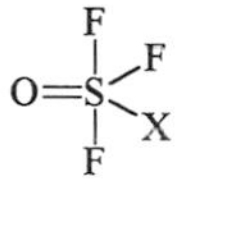

$O{=}SF_3X$ $O{=}SF_n(O{-}C_6H_4X)_{4-n}$

87, a, X=NMe$_2$
b, X=NEt$_2$
c, X=OPh

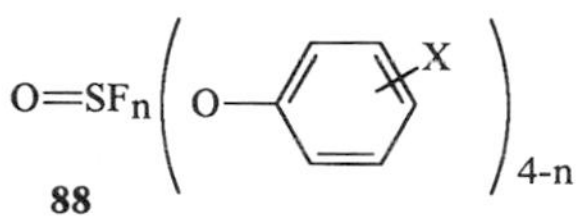

88

Diaryl or bis (perfluoroalkyl) sulfur oxydifluorides **89** were prepared by the direct fluorination of diarylsulfoxides with elemental fluorine[162] or by the reaction of chlorine monofluoride with bis (perfluoroalkyl) sulfoxides.[163] However, the relatively stable sulfurane oxide **90** was obtained as the single reaction product when a mixture of bis (trifluoromethyl) sulfoxide and trifluoromethyl hypochlorite was photolyzed at room temperature.[164] The sulfurane oxide **91** was produced in a ligand exchange reaction of **90** with lithium hexafluoroisopropylideneimine.[164]

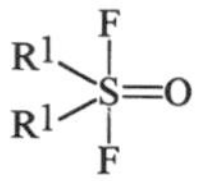

89, a, R^1=Ar
b, R^1=C_nF_{2n+1}

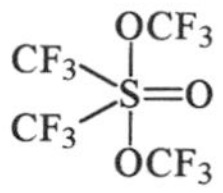

90

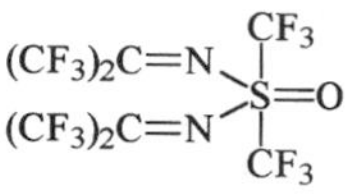

91

In addition to sulfurane oxides containing halogens, a few compounds with alkoxy substituents in the apical position of a trigonal bipyramidal structure have been isolated and fully characterized as stable species **18**, **27**, and **92–94**. As a rule they were prepared by the oxidation of the parent sulfuranes.[155,165,166]

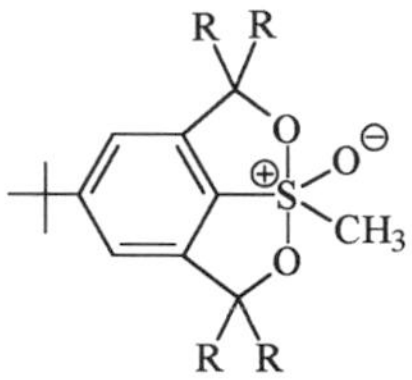

27, R=Me
18, R=CF_3

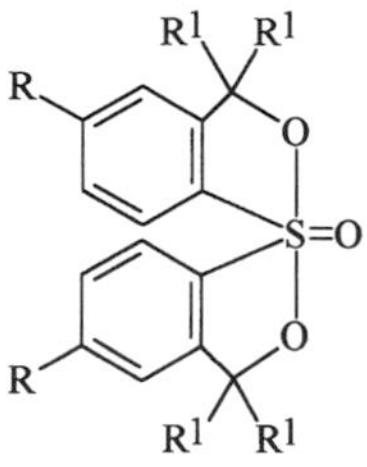

92, R=H, R^1=CF_3
93, R=t-Bu, R^1=CF_3
94, R=H, R^1=Me

A special type of sulfurane oxide is the dioxide salt **95**, the first example of an isolable intermediate postulated to explain an associative nucleophilic attack at the sulfonyl sulfur atom.[167]

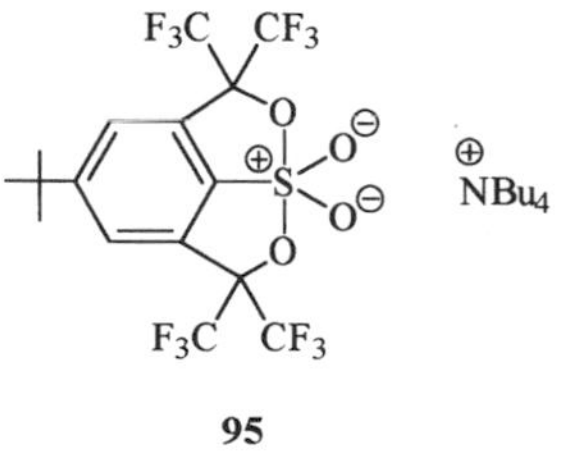

95

It is of interest to note that all stable sulfurane oxide analogs contain four halogen atoms (usually fluorine) bonded to the central sulfur atom.[11] One group of these compounds comprises sulfurane oxide analogs in which the equatorial S—O bond is replaced by an S—C bond, as in **96**, whereas the second group has an S—N bond, as in **97**.[168–169]

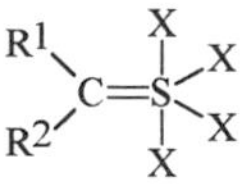

96

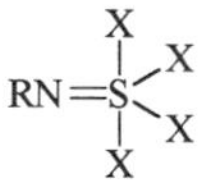

97

The persulfonium cation **10**, which may be considered as a special sulfurane oxide analog, has already been mentiond in this chapter (see Scheme 7.6).[19]

7.3.4 12-S-6 Persulfuranes

This name has been commonly accepted for hexacoordinated hexavalent sulfur compounds. The chemistry of persulfuranes, which began almost at the same time as the chemistry of sulfuranes (the parent member of this family, sulfur hexafluoride, was prepared five years earlier than sulfur tetrafluoride), has not been as extensively studied as the chemistry of tetracoordinated sulfuranes. Most of the stable persulfuranes comprise inorganic **98** and organic **99** derivatives of sulfur haxafluoride.

98, I=inorganic ligand

99, O=organic ligand

Because very detailed discussion on the synthesis and reactivity of these groups of hypervalent sulfur compounds can be found in Reference 11 and in Chapter 10, there is no need to include it here.

7.4 STEREOCHEMICAL ASPECTS OF THE CHEMISTRY OF SULFURANES

A compound having either a trigonal bipyramidal or tetragonal bipyramidal geometry can exist in enantiomeric forms when the number of different ligands is high enough to create chirality in such structures. Thus, in the case of 10-S-4 and 10-S-5 derivatives, all structures containing at least three different ligands can be chiral. Moreover, due to the topological properties of such molecules, chirality may still appear in the more symmetric spiro systems. Enantiomeric forms of such a derivative are pictured in Scheme 7.11.

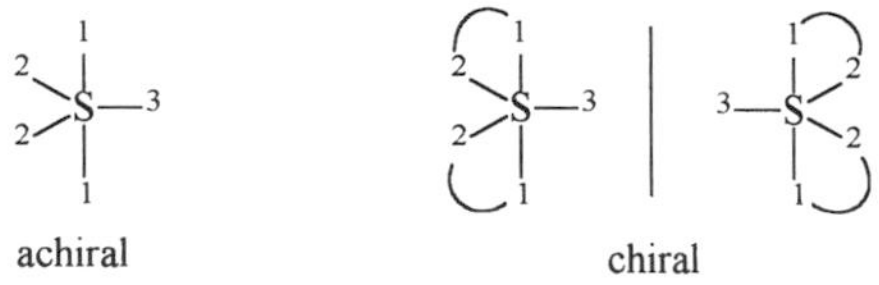

achiral chiral

Scheme 7.11

For this reason, with the exception of the dextrorotatory chlorosulfurane **100**,[42] all sulfuranes and sulfurane oxides which have, until now, been prepared as optically active species belong to this category of spiro derivatives. The sulfurane **100** was synthetized in 95% ee by treatment of the parent sulfoxide **101** with acetyl chloride (Eq. 7.26).[42]

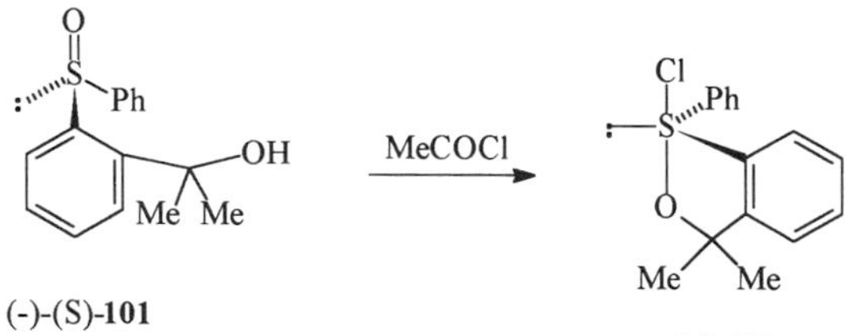

MeCOCl

(-)-(S)-**101** (+)-**100** (7.26)

All other optically active spirosulfuranes have been prepared by three different approaches: (1) stereoselective synthesis, (2) asymmetric synthesis, and (3) nonclassical resolution of racemic mixtures. The stereoselective dehydration of optically active sulfoxides **102** and **103** gave the corresponding spirosulfuranes **104** and **105** respectively (Eq. 7.27).[170]

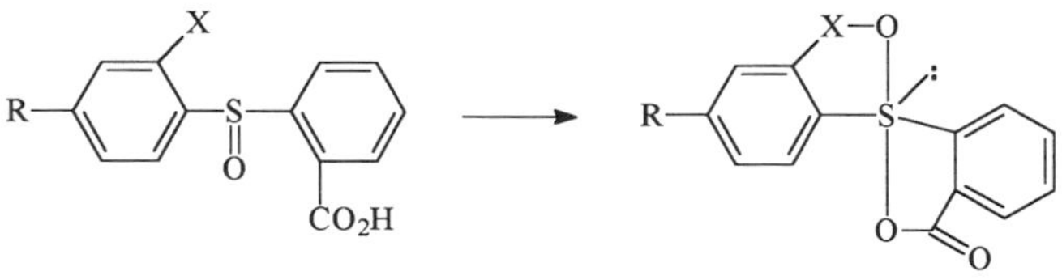

(7.27)

(-)-**102**, R=H, X=CH_2OH
(+)-**103**, R=NMe_2, X=COOH

(-)-**104**, R=H, X=CH_2
(-)-**105**, R=NMe_2, X=CO

The first asymmetric synthesis of optically active spirosulfuranes **53** and **55** is based on an asymmetric dehydration of either prochiral bis-hydroxysulfoxide **106** or the mono-tetra-*n*-butyloammonium or the potassium salts of the bis-hydroxysulfoxide **107** in the presence of an optically active acid (HA*) (Eq. 7.28).[171]

$\xrightarrow{HA^*}$ (+) or (-)-**53** (ee up to 75%) and (+) or (-)-**55** (ee up to 100%) (7.28)

106, R^1=Me, R^2=H, X=H
107, R^1=CF_3, R^2=t-Bu, X=H or n-Bu_4N

It is interesting to note that a substantial increase of the optical purity of spirosulfuranes **53** and **55** was achieved either by their partial dissolution in pentane (for **53**) or by recrystallization from petroleum ether (for **59**).

The enantiomers of the first optically active spirosulfurane **69** containing a tridentate ligand were obtained by the nonclassical resolution of racemic **69** with optically active 2.2-dihydroxy-1,1-binaphthol as a chiral resolving agent.[172]

Very recently, resolution of racemic spirosulfuranes was achieved on the analytical scale by standard chromatography using a chiral column.[173]

The first optically active spirosulfurane oxides **93** and **94** have been prepared by the stereoselective oxidation of the corresponding optically active spirosulfuranes **53** or **55**.[172b,c] All isolated optically active spirosulfuranes and their oxides have been found to be optically stable indefinitely at room temperature. They lose their optical activity at temperatures above 80°C. This process is slightly accelerated by protonic acid and becomes very fast for spirosulfuranes when racemization is carried out in the presence of traces of water.

The detailed kinetic studies on these proceses and discussion on mechanisms that could account for the thermal and catalyzed racemization of spirosulfuranes and their oxides can be found in a very recent paper.[174]

ACKNOWLEDGMENTS

Support for the preparation of this review was provided by grant from the State Committee for Scientific Research (KBN)-grants TO39A08508 and TO9A07714. The author would like to thank Prof. M. Mikołajczyk and Dr P. Łyżwa for their useful discussions and comments.

REFERENCES

1. (a) P. Laur, in *Sulfur in Organic and Inorganic Chemistry*, Vol. 3, A. Senning, Ed. (Marcel Dekker Inc., New York, 1972), p. 91; (b) S. Oae, *Organic Sulfur*

Chemistry: Structure and Mechanism (CRC Press, Boca Raton, Ann Arbor, Boston, London, 1991); (c) E. Block, *Reactions of Organosulfur Compounds* (Academic Press, New York, 1978); (d) D. N. Jones, Ed., *Comprehensive Organic Chemistry* (Pergamon, Oxford, 1979).

2. (a) C. W. Perkins, J. C. Martin, A. J. Arduengo, W. Lau, A. Alergia, J. K. Kochi, *J. Am. Chem. Soc.*, **102**, 7753 (1980); (b) J. I. Musher, *Angew. Chem.*, **81**, 68, (1969); *Angew. Chem.*, Int. Ed. Engl., **8**, 54 (1969); (c) J. I. Musher, in *Sulfur Research Trends*, R. F. Gould, Ed., Advances in Chemistry Series, Vol. 110 (ACS, Washington, D.C., 1972) p. 44; (d) V. B. Koutecky and J. I. Musher, *Theor. Chim. Acta*, **33**, 227 (1974).
3. B. M. Trost, *Top. Curr. Chem.*, **41**, 1 (1973).
4. H. O. House, *Modern Synthetic Reactons*, 2nd ed. (W. A. Benjamin Inc., New York, 1972).
5. L. Field, *Synthesis*, 101 (1972).
6. B. M. Trost, D. S. Melvin Jr., *Sulfur Ylides, Emerging Synthetic Intermediates* (Academic Press, New York, 1975).
7. R. A. Hayes, J. C. Martin, in *Organic Sulfur Chemistry, Theoretical and Experimental Advances*, F. Bernardi, J. G. Csizmadia, A. Mangini, Eds (Elsevier, Amsterdam, 1985), p. 408.
8. J. C. Martin, E. F. Perozzi, *Science*, **191**, 154, (1976).
9. J. M. Shreeve, in *Sulfur in Organic and Inorganic Chemistry,* A. Senning, Ed. (Marcel Dekker Inc., New York, 1982) p. 131.
10. N. Furukawa, in *Heteroatom Chemistry*, E. Block, Ed. (Verlag Chemie, Weinheim, 1990), p. 165.
11. J. Drabowicz, P. Łyżwa, M.Mikołajczyk, in *Supplement S: The Chemistry of Sulfur-Containing Functional Groups*, S.Patai, Z.Rappoport, Eds. (John Wiley & Sons, Chichester, 1993), pp. 799–956.
12. A. Michaelis, O. Schifferdecker, *Chem. Ber.*, **6**, 993 (1873); O. Ruff, *Chem. Ber.*, **37**, 4513 (1904).
13. A. J. Arduengo, III, E. M. Burgess, *J. Am. Chem. Soc.*, **99**, 2376 (1977).
14. (a) R. W. La Rochelle, B. M. Trost, *J. Am. Chem. Soc.*, **93**, 6077 (1971); (b) E. L. Muetterties and W. D. Phillips, *J. Chem. Phys.*, **46**, 2861 (1967); (c) R. L. Redington and C. V. Berney, *J. Chem. Phys.*, **43**, 2020 (1965); R. L. Redington and C. V. Berney, *Ibid.*, **46**, 2862 (1967); (d) I. W. Levin and W. C. Harris, *J. Chem. Phys.*, **55**, 3048 (1971); (e) G. W. Astrologes and J. C. Martin, *J. Am. Chem. Soc.*, **98**, 2895 (1976); (f) W. G. Klemperer, J. K. Krieger, M. D. McCreary, E. L. Muetterties, D. D. Traficante and G. M. Whitesides, *J. Am. Chem. Soc.*, **97**, 7023 (1975).
15. (a) R. S. Berry, *J. Chem. Phys.*, **32**, 933 (1960); (b) R. S. Berry, *Rev. Mod. Phys.*, **32**, 447 (1960).
16. a) I. Ugi, D. Marquarding, H. Klusacek, P. Gillespie, F. Ramirez, *Accounts Chem. Res.*, **4**, 288 (1971); (b) I. Ugi, D. Marquarding, H. Klusacek, G. Gokel, P. Gillespie, *Angew. Chem.*, Int. Ed. Engl., **9**, 703 (1970); (c) F. Ramirez and I. Ugi, *Advances in Physical Organic Chemistry*, Vol. 9, V. Gold, Ed (Academic Press, London, 1971), p. 25.

17. O. Takahashi, S. Sato, N. Furukawa, *17th International Symposium on the Organic Chemistry of Sulfur, Tsukuba, Japan, July 7–12, 1996, Book of Abstracts*, p. 106.
18. L. J. Adzima, J. C. Martin, *J. Org. Chem.*, **42**, 4006 (1977).
19. R. S. Michalak, J. C. Martin, *J. Am. Chem. Soc.*, **104**, 1683 (1982).
20. K. C. Hodges, D. Schomburg, J.-V. Weiss, R. Schmutzler, *J. Am. Chem. Soc.*, **99**, 6096 (1977).
21. M. R. Ross, Ph.D. Thesis, University of Illinois, Urbana-Champaign (1981).
22. J. C. Martin and M. R. Ross, unpublished results cited in ref. 7.
23. S. W. Benson, *Chem. Rev.*, **78**, 23 (1978).
24. J. A. Kerr, *Chem. Rev.*, **66**, 465 (1966).
25. J. C. Martin, R. J. Arhart, J. A. Franz, E. F. Perozzi, L. J. Kaplan, *Org. Synth.*, **57**, 22 (1977).
26. (a) J. C. Martin, R. J. Arhart, *J. Am. Chem. Soc.*, **93**, 4327 (**1971**); (b) M. Majewski and V. Snieckus, *Tetrahedron Lett.*, **23**, 1343 (1982); (c) W. Eschenmoser and C. H. Eugster, *Helv. Chim. Acta*, **58**, 1722 (1975); (d) E. G. Baggiolini, J. A. Iacobelli, B. M. Hennessy, M. R. Uskokovic, *J. Am. Chem. Soc.*, **104**, 2945 (1982); (e) D. A. Evans, D. J. Hart, P. A. Cain, *Ibid.*, **100**, 1548 (1978); (f) J. Wrobel, K. Takahashi, V. Honkan, G. Lannoye, J. M. Cook, S. H. Bertz, *J. Org. Chem.*, **48**, 139 (1983); (g) L. E. Friedrich and P. Y.-S. Lam, *Ibid.*, **46**, 306 (1981); (h) C. W. Funke, H. Cerfontain, *J. Chem. Soc. Perkin Trans II*, 1902 (1976).
27. R. J. Arhart, J. C. Martin, *J. Am. Chem. Soc.*, **94**, 5003 (1972).
28. J. C. Martin, J. A. Franz, R. J. Arhart, *J. Am. Chem. Soc.*, **96**, 4604 (1974).
29. (a) W. Eschenmoser, P. Uebelhart, C. H. Eugster, *Helv. Chim. Acta*, **62**, 2534 (1979); (b) W. Eschenmoser, C. H. Eugster, *Ibid.*, **61**, 822 (1978); (c) C. H. Eugster, *Pure and Appl. Chem.*, **51**, 463 (1979); (d) W. Eschenmoser, P. Vebelhart, C. H. Eugster, *Helv. Chim. Acta*, **65**, 353 (1982).
30. K. Ohkata, M. Ohnishi, K. Yoshinaga, K. Akiba, J. C. Rongione, J. C. Martin, *J. Am. Chem. Soc.*, **113**, 9270 (1991).
31. J. C. Rougione, J. C. Martin, *J. Am. Chem. Soc.*, **112**, 1637 (1990).
32. G. Wittig, H. Fritz, *Liebigs Ann. Chem.*, **577**, 39 (1952).
33. (a) B. M. Trost, S. D. Ziman, *J. Am. Chem. Soc.*, **93**, 3825 (1971); (b) B. M. Trost, R. C. Atkins, L. Hoffman, *J. Am. Chem. Soc.*, **95**, 1285 (1973); (c) B. M. Trost, H. C. Arndt, *J. Am. Chem. Soc.*, **95**, 5288 (1973).
34. B .M. Trost, W. L. Schinski, I. D. Mantz, *J. Am. Chem. Soc.*, **91**, 4320 (1969).
35. D. Harrington, J. Weston, J. Jacobus, K. Mislow, *J. Chem. Soc., Chem. Commun.*, 1079 (1972).
36. H. Hori, T. Kataoka, H. Shimizu, M. Miyagaki, *Chem. Pharm. Bull.*, **22**, 1711 (1974).
37. K. K. Andersen, S. A.Yeager, N. B. Peynircioglu, *Tetrahedron Lett.*, 2485 (1970).
38. W. A. Sheppard, *J. Am. Chem. Soc.*, **93**, 5597 (1971).
39. (a) S. Oae, T. Kawai, N. Furukawa, *Tetrahedron Lett.*, **25**, 69 (1984); (b) S. Oae, *Phosphorus and Sulfur*, **27**, 13 (1986); (c) T. Kawai, Y. Kodera, N. Furukawa, S. Oae, M. Ishida, T. Takeda, S. Wakabayashi, *Phosphorus and Sulfur*, **34**, 139 (1987); (d) S. Oae, T. Kawai, N. Furukawa, F. Iwasaki, *J. Chem. Soc., Perkin Trans. II*, 405 (1987); (e) S. Oae, T. Takeda, S. Wakabayashi, *Tetrahedron Lett.*,

29, 4445 (1988); (f) N. Furukawa, T. Shibutani, H. Fujihara, *Tetrahedron Lett.*, **27**, 3899 (1986); (g) N. Furukawa, T. Shibutani, H. Fujihara, *Tetrahedron Lett.*, **28**, 5845 (1987).

40. (a) T. Kitazume and J. M. Shreeve, *J. Am. Chem. Soc.*, **99**, 4194 (1977); (b) T. Kitazume and J. M. Shreeve, *Inorg. Chem.*, **17**, 2173 (1978).
41. C. W. Tullock, D. D. Coffman, E. L. Muetterties, *J. Am. Chem. Soc.*, **86**, 357 (1964).
42. J. C. Martin, T. M. Balthazor, *J. Am. Chem. Soc.*, **99**, 152 (1977).
43. J. L. Kice, *Adv. in Phys. Org. Chem.*, **17**, 65 (1980).
44. For reviews on this topic, see: (a) R. Gleiter, R. Gygax, *Top. Curr. Chem.*, **63**, 49 (1976); (b) N. Lozac'h, *Adv. Heterocyclic Chem.*, **13**, 161 (1971). See also: (c) R. Gleiter, J. Spanget-Larsen, *Top. Curr. Chem.*, **86**, 139 (1979).
45. P. H. W. Lau, J. C. Martin, *J. Am. Chem. Soc.*, **100**, 7077 (1978).
46. P. Livant, J. C. Martin, *J. Am. Chem. Soc.*, **99**, 576 (1977).
47. J. J. Liang, C.-L. Gu, M. L. Kacher, C. S. Foote, *J. Am. Chem. Soc.*, **105**, 4717 (1983).
48. Y. Watanabe, N. Kuriki, K. Ishiguro, Y. Sawaki, *J. Am. Chem. Soc.*, **113**, 2677 (1991).
49. E. L. Clennan, X.Chen, *J. Am. Chem. Soc.*, **111**, 5787 (1989).
50. E. L. Clennan, K. Yang, *J. Am. Chem. Soc.*, **112**, 4044 (1990).
51. E. L. Clennan, K. Yang, *Tetrahedron Lett.*, **34**, 1697 (1993).
52. P. R. Young, Li-shan Hsiech, *J. Am. Chem. Soc.*, **100**, 712 (1978).
53. (a) P.R.Young, M.Till, *J. Org. Chem.*, **47**, 1416 (1982); (b) P. R. Young, Li-shan Hsiech, *J. Am. Chem. Soc.*, **104**, 1159 (1982).
54. J. Takadashi Doi, W. K. Musker, *J. Am. Chem. Soc.*, **103**, 1159 (1981).
55. A. S. Hirschon, J. T. Doi, W. K. Musker, *J. Am. Chem. Soc.*, **104**, 725 (1982).
56. W. G. Bentrude, J. C. Martin, *J. Am. Chem. Soc.*, **84**, 1561 (1962).
57. I. B. Douglass, B. S. Farah, *J. Org. Chem.*, **23**, 330 (1958).
58. I. B. Douglass, *J. Org. Chem.*, **30**, 633 (1965).
59. E. N. Givens, H. Kwart, *J. Am. Chem. Soc.*, **90**, 378 (1968).
60. E. N. Givens, H. Kwart, *J. Am. Chem. Soc.*, **90**, 386 (1968).
61. G. Capozzi, G. Modena, L. Pasquato, in *The Chemistry of Sulfenic Acids and their Derivatives*, S. Patai, Ed. (Wiley, Chichester, 1990), p. 403.
62. V. M. Csizmadia, G. H. Schmid, P. G. Meey, I. G. Csizmadia, *J. Chem. Soc., Perkin Trans II*, 1019 (1977).
63. V. Calo, G. Scorrano, G. Modena, *J. Org. Chem.*, **34**, 2020 (1969).
64. V. Francen, H. J. Schmidt, C. Mertz, *Chem. Ber.*, **94**, 2942 (1961).
65. J. Bornstein, J. H. Supple, *Chem. Ind.*, 1333 (1960).
66. J. Botnstein, J. E. Shields, J. H. Supple, *J. Org. Chem.*, **32**, 1499 (1967).
67. H. Y. Kim, S. Oae, *Bull. Soc. Chem. Jpn.*, **42**, 1968 (1969).
68. G. H. Wiegand, W. E. McEwen, *J. Org. Chem.*, **33**, 2671 (1968).
69. K. Uemura, H. Matsuyama, W. Watanabe, H. Kobayashi, N. Kamigata, *J. Org. Chem.*, **54**, 2374 (1989).

70. K. Mislow, T. Simmons, J. J. Melillo, A. L. Ternay Jr., *J. Am. Chem. Soc.*, **86**, 1452 (1964).
71. H. Yoshida, T. Numata, S. Oae, *Bull. Soc. Chem. Jpn.*, **44**, 2875 (1971).
72. H. Kwart, H. Omura, *J. Am. Chem. Soc.*, **93**, 7250 (1971).
73. N. Kunieda, S. Oae, *Bull. Soc. Chem. Jpn.*, **46**, 1745 (1973).
74. S. Oae, M. Kise, *Tetrahedron Lett.*, 1409 (1967).
75. (a) S. Oae, M. Kise, *Bull. Chem. Soc. Jpn.*, **43**, 1416 (1970); (b) M. Kise, S. Oae, *Bull. Chem. Soc. Jpn.*, **43**, 1426 (1970).
76. J. Drabowicz, S. Oae, *Tetrahedron*, **34**, 63 (1978).
77. R. Curci, A. Giovine, G. Modena, *Tetrahedron*, **22**, 1235 (1966).
78. S. Oae, T. Takata, *Tetrahedron Lett.*, **21**, 3689 (1980).
79. T. Takata, Y. H. Kim, S.Oae, *Bull. Chem. Soc. Jpn.*, **54**, 1443 (1981).
80. J. L. Kice, A. R. Puls, *J. Am. Chem. Soc.*, **99**, 3455 (1977).
81. T. J. Wallace, J. J. Mahon, *J. Am. Chem. Soc.*, **86**, 4099 (1964).
82. S. Oae, *Organic Sulfur Chemistry: Structure and Mechanism* (CRC Press, Boca Raton, Ann Arbor, Boston, London, 1991), p. 209.
83. S. Oae, T. Yagihara, T. Okabe, *Tetrahedron*, **28**, 3203 (1972).
84. See ref. 82, p. 211.
85. S. Tamagaki, M. Mizuno, H. Yoshida, H. Hirota, S. Oae, *Bull. Chem. Soc. Jpn.*, **44**, 2456 (1971).
86. J. Drabowicz, S. Oae, *Synthesis*, 402 (1977).
87. J. Drabowicz, B. Dudziński, and M.Mikołajczyk, *Synlett*, 152 (1992).
88. T. Sagae, S. Ogawa, N. Furukawa, *Tetrahedron Lett.*, **34**, 4043 (1993).
89. C. A. Bunton, P. B. D. de la Mare, P. M. Greeseley, D. R. Llewellyn, N. H. Pratt, J. G. Tillett, *J. Chem. Soc.*, 4751 (1958).
90. J. L. Kice, G. Guaraldi, *J. Am. Chem. Soc.*, **89**, 4113 (1967).
91. M. Mikołajczyk, J. Drabowicz, H. Ślebocka-Tilk, *J. Am. Chem. Soc.*, **101**, 1302 (1979).
92. J. L. Kice, C. A. Walters, *J. Am. Chem. Soc.*, **94**, 590 (1972).
93. J. Drabowicz, *Phosphorus and Sulfur*, **31**, 123 (1987).
94. M. Mikołajczyk, J.Drabowicz, B.Bujnicki, *Tetrahedron Lett.*, **26**, 5699 (1985).
95. J. Drabowicz, M. Mikołajczyk, *Tetrahedron Lett.*, **26**, 5703 (1985).
96. P. G. Theobald, W. H. Okamura, *J. Org. Chem.*, **55**, 741 (1990).
97. J. Drabowicz, B. Dudziiski, M. Mikołajczyk, unpublished results.
98. S. Oae, M. Yokoyama, M. Kise, N. Furukawa, *Tetrahedron Lett.*, 4131 (1968).
99. (a) J. Day, D. J. Cram, *J. Am. Chem. Soc.*, **87**, 4398 (1965); (b) D. R. Rayner, D. M. von Schriltz, J. Day, D. J. Cram, *J. Am. Chem. Soc.*, **90**, 2721 (1968); (c) D. J. Cram, J. Day, D. R. Rayner, D. M. von Schriltz, D. J. Duchamp, D. C. Garwood, *J. Am. Chem. Soc.*, **92**, 7369 (1970).
100. H. Kwart, K. G. King, *d-Orbitals in the Chemistry of Silicon, Phosphorus and Sulfur* (Springer-Verlag, New York, 1977), p. 96.
101. T. J. Maricich, V. L. Hoffman, *J. Am. Chem. Soc.*, **96**, 7770 (1974).

102. T. Okuyama, in *The Chemistry of Sulfinic Acids, Esters and their Derivatives*, S. Patai, Ed. (John Wiley and Sons, Chichester, 1990), pp. 623–637.
103. T. Okuyama, in *17th International Symposium on the Organic Chemistry of Sulfur, Tsukuba, Japan, July 7-12, 1996, Abstract Book*, p. 102.
104. C. R. Johnson, R. A. Kirchhoff, *J. Org. Chem.*, **44**, 2280 (1979).
105. (a) P. R. Young, P. E. McMahon, *J. Am. Chem. Soc.*, **107**, 7572 (1985); (b) P. R. Young, H. C. Huang, *J. Am. Chem. Soc.*, **109**, 1810 (1987).
106. K. Kim, In Bae Jung, Y.Hae Kim, S.Oae, *Tetrahedron Lett.*, **30**, 1087 (1989).
107. N. Kunieda, S. Oae, *Int. J. Sulfur Chem.*, **8**, 40 (1973).
108. T. Akasaka, N. Furukawa, S. Oae, *Chem. Lett.*, 417 (1978).
109. D. C. Owsley, G. K. Helmkamp, M. F. Rettig, *J. Am. Chem. Soc.*, **91**, 5239 (1969).
110. J. C. Johnson, J. J. Rigau, *J. Am. Chem. Soc.*, **91**, 5398 (1969).
111. G. W. Wilson Jr., M. M. Y. Chang, *J. Am. Chem. Soc.*, **96**, 7533 (1974).
112. W. H. Sheppard, *Tetrahedron Lett.*, 875 (1971).
113. A. Schwöbel, G. Kresze, M. A. Perez, *Tetrahedron Lett.*, **23**, 1243 (1982).
114. S. Ogawa, Y. Matsunaga, S. Sato, T. Erata, N. Furukawa, *Tetrahedron Lett.*, **33**, 93 (1992).
115. S. Sato, N. Furukawa, *Tetrahedron Lett.*, **36**, 2803 (1995).
116. D. L. Chamberlain, N. Kharasch, *J. Am. Chem. Soc.*, **77**, 1041 (1955).
117. W. A. Sheppard, *J. Am. Chem. Soc.*, **84**, 3058 (1962).
118. R. M. Rosenberg, E. L. Muetterties, *Inorg. Chem.*, **1**, 756 (1962).
119. W. J. Middleton, US Patent U. 3111651, 1982; *Chem. Abstr.*, **96**, 180785g (1982).
120. (a) W. R. Hasek, W. C. Smith, V. A. Engelhart, *J. Am. Chem. Soc.*, **82**, 543 (1960); (b) L. N. Markovski, V. E. Pashnik, *J. Org. Chem.*, USSR (Engl. Transl.), **17**, 409 (1981).
121. K. Baum, *J. Am. Chem. Soc.*, **91**, 4594 (1969).
122. J. I. Darragh, D. W. A. Sharp, *Angew. Chem.*, **82**, 45, (1970); *Angew. Chem., Int. Ed. Engl.*, **9**, 73 (1970).
123. J. I. Darragh, S. F. Hossain, D. W. A.Sharp, *J. Chem. Soc., Dalton Trans.*, 218 (1975).
124. G. C. Demitras, P. A. Kent, A. G. MacDiarmid, *Chem. Ind.*, 1712 (1964).
125. G. C. Demitras, A. G. MacDiarmid, *Inorg. Chem.*, **6**, 1093 (1967).
126. W. J. Middleton, *J. Org. Chem.*, **40**, 574 (1975).
127. J. A. Gibson, D. G. Ibbot, A. F. Janzen, *Can. J. Chem.*, **51**, 3202 (1973).
128. C. von Braun, W. Dell, H. E. Sasse, M. L. Ziegler, *Z. Anorg. Allg. Chem.*, **450**, 139 (1979).
129. W. J. Middleton, US Patent 3888924, 1975; *Chem. Abstr.*, **83**, 78600d (1975).
130. L. N. Markovski, V. E. Pashnik, A. V. Kirsanov, *Zg. Org. Khim.*, **11**, 74 (1975).
131. W. Price, S. Smiles, *J. Chem. Soc.*, 2858 (1928).
132. A. H. Cowly, *J. Fluorine Chem.*, **7**, 333 (1976)
133. J. Adzima, E. N. Duesler, J. C. Martin, *J. Org. Chem.*, **42**, 4001 (1977).
134. A. K. Datta, P. D. Livant, *J. Org. Chem.*, **48**, 2445 (1983).

135. G. V. Röschenthaler, *Angew. Chem.*, **89**, 900 (1977); *Angew. Chem., Int., Engl.*, **16**, 862 (1977).
136. L. J. Adzima, J. C. Martin, *J. Org. Chem.*, **42**, 4006 (1977).
137. P. H. W.Lau, J. C. Martin, *J. C. S. Chem. Commun.*, 521 (1977).
138. P. H. W. Lau, J. C. Martin, *J. Am. Chem. Soc.*, **99**, 5490 (1977).
139. (a) T. Kitazume, J. M. Shreeve, *J. Am. Chem. Soc.*, **100**, 985 (1978); (b) D. T. Sauer, J. M. Shreeve, *J. Fluorine Chem.*, **1**, 1 (1971); (c) H. Oberhammer, R. C. Kumar, G. D. Kunerr, J. M. Shreeve, *Inorg. Chem.*, **20**, 3871 (1981).
140. W. Y. Lam, E. N. Deusler, J. C. Martin, *J. Am. Chem. Soc.*, **103**, 127 (1981).
141. I. Kapovits, J. Rabai, F. Ruff, A. Kucsman, *Tetrahedron*, **35**, 1869 (1979); I. Kapovits, J. Rabai, F. Ruff, A. Kucsman, B. Tanacs, *Tetrahedron*, **35**, 1875 (1979).
142. W. Walter, B. Krische, J. Voss, *J. Chem. Res. (S)*, 1332 (1978); *J. Chem. Res. (M)*, 4101 (1978).
143. G. E. Wilson, B. A. Belkind, *J. Am. Chem. Soc.*, **100**, 8124 (1978).
144. S. S. Campbell, D. B. Denney, D. Z. Denney, Li-Shang Shih, *J. Am. Chem. Soc.*, **97**, 3850 (1975).
145. S. Ogawa, N. Matsunaga, S. Sato, I. Iida, N. Furukawa, *J. Chem. Soc., Chem. Commun.*, 1141 (1992).
146. E. Ciuffarin, L. Senatore, H. Isola, *J. Chem. Soc., Perkin Trans. II*, 468 (1972).
147. E. Ciuffarin, L. Senatore, *Tetrahedron Lett.*, 1635 (1974).
148. W. Tagaki, T. Kurusu, S. Oae, *Bull. Soc. Chem. Jpn.*, **42**, 2894 (1969).
149. J. L. Kice, E. Legan, *J. Am. Chem. Soc.*, **95**, 3912 (1973).
150. S. Oae, Y. Kadoma, *Can. J. Chem.*, **64**, 1184 (1986).
151. M. Mikołajczyk, M.Gajl, *Tetrahedron Lett.*, 1325 (1975).
152. C. K. Ingold, J. A. Jessop, *J. Chem. Soc.*, 708 (1930).
153. I. W. Still, F. J. Ablenas, *J. Org. Chem.*, **48**, 1617 (1983).
154. I. B. Douglass, B. S. Farah, E. G. Thomas, *J. Org. Chem.*, **26**, 1996 (1961).
155. E. F. Perozzi, J. C. Martin, *J. Am. Chem. Soc.*, **94**, 5519 (1972).
156. A. M. El-Khagawa, R. M. Roberts, *J. Org. Chem.*, **50**, 3334 (1985).
157. German Patent R 100 449 DP 440/39 from 7.10.1937.
158. H. Jonas, *Z. Anorg. Allg. Chem.*, **265**, 273 (1951).
159. S. P. von Halasz, O. Glemser, *Chem. Ber.*, **103**, 594 (1970).
160. S. P. von Halasz, O. Glemser, M. F. Feser, *Chem. Ber.*, **104**, 1242 (1971).
161. R. Cramer, D. D. Coffman, *J. Org. Chem.*, **26**, 4164 (1961).
162. I. Ruppert, *Angew. Chem.*, **91** 941 (1979); *Angew. Chem., Int. Ed. Engl.*, **18**, 880 (1979).
163. D. T. Sauer, J. M. Shreeve, *Z.Anorg. Allg. Chem.*, **385**, 113 (1971).
164. J. M. Shreeve, T. Kitazume, *Inorg. Chem.*, **17**, 2173 (1978).
165. L. J. Adzima, J. C. Martin, *J. Am. Chem. Soc.*, **99**, 1657 (1977).
166. P. W. Lau, J. C. Martin, *J. Am. Chem. Soc.*, **99**, 5490 (1977).
167. C. W. Perkins, J. C. Martin, *J. Am. Chem. Soc.*, **105**, 1377 (1983).
168. G. Kleemann, K. Seppelt, *Angew. Chem., Int. Ed. Engl.*, **17**, 516 (1978).

169. A. Simon, E. M. Peters, D. Lenz, K. Seppelt, *Z. Anorg. Allg. Chem.*, **468**, 7 (1986).

170. I. Kapovits, *Phosphorus, Sulfur and Silicon*, **58**, 39 (1991).

171. (a) J. Drabowicz, J. C. Martin, in *The XIV International Symposium on the Organic Chemistry of Sulfur, Łodź, September 1990, Abstract Book*, BP-14; (b) J. Drabowicz, J. C. Martin, in *The 199th National Meeting of the American Chemical Society, Boston, April 1990, Abstract Book*, Part II, ORG-256, and manuscript in preparation.

172. (a) J. Drabowicz, J. C. Martin, *Tetrahedron: Asymmetry*, **4**, 297 (1993); (b) J. Drabowicz, J. C. Martin, in *The XV International Symposium on the Organic Chemistry of Sulfur, Caen, June 1992, Abstract Book*, OB12; (c) J. Drabowicz, J. C. Martin, *Phosphorus, Sulfur and Silicon*, **74**, 439 (1993).

173. S. Allenmark, *Tetrahedron: Asymmetry*, **3**, 2329 (1993).

174. J. Drabowicz, J. C. Martin, *Pure Appl. Chem.*, **68**, 951 (1996).

8 Structure and Reactivity of Hypervalent Chalcogen Compounds: Selenurane (Selane) and Tellurane (Tellane)

NAOMICHI FURUKAWA AND SOICHI SATO
Department of Chemistry, University of Tsukuba, Tsukuba, Ibaraki, Japan

8.1 GENERAL INTRODUCTION AND HISTORICAL BACKGROUND

The octet rule of Lewis has long been considered to be the sole central dogma to govern the chemical bonds of organic molecules which are composed normally from the second row elements; namely, carbon as a central atom and nitrogen, oxygen, hydrogen, and some other elements as auxiliaries.[1] Furthermore, orbital hybridization of *s* and *p* orbitals proposed by Pauling is also an important and general concept for explaining of the chemical bonds of organic compounds.[2] However, in the past several decades there have been a few unusual molecules reported which do not obey the octet rule, for example, SCl_4, SF_4, SF_6, and I_3^-.[3] The central atoms of these molecules are composed of elements below the third row and from Group 14 to Group 17 in the periodic table. The characteristic bonding modes for these molecules are decet[4] or even dodecet[5] and have long been considered as the few exceptions to the octet rule. Therefore, these molecules are called "unusual bonding molecules" *vis-à-vis* the octet rule. However, these decet and dodecet molecules have been well known in the inorganic compounds which use *d* and *f* orbitals for making chemical bonds. Therefore, these findings indicate the existence of a new field between the inorganic and organic regions, which we now call heteroatom chemistry.[6] There are two ways in which the chemical bond of SF_4 can be explained. One is the new hybridization using one 3*d* orbital of the sulfur atom with the normal $3sp^3$, resulting in the formation of the five coordinate bonds sp^3d required for the decet.[2] The other is the resonance among the conceivable ionic forms, that is, in the case of SF_4 it should be a resonance hybrid of $S^+F_3 \cdot F^-$, SF_4, and so

Chemistry of Hypervalent Compounds, edited by Kin-ya Akiba.
ISBN 0-471-24019-2

on. However, these propounded methods cannot illustrate correctly any molecular structures bearing an unusual bond. As a rational illustration of the bonding for the unusual valency, Rundle, earlier than Musher, had proposed the notation of "hypervalent bonding."[7] According to the molecular orbital calculation, *d*-orbital participation is not important for hybridization; instead, 3*s* and 3*p* orbitals are enough to make a decet and a dodecet of unusual molecules.[8] The chemical bonding used in the decet is called three-center–four-electron (3c–4e) bonding. Actually, the central atoms should have two types of chemical bonding, one is a 3c–4e bond (axial, or apical, bond) and the other is three sp^2 hybridized orbitals orthogonal to each other (equatorial bond); hence, the molecular structure should be a trigonal bipyramid (TBP). This hypothetical molecular structure agrees well with the crystal structure of the unusual bonding molecules. The typical molecular orbitals for the TBP structure are illustrated in Figure 8.1; the other unusual valency is the hexacoordinate square bipyramid (SBP), which has three identical 3c–4e bonds.

The dodecet species are considered to have three sets of 3c–4e bonds which are perpendicular to each other; hence, the corresponding molecule should be a octahedral structure (Oh). However, strictly speaking, the 3c–4e bonds in dodecet species differ from that of decet species. In decet species, the valence *s* orbital of the central atom is not involved in the 3c–4e bond. However, in the octahedral dodecet species, the *s* orbital is empty if only the *p* orbitals are utilized in description of the three 3c–4e bonds. Actually, the *s* orbital can be mixed with the nonbonding orbital in the Rundle–Musher 3c–4e model, that is, the nonbonding orbital splits into bonding and antibonding orbitals, as shown in Figure 8.2 (expanded Rundle–Musher model).[9] Therefore, the corresponding 3c–4e bond is stronger than that of the decet species. Generally, the hexacoordinated chalcogen species are more stable than the corresponding tetracoordinated species.

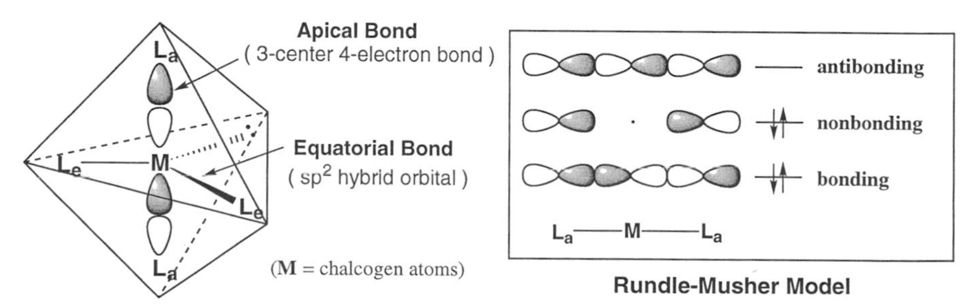

Figure 8.1 Trigonal bipyramid structure and Rundle–Musher model of 3c–4e bond.

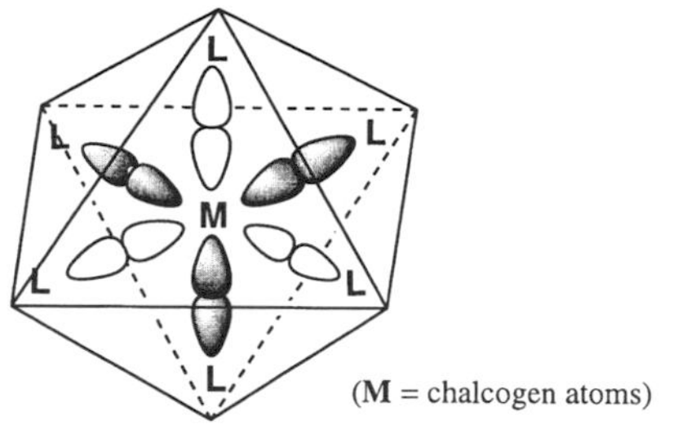

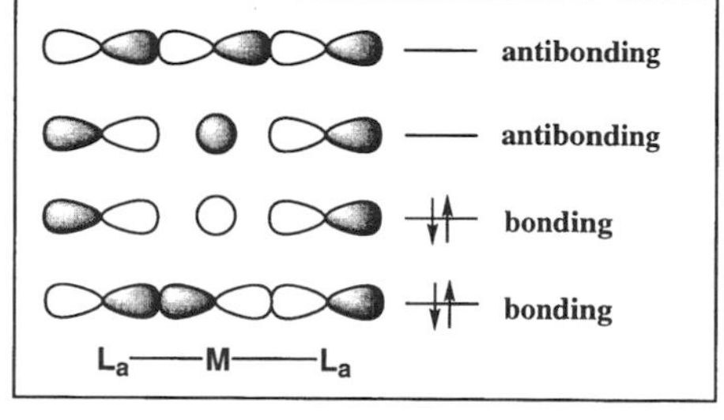

Figure 8.2 Octahedral structure and expanded Rundle–Musher model of 3c–4e bond.

In organic chemistry, these decet species have been conceived to be a transition state in the nucleophilic substitution reaction (S_N2) and not to exist as a real molecule. However, recently numerous heteroatom compounds bearing hypervalent structures have been detected and even synthesized, and the investigation of these molecules has been of considerable interest to many organic chemists. The hypervalent molecules are now known in the elements of below the third row and Group 14 to Group 17 elements in the periodic table. Among them, the chemistry of hypervalent compounds of the Group 16 elements, chalcogens (sulfur, selenium, and tellurium) has been most widely investigated and has been growing into a new field in organic chemistry. Since the pioneering works by Martin[10] and Kapovits[11] on the first synthesis of organosulfuranes (1,1′-spirobi(3H-2,1-benzoxathiole)-3,3′-dion and 3,3,3′,3′-tetramethyl-1,1′-spirobi[3H-2,1-benzoxathiole]), sulfuranes and persulfuranes have been extensively and systematically investigated.[12]

Their chemical and physical properties have been widely documented by Martin and his co-workers. According to Martin, molecular structures of hypervalent compounds or the arrangement of the ligands around the central elements (TBP) obey the following restrictions: (1) the Mutterties rule,[13] which states that the more electronegative ligands occupy the axial position, whereas electropositive ligands must be at the equatorial position. Lone pair electrons should be located at the equatorial position. (2) Five- and six-membered rings, which stabilize the hypervalent compounds, expand at both the axial and equatorial positions.[14]

As for the chemical properties of the hypervalent compounds, there are two typical reactions; one is the ligand-exchange reaction and the other is the ligand-coupling reaction. Pseudorotation or turnstile rotation[15] is also a typical stereochemical phenomenon observed in the hypervalent compounds having pentacoordinated structures.[16] In the mechanism of nucleophilic substitution reactions at the sulfur and selenium atoms, it has long been argued whether the reaction proceeds via a pentacoordinated species as an intermediate, unlike the S_N2 type substitution on the carbon atom, or via a simple S_N2 type process. In these reactions, the nucleophiles must come from the apical direction and the leaving group should also be eliminated from the apical position, that is, an **a**→**a** process resulting in the inversion of the stereochemical course. There are a few other possible combinations of attacking and leaving groups, namely **a**→**e** (or **e**→**a**) or **e**→**e** processes in the reactions in which the stereochemical results are either retention or inversion and hence quite different from that of an **a**→**a** process, as shown in Figure 8.3.[17]

In contrast to sulfur and selenium compounds, tellurium derivatives seem to show slightly different behavior. Tellurium gives various stable hypervalent compounds, and hence it has become one of the elements available for investigation of the hypervalent compounds.

According to the molecular orbital calculation,[18] important factors to be noted for the preparation of hypervalent compounds are as follows: (1) The smaller the electronegativity of the central element, the more stable the hypervalent compounds, that is, the 3c–4e bonding must be stabilized. (2) With regard to the ligand, the more electronegative elements or groups stabilize the hypervalent bond. (3) Lone pairs of electrons must play an important role in the formation and stabilization of the hypervalent bonds.

Therefore, a combination of electropositive elements and electronegative ligands provides the most stable hypervalent bonding. Thus, in the case of chalcogenuranes, tellurium fulfills these requirements and gives the most stable hypervalent molecules, followed by selenium and sulfur. As to the ligands, fluorine, chlorine, and oxygen must be appropriate candidates, together with their groups. As predicted by the valence bond treatment of Shaik and coworkers,[19] the lower the bond energy, the stronger the hypervalent bond formed. In fact, telluranes have been known to be more stable than selenuranes, and sulfuranes are the least stable. The various essential properties of chalcogen elements are summarized in Table 8.1.[20]

In organic sulfur chemistry, there is extensive indirect evidence to prove that the sulfurane is an intermediate in the substitution reaction. However, sulfuranes having electronegative ligands, such as in SF_4[21] and Ar_2SCl_2,[22] have been known; *p,p′*-dichlorophenyldichlorosulfurane was isolated and its crystal structure was determined to be TBP.

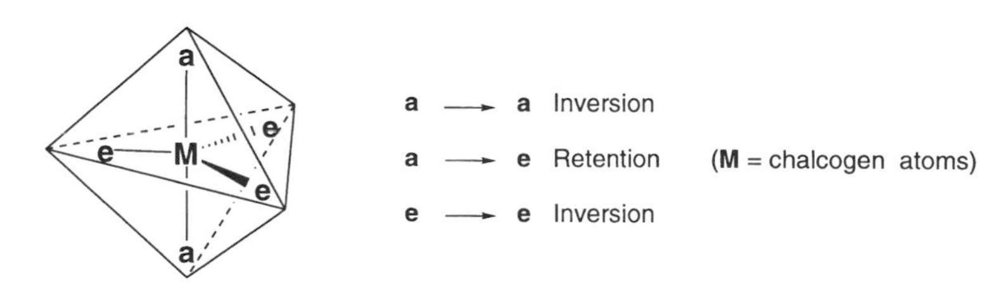

Figure 8.3 Stereochemical processes.

TABLE 8.1 Physical Properties of Chalcogen Elements

	Chalcogen Atoms			
	O	S	Se	Te
Electronegativity	3.5	2.5	2.4	2.1
Electron affinity	−7.28	−3.44	−4.21	—
Ionization potential	13.61	10.36	9.75	9.01
pK_a (aq. MH_2)	16	7.0	3.8	2.6
pK_a (aq. MH)	—	12.9	11.0	11.0
Bond length (Å)				
C(sp^3)—M	1.41	1.81	1.98	2.15
C(sp^2)=M	1.22	1.54	1.67	—

In the reaction of triphenylsulfonium chloride with phenyllithium or Grignard reagents, the products obtained are both biphenyl and diphenyl sulfide in quantitative yields. The mechanism for this reaction was investigated by Franzen and Mertz.[23] They described the reaction as proceeding via an initial formation of tetraphenylsulfurane that had not been detected. Wittig and co-workers found that similar treatment of triphenylselenonium salt with phenyllithium also afforded biphenyl and diphenyl selenide, but they were unable to detect tetraphenylselenurane. However, they found that tetraphenyltellurane could actually be isolated as a crystalline compound that undergoes various substitution reactions with nucleophiles and also ligand coupling reactions to result in the formation of biphenyl and diphenyl telluride quantitatively.[24] This is the first example of the isolation of a tellurane having four carbon ligands. After the more stable tellurane with four carbon ligands, bis(2,2′-biphenylylene)tellurane, was prepared by Hellwinkel and co-workers, a new aspect on the hypervalent compounds was developed.[25] They also isolated bis(2,2′-biphenylylene)selenurane, and quite recently Furukawa and co-workers have succeeded in synthesizing bis(2,2’-biphenylylene)sulfurane as stable crystals and they determined the structure by X-ray crystallographic analysis. They also succeeded in detecting the tetraphenylsulfurane and tetraphenylselenurane by a low-temperature nuclear magnetic resonance (NMR) technique. Thus, it has been acknowledged that tetraarylchalcogenuranes are common intermediates for the substitution reactions of triarylchalcogenium salts with organometallic reagents. Since the chemistry of sulfurane has been well documented by Martin, this review describes only the hypervalent compounds of tellurium and selenium. The chapter will be divided into two parts: (1) unstable hypervalent compounds such as ate-complexes of tellurium and selenium; (2) recent advancement in the study of the stable hypervalent compounds of selenium and tellurium. Several review articles have been published on these subjects.[26]

8.2 DETECTION OF UNSTABLE SELENURANE AND TELLURANE AS INTERMEDIATES

8.2.1 Detection of Tellurium and Selenium ate-Complexes [10—M—3(C3)] (M = Se, Te) Using Li–Te and Li–Se Exchange Reactions

Dicoordinated tellurides and selenides are, in general, thermally stable. They react readily with electrophiles such as an alkylating agent to give the corresponding onium salts. In contrast to the electrophilic reagents, tellurides and selenides do not undergo the substitution reaction with nucleophilic reagents. However, diaryl tellurides are known to react with Grignard reagents in the presence of Ni^{II} and Co^{II} salts as catalyst.[27] The mechanism for this reaction has been postulated to proceed via the formation of the metal complexes, which react further with other Grignard reagents to undergo exchange reactions; then a coupling reaction should take place to give the mixtures of the coupling products.

In the case of diaryl sulfides, though they do not react directly with Grignard reagents, coupling reactions have been conducted in the presence of metallic salts by Takei and co-workers. Since the electronegativity of tellurium is 2.1, the compounds of tellurium may undergo nucleophilic reactions or electron-accepting reactions. In fact, recently, the groups of Reich and Furukawa have found that diaryl tellurides undergo facile ligand-exchange reactions on treatment with organolithium reagents. The organotellurium–organolithium exchange reactions proceed via an initial formation of $Ar_3Te^-Li^+$ telluranes [10—Te—3(C3)] (ate-complex **1**), which can be detected via low-temperature ^1H-, ^{13}C-, ^7Li- and ^{125}Te-NMR spectroscopy. For example, in ^{125}Te NMR, diphenyl telluride shows the chemical shift at 670.4 ppm, which shifts upfield to 320.6 ppm when one equivalent of PhLi is added at −100C.[28] This behavior of the ^{125}Te NMR chemical shift indicates the unambiguous formation of a tellurane ate-complex, as shown in Scheme 8.1.[29]

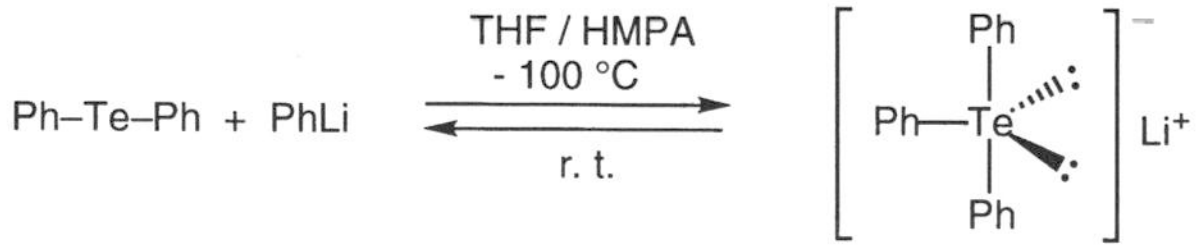

Ph–Te–Ph + PhLi ⇌ (THF / HMPA, −100 °C; r. t.) [Ph–Te(Ph)(Ph)]$^-$ Li$^+$ **1**

Scheme 8.1

Similarly, when phenyl pentafluorophenyl and 3,5-dichloro-2,4,6-trifluorophenyl tellurides were treated with pentafluorophenyllithium, ate-complexes **2** of telluride were observed via low-temperature ^{125}Te-, ^{19}F-, ^{13}C-NMR spectroscopy.[30] Interestingly, in these exchange reactions, the ligand exchange clearly takes place at the apical position and pseudorotation was not observed at all (Scheme 8.2).

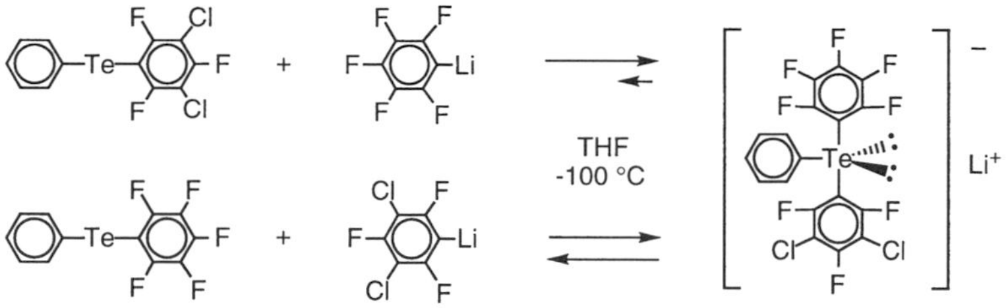

Scheme 8.2

These results reveal that electronegative ligands should be preferentially substituted at the apical positions in the tellurium ate-complexes.

Neither selenium nor sulfur ate-complexes of the tellurium analog have been detected or isolated. However, Reich and his co-workers have recently succeeded in detecting the formation of the selenium ate-complex of [10—Se—3(C3)] **3** by manipulating the selenide substrate, as shown in Scheme 8.3.[31]

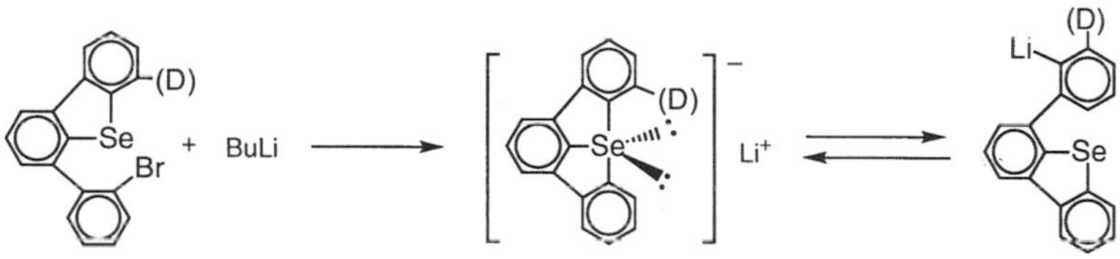

Scheme 8.3

Although the ate-complexes of both tellurium [10—Te—3(C3)] and selenium [10—Se—3(C3)] have been detected, no isolation of stable ate-complexes of tellurium and selenium has been reported.

8.2.2 Organic Synthesis via Tellurium ate-Complexes

Dicoordinated tellurium compounds undergo facile Te–Li exchange reactions via tellurium ate-complexes. Using this process, new organolithium reagents can be generated and applied for organic synthesis. Sonoda and his co-workers have extensively investigated the Te–Li metal exchange reactions using various tellurides such as telluroesters **4** and tellurocarbamates **5** and other tellurides bearing many functional groups. They reported that these compounds on treatment with alkyllithium undergo facile Te–Li exchange reactions, resulting in the generation of various functional anions such as acyl, carbamoyl, and thiocarbamoyl anions, which react with electrophiles *in situ* to give synthetically important products, as shown in Scheme 8.4. Although they did not detect the formation of tellurium ate-complexes as intermediates, the methods

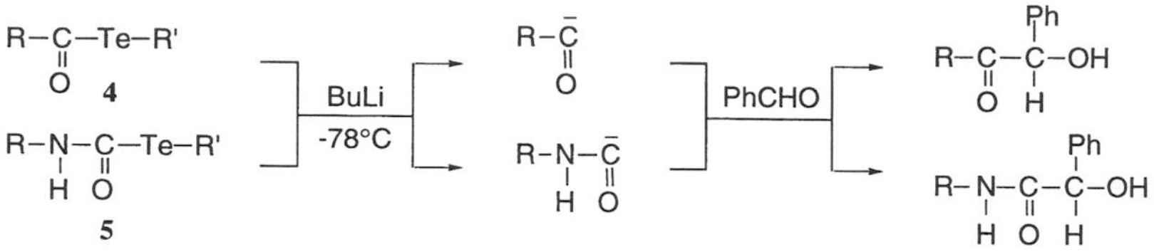

Scheme 8.4

are convenient and simple for the generation of these carbanions, which are difficult to prepare by other procedures.[32]

8.2.3 Ligand Exchange and Coupling Reactions of Chalcogen Oxides with Organometallic Reagents

Although sulfoxides have been well known to undergo simple ligand-exchange reactions on treatment with Grignard or organolithium reagents, there are only a few reports regarding the ligand-coupling processes, as described in Chapter 9. In contrast to the reactions of sulfoxides, there has been little investigation on the reactions using selenoxides and telluroxides with organometallic reagents. In the case of selenoxides, Ogawa and co-workers conducted the reaction of diphenyl selenoxide with phenylmagnesium bromide to give triphenylselenonium salt, which undergoes further reaction with PhMgBr to give finally diphenyl selenide and biphenyl (Scheme 8.5). These results demonstrate that the reactions of selenoxides with organometallic reagents proceed via the Se—O bond fission to give **6**, in marked contrast with the reaction of sulfoxide.[33]

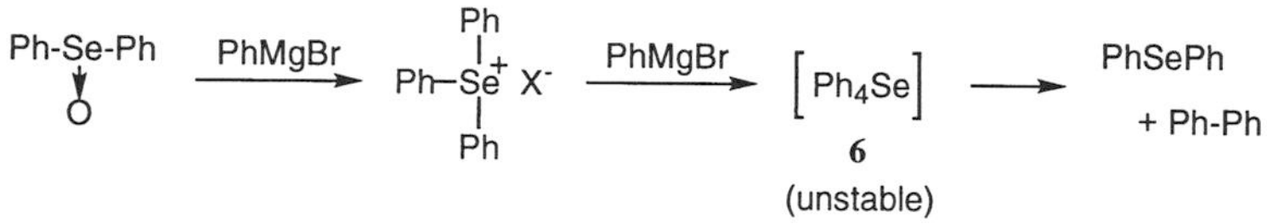

Scheme 8.5

Recently, Furukawa and co-workers repeated the reaction with diaryl selenoxides **7** with organolithium reagents and found that both the C—Se and Se—O bond fission take place competitively to afford the disproportionated selenoxides and selenides together with selenonium salts (Scheme 8.6).[34]

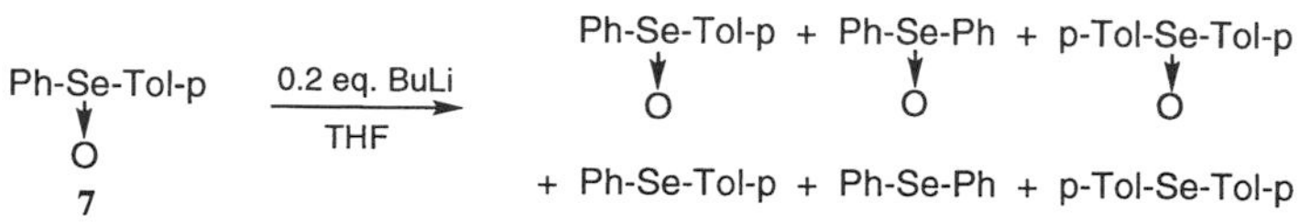

Scheme 8.6

As compared with the sulfoxides[35] and selenoxides,[36] the preparation of telluroxides by conventional synthetic methods using oxidizing agents is difficult, and hence only a few telluroxides are known.[37] In a few examples of the reactions of telluroxides with organometallic reagents, diaryl telluroxides are made to react with organolithium reagents. Unlike sulfoxides and selenoxides, diaryltelluroxides react with one equimolar amount of aryllithium to give triaryloxytelluranes **8** quantitatively (Scheme 8.7).[38]

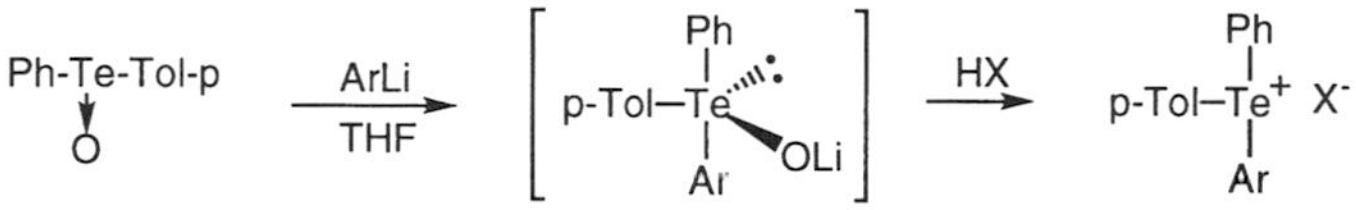

Ph-Te-Tol-p (Te→O) —ArLi, THF→ [p-Tol–Te(Ph)(Ar)(OLi)] **8** —HX→ p-Tol–Te⁺(Ph)(Ar) X⁻ **9**

Scheme 8.7

The structures of these oxytelluranes **8** have not been confirmed. However, on hydrolysis with aqueous HCl solution, these compounds are converted to the corresponding telluronium salts **9** bearing three different ligands, and hence this process is convenient for the preparation of unsymmetrically substituted telluronium salts. This different reaction mode between chalcogen oxides with organometallic reagents is accounted for in terms of the electronegativity and bond energies of M—O and M—C (M: S, Se, Te), as shown in Table 8.1; in the case of sulfoxide the C—S bond energy is smaller than that of S—O bond and hence organometallic reagents attack preferentially at the C—S σ^*-bond rather than the SO σ^*-bond. Meanwhile, in the case of telluroxides, the Te—O bond energy is smaller than that of the Te—C bond, and the tellurane structure **8** should be stabilized to give oxytelluranes or oxytelluronium salts.

8.2.4 Formation of Tetraarylchalcogenuranes [10—M—4(C4)] (M = Se, Te) and Ligand-Coupling Reactions

The hypervalent organochalcogen compounds containing only carbon ligands have been known to be unstable. Although some of the corresponding tellurane species [10—Te—4(C4)] are metastable and have been isolated, both sulfurane [10—S—4(C4)] and selenurane [10—Se—4(C4)] have been considered as intermediates or transition states for the reaction of the corresponding onium salts or oxides with various organometallic reagents. For a long time, these species were not detected or isolated.

In contrast to the tetraarylsulfuranes and selenuranes, Wittig and co-workers had already succeeded in the isolation of tetraphenyltellurane **10** from the reaction of tetrachlorotellurane and phenyllithium or phenyltrichlorotellurane with phenyllithium in high yield, having a melting point of 105–108°C (Scheme 8.8).[39]

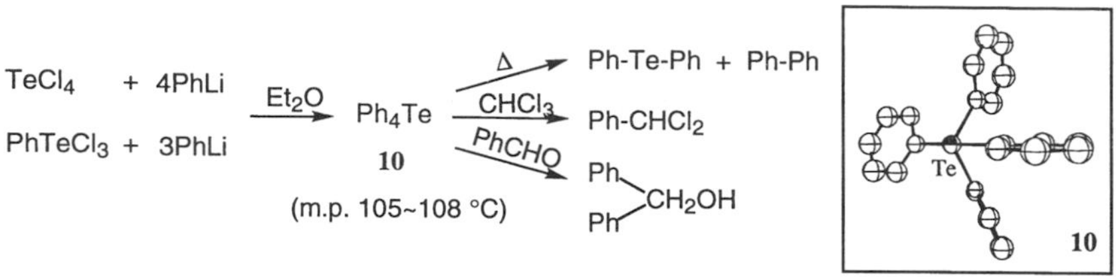

Scheme 8.8

They found that the tellurane **10** reacts with various reagents to undergo substitution reactions, and also it affords biphenyl and diphenyl telluride quantitatively on thermolysis. Therefore, this result indicates that the nucleophilic substitution reaction on the tellurium atom proceeds via two-step processes involving tetracoordinated tellurane as a real intermediate, which subsequently undergoes ligand-coupling or substitution reactions. Thereafter, Hellwinkel and co-workers also isolated the more stable bis(2,2'-biphenylylene)tellurane **11** as crystals.[40] Furthermore, they successfully achieved the isolation of bis(2,2′-biphenylylene)selenurane **12** as the first isolable selenurane having four carbon ligands, [10—Se—4(C4)], as shown in Scheme 8.9.[41]

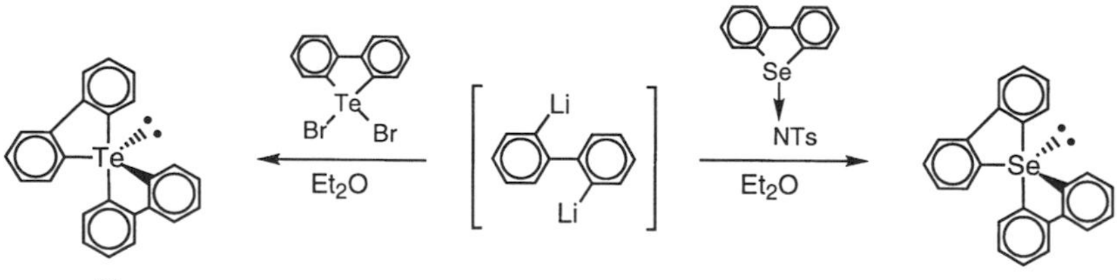

Scheme 8.9

Examination of the reactivity of selenurane **12** and tellurane **11** revealed that they undergo ring-opening reactions upon treatment with acids and chlorinated hydrocarbons such as dichloromethane and chloroform to give the selenonium and telluronium compounds. They also react with benzaldehyde to give diphenylmethyl alcohol, suggesting that both selenurane and tellurane bearing four-carbon atoms act as aryl anions. Recently, Barton and co-workers and Glover demonstrated that the reaction of tetraphenyltellurane with *t*-butylthiol gives biphenyl and di-*t*-butyl disulfide in quantitative yields. Interestingly, they revealed that tetraaryltelluranes undergo extremely facile ligand exchanges or disproportionation reactions, which are observed in the triaryltellurium ate-complexes, as described above.[42] The mechanism for this facile disproportionation has not been well elucidated, but they have postulated that the reaction proceeds via an ionic concerted manner and does not involve radicals, since no radical trapping agents work for the detection of phenyl radicals. These results represent only scattered data on the comparison of the reactivity of these hypervalent compounds.

Quite recently, however, Furukawa and co-workers have succeeded in the detection of both tetraphenylsulfurane **13**[43] and tetraphenylselenurane **14**[44] using low-temperature NMR spectroscopy. They found that **14** can exist at 0°C and even the formation of **13**, which has long been believed to be undetectable, was first confirmed by ^{1}H- and ^{13}C-NMR spectroscopy at around −40 to −105°C. The generation of **13** and **14** was performed by adding one or two equimolar amounts of phenyllithium to the corresponding chalcogenium salts or the chalcogen oxides in tetrahydrofuran solution in an NMR tube *in situ* (Scheme 8.10). Furthermore, they have reported that the formation of 2,2′-biphenylylene-diphenylsulfurane **15** and -selenurane **16** was detected by the same methods at low temperature; the isolation of a tellurane **17** was one example of tellurane bearing four carbon ligands.[45]

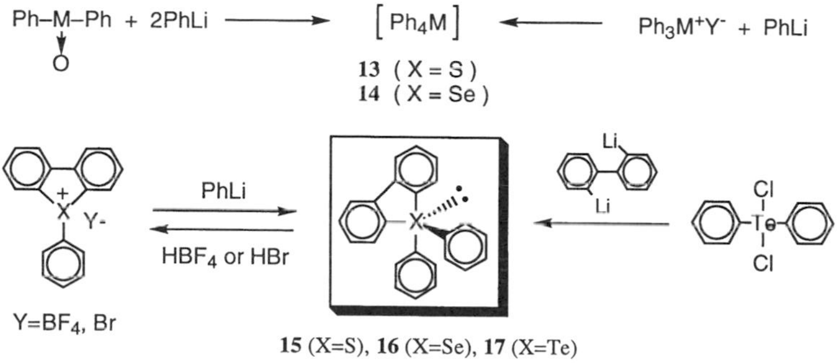

15 (X=S), **16** (X=Se), **17** (X=Te)

Scheme 8.10

The typical ^{1}H- and ^{13}C-NMR spectra at low temperature of **13** and **14**, together with that of stable tellurane **10**, are shown in Figure 8.4. These NMR spectra change to those of biphenyl and diphenyl chalcogenide upon heating to an appropriate temperature. Each chemical shift was determined by two-dimensional shift correlation (^{1}H—^{1}H— and ^{13}C—^{1}H—COSY) spectra.

Since tetraphenyltellurane **10** can be isolated as a crystalline material,[46] the stability values of these three phenyl chalcogenuranes are qualitatively in the following order: Te > Se > S. As a further example for comparison of the stability or reactivity of these chalcogenuranes, kinetic studies for the ligand-coupling reactions were conducted, since all three chalcogenuranes undergo the ligand-coupling reactions to give biphenyl and diphenyl chalcogenides quantitatively. The kinetic investigation was carried out quite readily with low-temperature ^{1}H- and ^{13}C-NMR spectroscopy. The kinetic results reveal that the facility of the ligand-coupling reactions is in the following expected order: S > Se > Te, which agrees well with the stability for the hypervalent bond of

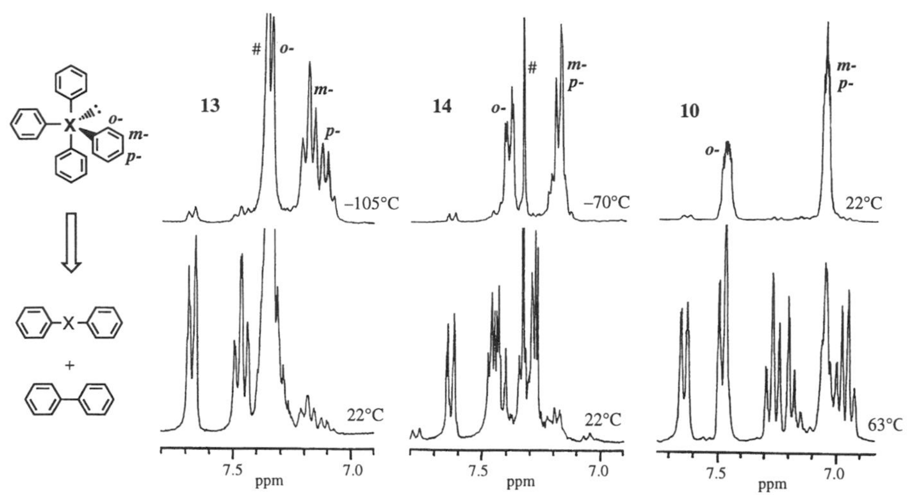

Figure 8.4 ^{1}H-NMR spectra of **10**, **13**, and **14**; initial (upper) and final (lower) spectra at present temperature.

these elements.[47] The first-order rate constants and activation parameters for ligand-coupling reactions are summarized in Table 8.2.

Although these tetraphenyl chalcogenuranes give ample information on the stability of the hypervalent bonds of chalcogenuranes, the question still remains whether it is possible to isolate sulfurane bearing four-carbon ligands. Many chemists have tackled this problem; our group have succeeded finally in isolating stable tetraaryl sulfurane **18** using dibenzothiophene monooxide and 2,2'-dilithiobiphenyl in the presence of trimethylsilyltriflate as an activation reagent of the sulfoxide group, as shown in Scheme 8.11.[48]

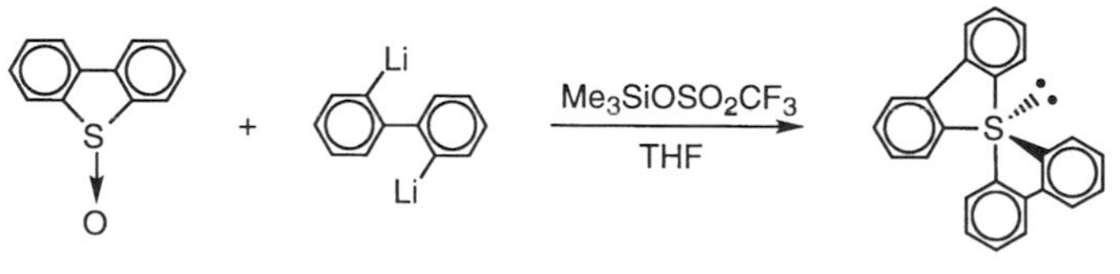

S, O + Li, Li $\xrightarrow[\text{THF}]{Me_3SiOSO_2CF_3}$ S **18**

Scheme 8.11

Furthermore, we have also determined the structure of **18** by X-ray crystallographic analysis. The compound has a nearly TBP structure. The ORTEP drawings of **18**[48] and the tellurium analogs **11**[49] are shown in Figure 8.5.

The X-ray crystallographic analysis of **18** and **11** (the structure of selenurane **12** has not been determined as yet) demonstrates clearly that they are nearly TBP structures and that the apical and equatorial bonds can be distinguished. However, in solution both apical and equatorial bonds seem to be averaged

TABLE 8.2 Activation Parameters of Ligand-Coupling Reaction of Tetraphenyl Chalcogen Compounds[a]

Compound (Solvent)	k_l (sec^{-1})		E_{act} ($kcal\ mol^{-1}$)	$\Delta G^{\ddagger}_{298}$ ($kcal\ mol^{-1}$)	$\Delta H^{\ddagger}$ ($kcal\ mol^{-1}$)	$\Delta S^{\ddagger}$ (eu)
13 Ph_4S(THF-d_8)	2.48×10^{-4} (−67°C)	1.22×10^{-4} (−72°C)	10.9	17.5	10.5	−23.5
	5.00×10^{-5} (−77°C)	3.25×10^{-5} (−82°C)				
14 Ph_4Se (THF-d_8)	2.20×10^{-4} (0°C)	1.30×10^{-4} (−5°C)	21.3	20.4	21.3	3.1
	5.31×10^{-5} (−11°C)	2.21×10^{-5} (−15°C)				
10 Ph_4Te (Toluene-d_8)	3.9×10^{-4} (84°C)	1.77×10^{-4} (74°C)	29.0	26.9	28.4	5.2
	3.00×10^{-5} (63°C)	8.26×10^{-6} (52°C)				

[a]Values shown are least-square treatments of Arrhenius and Eyring plots.

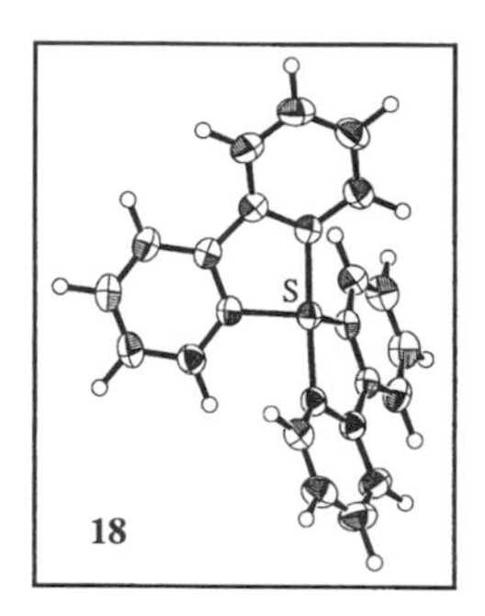

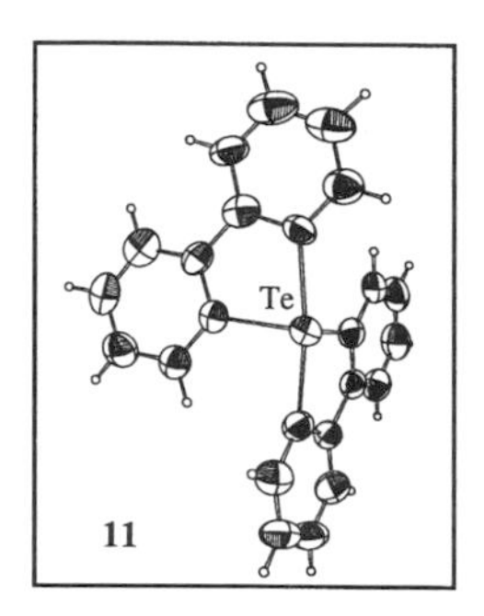

Figure 8.5 ORTEP drawins of **18** and **11**.

due to pseudorotation; NMR gives a single chalcogen—C bond at room temperature. At low temperature, however, the apical and equatorial bonds for selenurane **12** and tellurane **11** can be distinguished from each other by variable-temperature NMR spectroscopy. By elevating the temperature, the ^{1}H chemical shifts due to axial and equatorial C—H bonds in the aromatic rings or the ^{13}C-NMR chemical shifts coalesce to a peak shift, as shown in Figure 8.6. Hence, using this variable-temperature NMR technique, the energy barriers for the pseudorotation or turnstile rotation of both selenurane and tellurane have been determined. However, sulfurane always shows a single ^{1}H- or ^{13}C-NMR shift due to the phenyl rings at any temperature range, suggesting that **18** undergoes pseudorotation extremely rapidly or that it has a TBP structure in solution. The comparison of the pseudorotational ability between selenurane and tellurane shows that tellurane undergoes pseudorotation more readily than selenurane. The energy barriers of the two hypervalent compounds are about 13 kcal mol^{-1} (**12**) and 9 kcal mol^{-1} (**11**) respectively.[50]

Thus, the stability and pseudorotational ability of these two chalcogenuranes are opposite to each other. A kinetic study of the pseudorotational feasibility of selenurane and tellurane using tetraoxy derivatives has also been presented by Denney and his co-workers. They find that the trend is identical with that of tetraaryl derivatives (**19**, **20**), namely Te > Se.[51]

Another comparison of the reactivity of the chalcogenuranes has recently been performed by our group. We have investigated the reactions of 2,2′-bis(biphenylylene)chalcogenuranes (**11**, **12**, **18**) with several nucleophilic reagents. As is already known, both **11** and **12** undergo a facile ring-opening reaction with strong acids to give the corresponding onium salts.

In an extension of this study, when **12** and **18** were treated with alcohols and phenols, they gave *o*-alkoxy or -phenoxybiphenyl and diphenyl sulfide or selenide at room temperature in quantitative yields.[52] Following the reaction of **12** by ^{1}H- and ^{77}Se-NMR reveals that these chalcogenuranes react with hydroxy compounds to give the ring-opening compounds initially. These then undergo ligand-coupling or ipso-substitution reactions to give the coupling products, as shown in Scheme 8.12.[53]

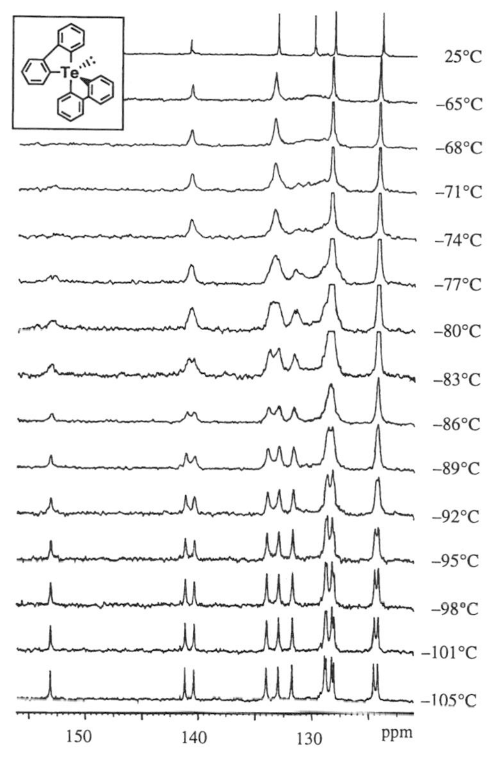

Figure 8.6 Variable-temperature ^{13}C-NMR spectra of **11** in THF-d_8.

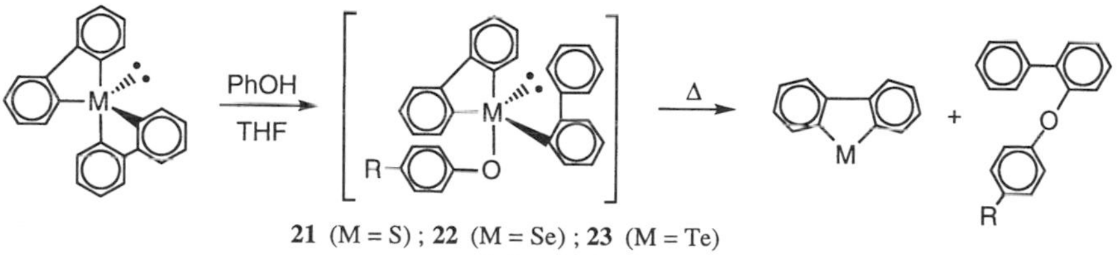

PhOH
THF
Δ
M
R
O
+

21 (M = S) ; **22** (M = Se) ; **23** (M = Te)

Scheme 8.12

Whether the intermediary compounds formed are oxysulfurane **21**(R = H) or -selenurane **22**(R = H) is a problem yet to be solved. Similarly, tellurane **11** also undergoes the ring-opening reaction on treatment with several phenols to afford the crystalline compounds **23**(**a**, R = Me; **b**, R = H; **c**, R = Cl; **d**, R = MeOCO; **e**, R = MeCO; **f**, R = NO_2; **g**, R = 2,4,6-Cl_3) after workup.[49] These compounds were subjected to X-ray crystallographic analysis, which revealed, as shown in Figure 8.7, a new oxytellurane having a slightly distorted TBP structure of [10—Te—4(C3O)]. This is also a first example of oxytellurane **23**

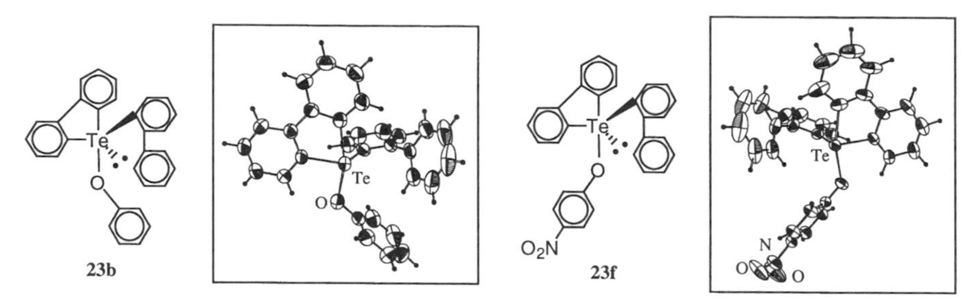

Figure 8.7 ORTEP drawing of **23**.

or typical telluronium salts having secondary bonding that is a centrosymmetric *O*-bridged dimer.[54] The hypervalency of the tellurium compounds **23** in solution was estimated by correlating their ^{125}Te-NMR chemical shifts and the Gutmann donor number, DN, employed in the experiments.

On the basis of examining the reaction patterns and reactivity orders for the chalcogenurane series, we concluded that sulfurane and selenurane also afford oxysulfurane and -selenurane as well as tellurane. Therefore, the chalcogenuranes bearing four-carbon ligands provide various new chalcogenuranes having numerous ligands. This promises the opening of a new frontier in the field of hypervalent compounds.

8.3 STABLE HYPERVALENT COMPOUNDS OF SELENIUM AND TELLURIUM

Numerous hypervalent compounds of selenium and tellurium have been reported in the literature. However, as predicted, the stability of the hypervalent bond depends both on the electronegativity of the central atom and on the ligands used. In the chalcogens, a heavier element provides more stable hypervalent bonds, and hence tellurium gives the most stable hypervalent compounds. Telluranes bearing numerous ligands have been produced as isolably stable derivatives, whereas a few selenuranes have been reported.[55] Moreover, investigations seem to be concentrated on the synthesis and structure and not many systematic reports have been made on treatment of the reactivity of the hypervalent bonds, except for a few restricted compounds such as $TeCl_4$. Therefore, this chapter treats mainly the preparation and structure of telluranes and selenuranes. X-ray crystallographic analysis is of course the most efficient procedure to analyze structures, whereas ^{77}Se- and ^{125}Te-NMR are also promising techniques for determining the structure of hypervalent compounds. The compounds in this section are classified according to Martin's nomenclature, the *N–X–L* (A_nB_m) coding system, in which *N* designates the number of valence electrons associated formally with a central atom *X* with *L* ligands (A and B are the elements). For example, sulfurane (SCl_4) and selenurane ($SeCl_4$) are represented as [10–S–4(Cl4)] and [10–Se–4(Cl4)].[56]

8.3.1 Type 10–M–3 (M = Se,Te)

Ate-complexes of tellurium and selenium, like [10–Te–3(C3)], [10–Se–3(C3)], belong to this category and have already been mentioned above. Stable π-hypervalent selenenium and telluronium species [10–M–3] (M = Se, Te) **24**[57], **25**[58], **26**[59], **27**[59], **28**[60], and **29**[60] are shown in Figure 8.8.

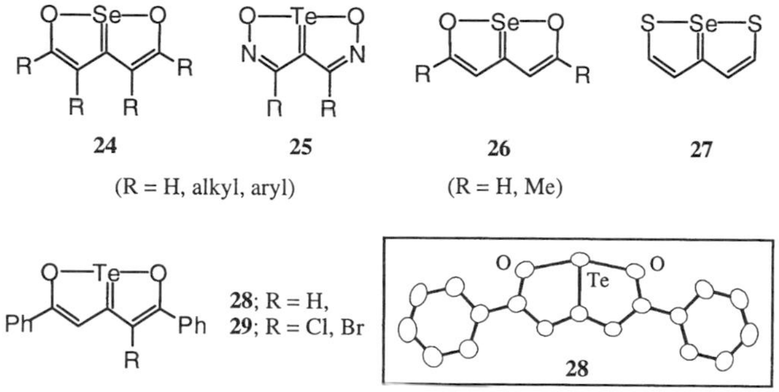

24 25 26 27
(R = H, alkyl, aryl) (R = H, Me)
28; R = H, 29; R = Cl, Br

Figure 8.8 Structure scheme and ORTEP view of type 10-M-3 (M = Se, Te).

Recently, the groups of Matsumura and Iwasaki have reported that the 12π-selenatetraazapentalene derivatives (π-selenurane) **30** containing a hypervalent selenium and two thiocarbonyl groups in their framework are synthesized in good yields by a convenient one-pot reaction.[61] The structures were determined by X-ray crystallographic analysis. The reactions of **30** with alkyl iodides and ω-bromoalkyl isothiocyanates gave novel selenatetraazapentalene derivatives, as shown in Scheme 8.13.

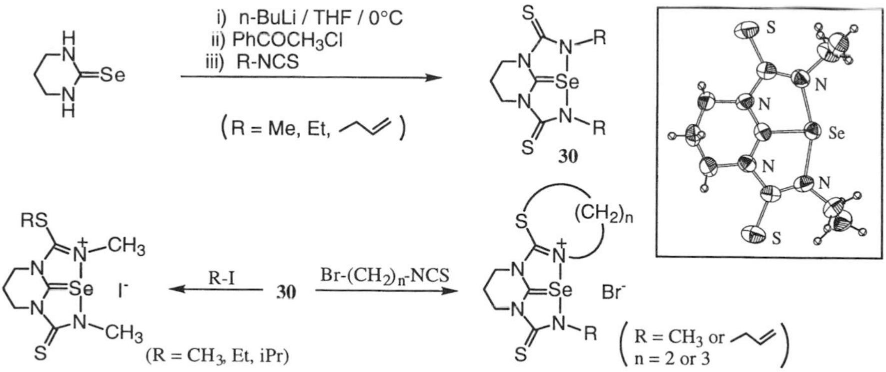

i) n-BuLi / THF / 0°C
ii) PhCOCH$_3$Cl
iii) R-NCS
(R = Me, Et, allyl)
30
R-I (R = CH$_3$, Et, iPr)
Br-(CH$_2$)$_n$-NCS
(R = CH$_3$ or allyl, n = 2 or 3)

Scheme 8.13

Our associates have reported the first isolation and crystal structure of heavier chalcogenium cations (RSe^+, RTe^+) stabilized by two neighboring amonio groups.[62] The reaction of a linear selenide **31** or telluride **32** with *t*-BuOCl or Br_2 gave the selenium **33** or tellurium cation **34**, without the formation of the corresponding selenoxide or telluroxide, in which its Se- or Te-dealkylation was induced by the neighboring group participation of the two nitrogen atoms. The structure of **33** was determined by X-ray diffraction analysis, as shown in Scheme 8.14.

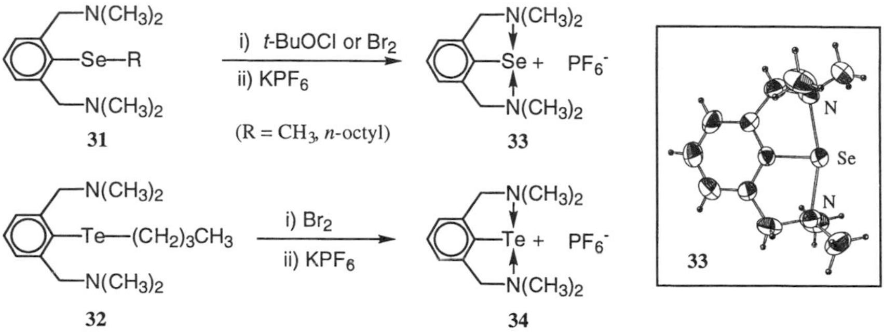

Scheme 8.14

8.3.2 Type 10–M–4 (Z4 and CZ3); (M = S, Se, Te; Z = halogen)

There are many stable hypervalent compounds belonging to this classification, in which Z is a halogen, such as SCl_4, SF_4, $SeCl_4$, $SeBr_4$, $TeCl_4$, $TeBr_4$, and TeI_4, and one carbon and three halogens, such as $ArTeCl_3$. More than 100 compounds were prepared and their structures were determined by X-ray crystallographic analysis. These compounds have been well documented in the reviews by Bergman and Sadekov, and hence are excluded from this review.[63]

$TeCl_4$, $PhTeCl_3$, and analogous bromine and iodine derivatives[64] are aggregates. X-ray analysis indicates that the crystalline structure is not genuine TBP, but more or less SP or a dimeric octahedral geometry.

Similarly, the selenium analogs of 10–M–4(Z4 and CZ3) are well known in the literature and their structures and reactivity are summarized in the reviews.[65]

8.3.3 Type 10–M–4(C2X2) (M = S, Se, Te; X = halogen, O, S)

A number of chalcogenuranes belong to this classification. In general, hypervalent compounds having two electronegative ligands such as halogen and oxygen are stable, and their structures can be analyzed by X-ray crystallography.[66] Chalcogenuranes [10–M–4(C2X2)] **35** are prepared simply by mixing the corresponding dicoordinated chalcogenides with halogen. Particularly, symmetrically substituted diaryl compounds bearing two halogen ligands are hypervalent in character, except in the cases of combinations of sulfide-I_2 and -Br_2, and selenide-I_2, which are known to be molecular complexes like charge-transfer complexes. Otherwise, Friedel–Crafts-type reactions are employed for the synthesis, as shown in Scheme 8.15.

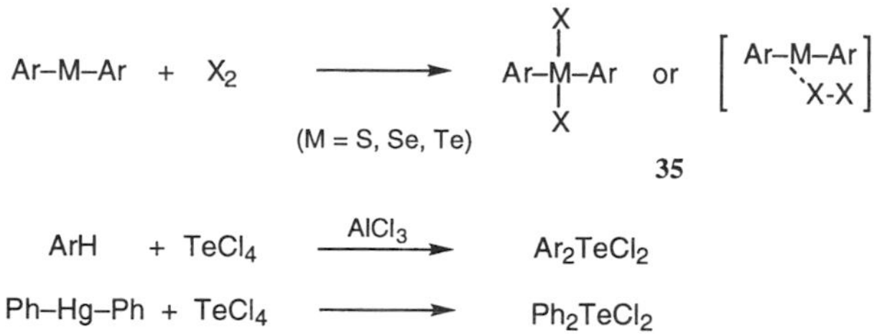

$$\text{Ar–M–Ar} + X_2 \xrightarrow[\text{(M = S, Se, Te)}]{} \text{Ar–M(X)}_2\text{–Ar} \;\; \text{or} \;\; [\text{Ar–M–Ar} \cdots \text{X–X}]$$

35

$$\text{ArH} + TeCl_4 \xrightarrow{AlCl_3} Ar_2TeCl_2$$

$$\text{Ph–Hg–Ph} + TeCl_4 \longrightarrow Ph_2TeCl_2$$

Scheme 8.15

The compounds having the general formula Ar_2MX_2 **35** are classified as shown in Table 8.3.[67]

TABLE 8.3 Hypervalent Compounds and Molecular Complexes[a]

Elements, **M** (Electronegativity)	Halogens, **X** (Electronegativity)		
	Cl(2.80)	Br(2.74)	I(2.21)
O(3.50)	—	MC	MC
S(2.48)	TBP	—	MC
Se(2.44)	TBP	TBP	MC
Te(2.10)	TBP	TBP	TBP

[a]TBP: trigonal bipyramid (hypervalent compound); MC: molecular complex.

The preparation of fluorine analogs **36**[68] of chalcogenuranes is carried out using suitable fluorinating reagents such as SF_4 and HF[69] with dicoordinated chalcogenides or chalcogenide monooxides, as shown in Scheme 8.16.[70]

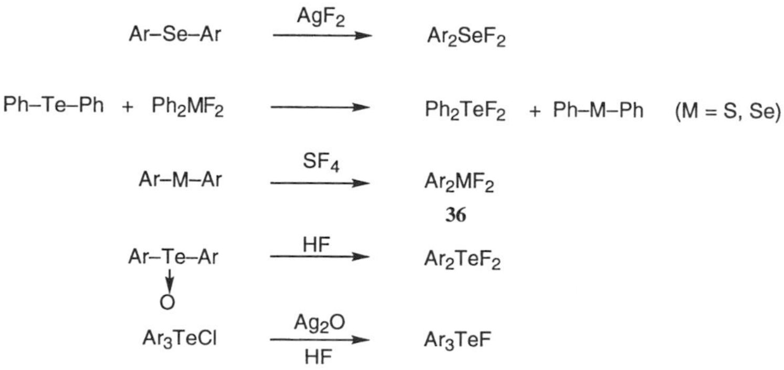

$$\text{Ar–Se–Ar} \xrightarrow{AgF_2} Ar_2SeF_2$$

$$\text{Ph–Te–Ph} + Ph_2MF_2 \longrightarrow Ph_2TeF_2 + \text{Ph–M–Ph} \quad (\text{M = S, Se})$$

$$\text{Ar–M–Ar} \xrightarrow{SF_4} Ar_2MF_2 \; \textbf{36}$$

$$\text{Ar–Te(}\rightarrow\text{O)–Ar} \xrightarrow{HF} Ar_2TeF_2$$

$$Ar_3TeCl \xrightarrow[HF]{Ag_2O} Ar_3TeF$$

Scheme 8-16

Similarly, preparation of several other diaryltelluranes **37** and selenuranes **38** having pseudohalogens such as —SCN[71] and —NCS[72] as ligands are also described, as shown in Scheme 8.17.

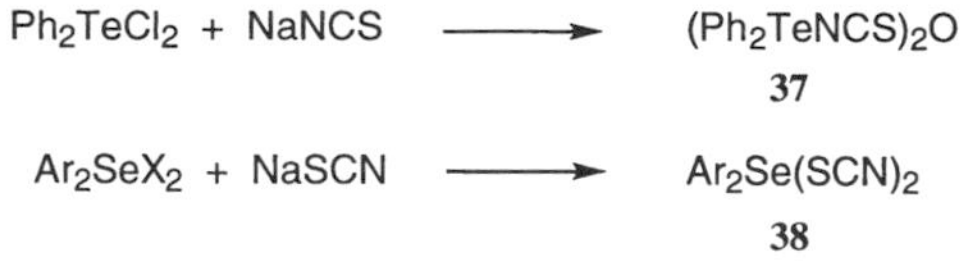

Ph2TeCl2 + NaNCS
(Ph2TeNCS)2O
37
Ar2SeX2 + NaSCN
Ar2Se(SCN)2
38

Scheme 8.17

These halogen-containing chalcogenuranes are readily transformed to the dihydroxy derivatives. However, many dialkoxy or diacyloxy selenuranes and telluranes having general structures $R_2Te(OR')_2$ **39**[73] or $R_2M(OCOR')_2$ **40** (M = Se, Te) have been prepared by the procedures shown in Scheme 8.18.

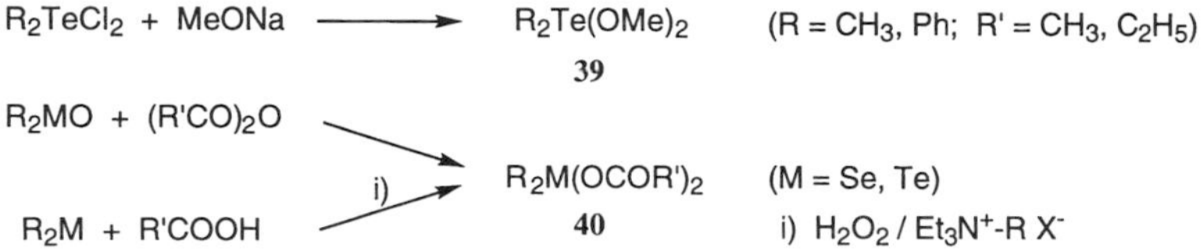

R2TeCl2 + MeONa
R2Te(OMe)2
39
(R = CH3, Ph; R' = CH3, C2H5)
R2MO + (R'CO)2O
R2M + R'COOH
i)
R2M(OCOR')2
40
(M = Se, Te)
i) H2O2 / Et3N+-R X-

Scheme 8.18

Recently, the bis(2,6-dimethylphenyl)telluranes **41** and **42** have been synthesized.[74] The ^{1}H-NMR spectrum of **41a** shows the existence of two broad methyl signals at 2.61 and 2.94 ppm. However, these signals coalesce to a singlet peak on elevating the temperature. The X-ray crystallographic analysis of **41a** indicates that the structure of **41a** has two apical bromine atoms and that two aryl ligands occupy the equatorial ligands, as shown in Figure 8.9. This result suggests that the free rotation at the tellurium atom in **41a** should be restricted at room temperature. On the basis of variable-temperature NMR the free energy for the free rotation of **41a** ($\Delta G^{\ddagger}$) was calculated to be 15.2 kcal

X
Te
R
X
OCOY
Te
R
OCOX
Br
Te
Br
41a

41a: R = H, X = Br
41b: R = H, X = Cl
41c: R = H, X = F
41d: R = Me, X = Br

42a: R = H, X = Y = CH_3
42b: R = H, X = Y = CF_3
42c: R = Me, X = Y = CH_3
42d: R = Me, X = Y = CF_3
42e: R = Me, X = Y = C_2H_5
42f: R = Me, X = Y = $C(CH_3)_3$
42g: R = Me, X = CH_3, Y = CF_3

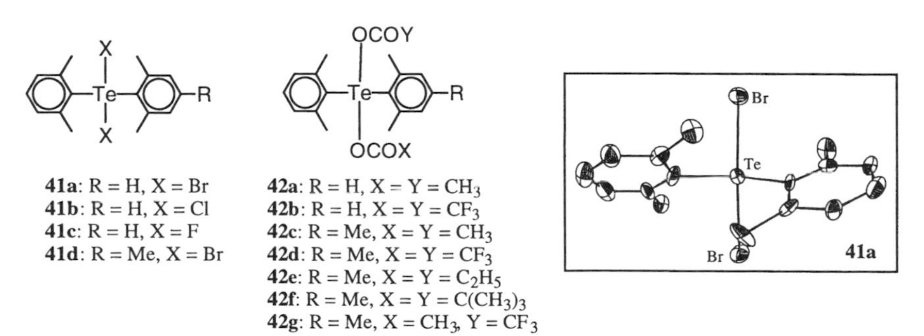

Figure 8.9 Structure scheme and ORTEP view of **41a**.

mol^{-1} (at 44.8°C). Similarly, structures of other dihalotelluranes **41b–d** and **42a–g** have demonstrated bearing conformationally restricted structures from free rotation of the 2,6-dimethyl groups by locking the two apical bromine ligands.

It is of interest to note that the tellurane **42g** is the first tetrasubstituted tellurane bearing a chiral center, though we could not succeed in separating **42g** into the individual enantiomers.

Although dialkoxy sulfurane **43** and selenurane **44** were prepared by Kapovits and Kalman[10] and Dahlen and Lindgren,[75] as shown in Figure 8.10, the preparation of dialkoxy tellurane derivative **45** has not been reported until quite recently.

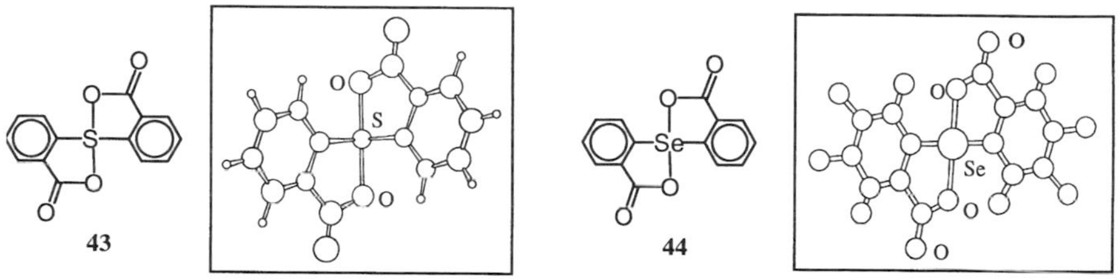

Figure 8.10 Structure scheme and ORTEP view of **43** and **44**.

Tellurane **45** was synthesized directly from anthranilic acid after diazotization and treatment with Na_2Te.[76] The intermediary formation of 2,2′-dicarboxylic telluride was not observed, in marked contrast to the syntheses of **43** and **44**. Tellurane **45** under by X-ray analysis was found to have a nearly TBP structure with a C2 molecular symmetry, as shown in Scheme 8.19. All the structures of hypervalent spiro-chalcogenuranes **43–45** have been determined by X-ray crystallographic analysis.

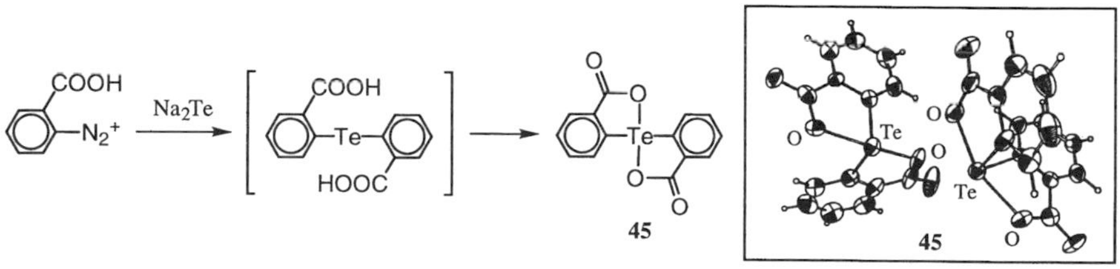

Scheme 8.19

The spiro-acyloxy tellurane **45** is quite stable against moisture and does not form the corresponding telluroxide even on heating with H_2O, in marked contrast to the sulfur analog. This unusual stability of the tellurane is accounted for in terms of the annelation effect provided by the five-membered ring structure. However, the spirotellurane **45** undergoes reactions with nucleophilic reagents such as sodium hydroxide, Laweson's reagent, and $LiAlH_4$ to afford the interesting tellurane derivatives **46**, **47**, **48**, and **49**.[77] Such reactions are illustrated in Scheme 8.20.

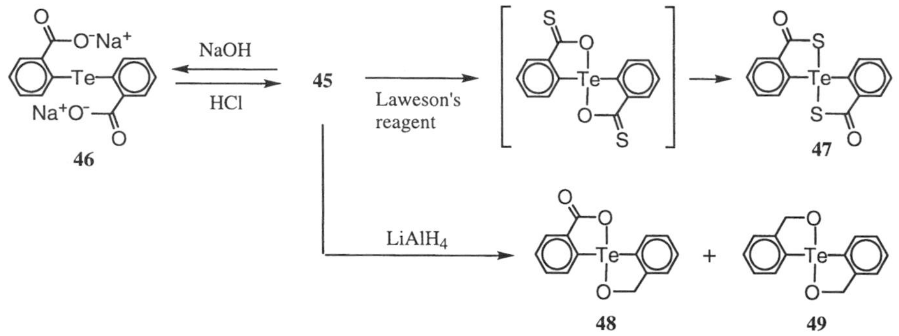

Scheme 8.20

Dialkoxysulfuranes are generally unstable and difficult to isolate. One should use ligands bearing strong electron-withdrawing fluorinated alcohols or the sulfur atom should be annelated at the ortho positions of the benzene ring, as was done for those first prepared by Martin and his co-workers.[78] Their manipulation of this enormous stabilizing ability of the ligands, as compared to the open chain analogs, has been utilized widely in the preparation of many heteroatom hypervalent compounds and is now referred to as Martin ligands **50**. It is shown typically in Figure 8.11.

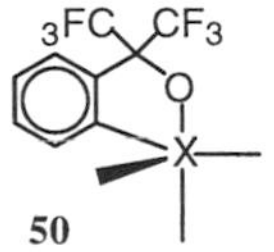

i) Five-membered ring annelation
ii) Gem-trifluoromethyl groups
iii) Apical alkoxy group of highly effective electronegativity

Figure 8.11 Martin ligand.

A few oxyselenuranes and -telluranes have been prepared; as typical examples, the selenuranes **51**[79] and telluranes **52**,[80] have been reported by the groups of Reich and Martin, are shown in Scheme 8.21.

Koizumi and co-workers have reported the first synthesis of optically pure haloselenuranes **53** and -tellurane **54**.[81] They have observed a complete retention of the configuration in the interconversion reactions of haloselenuranes **53**, in the corresponding selenoxide **55**, and in the nucleophilic substitution reactions of **53** and **54**, as shown in Scheme 8.22.

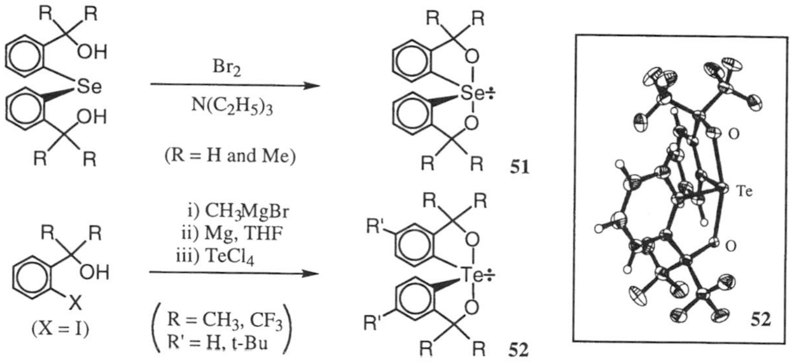

Scheme 8.21

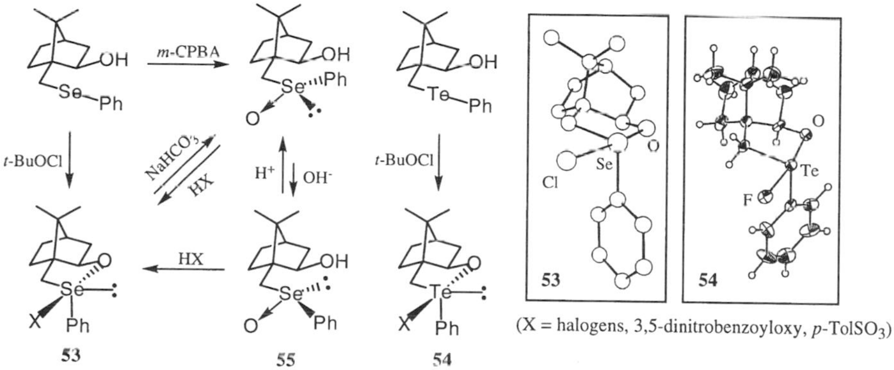

(X = halogens, 3,5-dinitrobenzoyloxy, p-TolSO$_3$)

Scheme 8.22

Other stable spiro-selenuranes **56**[82] and **57**[83] of this type are shown in Figure 8.12.

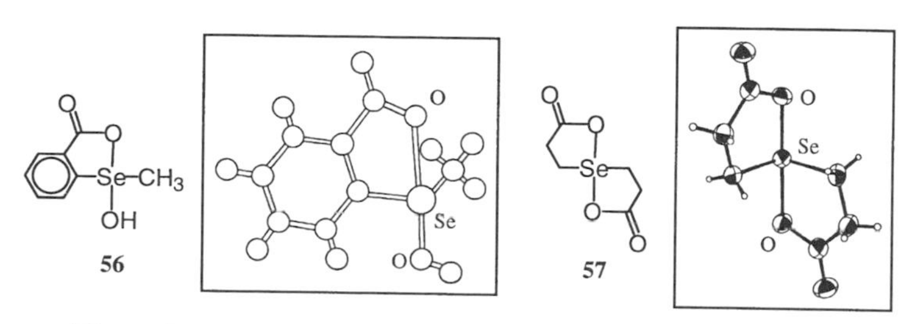

Figure 8.12 Structure scheme and ORTEP view of **56** and **57**.

Recently, a diphenoxytellurane **58**, where R' is a phenol moiety bearing an *o*-azomethine functional group, has been reported.[84] This tellurane is rather stable to moisture, unlike the other general alkoxy analogs, due to the presence of bulky *o*-azomethine substituents, which prevent the attack of H_2O. The X-ray crystallographic analysis of the tellurane **58** demonstrates that the structure is a slightly distorted TBP and coordinated by two azomethine nitrogen atoms. In other words, **58** belongs to the hexacoordinated tellurane type annelated by five-membered chelate rings. The structure is shown in Scheme 8.23.

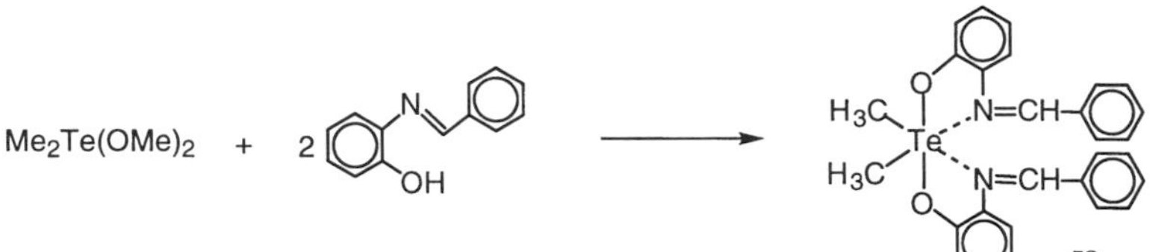

$Me_2Te(OMe)_2$ + 2 [o-hydroxyphenyl azomethine] → **58**

Scheme 8.23

As to the thio-analogs of telluranes **59**, Wieber and Kauzinger tried to convert the dioxytelluranes to dithiatelluranes **60** using dioxytellurane with bisdithiols.[85] However, although they did not succeed in the isolation of dithiatelluranes [10–Te–4(C2S2)] **60**, the coupling reaction proceeded to result in the formation of disulfide and telluride, as shown in Scheme 8.24. They only obtained dithiatelluranes **61** by using sodium xanthogenates,[86] as shown in Scheme 8.24.

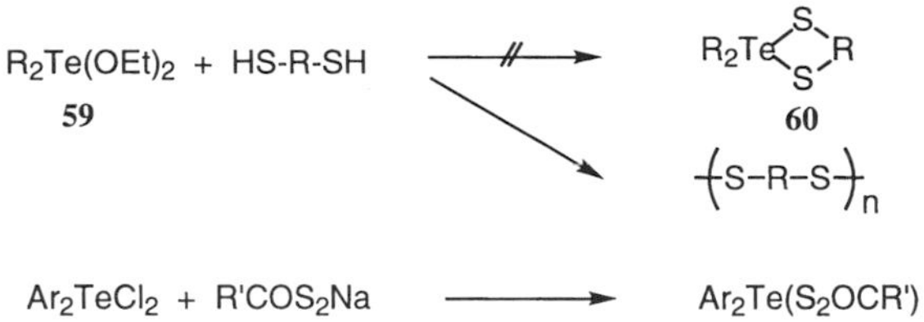

$R_2Te(OEt)_2$ + HS-R-SH (**59**) ⇸ $R_2Te(S_2R)$ (**60**)

→ $-(S-R-S)_n-$

Ar_2TeCl_2 + $R'COS_2Na$ → $Ar_2Te(S_2OCR')$ (**61**)

Scheme 8.24

In the course of the studies on the transannular interaction[87] between heteroatoms, our group have recently reported some interesting new stable azasulfuranes,[88] sulfurane dications bearing two sulfur atoms,[89] and selenurane dications bearing two sulfur and selenium atoms as ligands which are annelated in the two five-membered diazocine skeleton. As shown in Scheme 8.25, the central sulfur atom of the trissulfide **62** can be converted to the tetracoordinated sulfurane [10–S–4(C2S2)] **63** by employing suitable one-electron oxidants such as $NOBF_4$ or by dissolving the monooxide **64** into concentrated H_2SO_4.[89]

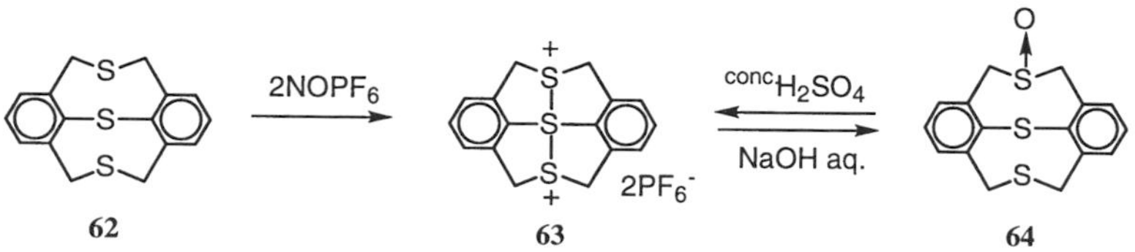

Scheme 8.25

The conformation of the trissulfide **62** is a twin-chair (TC) form which is transformed instantaneously to the twin-boat (TB) form on dissolution into concentrated H_2SO_4 via a transannular interaction between the three sulfur atoms. *Ab initio* MO calculation using STO-3G$^{(*)}$ as a basis set reveals also that the most stable conformations of **62** and **64** are TC and that of **63** is TB, respectively.

Furthermore, the trisselenide analog **65** and its derivative **66** behave similarly to **62**. Formation of the hypervalent bond between the three selenium atoms in **67** has been confirmed by examining the satellite peak of the ^{77}Se–^{77}Se coupling on NMR.[90] Also, X-ray crystallographic analyses of both **65**(neutral) and **67**(dication), in which X = Se, were successfully performed. The results support the hypothesis described in the case of trissulfide **62** and dication **63**: compound **65**, **66** has a TC conformer, whereas selenurane **67**, **68** has a TB structure, in which the central selenium atom is a slightly distorted TBP structure, as shown in Scheme 8.26.

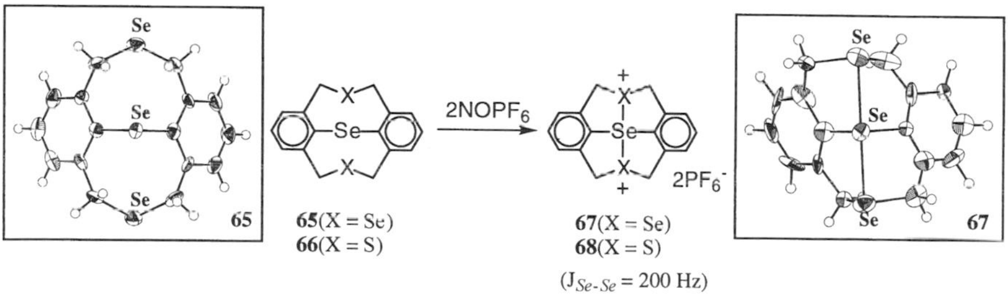

Scheme 8.26

This group has succeeded in the preparation of several other new selenuranes **69**,[91] **70**,[92], **71**,[93] and **72**[94] bearing nitrogen and/or halogen as ligands, as shown in Scheme 8.27.

Ryan and Harpp have also reported the detection of a new type of selenurane **73** having nitrogen and chlorine, on the treatment of bisdiallyl selenide with *N*-chlorobenzotriazole, by examining ^{77}Se-NMR at low temperature, as illustrated by Scheme 8.28.[95]

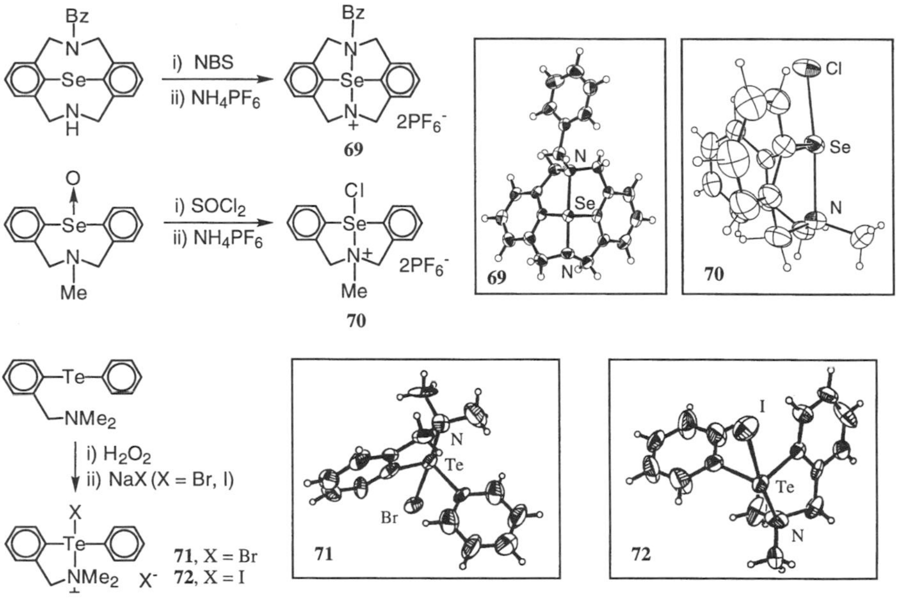

Scheme 8.27

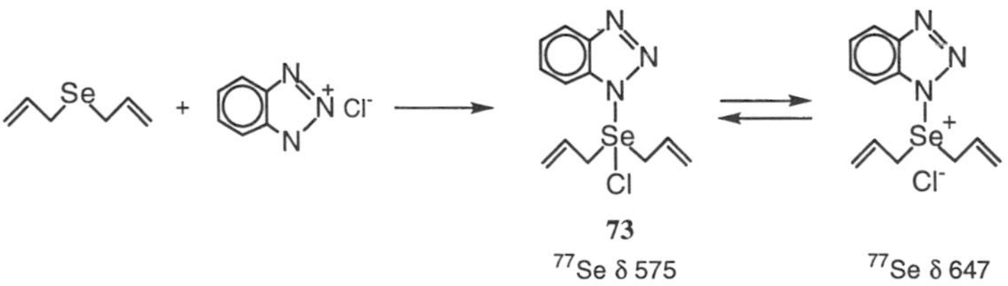

Scheme 8.28

Pinto and co-workers reported that the spontaneous electron-transfer reaction of the selenium coronand $[Cu(16Se4)]\text{-}[SO_3CF_3]_2$ **74** in organic solvents gave Cu(I) as well as the intermediate radical cation $[16Se4]^{\cdot +}$ and the stable hypervalent dication $[16Se4]^{2+}([SO_3CF_3]^-)_2$ **75** having a weak secondary Se—Se bonding, as shown in Scheme 8.29.[96]

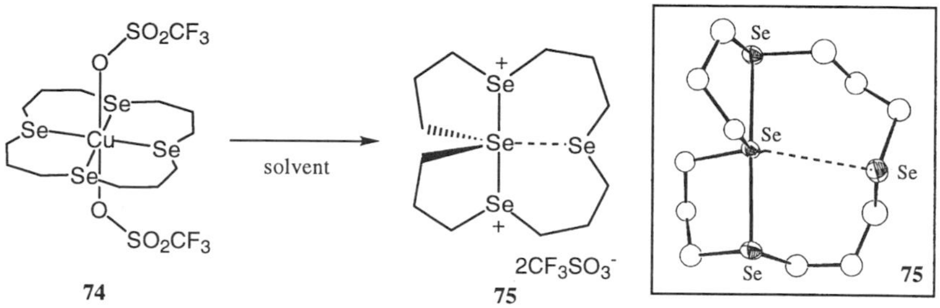

74 → (solvent) 75 $2CF_3SO_3^-$

Scheme 8.29

8.3.4 Type 10-M-4(C3X) (M = Se, Te; X = O, N, S)

As shown previously, the compounds belonging to this classification are in general the onium salts such as sulfonium, selenonium, and telluronium salts, and are essentially ionic in character.

Several preparative processes are known. In the case of unsymmetric triarylsulfonium salts, preparations should preferably be carried out using diaryl sulfoxides with aryl Grignard reagents in the presence of activating reagents such as Meerwein reagent, but the use of trimethylsilyl trifluorosulfonate is recommended.[97] This method can be applied especially for the synthesis of the salts having not only three different aryl groups but also aryl alkyl moieties. In the absence of this activating reagent, disproportionation of the sulfoxides employed for the synthesis has been observed, as described in the previous section.

However, we found that 2,2′bis(biphenylylene)tellurane **11** undergoes ring-opening reactions on treatment with alkoxides, phenoxides, and thiolates to give the corresponding oxytelluranes [10–Te–4(C3O)] **23**, as described previously.[53,54]

These results suggest that the telluronium and selenonium salts would have hypervalent structures in a crystalline state, whereas in solution, especially in polar solvents, it might be an onium structure. The electroconductivity values of several telluronium compounds reveal that they also should have an ionic character.

8.3.5 Type 10-M-4(X4) (M = S, Se, Te ; X = C, O, S, Se)

Sulfuranes having four carbon ligands have been described as intermediates in the substitution reactions at the sulfur atom. By similar procedures, nine tetra-arylic chalcogenuranes bearing four carbon ligands have been prepared, as described in the previous section (Figure 8.13).

Tetraalkoxysulfuranes were reported by Wilson and Belkind.[98] Many tetraalkoxyselenuranes and -telluranes have been prepared by Denney and co-workers[50] and Day and Holmes.[99] X-ray structural analysis results

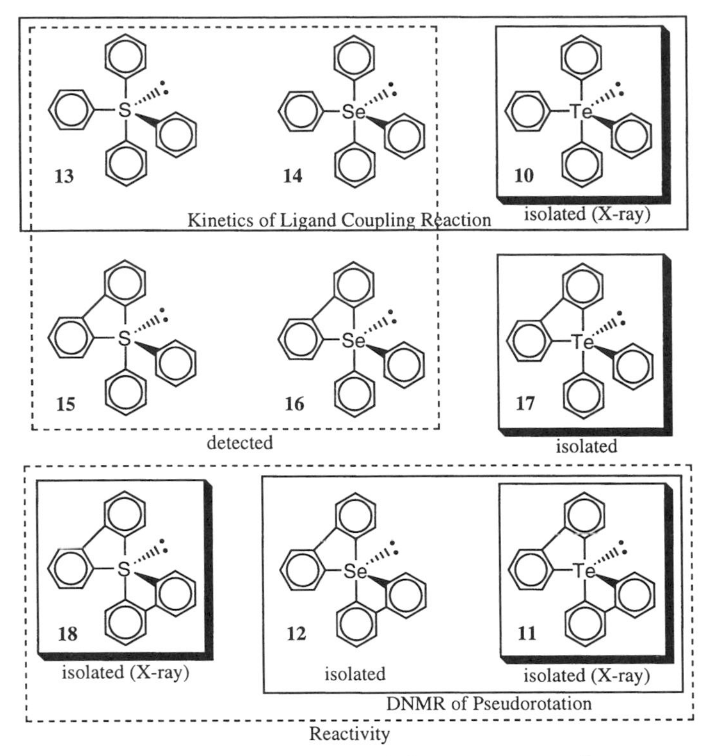

Figure 8.13 Tetraaryl chalcogen compounds.

of both tetraalkoxyselenuranes **19** and -telluranes **20** are shown in Scheme 8.30.[99]

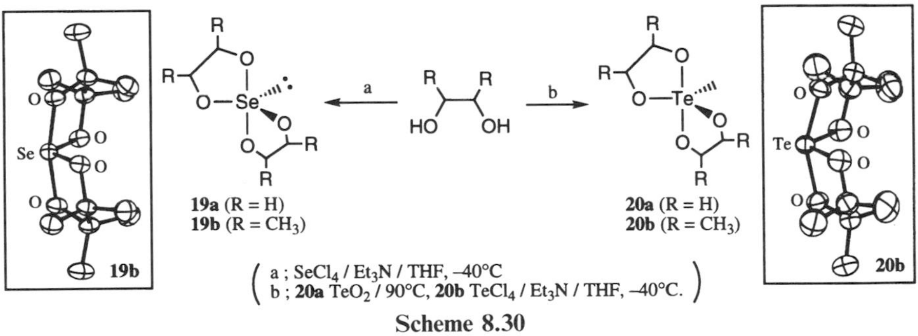

19a (R = H)
19b (R = CH_3)

20a (R = H)
20b (R = CH_3)

(a ; $SeCl_4$ / Et_3N / THF, –40°C
b ; **20a** TeO_2 / 90°C, **20b** $TeCl_4$ / Et_3N / THF, –40°C.)

Scheme 8.30

Tetrathiatelluranes **76** and **77** have been reported by Klar in the reactions of $TeCl_4$ and various thiols in ethanol, but these telluranes do not represent the correct structures, and Stukalo and co-workers concluded that the tetrathiatelluranes reported by Klar are all 1:1 mixtures of the disulfides and dithiatellurides, as shown in Scheme 8.31.[100]

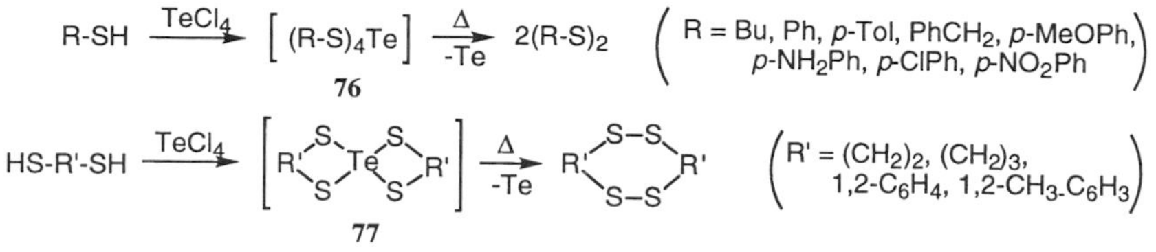

$$\text{R-SH} \xrightarrow{TeCl_4} [(\text{R-S})_4\text{Te}]\ (\mathbf{76}) \xrightarrow[-Te]{\Delta} 2(\text{R-S})_2 \quad (\text{R = Bu, Ph, } p\text{-Tol, PhCH}_2, p\text{-MeOPh, } p\text{-NH}_2\text{Ph, } p\text{-ClPh, } p\text{-NO}_2\text{Ph})$$

$$\text{HS-R'-SH} \xrightarrow{TeCl_4} [\text{R'(S)}_2\text{Te(S)}_2\text{R'}]\ (\mathbf{77}) \xrightarrow[-Te]{\Delta} \text{R'(S-S)}_2\text{R'} \quad (\text{R' = (CH}_2)_2, (\text{CH}_2)_3, \text{1,2-C}_6\text{H}_4, \text{1,2-CH}_3\text{-C}_6\text{H}_3)$$

Scheme 8.31

Selenuranes having four selenium ligands were reported initially by Husebye in 1970. The structure was determined by X-ray crystallographic analysis, as presented in Figure 8.14 (R = Et).[101] Recently, Kobayashi and co-workers reported the ^{77}Se-NMR of the selenurane [bis(diisobutylselenocarbamoyl)-triselenide] (R = *i*-Pr) **78**. At low temperature, ^{77}Se–^{77}Se couplings are observed as satellite peaks; hence they concluded that the compound **78** is actually a selenurane structure in solution.[102]

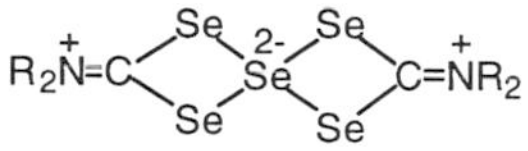

$$R_2\overset{+}{N}{=}C(Se)_2Se^{2-}(Se)_2C{=}\overset{+}{N}R_2$$

78

Figure 8.14 Structure scheme of **78**.

Recently, Devillanova and co-workers have reported the electrochemical synthesis of tetrakis[*N*-methylbenzothiazole-2(3H)-selone]selenium(2+) tetrafluoroborate **79**. It shows an uncommon dication containing a non-cyclic square-planar Se_5 framework, as shown in Figure 8.15.[103] The structure of **79** was determined by X-ray crystallographic analysis and MO calculation of a smaller model system.

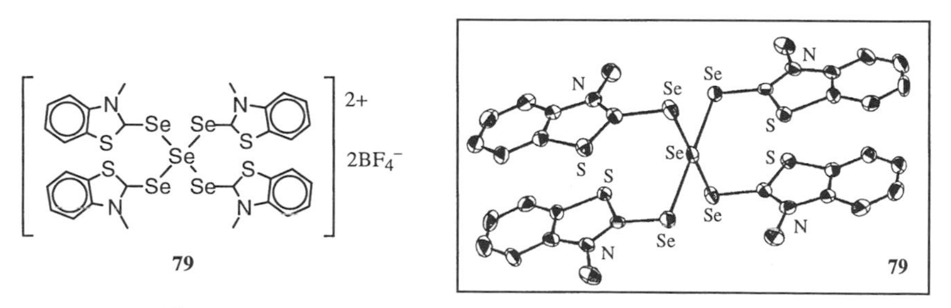

Figure 8.15 Structure Scheme amd ORTEP View of **79**.

8.3.6 Penta- and Hexacoordinated Selenium and Tellurium Species. Type (10–M–5, 12–M–6)

Tetracoordinated hypervalent selenuranes and telluranes are well known. Standard procedures for their preparation are described in the previous sections. In contrast, higher-coordinated selenium and tellurium compounds are quite rare, though a few hexacoordinated σ-perchalcogenuranes like MX_6 [12–M–6(X6), M = S, Se, Te] are known. For example, perfluorosulfurane and -selenurane (SF_6 and SeF_6) are well known as inorganic compounds. However, most of the known perchalcogenuranes are those of the tellurium derivatives. The groups of Seppelt and Schrobilgen have reported the inorganic pertellurane species having fluorine atoms or pentafluorotellurium(VI) oxide groups as the ligands described in Chapter 10.

The first hexavalent organopertellurane(VI) species [12–Te–6(C2O2F2)] **80**(*trans*) and **81**(*cis*) have been reported by Martin and co-workers,[80] as shown in Scheme 8.32. These pertelluranes **81** are generated by the oxidative addition of the corresponding tetravalent chalcogenurane [10–Te–4(C2O2)] **52** with BrF_3 and react with more Lewis acid to give the *cis*-pertellurane **81**. The structures of **81** were determined to be octahedral with *cis* configurations by X-ray crystallographic analysis.

52a : R = Me, R' = H
52b : R = CF_3, R' = *t*-butyl

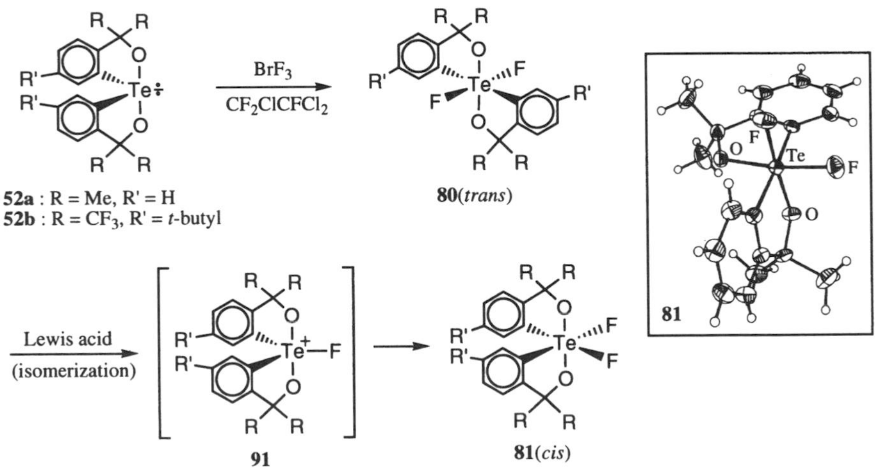

Scheme 8.32

They also reported the synthesis of the tellurane *Te*-oxide dimer [12–Te–6(C2O4)]$_2$ **82**, as a first bis(organopertellurane) which was prepared by the oxidation of the tellurium(IV) **52** using ozone, as shown in Scheme 8.33.[80] However, the structure of this tellurane *Te*-oxide dimer species **82** has only been characterized by spectroscopic experiments and has not been determined by X-ray crystallographic analysis.

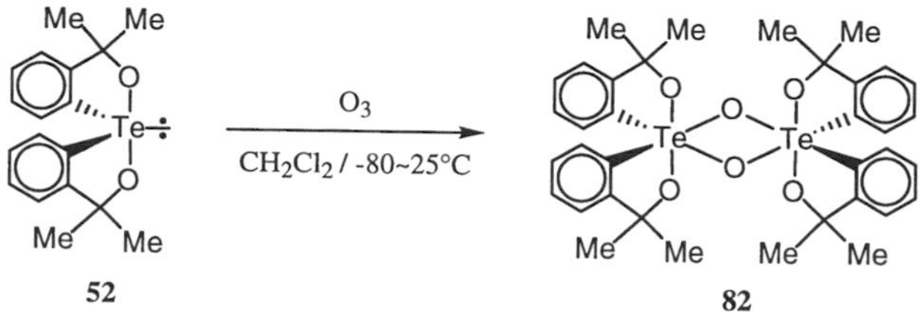

Scheme 8.33

Recently, the syntheses of hexamethylpertellurane [12–Te–6(C6)] **83** and tetramethyldifluoropertellurane [12–Te–6(C4F2)] **84** were reported by Ahmed and Morrison.[104] In addition, the molecular structural analysis of $(CH_3)_6Te$ was conducted by gas-phase electron diffraction[105] and its theoretical MO calculation was presented by Schaefer and co-workers[106] and Marsden and Smart,[107] as shown in Scheme 8.34.

$$Me_4Te + XeF_2 \xrightarrow[CH_3CN]{-30\ °C} \underset{\mathbf{84}}{Me_4TeF_2} \xrightarrow[Et_2O]{ZnMe_2} \underset{\mathbf{83}}{Me_6Te}$$

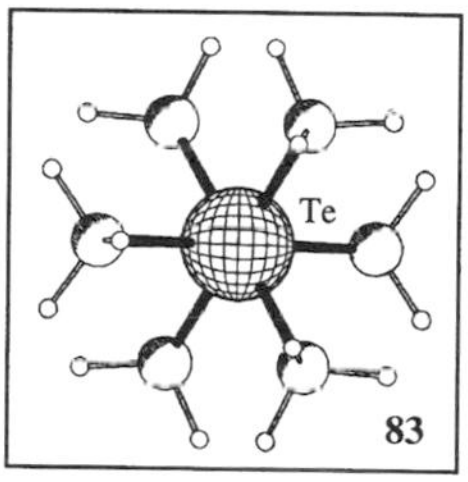

Scheme 8.34

The groups of Akiba and Seppelt have reported that the reaction of LiC_6H_4-4-CF_3 with $TeCl_4$ gives hexakis(4-trifluorophenyl)pertellurane **85** and the reaction of $F_2Te(C_6H_5)_4$ with PhLi also gives hexaphenylpertellurane **86**; this is the first isolation of hexaarylpertellurane.[108] The molecular structures of both **85** and **86** were determined by X-ray crystallographic analysis, as shown in Scheme 8.35.

$$2TeCl_4 + 8ArLi \xrightarrow[-8LiCl]{} \underset{\mathbf{85}}{Ar_6Te} + Ar_2Te$$

(Ar = 4-CF_3-Ph)

$$Ph_4TeF_2 + 2.2PhLi \xrightarrow[-2LiF]{} \underset{\mathbf{86}}{Ph_6Te}$$

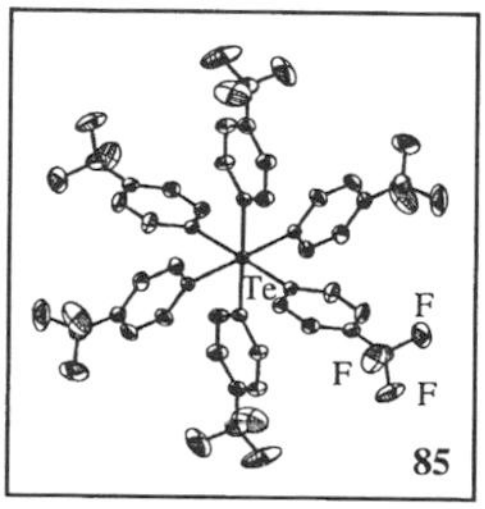

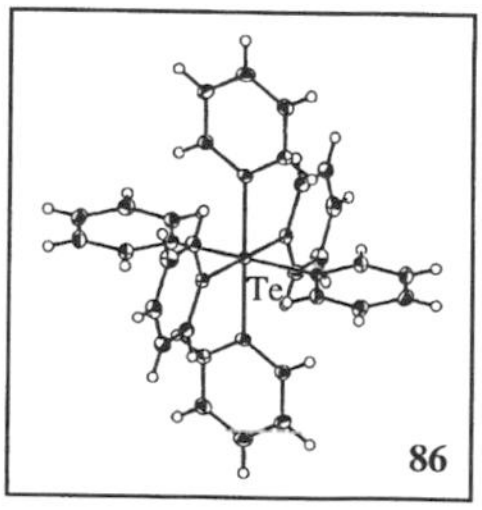

Scheme 8.35

Recently, our associates have reported that bis(2,2′-biphenylylene)tellurane **11** reacts with sulfuryl chloride to give bis(2,2′-biphenylylene)dichloropertellurane [12–Te–6(C4C12)] **87**, as shown in Scheme 8.36.[109] The structure of this pertellurane was determined by X-ray crystallographic analysis, revealing that it has a distorted-octahedral geometry and a *cis* configuration with respect to the coordinated chlorine atoms.

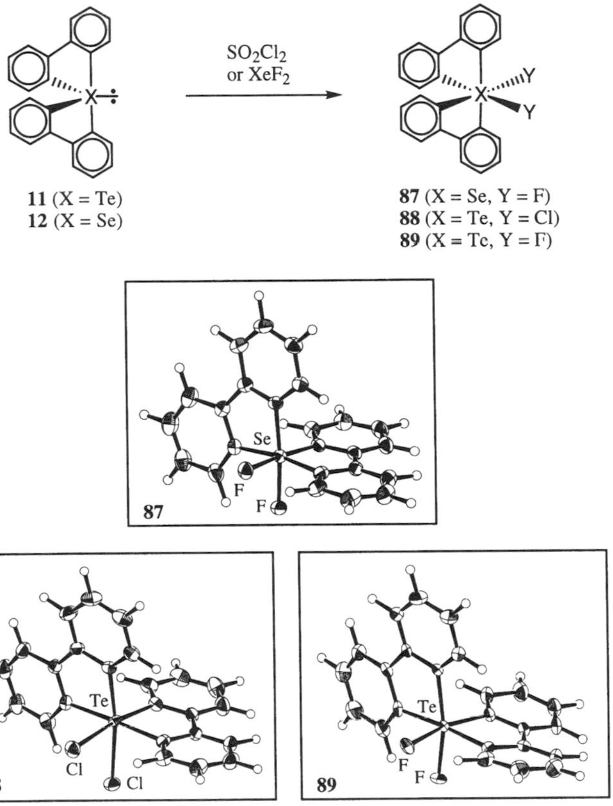
SO_2Cl_2
or XeF_2
X
Y
Y
11 (X = Te)
12 (X = Se)
87 (X = Se, Y = F)
88 (X = Te, Y = Cl)
89 (X = Te, Y = F)
Se
F
F
87
Te
Cl
Cl
88
Te
F
F
89

Scheme 8.36

Furthermore, they reported that bis(2,2′-biphenylylene)tellurane **11** reacts with ozone at low temperature to afford diastereoselectively the *O*-bridged tellurane *Te*-oxide dimer **90** having a bis(octahedral) configuration.[110] Compound **88** has been characterized by elemental and spectroscopic (IR and FABMS, ^{1}H-, ^{13}C-, and ^{125}Te-NMR) analyses. The crystal and molecular structure analysis of **88** has been confirmed to be composed of the ΛΛ and ΔΔ diastereomeric isomers, as shown in Scheme 8.37.

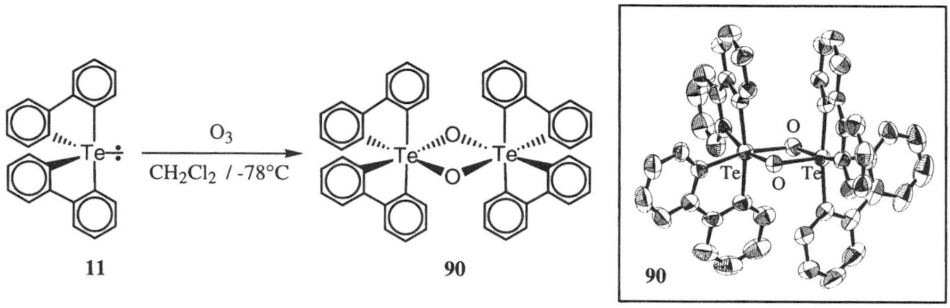

Scheme 8.37

Only few σ-pentacoordinated chalcogen species, like MX_5, have been reported, as compared with the tetra- or hexacoordinated hypervalent chalcogen species. There are three kinds of pentacoordinated chalcogen species, namely, perchalcogenium cation species $[(MX_5)^+]$ [10–M–5(X5), M = S, Se, Te], ate-complex perchalcogen species $[(MX_5)^-]$ [12–M–5(X5), M = S, Se, Te], and neutral perchalcogen oxides $[MX_4O]$.

A number of inorganoperchalcogenium compounds have been reported by Seppelt and co-workers (Chapter 10). The first σ-organoperchalcogenium species [10–Te–5(C2O2F)] have been reported by Martin and co-workers[80] (Scheme 8.34). However, these compounds were only predicted as the intermediates in the reaction of the corresponding tellurane(IV) with BrF_3 and the equilibrium between *trans*- **80** and *cis*-pertellurane **81** in the presence of Lewis acid.

The ate-complex perchalcogen species are generally unstable and have been reported to be detected as intermediates in the reactions of the corresponding tellurane(IV) with nucleophiles. Wittig and co-workers investigated the reaction of tetraphenyltellurane(IV) **10** with phenyllithium and found some traces of the adduct formation ($Ph_5Te^+Li^-$) **91**, but no definite experimental evidence could be obtained.[38] Recently, Reich and co-workers observed that the tetraphenyltellurane(IV) **10** reacts with phenyllithium at −120°C to give the lithium pentaphenyl tellurium(V) **91** ate-complex directly as an unstable intermediate, detected by variable-temperature ^{125}Te-NMR experiments, as shown in Figure 8.16.[111]

Detty and co-workers have reported that the reactions of the π-telluranes **28** described above with halogens give π-pertelluranes [12—Te—5] **92** (at low temperature) and **93** (at higher temperature). The π-pertelluranes function as mild oxidants for thiols, selenol, and tellurides, and are reduced by these reagents and by hydrazine to π-telluranes **28** or **29**.[60] The structure of **92** was assigned by X-ray crystallographic analysis, as shown in Scheme 8.38.

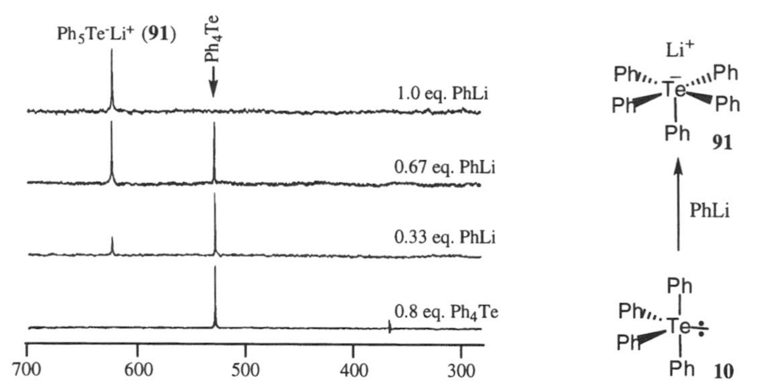

Figure 8.16 ^{125}Te-NMR spectra of titration of Ph_4Te (**10**) with PhLi at -120C.

Scheme 8.38

REFERENCES

1. J. March, *Advanced Organic Chemistry* (John Wiley & Sons, New York, 1985).
2. L. Pauling, *The Nature of Chemical Bond* (Cornell University Press, New York, 1960).
3. (a) A. Michaelis, O. Schifferdecker, *Chem. Ber.*, **6**, 993 (1873). (b) O. Ruff, A. Heizelmann, *Z. anorg. allg. Chem.*, **72**, 63 (1911).
4. E. A. Innes, I. G. Csizmadia, Y. Kanada, *J. Mol. Structure (Theochem.)*, **186**, 1 (1989).
5. E. A. Robinson, *J. Mol. Structure (Theochem.)*, **186**, 9 (1989).
6. (a) E. Block, *Heteroatom Chemistry* (VCH, New York, 1990). (b) W. E. McEwen, *Heteroatom Chem.*, VCH, International journal since 1990.

7. (a) R. J. Hatch, R. E. Rundle, *J. Am. Chem. Soc.*, **73**, 4321 (1951). (b) R. E. Rundle, *ibid.*, **85**, 112 (1963). (c) J. I. Musher, *Angew. Chem., Int. Ed. Engl,* **8**, 54 (1969).
8. (a) M. M. L. Cheu, R. Hoffmamm, *J. Am. Chem. Soc.*, **98**, 1647 (1976). (b) A. E. Reed, P. von R. Schleyer, *J. Am. Chem. Soc.*, **112**, 1434 (1990). (c) W. Kutzelnigg, *Angew. Chem., Int. Ed. Engl.*, **23**, 272 (1984).
9. (a) A. E. Reed, F. Weinhold, *J. Am. Chem. Soc.*, **108**, 3586 (1986). (b) S. Sato, T. Yamashita, O. Takahashi, N. Furukawa, M. Yokoyama, K. Yamaguchi, *Tetrahedron*, **53**, 12183 (1997).
10. J. C. Martin, R. Arhart, *J. Am. Chem. Soc.*, **93**, 2339 (1971).
11. I. Kapovits, A. Kalman, *J. Chem. Soc. Chem. Commun.*, 649 (1971).
12. (a) R. A. Hayes, J. C. Martin, *Organic Sulfur Chemistry* (Elsevier, Amsterdam, 1985), Chapt. 8, p. 408. (b) J. C. Martin, E. F. Perozzi, *Science*, **191**, 154 (1976).
13. E. L. Mutterties, R. A. Schunn, *Quat. Rev.*, **20**, 245 (1966).
14. (a) R. A. Hayes, J. C. Martin, *Organic Sulfur Chemistry. Theoretical and Experimental Advances* (Elsevier, Amsterdam, 1985), Chapt. 8, p. 408.
15. (a) I. Ugi, F. Ramirez *et al.*, *Acc. Chem. Res.*, **4**, 288 (1971). (b) R. S. Berry, *J. Chem. Phys.*, **32**, 933 (1960).
16. (a) R. R. Holmes, *Acc. Chem. Res.*, **12**, 257 (1979); R. Luckenbach, *Dynamic Stereochemistry of the Pentaco-ordinated Phosphorus and Related Elements* (George Thieme, Stuttgart, 1973).
17. S. Oae, J. T. Doi, *Organic Sulfur Chemistry: Structure and Mechanism* (CRC Press, Boston, 1991), Chapt. 4, p. 119.
18. M. M. L. Chen, R. Hoffmann, *J. Am. Chem. Soc.*, **98**, 1647 (1976).
19. (a) A. Pross, S. S. Shaik, *Acc. Chem. Res.*, **16**, 363 (1983). (b) G. Sini, G. Ohanessian, P. C. Hiberty, S. S. Shaik, *J. Am. Chem. Soc.*, **112**, 1407 (1990).
20. C. Paulmier, *Selenium Reagents and Intermediates in Organic Synthesis* (Pergamon, Oxford, 1986).
21. M. W. Tolles, W. D. Gwinn, *J. Chem. Phys.*, **36**, 119 (1962).
22. N. C. Baezinger, R. E. Buckles, R. J. Maner, T. D. Simpson, *J. Am. Chem. Soc.*, **91**, 5749 (1969).
23. V. Franzen, C. Mertz, *Justus Liebigs Ann. Chem.*, **643**, 24 (1961).
24. G. Wittig, H. Fritz, *Justus Liebigs Ann. Chem.*, **577**, 39 (1952).
25. D. Hellwinkel, *Ann. New York Acad. Sci.*, **192**, 158 (1972).
26. (a) D. L. Klayman, W. H. H. Günther, *Organic Selenium Compounds* (John Wiley & Sons, New York, 1973). (b) I. Hargittai, B. Rozsondai, *The Chemistry of Organic Selenium and Tellurium Compounds,* S. Patai and Z. Rappoport, Eds (John Wiley & Sons, New York, 1986), Vol. 1, Chapt. 3. (c) J. Bergmann, L. Engman, J. Sidien, *ibid.*, Chapter 14. (d) K. J. Irgolic, *The Organic Chemistry of Tellurium* (Gordon and Breach, New York, 1974). (e) N. Furukawa, S. Sato, in *RODD'S Chemistry of Carbon Compounds,* 2nd Edition, M. Sainsbury, Ed. (Elsevier Science, Amsterdam, 1996), Vol. III, pp. 469–520.
27. S. Uemura, S. I. Fukuzawa, S. R. Patil, *J. Organomet. Chem.*, **243**, 9 (1983).
28. S. Ogawa, Y. Masutomi, N. Furukawa, T. Erata, *Heteroatom Chem.*, **3**, 423 (1992).

29. (a) H. J. Reich, D. P. Green, N. H. Phillips, *J. Am. Chem. Soc.*, **113**, 414 (1991). (b) H. J. Reich, D. P. Green, N. H. Phillips, J. P. Borst, I. J. Reich, *Phosphorus, Sulfur, and Silicon*, **67**, 83 (1992).
30. S. Ogawa, Y. Masutomi, T. Erata, N. Furukawa, *Chem. Lett.*, 2471 (1992).
31. H. J. Reich, B. O. Gudmundson, D. R. Dykstra, *J. Am. Chem. Soc.*, **114**, 7937 (1992).
32. (a) T. Hiiro, N. Kambe, A. Ogawa, N. Sonoda, *Angew. Chem., Int. Ed. Engl.*, **26**, 1187 (1987). (b) N. Sonoda *et al.*, *Organometallics*, **9**, 1355 (1990); *Synth. Commun*, **20**, 1703 (1990); *J. Am. Chem. Soc.*, **112**, 455 (1990).
33. Y. Iwata, M. Aragi, M. Sugiyama, K. Matsui, Y. Ishii, M. Ogawa, *Bull. Chem. Soc. Jpn.*, **54**, 2065 (1981).
34. N. Furukawa, S. Ogawa, K. Matsumura, H. Fujihara, *J. Org. Chem.*, **56**, 6341 (1991).
35. S. Patai, Z. Rappoport, C. J. M. Stirling, Eds., *The Chemistry of Sulphoxides and Sulphones* (John Wiley & Sons, New York, 1988).
36. C. Paulmier, *Selenium Reagents and Intermediates in Organic Synthesis* (Pergamon Press, Oxford, 1986), Chapt. V, p.124.
37. M. R. Detty, *J. Org. Chem.*, **45**, 274 (1980).
38. N. Furukawa, Y. Masutomi, *Chem. Lett.*, to be published.
39. G. Wittig, H. Fritz, *Justus Liebigs Ann. Chem.*, **577**, 39 (1952).
40. D. Hellwinkel, G. Fahrbach, *Justus Liebigs Ann. Chem.*, **712**, 1 (1968).
41. D. Hellwinkel, G. Fahrbach, *Justus Liebigs Ann. Chem.*, **715**, 68 (1968).
42. (a) D. H. R. Barton, S. A. Glover, S. V. Ley, *J. Chem. Soc. Chem. Commun.*, 266 (1977). (b) S. A. Glover, *J. Chem. Soc. Perkin I,* 1338 (1980).
43. S. Ogawa, Y. Matsunaga, S. Sato, T. Erata, N. Furukawa, *Tetrahedron Lett.*, **33**, 93 (1992).
44. S. Ogawa, S. Sato, T. Erata, N. Furukawa, *Tetrahedron Lett.*, **32**, 3179 (1991).
45. S. Sato, N. Furukawa, *Tetrahedron Lett.*, **36**, 2803 (1995).
46. I. C. S. Smith, J.-S. Lee, D. D. Titus, R. F. Ziolo, *Organometallics*, **1**, 350 (1982).
47. S. Ogawa, S. Sato, N. Furukawa, *Tetrahedron Lett.*, **33**, 7925 (1992).
48. S. Ogawa, Y. Matsunaga, S. Sato, I. Iida, N. Furukawa, *J. Chem. Soc. Chem. Commun.*, 1141 (1992).
49. S. Sato, N. Kondo, N. Furukawa, *Organometallics*, **13**, 3393 (1994).
50. S. Ogawa, S. Sato, T. Erata, N. Furukawa, *Tetrahedron Lett.*, **33**, 1915 (1992).
51. D. B. Denney, D. Z. Denney, P. J. Hammond, *J. Am. Chem. Soc.*, **103**, 2340 (1981).
52. N. Furukawa, Y. Matsunaga, S. Sato, *Synlett.*, 655 (1993).
53. S. Sato, N. Furukawa, *Chem. Lett.*, 889 (1994).
54. S. Sato, N. Kondo, N. Furukawa, *Organometallics*, **14**, 5393 (1995).
55. (a) J. Bergman, L. Engman, J. Siden, *The Chemistry of Organic Selenium and Tellurium Compounds,* Vol. 1 (John Wiley & Sons, Chichester, 1986), Chapt. 14, p. 517. (b) T. B. Rauchfuss, *ibid.*, Vol. 2, 1987, Chapt. 6, p. 339.
56. C. W. Perkins, J. C. Martin, A. J. Arduengo III, W. Law, A. Algeria, J. K. Kochi, *J. Am. Chem. Soc.*, **102**, 7753 (1980).

57. D. H. Reid, R. G. Webster, *J. Chem. Soc. Perkin Trans.*, **1**, 775 (1975).
58. M. Perrrier, J. Vialle, *Bull. Soc. Chim. Fr.*, 4591 (1971).
59. D. H. Reid, R. G. Webster, *J. Chem. Soc., Chem. Commun.*, 3187 (1971).
60. M. R. Detty, H. R. Luss, *J. Org. Chem.*, **48**, 5149 (1983).
61. (a) F. Iwasaki, H. Murakami, N. Yamazaki, M. Yasui, M. Tomura, N. Matsumura, *Acta. Cryst.*, **C47**, 998 (1991). (b) N. Matsumura, M. Tomura, K. Inazu, H. Inoue, N. Yamazaki, F. Iwasaki, *Phosphorus, Sulfur, and Silicon*, **67**, 135 (1992).
62. H. Fujihara, H. Mima, N. Furukawa, *J. Am. Chem. Soc.*, **117**, 10153 (1995).
63. (a) J. Bergman, *Tetrahedron*, **28**, 3323 (1972). (b) I. D. Sadekov, A. Ya. Bushkov, I. Minkin, *Russ. Chem. Rev.*, **48**, 343 (1979). (c) K. J. Irgolic, *The Organic Chemistry of Tellurium* (Gordon and Breach, New York, 1974).
64. P. H. Bird, V. Kumar, B. C. Pant, *Inorg. Chem.*, **19**, 2487 (1980).
65. R. V. Mitcham, B. Lee, K. B. Mertes, R. F. Ziolo, *Inorg. Chem.*, **18**, 3498 (1979).
66. Y. Okamoto, K. L. Chellappa, R. Horsony, *J. Org. Chem.*, **38**, 3172 (1973).
67. (a) N. W. Alcock, W. D. Harrison, *J. Chem. Soc., Dalton Trans,* 251 (1983). (b) C. S. Mancinelly, D. D. Titus, R. F. Ziolo, *J. Org. Chem.*, **113**, 140 (1977).
68. K. J. Wynne, *Inorg. Chem.*, **9**, 299 (1970).
69. S. Herberg et al., *Z. anorg. allg. Chem.*, **492,** 95. (1982).; *ibid.*, **494**, 151, 159 (1982).
70. I. D. Sadaekov, A. Y. Bushukov, V. L. Pavlova, Y. L. Yureva, V. I. Minkin, *Zh. Obshch. Khim.*, **46**, 1660 (1976).; *ibid.*, **47**, 1305 (1977).
71. R. Paetzold, U. Lindner, *Z. anorg. allg. Chem.*, **350**, 295 (1967).
72. C. S. Mancinelli, D. D. Titus, R. F. Ziolo, *J. Organomet. Chem.*, **140**, 113 (1977).
73. M. Wieber, E. Kauzinger, *J. Organomet. Chem.*, **129**, 339 (1977).
74. Y. Takaguchi, H. Fujihara, N. Fujihara, *J. Organomet. Chem.*, **498**, 49 (1995).
75. B. Dahlen, B. Lindgren, *Acta. Chem. Scand.*, **27**, 2218 (1973).
76. Y. Takaguchi, H. Fujihara, N. Furukawa, *Heteroatom Chem.*, **6**, 481 (1995).
77. (a) Y. Takaguchi, H. Fujihara, N. Furukawa, *Organometallics*, **15**, 1913 (1996). (b) Y. Takaguchi, N. Furukawa, *Chem. Lett.*, **18**, 365 (1996).
78. J. C. Martin, E. F. Perozzi, *Science*, **191**, 154 (1976).
79. H. J. Reich, *J. Am. Chem. Soc.*, **95**, 964 (1973).
80. R. S. Michalak, S. R. Wilson, J. C. Martin, *J. Am. Chem. Soc.*, **106**, 7529 (1984).
81. (a) T. Takahashi, N. Kurose, S. Kawanami, Y. Arai, T. Koizumi, *J. Org. Chem.*, **59**, 3262 (1994). (b) T. Takahashi, J. Zhang, N. Kurose, S. Takahashi, T. Koizumi, *Tetrahedron Assymmetry*, **7**, 2797 (1996).
82. B. Dahlen, *Acta. Cryst.*, **B29**, 595 (1973).
83. B. Dahlen, B. Lindgren, *Acta. Chem. Scand.*, **A33**, 403 (1979).
84. V. I. Minkin, A. A. Maksimenko, G. K. Mehrota, A. G. Maslakow, Yu. T. Struchkov, D. S. Yufit, *J. Organomet. Chem.*, **348**, 63 (1988).
85. M. Wieber, E. Kauzinger, *J. Organomet. Chem.*, **129**, 339 (1977).
86. M. Wieber, E. Schmidt, C. Burschka, *Z. anorg. allg. Chem.*, **525**, 127 (1985).
87. H. Fujihara, N. Furukawa, *J. Mol. Structure (Theochem)*, **186**, 261 (1989).
88. H. Fujihara, N. Oi, T. Erata, N. Furukawa, *Tetrahedron Lett.*, **31**, 1019 (1990).

89. H. Fujihara, J.-J. Chiu, N. Furukawa, *J. Am. Chem. Soc.*, **110**, 1280 (1988).
90. H. Fujihara, H. Mima, T. Erata, N. Furukawa, *J. Am. Chem. Soc.*, **114**, 3117 (1992).
91. H. Mima, H. Fujihara, N. Furukawa, *J. Am. Chem. Soc.*, **118**, 1783 (1996).
92. H. Fujihara, H. Mima, M. Ikemori, N. Furukawa, *J. Am. Chem. Soc.*, **113**, 6337 (1991).
93. M. R. Detty, A. E. Friedman, M. McMillan, *Organometallics*, **13**, 3338 (1994).
94. M. R. Detty, A. J. Williams, J. M. Hewitt, M. McMillan, *Organometallics*, **14**, 5258, (1995).
95. M. O. Ryan, D. N. Harpp, *Tetrahedron Lett.*, **33**, 2129 (1992).
96. R. J. Batchelor, F. W. B. Einstein, I. D. Gay, J. -H. Gu, B. M. Pinto, X. -M. Zhou, *J. Am. Chem. Soc.*, **112**, 3706 (1996).
97. R. D. Miller, A. F. Renaldo, H. Ito, *J. Org. Chem.*, **53**, 5571 (1988).
98. G. E. Wilson, Jr., B. A. Belkind, *J. Am. Chem. Soc.*, **100**, 8124 (1978).
99. (a) R. O. Day, R. R. Holmes, *Inorg. Chem.*, **20**, 3071 (1981). (b) R. O. Day, R. R. Holmes, *Phosphorus, Sulfur, and Silicon,* **98**, 241 (1995).
100. (a) B. Nakhdjavan, G. Klar, *Libigs Ann. Chem.*, 1683 (1977). (b) E. A. Stukalo, E. M. Yuréva, L. N. Narkovskii, *Zhur. Obsch. Khim.*, **19**, 343 (1983).
101. S. Husebye, G. H. Mandensen, *Acta. Chem. Scand.*, **24**, 2273 (1970).
102. Y. Mazaki, K. Kobayashi, *Tetrahedron Lett.*, **30**, 2813 (1989).
103. C. Adamo, F. Demartin, P. Deplano, F. A. Devillanova, F. Isaia, F. Lelj, V. Lippolis, P. Lukes, M. L. Mercuri, *J. Chem. Soc., Chem. Commun.*, 873 (1996).
104. L. Ahmed, J. A. Morrison, *J. Am. Chem. Soc.*, **112**, 7411 (1990).
105. A. Haaland, H. P. Verne, H. V. Volden, J. A. Morrison, *J. Am. Chem. Soc.*, **117**, 7554 (1995).
106. J. E. Fowler, T. P. Hamilton, H. F. Schaefer III, *J. Am. Chem. Soc.*, **115**, 4155 (1993).
107. C. J. Marsden, B. A. Smart, *Organometallics*, **14**, 5399 (1995).
108. M. Minoura, T. Sagami, K.-y. Akiba, C. Modrakowski, A. Sudau, K. Seppelt, S. Wallenhauer, *Angew. Chem. Int. Ed. Engl.*, **35**, 2660 (1996).
109. S. Sato, T. Yamashita, E. Horn, N. Furukawa, *Organometallics*, **15**, 3256 (1996).
110. S. Sato, T. Ueminami, E. Horn, N. Furukawa, *J. Organomet. Chem.*, **543**, 77 (1997).
111. H. J. Reich, D. P. Green, N. H. Phillips, J. P. Borst, I. L. Reich, *Phosphorus, Sulfur, and Silicon,* **67**, 83 (1992).

9 Ligand-Coupling Reactions of Compounds of Group 15 and Group 16 Elements

YOHSUKE YAMAMOTO AND KIN-YA AKIBA
Department of Chemistry, Faculty of Science, Hiroshima University, Higashi-Hiroshima, Japan

9.1 INTRODUCTION

The ligand-coupling reaction (LCR) is one of the fundamental reactions of hypervalent organic molecules of main group elements. The phenomenon is described as reductive elimination[1] in transition metal compounds, but the term LCR has been preferentially used for concerted reductive coupling of hypervalent organic compounds. This usage reflects the fact that the term "reductive elimination" has already been used in organic chemistry for olefin formation by reductively eliminating reactions of leaving groups.[2] There should be three possible types of LCR when the compounds have two kinds of bonds. For example, in 10-M-5[3] or 10-M-4[3] compounds there are apical (axial) and equatorial bonds in trigonal bipyramidal (TBP) or pseudo-trigonal bipyramidal (Ψ-TBP) structures, and axial and basal bonds are present in square pyramidal (SP) or pseudo-square pyramidal (Ψ-SP) structures.[4] Therefore, special attention has been paid to the selectivity of the LCR, that is, which kind of ligand will take part in the LCR of 10-M-5 or 10-M-4 compounds (Scheme 9.1).[5] In this review we will focus on LCRs from Group 15 element (10-M-5) and Group 16 element (10-M-4) compounds. Several reviews by Oae show various experimental examples of LCR.[2]

Chemistry of Hypervalent Compounds, edited by Kin-ya Akiba.
ISBN 0-471-24019-2

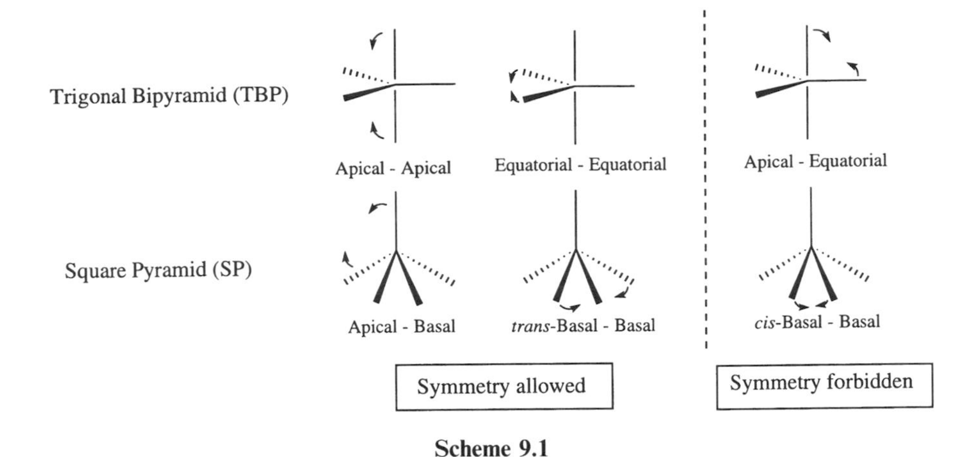

Scheme 9.1

9.2 LCRS FROM GROUP 15 ELEMENT COMPOUNDS (10-M-5)

9.2.1 Theoretical Investigations

Studies of pentacoordinate Group 15 elements (10-M-5 compounds) include the theoretical investigation of the mechanism of the LCR of phosphorane (PH_5). This was first carried out by Hoffmann and co-workers.[5] They concluded that LCRs between the apical–apical ligands and the equatorial–equatorial ligands from TBP structures were symmetry allowed, and those between the apical–equatorial ligands were forbidden (Scheme 9.1). Although the symmetry-forbidden, highly polarized apical–equatorial coupling process was once reported to be favored by *ab initio* calculation,[6] recent more sophisticated calculations by Kutzelnigg[7] and Morokuma[8] essentially supported the conclusion by Hoffmann. The equatorial–equatorial coupling was calculated to be the lowest energy process for PH_5,[7–9] AsH_5,[8] and SbH_5[8] although the apical–equatorial process was calculated to be favored for BiH_5 (Scheme 9.2).[8] The barriers of the preferred process for the H_2 elimination reaction from MH_5 (M = P, As, Sb, Bi) were 32.6 kcal mol^{-1} (PH_5), 33.3 kcal mol^{-1} (AsH_5), 38.1 kcal mol^{-1} (SbH_5), and 34.0 kcal mol^{-1} (BiH_5) by the MP2/ECP calculation. The net charges at the MP2/ECP transition states for the H_2 elimination by use of natural population analysis are shown in Scheme 9.2.[8] The highly polarized nature of the apical–equatorial process is evident, based on the charges in the transition state for BiH_5 in comparison with other transition states for PH_5, AsH_5, and SbH_5.

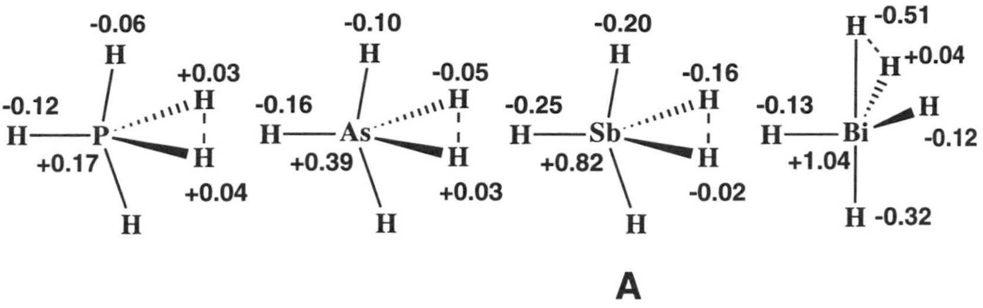

A

Scheme 9.2

9.2.2 Backgrounds of Experimental Investigations

Experimental investigations on the selectivity of LCRs are very difficult because the following requirements have to be fulfilled in order to get any conclusions about the selectivity of LCRs: (1) LCRs should take place in a concerted manner; (2) reverse reactions from the products should not take place; (3) the system should be sterically unbiased (acyclic); (4) structures of starting compounds are clearly determined (hopefully by X-ray analysis). In addition if one chooses sterically unbiased systems, barriers of Berry pseudorotation (BPR), which exchanges apical and equatorial ligands in the TBP structures, have been shown to be very low.[4,10] Besides the unavoidable, difficult problems caused by the fast BPR, there had been no reports which fulfilled all of the above requirements until we investigated LCRs of mixed pentaarylantimony compounds (*vide infra*), although various examples had been reported, such as sterically rigid (biased) bis(biphenylene)methylphosphorane[11] (Scheme 9.3). Among these examples, Barton and co-workers reported LCRs for the Bi(V) compounds and showed evidence that the LCR of Bi(V) reagents took place concertedly, based on the phenyl radical trapping experiments, which did not show any detectable compounds formed from phenyl radicals (Scheme 9.4).[12] The system is quite interesting and provides a useful application of the LCR to organic synthesis (phenylation), but unfortunately the system is not suitable for investigations on the selectivity of ligands because the transition states for the reaction are unclear: we do not know whether the LCR took place via cyclic (**1**) or acyclic (**2**) transition states.

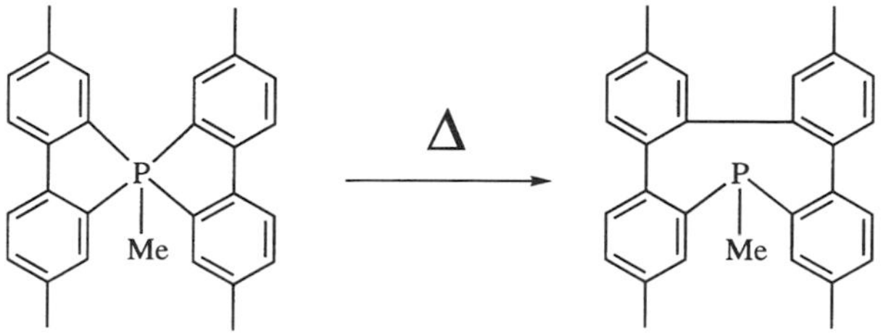

Scheme 9.3

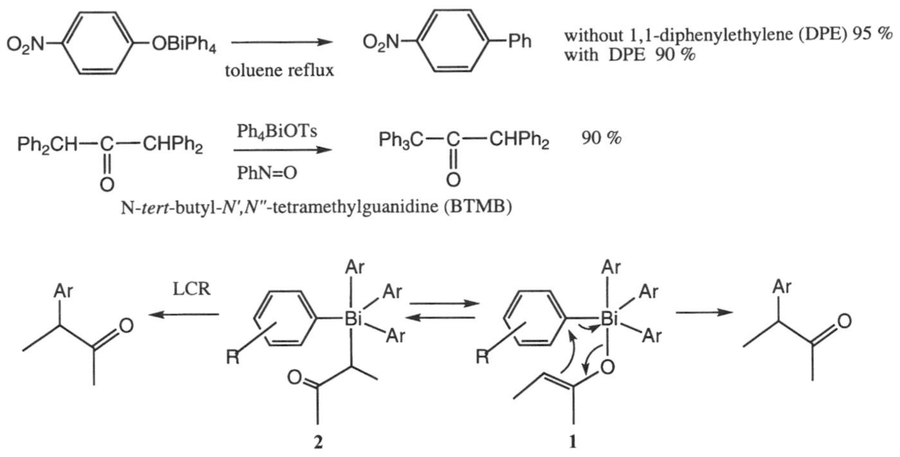

$O_2N-C_6H_4-OBiPh_4 \xrightarrow{\text{toluene reflux}} O_2N-C_6H_4-Ph$ without 1,1-diphenylethylene (DPE) 95 % with DPE 90 %

$Ph_2CH-C(=O)-CHPh_2 \xrightarrow[PhN=O]{Ph_4BiOTs} Ph_3C-C(=O)-CHPh_2$ 90 %

N-*tert*-butyl-*N′*,*N″*-tetramethylguanidine (BTMB)

LCR

Ar Ar Ar Bi R O 2

Ar Ar Ar Bi R O 1

Scheme 9.4

Generally speaking, R_3MX_2, R_4MX (X = halogen), $R_4M(OR)$, and $R_4M(SR)$ systems are not suitable for investigation of the selectivity of the LCR experimentally, because reversible reactions have been reported for the reactions of R_3MX_2 to form R_3M and X_2.[13] The strong ionic nature of the M—X bond has been confirmed in R_4MX,[14] and radical processes have been reported to take place in $Ph_4Sb(OMe)$[15] and $Ph_4Sb(SAr)$.[16] Thus, suitable compounds for investigation on the selectivity of the LCR should be pentakis (carbon-substituted) Group 15 element compounds, since the reverse reactions from the LCR products such as bialkyl or biaryls and trialkyl- or triaryl-group 15 element compounds to form pentaalkyl- or pentaarylantimony compounds should not take place. Among the cases for the sterically unbiased (acyclic) system thermolysis of Group 15 element compounds, those bearing five carbon substituents such as Ph_5P,[17] Ph_5As,[18] Ph_5Sb,[18] Ph_5Bi,[19] Me_5As,[20] or Et_5Sb[21] have been reported. However, thermolysis of Me_5As or Et_5Sb in benzene gave methane, ethane, or ethylene, indicating that thermolysis took place via radical processes (Scheme 9.5).[20] In addition, thermolysis of Ph_5P, Ph_5As, and Ph_5Bi in benzene gave biphenyl and the corresponding Ph_3X (X = P, As, Bi), but the yields are not quantitative; thus, radical decomposition processes played some roles in these compounds. Only in the case of Ph_5Sb has the concertedness of the LCR been strictly confirmed, by McEwen and co-workers.[22] The LCR of ^{14}C-labelled pentaphenylantimony in unlabelled benzene afforded biphenyl bearing ^{14}C-label for both phenyl rings and ^{14}C-labelled triphenylantimony, without either ^{14}C-labelled benzene or biphenyl having a ^{14}C-label for one of the two phenyl rings being formed. Thus the LCR of Ph_5Sb in benzene was strictly confirmed to take place concertedly (Scheme 9.6), although the LCR of Ph_5Sb in CCl_4 was reported to take place through a radical process.[23,24]

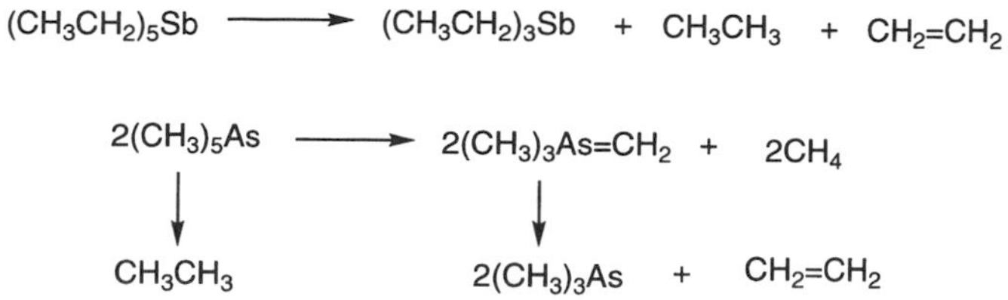

$$(CH_3CH_2)_5Sb \longrightarrow (CH_3CH_2)_3Sb + CH_3CH_3 + CH_2{=}CH_2$$

$$2(CH_3)_5As \longrightarrow 2(CH_3)_3As{=}CH_2 + 2CH_4$$

$$2(CH_3)_5As \downarrow CH_3CH_3 \qquad 2(CH_3)_3As{=}CH_2 \downarrow 2(CH_3)_3As + CH_2{=}CH_2$$

Scheme 9.5

$$Ph^*{}_5Sb \xrightarrow{PhH} Ph^*{}_3Sb + Ph^*\text{-}Ph^* \qquad * : {}^{14}C$$

no Ph*H

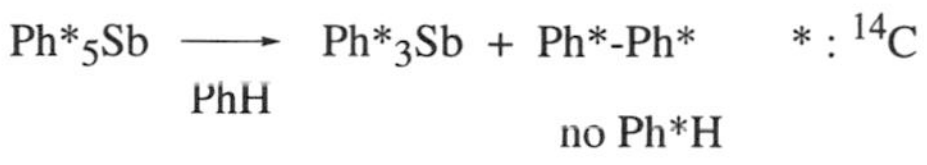

Scheme 9.6

In 1987, we reported LCRs from triarylbis(phenylethynyl)antimony(V) compounds (**3**: aryl = p-$CH_3C_6H_4$; **4**: aryl = Ph; **5**: aryl = p-ClC_6H_4) in benzene for this purpose and found that no biaryls were formed from the LCR of **3**–**5** (Scheme 9.7). The ratio of aryl–alkynyl coupling increased as the aryl group became more electronegative. The result was interpreted based on the X-ray structure of trimethyldipropynylantimony with the two propynyl substituent at the apical positions and on some well-known facts[4] about the apicophilicity (isomers with more electronegative substituents in the apical positions are preferred). Thus, the fast equilibrium among B–D takes place and the equilibrium shifts gradually to the right in Scheme 9.8 as the aryl group becomes more electronegative. In accordance with the experimental results, especially based on the absence of Ar—Ar in the products, it can be concluded that the LCR takes place via an apical–apical path. However, the instability of **3**–**5** prevented detailed examination.[25]

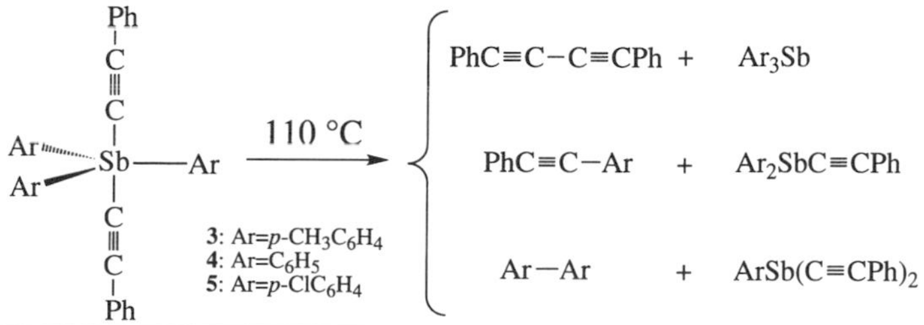

$$Ar_3Sb(C{\equiv}CPh)_2 \xrightarrow{110\,^{\circ}C} \begin{cases} PhC{\equiv}C{-}C{\equiv}CPh + Ar_3Sb \\ PhC{\equiv}C{-}Ar + Ar_2SbC{\equiv}CPh \\ Ar{-}Ar + ArSb(C{\equiv}CPh)_2 \end{cases}$$

3: Ar=p-$CH_3C_6H_4$
4: Ar=C_6H_5
5: Ar=p-ClC_6H_4

Ar	Ratio PhC≡C−C≡CPh	:	PhC≡C−Ar	:	Ar−Ar	Total Yield (%)[a]
p-$CH_3C_6H_4$	76	:	24	:	0	93
Ph	66	:	34	:	0	99
p-ClC_6H_4	50	:	50	:	0	64

a : GLC yield

Scheme 9.7

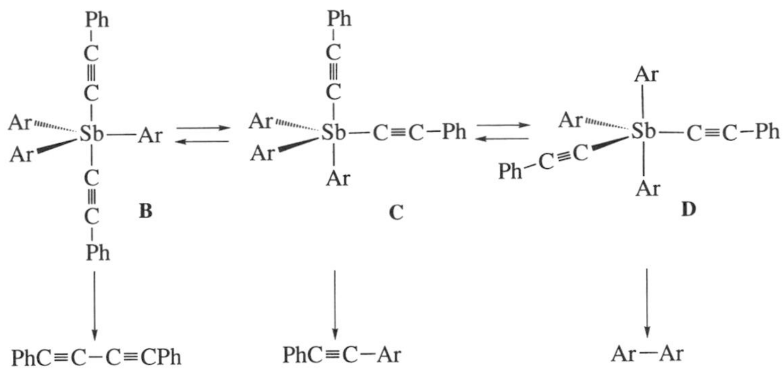

Scheme 9.8

9.2.3 Selectivity of LCR of Mixed Pentaarylantimony (V) Compounds

We chose mixed pentaarylantimony(V) species, $Ar_nTol_{5-n}Sb$ ($n = 1$–4: Ar = *p*-$CF_3C_6H_4$, Tol = *p*-$CH_3C_6H_4$), for which synthetic procedures had to be established. The structures in solid states and in solution were investigated by X-ray analyses and ^{13}C variable temperature NMR measurements.[26] Ligand-exchange reactions (LERs: Scheme 9.9), which took place at much lower temperatures (ca. 60°C) than LCRs (160–200°C) in solution,[27] made examination of the selectivity of the LCR very complex. The experimental investigations of the selectivities from these compounds were carried out by use of the catalysts such as $Cu(acac)_2$[28] and Li^+TFPB^- (TFPB: $B(3,5\text{-}(CF_3)_2C_6H_3)_4$).[29] The activities of these catalysts were examined by heating the benzene-d_6 solution (or suspension) of Tol_5Sb (**6**) or Ar_5Sb (**11**) with or without catalysts in sealed NMR tubes at 165°C, and the results are shown in Table 9.1. The LCRs from **6** and **11** were drastically accelerated by $Cu(acac)_2$, with a complete reaction occurring at 165°C within 5 min. The activity of Li^+TFPB^- was less than that of $Cu(acac)_2$; however, a large acceleration effect was observed.

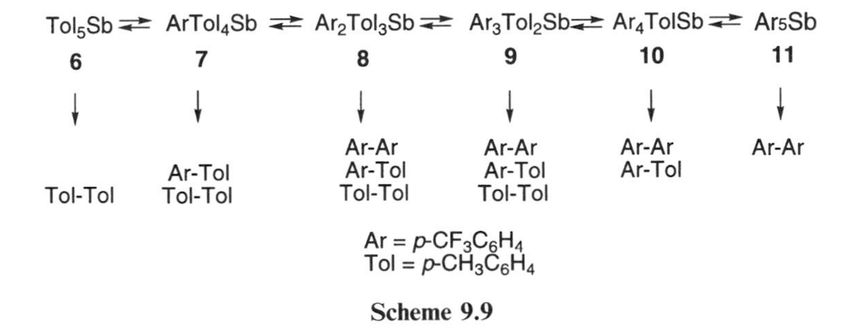

Scheme 9.9

The LCRs catalyzed by Li^+TFPB^- and $Cu(acac)_2$ were applied for mixed pentaarylantimony compounds (**7**–**10**). Experimental ratios of biaryls and

TABLE 9.1 Catalyzed LCRs from Tol_5Sb (6) and Ar_5Sb (11)

Catalysts	Conversion (%) of the Reaction[a]	
	Tol_5Sb (**6**)	Ar_5Sb (**11**)
Nothing	23	No reaction (19 h)
LiBAr′ ($Ar' = 3,5\text{-}(CF_3)_2C_6H_3$)	91	16
$Cu(acac)_2$	100	100

[a]After 30 min at 165°C in C_6D_6.

triarylantimony compounds were determined by GC and by relative integral intensities of ^{19}F NMR. The results are shown in Tables 9.2 and 9.3. It is clear that the selectivities of LCRs from **7–10** catalyzed by Li^+TFPB^- and $Cu(acac)_2$ are quite similar, and bitolyl was not formed from Ar_2Tol_3Sb (**8**), Ar_3Tol_2Sb (**9**), and Ar_4TolSb (**10**). Although bitolyl was obtained in the LCR from $ArTol_4Sb$ (**7**) we had already found that the rate of the LER from **7** was exceptionally fast.

The selectivities of the LCRs from each $Ar_nTol_{5-n}Sb$ could be obtained with the help of catalysts in solution; however, LERs could still be competitive and the coordination effects of the catalysts on the selectivities were unclear. After several trials, flash vacuum thermolysis (FVT) could finally be used to prevent the LER and the conclusive selectivities of the LCRs from each of $Ar_nTol_{5-n}Sb$ ($n = 1$–4) were obtained. The apparatus is shown in Figure 9.1. A pure mixed pentaarylantimony compound (10–15 mg) is placed at the head of a quartz tube in which "fillings" are packed in the center. These are made of crushed quartz glass and should be pretreated by "coating" to avoid protonolysis and any catalyzed reactions on the surface of fillings. Coating means that the fillings are boiled in a benzene solution of a mixture of Ar_5Sb, Tol_5Sb, Ar_3Sb, Tol_3Sb, or Tol_2 overnight; the fillings are washed out thoroughly with acetone and deionized water and then dried. When the dried fillings were heated in the quartz tube at 300°C for several hours in a high vacuum (10^{-3} Torr) under a gentle stream of argon, any organic compound could not be detected in a liquid N_2 trap, which certified "cleanness of the fillings."

The FVT tube was evacuated first to $1\text{–}2 \times 10^{-5}$ Torr with an oil diffusion pump and the pressure was kept to $1\text{–}2 \times 10^{-3}$ Torr under a gentle and constant stream of argon throughout the pyrolysis. The starting material was heated up to 120–125°C in solid by a ribbon heater ("preheating") at the head of the quartz tube, and the sublimed starting material was pyrolyzed in the oven (300°C) through the fillings to give LCR products, which were trapped by liquid N_2. The reaction period was 30 min and the obtained products were analyzed by GC. About 5–10% of the starting material was

TABLE 9.2 LCR Catalyzed by $Cu(acac)_2$

Sample	Mol Ratio Cat./Sample	Statistical Ratio			Experimental Ratio			LER (%)
		Ar—Ar:	Ar—Tol:	Tol—Tol	Ar—Ar:	Ar—Tol:	Tol—Tol	
Ar_4TolSb (**10**)	0.83	60	40	0	69	31	0	24
Ar_3Tol_2Sb (**8**)	0.36	30	60	10	50	50	0	18
	1.0				51	49	0	15
Ar_2Tol_3Sb (**8**)	0.33	10	60	30	34	66	0	23
	1.8				33	67	0	20
$ArTol_4Sb$ (**7**)	1.2	0	40	60	12	69	19	31

Conditions: in C_6D_6 at 165°C. 5 min. Concentration of the sample 0.0072–0.044 mol^{-1}.

TABLE 9.3 LCR Catalyzed by LiTFPB (TFPB: $B(3,5\text{-}(CF_3)_2C_6H_3)_4$)

Sample	Mol Ratio Cat./Sample	Statistical Ratio			Experimental Ratio			LER (%)
		Ar—Ar:	Ar—Tol:	Tol—Tol	Ar—Ar:	Ar—Tol:	Tol—Tol	
Ar_4TolSb (**10**) (3.5 h)	0.11	60	40	0	73	27	0	
Ar_3Tol_2Sb (**9**)	0.20	30	60	10	51	49	0	
(10 min)	0.088				50	50	0	12
Ar_2Tol_3Sb (**8**)	0.10	10	60	30	31	69	0	12
(10 min)	0.13				32	68	0	15
$ArTol_4Sb$ (**7**) (10 min)	0.089	0	40	60	10	76	14	24

Conditions: in C_6D_6 at 165°C. Concentration of the sample 0.0011–0.034 mol^{-1}.

TABLE 9.4 LCR by Flash Vacuum Thermolysis (FVT)

Sample	Statistical Ratio			Experimental Ratio		
	Ar—Ar:	Ar—Tol:	Tol—Tol	Ar—Ar:	Ar—Tol:	Tol—Tol
Ar_4TolSb (**10**)	60	40	0	76 ± 2.3	24 ± 2.3	0
Ar_3Tol_2Sb (**9**)	30	60	10	58 ± 1.9	42 ± 1.9	0
Ar_2Tol_3Sb (**8**)	10	60	30	36 ± 1.5	64 ± 1.5	0
Ar_2Tol_4Sb (**7**)	0	40	60	19	67	14

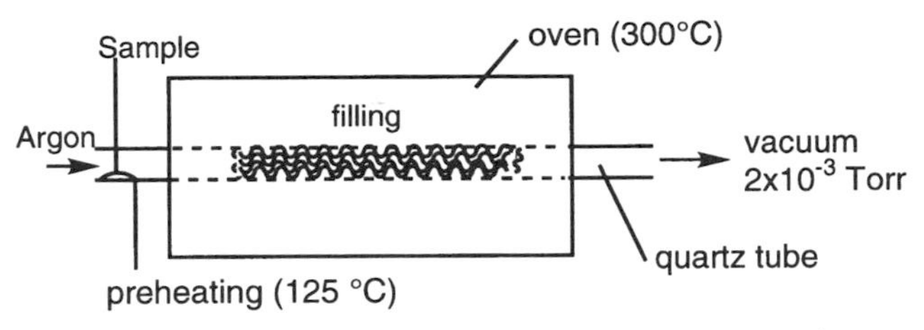

Figure 9.1 Apparatus of flash vacuum thermolysis.

sublimed to the oven, and the residual sample which did not vaporize was analyzed by NMR in order to estimate the extent of the LER. Under these conditions, the extent of the LER was determined to be less than 5%, except for $ArTol_4Sb$, in which the rate of the LER was exceptionally fast, as described in the previous section. The results of FVT are shown in Table 9.4 as the yields of biaryls, together with statistical ratios. FVT experiments were carried out at least four times for all compounds (except for $ArTol_4Sb$). The data show an average of all runs and are presented with probable errors. The data of $ArTol_4Sb$ are an average of two runs. $ArTol_4Sb$ (**7**) is an exception, from which Ar–Ar (19%) and Tol–Tol (14%) were obtained, probably due to very fast LER. Tol–Tol was not detected at all from Ar_2Tol_3Sb (**8**), Ar_3Tol_2Sb (**9**), and Ar_4TolSb (**10**).

The fact that Tol–Tol could not be detected at all from Ar_2Tol_3Sb (**8**) and Ar_3Tol_2Sb (**9**) rules out the possibility of the LCR taking place via equatorial–equatorial coupling, because **8** and **9** should give Tol-Tol if such a possibility were real. Apical–equatorial coupling may certainly be not the case because **8** and **9** should give Tol–Tol from a stereoisomer, which should be present to a certain extent by BPR. There is an apparent selectivity that the more electronegative p-$CF_3C_6H_4$ group preferred for LCR to afford Ar–Ar and Ar–Tol at the expense of Tol–Tol (Scheme 9.10). The results of selective LCRs in the presence of catalysts (Tables 9.2 and 9.3) are essentially the same as those by FVT. This can be interpreted that the catalyzed LCR takes place faster than the LER and the the LCR goes through by the apical–apical coupling. This trend was also true in the reaction of triarylbis(phenylethynyl)antimony compounds (**3**–**5**) (Scheme 9.7).

9.2.4 Conclusion

These experimental results on the LCR can be interpreted by invoking the idea that the apical–apical coupling is the sole reaction path in stereoisomers which are inevitably generated by very fast BPR.[30] That is, once the apical substituents start a bending motion for the LCR, the combination of the substituents is kept through the transition state to the final products with the conservation of the momentum (Scheme 9.11); we propose to call this the *memory effect*. The

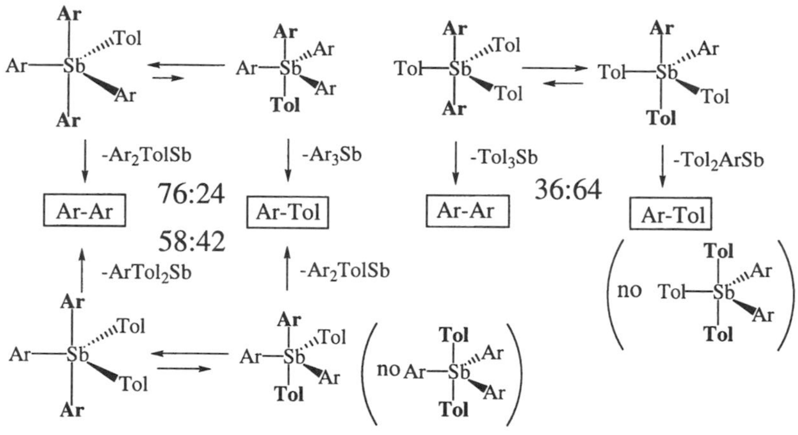

Scheme 9.10 Apical–apical coupling from pentaarylantimony compounds.

memory effect is intuitively acceptable because the hypervalent bond (3c–4e) is weak and polarized (polarizable) and should be easier to get into a bending motion than equatorial sp^2 bond. Moreover, the more electronegative substituents preferably occupy the apical positions. The definite preference of the more electronegative substituents to participate in the LCR seems to be general. This kind of discussion on the selectivity of some organic reactions has been described by Carpenter, who stated that reacting molecules or reacting positions in a molecule tended to take a trajectory (from reactants, through the intermediate, and on to the product) that is closest to the straight-line path, where conservation of momentum plays an important role.[31] Recent advanced *ab initio* calculation predicted that the LCR from SbH_5 took place with coupling of the two equatorial substituents through the transition state A, as shown in Scheme 9.2.[8] The charge distribution of A showed that the electron density of the leaving dihydrogens in the equatorial positions was less than that of other hydrogens. The apparent contradiction between the theoretical study and the present experimental result may be due to several reasons, and the neglect of *p*-orbitals in the calculation should be one of them. Further theoretical study is necessary to make clear the essential reason for the present contradiction.

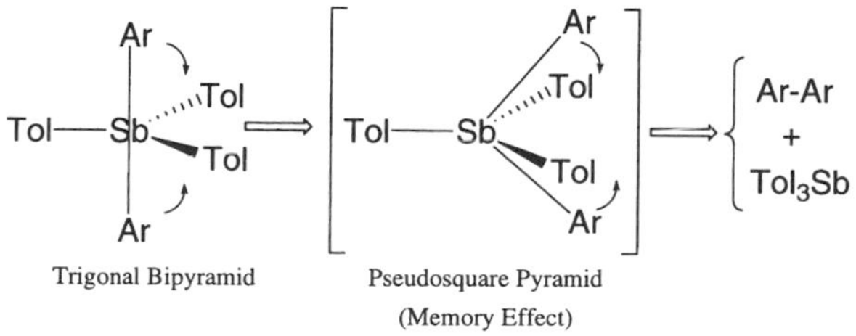

Trigonal Bipyramid

Pseudosquare Pyramid (Memory Effect)

Scheme 9.11

9.3 LCRs FROM GROUP 16 ELEMENT COMPOUNDS

Studies of Group 16 elements include theoretical investigations on the mechanism of LCRs of λ^4-sulfane (SH_4), λ^4-selenane (SeH_4), λ^4-tellane (TeH_4) by several groups. A first report by Schaefer suggested the symmetrically allowed equatorial–equatorial coupling process (42 kcal mol^{-1})[32] for SH_4, but Morokuma and co-workers recently reported that the LCR between the ionic apical–equatorial ligands was energetically favorable in all cases (SH_4, SeH_4, TeH_4).[33] The comparison of the calculated energies of LCRs from apical–equatorial coupling with those from equatorial–equatorial coupling are shown in Scheme 9.12, together with the net charges in the transition states.

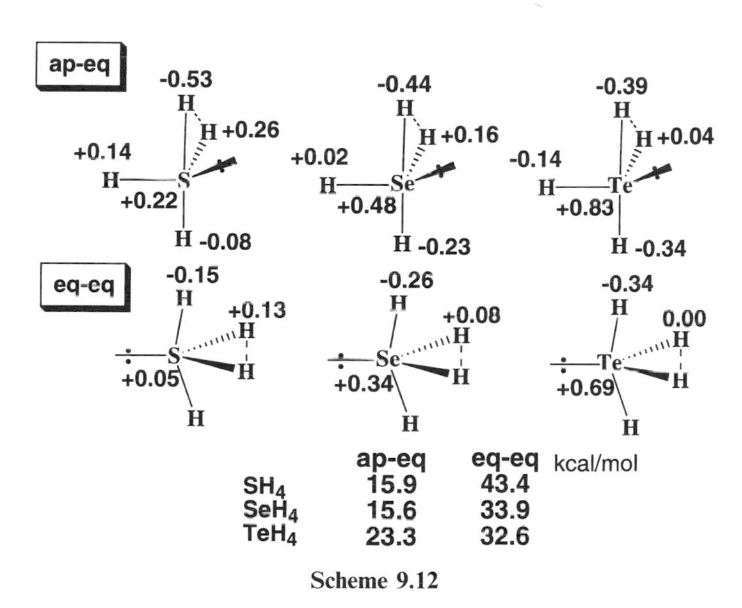

Scheme 9.12

The calculation suggests that highly polarized apical–equatorial coupling from 10-M-4 Group 16 element compounds is favored, probably because of the presence of lone pair electrons at the central atom. Experimental investigations on LCRs from λ^4-sulfane,[34] λ^4-selenane,[35] and λ^4-tellane[36] have been reported. Oae reported the concerted nature of LCRs from 10-S-4 sulfanes, which are intermediates in the reaction of 2-pyridyl sulfoxide with methylmagnesium bromide. The stereochemistry around the carbon atom in the coupled products was completely retained (Scheme 9.13).[37] In the thermolysis of Ph_4Te, the yields of biphenyl and diphenyl telluride from decomposition at 140°C in triethylsilane and styrene did not differ significantly from those of the neat reaction.[36] Thus, the concerted nature of the LCR from Ph_4Te was confirmed (Scheme 9.14).

Interestingly, they investigated LCRs from the mixture of Ph_4Te and (p-$CH_3C_6H_4$)$_4$Te and found that the LERs took place prior to the LCRs and mixed biaryl and mixed diaryl telluride were formed (Scheme 9.15). The situation is quite similar to the reactions of $Ar_nTol_{5-n}Sb$ ($n = 1$–4: Ar = p-$CF_3C_6H_4$, Tol = p-$CH_3C_6H_4$) described above (Scheme 9.9), but in the tellurium case almost a statistical distribution of the organic groups was observed, probably because the electronic effects of the phenyl and p-methylphenyl group are not quite different. Some of the reactions have been confirmed to take place concertedly, but the selectivity of LCRs in 10-M-4 compounds (M: Group 16 element) has been difficult to conclude. Further study on the LCRs of mixed acyclic Group 16 element compounds such as $Ar_nTol_{4-n}Te$ ($n = 1$–3: Ar = p-$CF_3C_6H_4$, Tol = p-$CH_3C_6H_4$) will clarify the point.

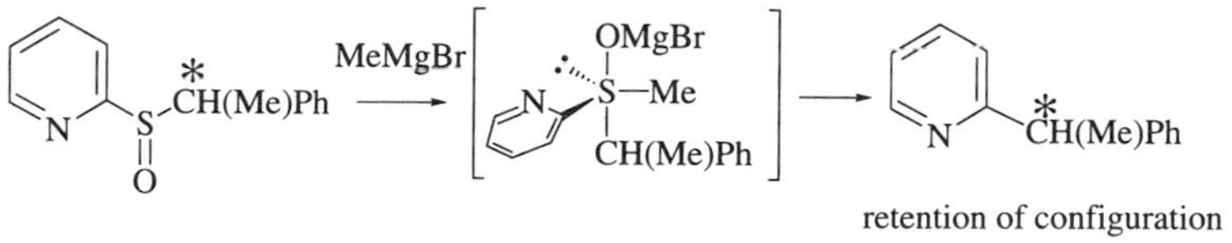

MeMgBr
*CH(Me)Ph
OMgBr
S—Me
CH(Me)Ph
*CH(Me)Ph
retention of configuration

Scheme 9.13

$Ph_4Te \xrightarrow{140\ °C} Ph\text{-}Ph + Ph_2Te$

	Ph-Ph %	Ph_2Te %
no solvent	89	92
Et_3SiH (N_2)	90	80
styrene (N_2)	95	103

Scheme 9.14

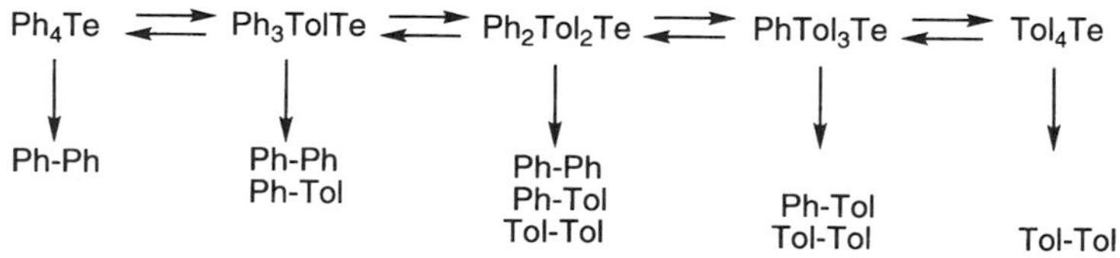

$Ph_4Te \rightleftarrows Ph_3TolTe \rightleftarrows Ph_2Tol_2Te \rightleftarrows PhTol_3Te \rightleftarrows Tol_4Te$

Ph_4Te → Ph-Ph
Ph_3TolTe → Ph-Ph, Ph-Tol
Ph_2Tol_2Te → Ph-Ph, Ph-Tol, Tol-Tol
$PhTol_3Te$ → Ph-Tol, Tol-Tol
Tol_4Te → Tol-Tol

Scheme 9.15

REFERENCES

1. For recent reviews: J. M. Brown, N. A. Cooley, *Chem. Rev.*, **88**, 1031 (1988). R. D. Pike, *Mech. Inorg. Organomet. React.*, **8**, 459 (1994).
2. S. Oae, Y. Uchida, *Acc. Chem. Res.*, **24**, 202 (1991). S. Oae, N. Furukawa, *Adv. Heterocycl. Chem.*, **48**, 1 (1990). S. Oae, *Rev. Heteroatom Chem.*, **1**, 304 (1988). S. Oae, Y. Uchida, *Rev. Heteroatom Chem.*, **2**, 76 (1989).
3. The N—X—L designation has been previously described: *N* valence electrons about a central atom X with *L* ligands. C. W. Perkins, J. C. Martin, A. J. Arduengo, III., W. Lau, A. Alegria, K. Kochi, *J. Am. Chem. Soc.*, **102**, 7753 (1980).
4. J. Drabowicz, P. Lyzwa, M. Mikolajczyk, *Supplement S: The Chemistry of Sulphur-Containing Functional Groups*, S. Patai, Z. Rapppoport, Eds (John Wiley & Sons, 1993, Chapter 15, p. 799). R. A. Hayes, J. C. Martin, *Organic Sulfur Chemistry. Theoretical and Experimental Advances*, F. Bernardi, I. G. Csizmadia, A. Mangani, Eds (Elsevier, 1985, Chapter 8, p. 408). R. Burgada and R. Setton, *The Chemistry of Organophosphorus Compounds*, S. Patai, F. R. Hartley, Eds (John Wiley & Sons, 1994, Chapter 3, p. 185); R. R. Holmes, *Pentacoordinated Phosphorus*, ACS Monograph Series 175 and 176 (American Chemical Society, Washington, DC, 1980), Vols 1 and 2.
5. R. Hoffmann, J. M. Howell, E. L. Muetterties, *J. Am. Chem. Soc.*, **94**, 3407 (1972). R. G. Pearson, *Symmetry Rules for Chemical Reactions: Orbital Topology and Elementary Processes* (John Wiley & Sons, 1976, pp. 270, 286).
6. J. M. Howell, *J. Am. Chem. Soc.*, **99**, 7447 (1977).
7. W. Kutzelnigg, J. Wasilewski, *J. Am. Chem. Soc.*, **104**, 953 (1982).
8. J. Moc, K. Morokuma., *J. Am. Chem. Soc.*, **117**, 11790 (1995).
9. P. Kolandaivel, R. Kumaresan, *J. Mol. Struct.*, **337**, 225 (1995). A. E. Reed, P. v. R. Schleyer, *Chem. Phys. Lett.*, **133**, 553 (1987).
10. H. Wasada, K. Hirao, *J. Am. Chem. Soc.*, **114**, 16 (1992).
11. D. Hellwinkel, W. Lindner, *Chem. Ber.*, **109**, 1497 (1976).
12. D. H. R. Barton, J.-P. Finet, C. Giannotti, F. Halley, *Tetrahedron*, **44**, 4483 (1988). R. A. Abramovitch, D. H. R. Barton, and J.-P. Finet, *Tetrahedron*, **44**, 3039 (1988). D. H. R. Barton, N. Y. Bhatnagar, J.-P. Finet, W. B. Motherwell, *Tetrahedron*, **42**, 3111 (1986). D. H. R. Barton, J.-P. Finet, C. Giannotti, F. Halley, *J. Chem. Soc., Perkin Trans. I* , 241 (1987). J.-P. Finet, *Chem. Rev.*, **89**, 1487 (1989). D. H. R. Barton, S. A. Glover, S. V. Ley, *J. Chem. Soc., Chem. Commun.*, 266 (1977).
13. H. Hartmann, H. Kühl, *Angew. Chem.*, **68**, 619 (1956). *Z. Anorg. Chem.*, **312**, 186 (1961). T. Morgan and G. R. Davies, *Proc. Royal Soc.*, **110**, 523 (1926). H. J. Breunig, W. Kanig, *Phosphorus Sulfur*, **12**, 149 (1982).
14. C. A. McAuliffe, A. G. Mackie, *The Chemistry of Organic Arsenic, Antimony, and Bismuth Compounds*, S. Patai, Ed. (John Wiley & Sons, 1994), Chapter 13, p. 527. J. L. Wardell, *Comprehensive Organometallic Chemistry*, Vol. 2 (1982), Chapter 13, p. 681. J. L. Wardell, *Comprehensive Organometallic Chemistry II*, Vol. 2 (1994), Chapter 8, p. 321.
15. G. A. Razubaev and N. A. Osanova, *J. Organometal. Chem.*, **38**, 77 (1972).
16. B. S. Bedi, D. W. Grant, L. Tewnion, and J. L. Wardell, *J. Organometal. Chem.*, **239**, 251 (1982).

17. G. Wittig and M. Rieber, *Liebigs Ann. Chem.*, **562**, 187 (1949).
18. G. Wittig and K. Clauss, *Liebigs Ann. Chem.*, **577**, 26 (1952).
19. G. Wittig and K. Clauss, *Liebigs Ann. Chem.*, **578**, 136 (1952).
20. K.-H. Mitschke and H. Schmidbaur, *Chem. Ber.*, **106**, 3645 (1973).
21. Y. Takashi, *J. Organometal. Chem.*, **8**, 225 (1967).
22. K. Shen, W. E. McEwen, and A. P. Wolf, *J. Am. Chem. Soc.*, **91**, 1283 (1969).
23. W. E. McEwen and C. T. Lin, *Phosphorus*, **4**, 91 (1974).
24. G. A. Razubaev, N. A. Osanova, N. P. Shulaev, and B. M. Tsigin, *J. Gen. Chem. USSR (engl. Transl.)*, **30**, 3203 (1960).
25. K.-y. Akiba, T. Okinaka, M. Nakatani, and Y. Yamamoto, *Tetrahedron Lett.*, **28**, 3367 (1987).
26. Y. Yamamoto, G. Schröder, Y. Mimura, T. Okinaka, A. Hasuoka, M. Watanabe, T. Matsuzaki, F. Kondo, S. Kojima, K.-y. Akiba, to be published.
27. K.-y. Akiba, G. Schröder, M. Watanabe, Y. Mimura, T. Okinaka, Y. Yamamoto, to be published.
28. V. A. Dodonov, O. P. Bolotova, A. V. Gushchin, *J. Gen. Chem. USSR (engl. Transl.)*, **58**, 711 (1988).
29. J. Nie, T. Sonoda, H. Kobayashi, private communication.
30. K.-y. Akiba, *Pure Appl. Chem.*, **68**, 837 (1996).
31. B. K. Carpenter, *J. Am. Chem. Soc.*, **107**, 5730 (1985). B. A. Lyons, J. Pfeifer, B. K. Carpenter, *J. Am. Chem. Soc.*, **113**, 9006 (1991).
32. Y. Yoshioka, J. D. Goddard, H. F. Schaefer, III., *J. Chem. Phys.*, **74**, 1855 (1981).
33. J. Moc, A. E. Dorigo, K. Morokuma, *Chem. Phys. Lett.*, **204**, 65 (1993). J. Moc and K. Morokuma, *Inorg. Chem.*, **33**, 551 (1994).
34. M. J. Bogdanowicz, B. M. Trost, *Tetrahedron Lett.*, 887 (1970). B. M. Trost, M. J. Bogdanowicz, *J. Am. Chem. Soc.*, **95**, 5298 (1973).
35. D. Hellwinkel, *N. Y. Acad. Sci.*, **192**, 158 (1972).
36. D. H. R. Barton, S. A. Glover, S. V. Ley, *J. Chem. Soc., Chem. Commun.*, 266 (1977). S. A. Glover, *J. Chem. Soc., Perkin Trans. I*, 1338 (1980).
37. S. Oae, T. Kawai, N. Furukawa, *Tetrahedron Lett.*, **25**, 69 (1984). S. Oac, T. Kawai, N. Furukawa, F. Iwasaki, *J. Chem. Soc., Perkin Trans. II*, 405 (1987).

10 The —SF_5, —SeF_5, and —TeF_5 Groups in Organic Chemistry

DIETER LENTZ AND KONRAD SEPPELT
Institut für Anorganische und Analytische Chemie der Freien Universität Berlin, Berlin, Germany

10.1 INTRODUCTION

SF_6 is the obvious model for a hypervalent molecule. Without embarking on a discussion of the nature of the bonds in SF_6, it still can be said that its extreme kinetic stability should have called into question many decades ago all ideas of an octet rule for heavier main group elements. But the kinetic stability of SF_6 is also a disadvantage in its chemistry. Although there are many uses in technology for SF_6 that in one way or the other are all based on its kinetic stability, its stability prevents it from being a useful reagent in chemical reactions. Only the first attack on SF_6 is strongly sterically hindered and requires in general high activation energy. Any subsequent steps will proceed more rapidly, resulting inevitably in complete destruction of the SF_6 molecule.

So organic derivatives of SF_6 (R—SF_5, R_2SF_4, etc.) cannot be prepared directly from SF_6. But the question remains, how many of the special features of SF_6 are found in such derivatives compounds, after they have been made on along painfully difficult routes.

A first survey showed that in spite of the problems of preparation, the total number of well-characterized R—SF_5 compounds is already very large. Our focus will be on preparation of R—SF_5 compounds, and on the differences that the SF_5 substituent can make in organic chemistry.

There exists a recent review that covers very thoroughly the chemistry of SF_5 -substituted alkenes and alkynes.[1] Also we would like to draw attention to a review that deals with R—SF_3 compounds, among other topics, but these have a totally different chemistry.[2]

Chemistry of Hypervalent Compounds, edited by Kin-ya Akiba.
ISBN 0-471-24019-2

10.2 PREPARATION OF SF_5 ORGANIC COMPOUNDS

10.2.1 By Fluorination of Organic Sulfur Compounds

Similar to the situation in carbon–fluorine chemistry, the choice of the fluorinating agent for preparing organic SF_5 compounds is difficult, mainly because there is none that works satisfactorily, if simplicity and yield are considered. Many of the fluorinating agents (ClF, AgF_2, CoF_3) have to be prepared by use of elemental fluorine, so the simplest way could be using elemental fluorine itself. Indeed the very first compound, CF_3—SF_5, has been made from CH_3—SH or CS_2 and F_2.[3,4] The compound F_5S—CN was said to be obtained from CH_3—SCN and F_2,[5] but recently it was shown that the material obtained was CF_3—N=SF_2. Fluorination of $(SCN)_2$ or F_3SCN afforded SF_5—CN, finally.[6,7] (The isomer SF_5—NC is now also known.[8]) The methods to control the reaction with elemental fluorine by careful control of the temperature and dilution with helium made it possible to obtain CF_3—SF_5, $CF_2(SF_5)_5$[4,9,10] and even $CF_2(SF_3)_2$[9,10] from CS_2, as well as perfluoroalkyl sulfur pentafluorides with branched or longer chained alkyl groups.[11] Use of ClF instead of F_2 produced $CF_2(SF_5)_2$[10] from CS_2 as well as CF_3—SO_2—CF_2—SF_5[12] from O_2S—C—S—CF_2 and small amounts of perfluoroalkyl-SF_5 compounds from perfluoroalkylsulfenyl chlorides, perfluorodialkyl disulfides.[13] But in reactions with ClF, the main products are R—SF_4Cl compounds.

$$CH_3{-}SH \xrightarrow{F_2} CF_3{-}SF_5$$

$$CS_2 \xrightarrow{F_2} CF_2(SF_5)_2 + \cdots$$

$$(SCN)_2 \text{ or } F_3SCN \xrightarrow{F_2} F_5S{-}C{\equiv}N$$

$$R{-}C_6H_4{-}S{-}S{-}C_6H_4{-}R \xrightarrow{AgF_2} R{-}C_6H_4{-}SF_5$$

Scheme 10.1

CoF_3 converts CH_3—SH into CF_3—SF_5 and CHF_2SF_5[14] and also converts CF_3—S—S—CF_3 into CF_3—SF_5.[15] AgF_2 converts $(CF_3)_2N$—S—CH_3 into CH_2F—SF_5[16] and CF_2Cl—S—Cl into CF_3—SF_5.[17] Of importance is the reaction of AgF_2 with mono- or disubstituted aromatic disulfides, since it is one of the rare pathways into the aromatic-SF_5 chemistry[18-20] (see Section 10.5.5).

The well-known electrofluorination process has also been successfully used for generating SF_5 compounds. With few exceptions, only perfluorinated compounds have been obtained: CF_3—SF_5,[21,22,23,24,28,29] C_2F_5—SF_5,[25,26] C_3F_7—SF_5,[25] C_4F_9—SF_5,[25,30] C_5F_{11}—SF_5,[30] i-C_3F_7—SF_5,[31] $(CF_3)_2CF$—CF_2—SF_5,[31] i-C_5F_{11}—SF_5,[30] $C_2F_5CF(CF_3)$—SF_5,[31] $(C_2F_5)_2N$—CF_2—

CF_2—SF_5,[25,30] $(C_2F_5)_2N$—CF_2—CF_2—CF_2—SF_5,[25] HOOC—CF_2—SF_5,[23] $CF_2(SF_5)_2$,[22,24] C_6F_{13}—SF_5,[27] c-C_6F_{11}—SF_5,[27] SF_5—$(CF_2)_n$—SF_5 ($n=2$–5),[30] SF_5—CF_2—CF_2—O—CF_2—CF_2—SF_5,[30] and C_2F_5O—CF_2—CF_2SF_5.[30]

Only recently have the first partially fluorinated compounds been detected in the reaction mixtures: C_2F_5—CH_2—SF_5,[31] C_2H_5—SF_5,[32] CHF_2—CH_2—SF_5,[32] CHF_2—CHF—SF_5,[32] CH_2F—CHF—SF_5,[32] and CF_3—CH_2—SF_5,[32] but all of these in low yields only.

10.2.2 Introduction of the SF_5 Group by Means of S_2F_{10} or SF_5OF

S_2F_{10} may seem to be a very useful reagent for introducing the SF_5 group, because it is known to decompose into two free radicals $\cdot SF_5$. However, it must first be prepared from SF_5Cl, which is a better reagent for the same purpose. Also S_2F_{10} is very toxic and not detectable by smell, so this compound has been largely disregarded for any chemical reactions, in spite of the fact that S_2F_{10} is also unique for theoretical reasons. It reacts with $(COF)_2$, forming F—CO—SF_5 in low yields.[33] In two cases, double SF_5 addition was achieved across CC double bonds: F_5S—CClH—CH_2—SF_5 and F_5S—$CF(CF_3)$—CF_2—SF_5.[34] Reaction with benzene resulted in small amounts of C_6H_5—SF_5.[34] In another study, the reaction with $ClCH{=}CH_2$ gave only CH_2F—CHCl—SF_5, besides some product of a vinyl chloride dimer. Vinyl fluoride gives only products of dimerized material.[35] Tetrafluoroethylene reacts with S_2F_{10} and I_2 under formation of telomeric I—$(C_2F_4)_n$—SF_5 compounds.[36] Only small amounts of F—CO—SF_5 are formed by photochemical reactions between SF_5—OF and CO.[37]

10.2.3 Introduction of the SF_5 Group by Means of SF_5Cl

The addition reaction of SF_5Cl (or SF_5Br) to alkenes and alkynes has some general characteristics. The reaction dates back to 1959. But there are shortcomings: SF_5Cl is normally not commercially available, and its preparation requires handling of SF_4 and metal autoclave equipment.[38] Quite a number of alkenes have been tried: ethene, propene, butadiene, isobutene, cyclohexene, styrene, vinyl chloride,[39–40] C_2F_4,[41] C_2F_3H,[41–43] C_2F_3Cl,[41] $CH_2{=}CF_2$,[43] and C_3F_6.[41,43] The highly fluorinated alkenes give also some telomeric products.[41–43] Also vinyl silanes,[44] $CH_2{=}C{=}O$,[45–47] various unsaturated esters,[48] $F_2C{=}C{=}CF_2$,[49] $H_2C{=}C{=}CH_2$,[50] perfluorovinylamines,[51] and possibly others have been tried. The most recent reaction is that of SF_5Cl and 3-cyclopenten-1-ol or norbornene, which after intermediate steps results in pentafluoro-λ^6-sulfanyl-cyclopentadiene.[52]

The addition reaction has obviously a stepwise free radical chain mechanism. This is supported by a kinetic investigation of the SF_5Cl/C_2H_4 reaction.[53] Many of these reactions require irradiation or the addition of benzoyl peroxide. The direction of the reaction is guided by the most stable free radical that is first formed between $\cdot SF_5$ and $>C{=}C<$; examples are shown in the literature.[39,43,45–47,54]

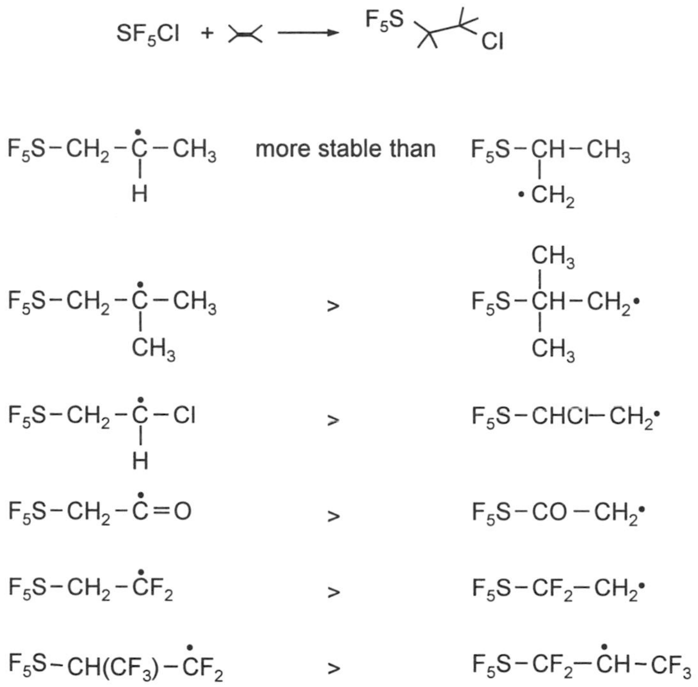

$$SF_5Cl + (CH_3)_2C{=}C(CH_3)_2 \longrightarrow F_5S{-}C(CH_3)_2{-}C(CH_3)_2{-}Cl$$

$$F_5S{-}CH_2{-}\dot{C}H{-}CH_3 \quad \text{more stable than} \quad F_5S{-}CH(\dot{C}H_2){-}CH_3$$

$$F_5S{-}CH_2{-}\dot{C}(CH_3){-}CH_3 \quad > \quad F_5S{-}C(CH_3)_2{-}CH_2\cdot$$

$$F_5S{-}CH_2{-}\dot{C}H{-}Cl \quad > \quad F_5S{-}CHCl{-}CH_2\cdot$$

$$F_5S{-}CH_2{-}\dot{C}{=}O \quad > \quad F_5S{-}CO{-}CH_2\cdot$$

$$F_5S{-}CH_2{-}\dot{C}F_2 \quad > \quad F_5S{-}CF_2{-}CH_2\cdot$$

$$F_5S{-}CH(CF_3){-}\dot{C}F_2 \quad > \quad F_5S{-}CF_2{-}\dot{C}H{-}CF_3$$

Scheme 10.2

The regiospecificity of the radical reactions prevails even if a sterically very crowded product is formed. Highly fluorinated alkenes like $F_2C{=}CHF$[41–43] or $F_2C{=}CF{-}CF_3$,[41,43] however, give both possible addition products. Electron-richer alkenes react more rapidly and with better yields with the thought-to-be electrophilic free radical SF_5.[48] The electron affinity of SF_5 is given as 84.4(10) or 78.2(35) kcal mol^{-1}, depending on the method.[55] This value is similar to that of chlorine, which has the highest electron affinity of any atom.

SF_5Cl additions occur also to alkynes: C_2H_2,[39,56] $CH_3{-}C{\equiv}CH$,[39,57] $CH_3{-}C{\equiv}C{-}OR$,[58] and $HC{\equiv}C{-}OR$ ($R{=}CH_3$, C_2H_5).[59] Again the addition follows the principle of the most stable SF_5-containing radical intermediate. The SF_5Cl/C_2H_2 reaction is regiospecific, resulting in only the *E* isomer;[56] for other alkynes no information is available.

10.2.4 Introduction of the SF_5 Group by Means of SF_5Br

SF_5Br adds in some cases when SF_5Cl fails, although there also are reported cases where SF_5Br fails and SF_5 Cl adds successfully.[59] There is of course the problem of availability: SF_5Br can be made from S_2F_{10},[60] the latter from SF_5Cl. The alternative route from SF_4 and unstable BrF is certainly not easier to handle.[50,61,62]

Addition reactions with $CH_2{=}CHF$,[63] $CH_2{=}CF_2$,[63,64] $CHF{=}CF_2$[63,64] and $ClCF{=}CF_2$[63] proceed smoothly and in the same direction as with SF_5Cl. Later this work was extended to $CH_2{=}CFCl$, $CH_2{=}CH{-}CF_3$, $CHF{=}CHCl$, $CHF{=}CFCl$, and $CF_2{=}CClH$.[65] It was also found that excess of fluoroalkenes results in telomerization along with the SF_5Br addition, for instance $SF_5{-}(CH_2{-}CF_2)_8Br$.[66] $CH_2{=}CH{-}SF_5$ adds a second SF_5Br, forming $SF_5{-}CH_2{-}CHBr{-}SF_5$.[67]

Of special interest is SF_5Br in the alkyne series. With SF_5Cl it takes four steps for the formation of $F_5S{-}C{\equiv}C{-}H$; use of SF_5Br requires only one addition to C_2H_2 and one elimination.[62,68] In two more steps, $F_5SC{\equiv}CH$ can be converted to $F_5S{-}C{\equiv}C{-}SF_5$ where SF_5Br is added first.[69] These kinds of reactions have been extended to $CH_3{-}C{\equiv}CH$, $CF_3{-}C{\equiv}CH$,[70] and $HC{\equiv}C{-}C{\equiv}CH$,[71] resulting in the corresponding SF_5 substituted alkynes. The $SF_5Br/CH_3{-}C{\equiv}CH$ reaction yields a single isomer, with $CF_3{-}C{\equiv}CH$ two isomeric propenes are formed.[70] The SF_5Br addition reaction is obviously also a radical one, as already indicated by the regiospecificity. Reactions are quicker if irradiated, but there is no marked difference in products, namely threo- and erythro, if added to *cis*- and *trans*-$CHF{=}CHF$.[72] Finally there is also one report on an α-addition of SF_5Br to $CF_3{-}NC$, forming $CF_3{-}N{=}CBr{-}SF_5$.[73]

10.2.5 Other Syntheses of the SF_5-Group-Containing Compounds

It is, of course, possible to obtain $C{-}SF_5$ compounds from unsaturated precursors, like alkylidene sulfur tetrafluorides, $R_2C{=}SF_4$, or alkylidyne sulfur trifluorides, $R{-}C{\equiv}SF_3$, by addition of HF or other fluorides.[74] This, however, is not a practical way, since these double-bonded compounds are always made from $C{-}SF_5$ compounds by elimination, and there is no SF_5 compound known that can be made only via such a route. We will discuss the formation reactions of $>C{=}SF_4$ and $-C{\equiv}SF_3$ compounds briefly in the section on chemistry at the $C{-}SF_5$ group.

One theoretically feasible reaction has so far not worked successfully: the nucleophilic displacement by SF_5^-. This anion is a well-characterized, square pyramidal entity.[75–77] Either the reaction has not been tried, or attempts were unsuccessful. It can indeed be expected that primary attack occurs on the fluorine atoms, since the electron-rich areas of the SF_5^- anion are certainly the fluorine atoms and not the sulfur atom. But the backward reaction seems to work: SF_5-containing organic fluoroanions undergo an SF_4 elimination reaction, as will be discussed later.[77]

10.3 PHYSICAL CHEMISTRY OF ORGANIC SF_5 COMPOUNDS

The structure of all SF_5-containing compounds is derived from the octahedral environment around sulfur, which is only slightly perturbed.

Table 10.1 shows a listing of a few chosen results of structure determinations by single crystal X-ray and gaseous electron diffraction. In almost all cases, the axial (*trans* to carbon) bonded S—F bond is shorter than the four equatorial S—F bonds, though the difference is small, and for sterical considerations meaningless. The S—C bond length varies; it is shorter (even below 1.73 Å) if the carbon atom is triple-bonded ($HC{\equiv}C{-}SF_5$, $F_3S{\equiv}C{-}SF_5$), intermediate if bonded to double-bonded carbon atoms, and long in case of single-bonded carbon atoms. Especially if the carbon atoms are also fluorinated, values above 1.90 Å for C—S are reached. These trends are in full accord with conventional theories (e.g., hybridization).

Bond angles are close to 90°: almost always the $F_{ax}{-}S{-}F_{eq}$ angles are a little below 90°, the $C{-}S{-}F_{eq}$ angles a little larger than 90°. It is important to know that the F atoms will not give way very much in cases of sterical crowding. The rigidity of the bonds within the SF_5 group is very characteristic, indicating a very bulky group. If one considers the S—C—S angles in $F_5S{-}CH_2{-}SF_5$ (126.1(8)°) and $F_5S{-}CF_2{-}SF_5$ (124.3(8)°), where the slightly smaller angle in the latter case is a result of the slightly longer C—S bonds, the sterical interaction between the two SF_5 groups is impressive. For comparison, the central C—C—C angle in Di-*t*-butylmethane is 128°.[93] Therefore we conclude that SF_5 is just a little less bulky that the *t*-butyl groups. For electronic reasons SF_5 is often compared to CF_3. Sterically, CF_3 groups are not particularly bulky. Using definitions similar to those above, not much material is known here: $CF_3{-}CFI{-}CF_3$ has a C—C—C angle of 113.2°, but the influence of the large iodine atom should be considered.[94] In $CF_3{-}CO{-}CF_3$ the angle is 121.4°.[95] A search in the Cambridge crystal structure data file for $F_3C{-}C{-}CF_3$ fragments resulted in a minimum C—C—C bond angle of 95.6°, and a maximum of 124.7° for thc perfluoro–iso–propyl silver.[96,97]

The bulkiness of two SF_5 groups is visible in the sterical hindrance of the —CF_3, —CF_2Cl, or —COF rotation in $(SF_5)_2CH{-}CF_3$, $(SF_5)_2CH{-}CF_2Cl$ and $(SF_5)_2CH{-}COF$. The low-temperature limit nuclear magnetic resonance (NMR) spectra of $(SF_5)_2CH{-}CF_2Cl$ are observed at −60°C, resulting in two rotational isomers. The barrier towards free rotation of the CF_3 group in $(SF_5)_2CH{-}CF_3$ was found to be 44.8 kJ mol^{-1}, probably the highest value known so far for any CF_3 group.[98] In no case has it been possible to freeze out the rotation of the C—S bond, which would result in chemically and magnetically different equatorial fluorine atoms on sulfur. In spite of the fact that gas-phase structures of $F_5S{-}CH_2{-}SF_5$ and $F_5S{-}CF_2{-}SF_5$ show distinct orientations of the equatorial fluorine atoms, the close to cylindrical fourfold symmetry of the group prevents detectability on the NMR time scale.[98] Due to the bulkiness of the SF_5 groups,

TABLE 10.1 Structural Data of SF_5-Containing Molecules

Compound	S—C (Å)	S—F (*cis*) (Å)	S—F (*trans*) (Å)	Other Important Bond Lengths (Å) or Angles (deg.)	F(*cis*)—S—F(*trans*) (deg.)	Method, Literature
$(F_5S)_2CH_2$	1.832(6)	1.570(4)	1.578(13)	S—C—S 126.1(8)	88.2(2)	ED[78]
$(F_5S)_2CF_2$	1.908(7)	1.566(4)	1.544(10)	S—C—S 124.3(7) F—C—F 103.0(15) S—C—F 106.9(5) 110.1(29)	89.6(2)	ED[79]
$F_5S—CH_3$	1.793(8)					ED[78]
$F_5S—CF_3$	1.887(8)	1.572(2)	1.562(7)	C—F 1.319(2) F—C—F 108.7(5)	89.5(2)	ED, MW[80]
$F_5S—CF{=}CF_2$	1.784(8)	1.571(3)	1.537(11)	S—C—C 126.5(8)	88.9(3)	ED[78]
$F_5S—C{\equiv}CH$	1.727(5)	1.574(2)	1.576(2)	C≡C 1.202(4)	88.9(2)	ED[81]
$F_5S—C{\equiv}N$	1.765(5)	1.566(6)	1.558(6)	C≡N 1.152(5)	90.1(2)	ED, MW[82]
$F_5S—C{\equiv}N$	1.757(2)	1.524–1.548(2)	1.550(2)	C≡N 1.141(3)	90.0(1)	X-ray[83]
$F_5S—C{\equiv}SF_3$	1.699(12)	1.559(2) mean		C≡S 1.401(9) S—C—S 159(3) F—S—F 93.9(6) C—S—F 122.4(6)	88.6(3)	ED[84]
$F_5S—C{\equiv}SF_3$	1.682(4)			C≡S 1.392(4) S—C—S 180 by lattice symmetry S—F 1.534(3) C—S—F 122.7(2)	93.5(1)	X-ray[85]
$F_5S—C({=}SF_2)—C(F){=}N—C_6F_5$	1.797(5)	1.566–1.570(3)	1.573(3)	C=S 1.601(5)		X-ray[86]
$(\eta^2\text{-}F_5S—C{\equiv}C—SF_5)Co_2(CO)_6$	1.780(5) 1.784(6)	1.50–1.58		Co—C 1.916-1.931(4) C≡C 1.364(8)		X-ray[87]

TABLE 10.1 Structural Data of SF_5-Containing Molecules (*continued*)

Compound	S—C (Å)	S—F (*cis*) (Å)	S—F (*trans*) (Å)	Other Important Bond Lengths (Å) or Angles (deg.)	F(*cis*)—S—F(*trans*) (deg.)	Method, Literature
$F_5S—CH_2$-$Co(CO)_4$	1.816(7)	1.57–159		Co—C 2.026(6) Co-C-S 122.4(3)		X-ray[87]
$F_5S—CH(SO_2F)—C(O)—NEt_2$	1.856(5)	1.532—1.553(4)	1.568(3)	C—S (SO_2F) 1.801(5)	87.7–88.9(2)	X-ray[88]
$[F_5S—CF{=}CFPMe_3]^+$ BF_4^-	1.818(7)	1.557–1.573(6)	1.560(5)	C=C 1.321(11) C—P 1.764–1.807(7) S—C—C 126.7(6)	88.6(3)–90.6(3)	X-ray[89]
$(F_5S—CH_2CHO)_3$	1.789(6)	1.552–1.581(4)	1.574(6)	C-O 1.387(6), 1.435(4)	87.2 -88.5(3)	X-ray[90]
$[Et_3NH][F_5S—C(SO_2F)C(O)OCHMe_2]$	1.770(4)	157.1–158.4(3)	1.59.5(3)	$FO_2S—C$ 1.679(4)		X-ray[91]
$(F_5S—CH_2)_2AsFO$ AsF_5				As—O 1.647(5), 1.840(5) As-C—1.905(8), 1.884(8)		X-ray[92]
$F_5S—CH(CF_3)—AsF_4$				As—C 1.970(7) As—F(ax) 1.726(5), 1.737(5) As—F(eq.) 1.668(5), 1.683(5)		X-ray[92]

the electron paramagnetic resonance (EPR) spectrum of the $(SF_5)_2C{-}CF_3$ radical again shows the effects of a hindered rotation around the C—C single bond.[99]

Much less is known about the electronic properties of an organic SF_5 group. From C-1s X-ray photoelectron spectra on $F_3C{-}C{\equiv}C{-}CF_3$, $F_3C{-}C{\equiv}C{-}SF_5$, and $F_5S{-}C{\equiv}C{-}SF_5$ it is found that the alkyne carbon atoms all have the same ionization potentials, so it can be concluded that the electron-withdrawing effects of the CF_3 and SF_5 ligands are very similar.[100] Semiempirical calculations indicate a slightly larger electron-withdrawing effect for CF_3, however.[100] In summary it may be stated that the SF_5 group is (almost) as bulky as a *t*-butyl group, and (almost) as electron-withdrawing as a CF_3 group.

One major advantage of all SF_5-containing compounds is that they can be detected easily by means of ^{19}F NMR spectroscopy.[101,102] The inevitable ab_4-type spectra are often very characteristic, and are often resolved into second-order splitting. Only in cases when the chemical shift of axial and the four basal fluorine atoms coincide accidentally, does the ^{19}F NMR spectra degenerate into a single line; see Fig. 10.1. With the advent of high-field NMR spectroscopy this case has become very rare. Characterization of any given SF_5 compounds nowadays includes the ^{19}F NMR data. Although the coupling constant $^2J(F_{ax}{-}F_{eq})$ is always in the vicinity of 145 Hz, chemical shifts $\delta(F_{ax})$ and $\delta(F_{eq})$ vary. Chemical shifts are positive[103] with respect to $CFCl_3$, and $\delta(F_{ax})$ occurs in general at higher frequencies than $\delta(F_{eq})$. This sequence is only inverted when a strong anisotropic paramagnetic influence exists, as in F_5S-alkynes. Couplings between equatorial fluorine atoms and protons or fluorine atoms of the organic parts are regularly observed, and are helpful in structural investigations. Coupling constants of the same atoms to the axial fluorine atoms are usually smaller by a factor of 10, or are not resolved. Chemical shifts of $\delta(F_{ax})$ and $\delta(F_{eq})$ have been correlated to electronegativity scales, or to Hammett σ-factors and Taft's inductive and resonance parameters in aromatic systems.[102] It is interesting to note that the F_5S and F_3C groups obviously have a similar influence on the ^{31}P chemical shifts in the phosphonates $XCF{=}CFP(O)(OR)_2$ ($X = F_5S$ and F_3C).[104] There is no information available about what kind of substituent in aromatic compounds an SF_5 group might be. However, from the assumed electronegativity and the nonexistence of nonbonding electron pairs on sulfur, one can deduce that an SF_5 group would exhibit an -I, -M effect. Also, little is known about the biological activity of organic SF_5 compounds. However, the extreme toxicity of S_2F_{10} as compared to the biological inertness of SF_6 sets the frame. $H_2C{=}CH{-}SF_5$ if breathed is toxic, while $Cl{-}CH_2{-}CH_2{-}SF_5$ or $Cl{-}(CF_2)_n{-}SF_5$ (n = 2,4,6) are not.[105] $F_5S{-}CH_2{-}COOH$ smells like isobutyric acid, so the human nose mistakenly identifies the F_5S group as a branched aliphatic group.

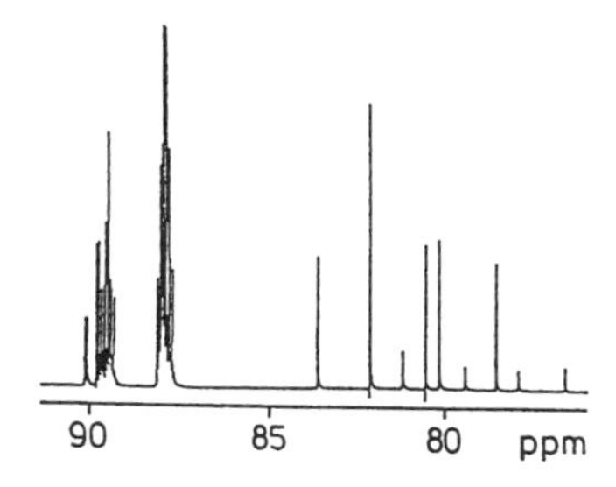

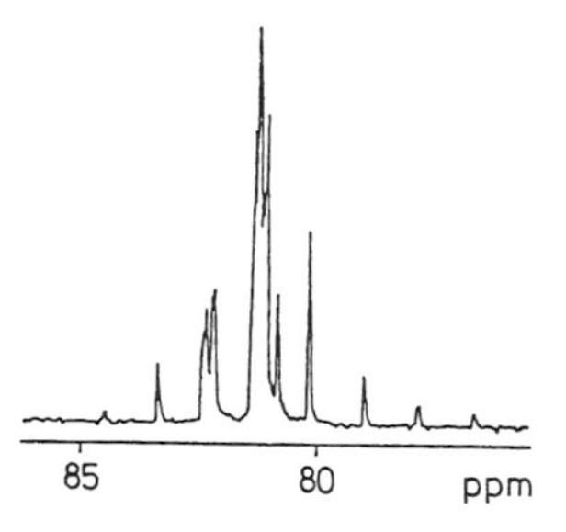

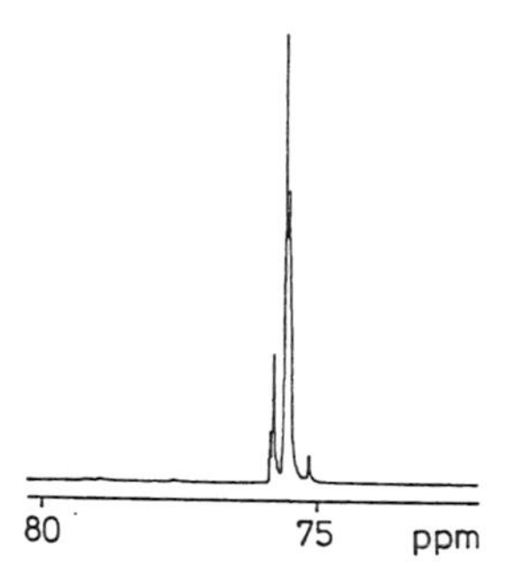

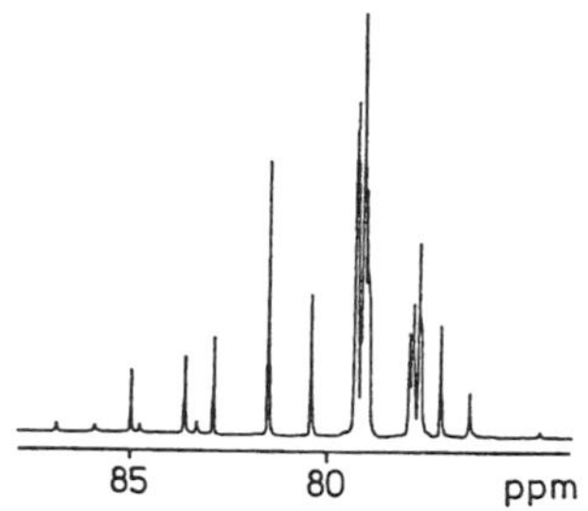

Figure 10.1 ^{19}F NMR spectra of the ketenes (from top to bottom) $F_5S—HC{=}C{=}O$, $F_5S—BrC{=}C{=}O$, $F_5S—ClC{=}C{=}O$ and $F_5S—(H_3C)C{=}C{=}O$. The *a* part of the ab_4 spectra seems to move from right to left (low to high frequencies) in this sequence. (Taken from Ref. 136 with permission of the copyright owner.)

10.4 CHEMICAL REACTIONS AT THE SF_5 GROUP

SF_5-containing compounds are in general quite stable: CF_3—SF_5 is an extreme case. There is no thermal decomposition up to 450°C[106] and no reaction with NaOH at ambient temperature.[3] With alkali metals, rapid reaction occurs only at dull red heat.[3] Of course, basic attack would increase the coordination number beyond 6 in the transition state, which is very unlikely. On the contrary, attack by Lewis acids is observed with varying ease: CH_3—SF_5 reacts with SbF_5/SO_2 under formation of the CH_3—SF_4^+ ion; under similar conditions CH_2C=CH—SF_4^+, C_2H_5—SF_4^+, and a few others are formed.[50,107] HC≡C—SF_4^+ is only formed in the less basic solvent SO_2FCl. F_5S—C≡C—SF_5 does not react even with neat SbF_5 at room temperature. Fluoride abstraction is also the principal reaction for preparing R_2C=SF_4 and R—C≡SF_3 compounds.[74] α-lithiated SF_5 compounds like Li—CH_2—SF_5 will undergo spontaneous loss of LiF and formation of H_2C=SF_4 even at −70°C.[74] The ketene F_5S—CH=C=O isomerizes to F_4S=CH—COF at 270–290°C.[108] CF_2=CH—SF_5 is in equilibrium with F_3C—CH=SF_4 at 100°C.[109] CF_3—CH_2—SF_5 and F_5S—CH_2—SF_5 will even lose 2 mol of HF if passed over KOH at elevated temperatures, forming CF_3—C≡SF_3[74,110] and F_5S—C≡SF_3.[85] In certain cases, mild bases such as $(C_2H_5)_3N{\cdot}BF_3$ promote HF elimination.[111]

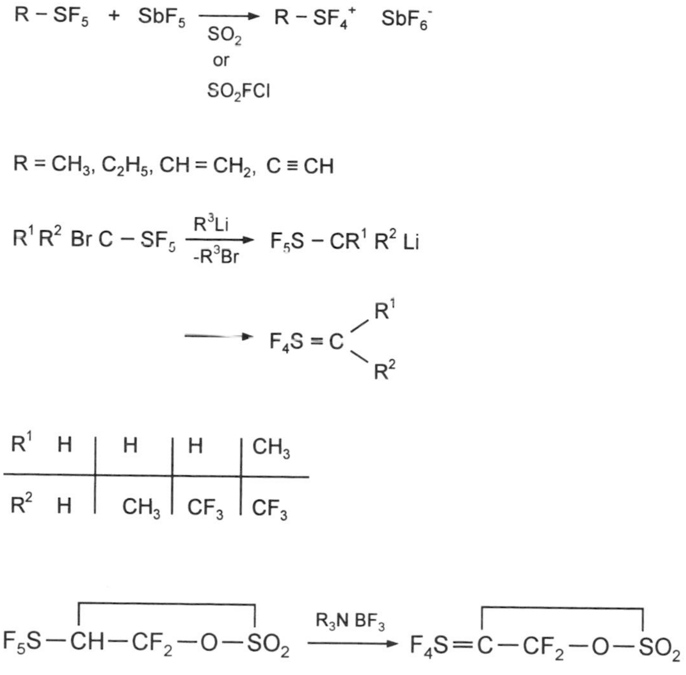

$$R-SF_5 + SbF_5 \xrightarrow[SO_2 \text{ or } SO_2FCl]{} R-SF_4^+ \; SbF_6^-$$

$R = CH_3, C_2H_5, CH=CH_2, C\equiv CH$

$$R^1R^2BrC-SF_5 \xrightarrow[-R^3Br]{R^3Li} F_5S-CR^1R^2Li \longrightarrow F_4S=CR^1R^2$$

R^1	H	H	H	CH_3
R^2	H	CH_3	CF_3	CF_3

$$F_5S-CH-CF_2-O-SO_2 \xrightarrow{R_3N\,BF_3} F_4S=C-CF_2-O-SO_2$$

(ring: CH/C bonded to SO_2)

Scheme 10.3

In the case of F_5S—$CH(SO_2F)$—$COOCH_3$ the same base produces $F_2S(O)$=$C(SO_2F)$—COF.[112] The first step is also thought to be a simple HF elimination, followed by oxygen transfer in a four-membered ring transient species.

Loss of an entire SF_5 group by basic attack (formally HSF_5 elimination) is occasionally observed for F_5S—CHBr—CH_2—SF_5, yielding F_5S—CBr=CH_2.[67] Similarly, F_5S—$C(CF_3)_2^-$ decomposes slowly into SF_5^- and the carbene dimer $(F_3C)_2C$=$C(CF_3)_2$.[77,99]

10.4.1 SF_5-Substituted Alkanes

In contrast to the numerous compounds containing halogenated alkyl groups, which are often the primary reaction products of SF_5Cl and SF_5Br addition reactions (and have in part been discussed in Section 10.2.4), there exist only CH_3—SF_5 and C_2H_5—SF_5. The first has been obtained by reduction of Br—CH_2—SF_5 using $Zn/H^+/H_2O$ or by HF addition to H_2C=SF_4.[46,47] C_2H_5—SF_5 is a minor product of the electrochemical fluorination of C_2H_5—SH.[32] It can be prepared in preparative amounts by the reduction of Tos—O—CH_2—CH_2—SF_5 with $LiAlH_4$.[50] No reactions at the alkyl group have been reported. In fact these compounds are very stable, for instance C_2H_5—SF_5 can be refluxed with *n*-C_4H_9—Li.[50]

$$F_5S-CH_2Br \xrightarrow{Zn\,/\,H^+} F_5S-CH_3$$

$$F_5S-CH_2-CH_2-OTos \xrightarrow{LiAlH_4} F_5S-C_2H_5$$

Scheme 10.4

10.4.2 SF_5-Substituted Alkenes

Dehydrohalogenation of halogenated alkanes is the dominant reaction for SF_5-alkene preparation.[39,43,50,52,54,56,57,64,68,98,109,113] Elimination is always enforced by basic conditions (OH^-, CO_3^{2-}). The yields are often modest, since loss of a whole SF_5 group (formally elimination of HSF_5 = $HF + SF_4$) is competing with the dehydrohalogenation.

Eliminations will preferentially yield alkenes with the double bond in α-position to the SF_5 group.[39,50,52] (For a single exception see literature.[39]) The reaction is thought to be largely a *syn* elimination, since in the one case investigated *erythro* and *threo* diastereomers gave alkenes with *cis–trans* ratios of 4:1 and 0.4:1.[72] Alternatively, SF_5 alkenes have often be prepared by adding SF_5Cl or SF_5Br on alkynes, as has been described above.[39,56,57,58,64,68,69,70,71,114] Only one of the two possible products (*cis* versus *trans* addition) is

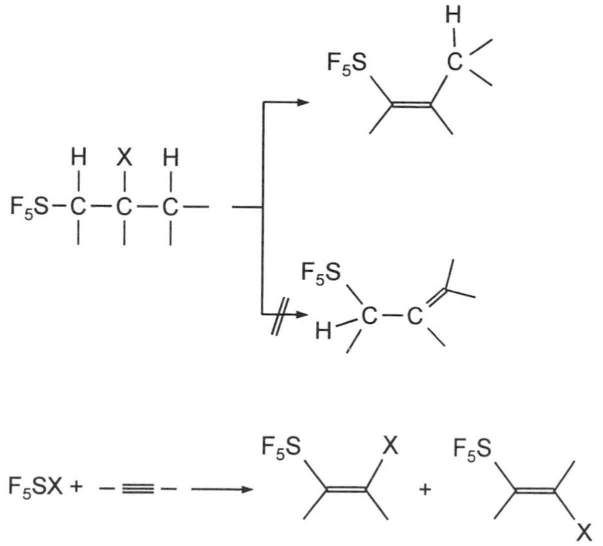

Scheme 10.5

observed.[56,57,65,69,114] In the case of SF_5—CH=CHCl, this has been shown to be the *E* isomer.[56] Addition of SF_5Br to CF_3—C≡CH gives both isomers, however.[70]

Elimination of H_2O by means of P_2O_5, and of C_2H_4 in a retro-Diels–Alder reaction has been used for generating cyclopentadienyl-SF_5 compounds.[52] The Diels–Alder reaction between SF_5—C≡CH and butadienes affords 1,4-cyclohexadienes.[56]

10.4.3 Reactions of SF_5-Substituted Alkenes

1, 2 additions across the double bond of SF_5-substituted alkenes have been performed with HF (via KF/formamide),[43] HBr,[58] COF_2 (with CsF catalyst),[115] CF_3OOCF_3,[116] F_5SOOSF_5,[116] BrF,[117] F_3S—CF_3 and F_3S—CFCl—CF_3 (with CsF catalyst),[118] NOF_3 (with AsF_5 catalyst),[119] NOF (by means of a NO/NOF_3 mixture),[120] SO_3,[64,121] IF (by means of a I_2/IF_5 mixture and AlI_3 catalyst),[122] SF_4 (with CsF catalyst),[123] Br_2,[56,68,114] Cl_2,[39,113] CF_3OCl,[113] F^- (by means of CsF),[77,99,123] SF_5Cl,[98] and H_2O.[98,124] In most cases an anionic mechanism is to be assumed; addition of an anionic group (often F^-) occurs first, especially in the cases of highly fluorinated alkenes like SF_5—CF=CF_2. The anion SF_5—$C(CF_3)_2^-$ has been observed by NMR methods.[77] A radical mechanism is certain for additions like CF_3—O—O—CF_3 or SF_5Cl. F_5S—CH=CF_2 and F_5S—CF=CF_2 can be epoxidated with NaOCl under phase transfer conditions.[125] Ozonization of F_5S—CF=CF_2 gives F_5S—O—CF_2—COF with 67% yield.[126] NSCl (supplied in form of $(NSCl)_3$) will add across the C—C bond in F_5S—CF=CF_2, forming an SF_5-substituted aziridine.[127] SF_5—

CH═CF_2 is in equilibrium with F_4S═CH—CF_3.[109] This observation is so far unique and makes it possible to establish the C═S double-bond energy.

Elimination reactions on SF_5-alkenes afford SF_5-alkynes and will be discussed there. Phosphorous III compounds like $(CH_3)_3P$[89,128] or $(C_2H_5O)_2POSi(CH_3)_3$[104] insert into the CF bond *trans* to the SF_5 group in F_5S—CF═CF_2. The $P(CH_3)_3$ can then be cleaved off by means of I_2 under formation of SF_5—CF═CF—I.[89,128] Thermal decomposition of the phosphorane or hydrolysis yields F_5S—CF═CFH.[128] The chlorine atom can be replaced by OCH_3 in SF_5CH═CHCl by using $Na^+OCH_3^-$.[124]

Finally, 1- and 2-SF_5-cyclopentadiene (the 5-SF_5-cyclopentadiene is not observed), prepared by two different adition–elimination routes, will serve as classical cyclopentadienyl ligands in organometallic chemistry (see below). They also dimerize reversibly to a single dimer, out of 32 theoretically possible isomers.[52] This finding can only be explained by a combination of steric and electronic effects of the large SF_5 groups.

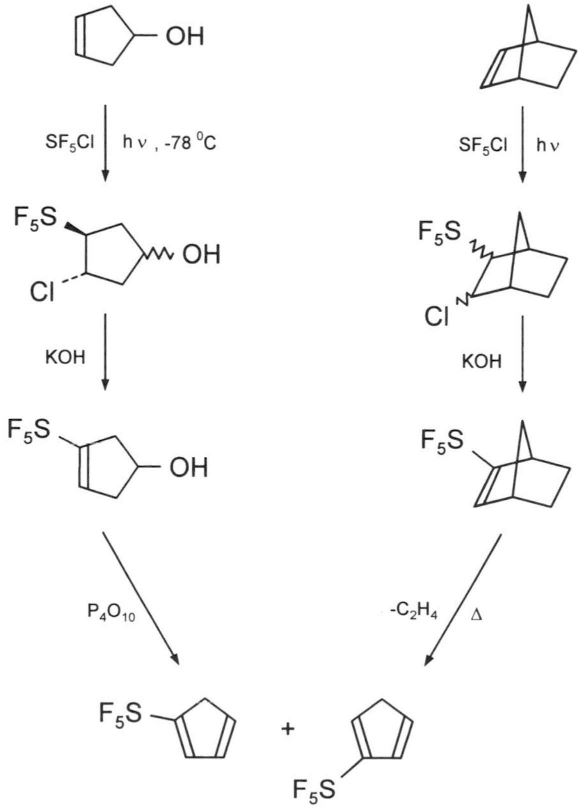

Scheme 10.6

The anions $SF_5—C(CF_3)_2^-$ and $(SF_5)_2C—CF_3^-$, made by reaction of $SF_5—C(CF_3)CF_2$ or $(SF_5)_2C{=}CF_2$ and CsF/CH_3CN are stable at room temperature. Furthermore, they can be electrochemically oxidized to the free radicals $SF_5—C(CF_3)_2$ and $(SF_5)_2C—CF_3$.[99] $SF_5—CF{=}CF_2$ with CsF will form a dimer of unknown structure.[115]

10.4.4 SF_5-Substituted Alkynes

In every case, SF_5-substituted alkynes are prepared by elimination from SF_5-alkene precursors. The total number of such alkynes is quite limited: $F_5S—C{\equiv}C—R$ (R = —H),[56,57,62,68] CH_3,[39,57] —CF_3,[69] —SF_5,[69] —$Si(CH_3)_3$,[129] —Cl,[129] —Br,[129] I,[129] —HgC_6H_5,[129] —MgI,[129], and —Li.[129] Butadiyne adds SF_5Br once and twice, and complete dehydrobromination affords $SF_5—C{\equiv}C—C{\equiv}C—H$ and $SF_5—C{\equiv}C—C{\equiv}C—SF_5$.[71]

The SF_5 group shows in the ^{19}F NMR spectrum a characteristic inversion of the axial and equatorial chemical shifts, as compared to most other SF_5 compounds. This is certainly a result of the grossly anisotropic electron distribution of the $C{\equiv}C$ triple bond.[62,68] Similar observations can be made for cumulated compounds like SF_5-substituted ketenes. Some compounds have been added to the triple bond of the alkynes. Base-catalyzed addition of methanol gives the normal stereospecific Z-product.[56]

1,3-Cycloaddition of CH_2N_2 to $SF_5—C{\equiv}C—H$ gives two different pyrazoles.[56] Diels–Alder cyclization between $F_5S—C{\equiv}C—H$ and butadiene or 2,3-dimethylbutadiene give cyclohexadienes, which in turn have been dehydrogenerated to SF_5-substituted benzenes.[56]

The reactions of SF_5-alkynes in organometallic chemistry are important (see Section 10.4.9).

10.4.5 SF_5-Substituted Benzenes

It is a weakness of this entire field of chemistry that aromatic compounds are so difficult to prepare, and can be obtained only by routes that cannot be generalized. $C_6H_5—SF_5$ has been made in low yields by fluorination of $C_6H_5—S—S—C_6H_5$ with AgF_2,[18] or by Diels–Alder reaction of $F_5S—C{\equiv}C—H$ to butadiene and subsequent catalytic dehydrogenation.[56] The former reaction has been extended mostly to NO_2-substituted benzenes.[18,19,20] Very recently, the fluorination of certain aromatic thiols or disulfides (with nitro or fluorine substituents) with nitrogen-diluted fluorine in acetonitrile was reported.[130] The nitro groups can be converted by standard reactions into NH_2, and OH groups. The only other Diels–Alder reaction reported is with 2,3-dimethylbutadiene.[56] The deactivation of the $C{\equiv}C$ triple bond by sterical or electronic effects makes this reaction unattractive for general use. $SF_5—C{\equiv}C—H$ can be cyclized to 1,3,5-tris-pentafluoro-λ^6-sulfanyl benzene by UV irradiation in low (19%) yield.[129] The $Co_2(CO)_8$-mediated cyclisation of $SF_5—C{\equiv}C—H$ gave a good yield (88%) of 1,2,4-trispentafluoro-λ^6-sulfanyl benzene.[129]

These reactions seem to be limited to H—C≡C—SF_5, since further substitution will sterically hinder the formation of the necessary cobalt complex intermediates.[129,131]

10.4.6 SF_5-Alcohols, -Aldehydes, and -Carboxylic Acids

If the CF_3 group and the SF_5 group are considered as somewhat similar, then the most simple alcohols would be CF_3OH and F_5SOH. Both compounds exist, are alike in their instability towards loss of HF, and are rather acidic.[132–134] The stable CF_3—CH_2—OH has so far no counterpart SF_5—CH_2—OH, but there is no obvious reason why this compound should not exist. SF_5—CH_2—CH_2—OH is the first and only simple alcohol known in this series,[50] besides an ene-ol F_5S—CH=CH—CH_2OH.[48]

F_5S-COF was made from F_5SOF or S_2F_{10} in low yields, and is so far not fully characterized.[33,37] F_5S—CO—CF_3, made by oxidation of F_5S—CHBr—CF_3, is stable up to 80°C, then decomposes to CF_3—COF and SF_4.[117] The simplest aldehyde F_5S—CH_2—CHO has been prepared by different routes.[45,90,124]

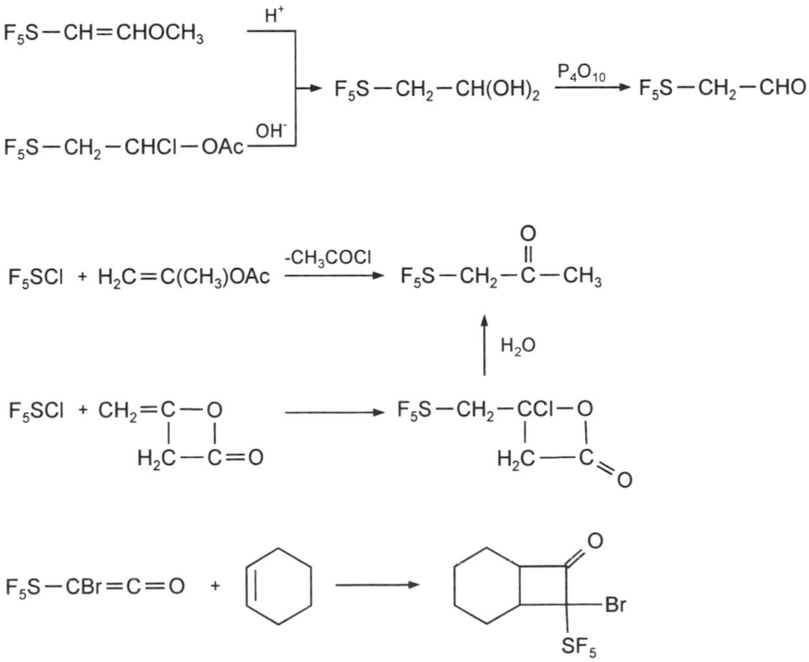

$$F_5S-CH=CHOCH_3 \xrightarrow{H^+} F_5S-CH_2-CH(OH)_2 \xrightarrow{P_4O_{10}} F_5S-CH_2-CHO$$

$$F_5S-CH_2-CHCl-OAc \xrightarrow{OH^-} F_5S-CH_2-CH(OH)_2$$

$$F_5SCl + H_2C=C(CH_3)OAc \xrightarrow{-CH_3COCl} F_5S-CH_2-C(=O)-CH_3$$

$$F_5SCl + CH_2=C-O\ (\text{ring: } H_2C-C=O) \longrightarrow F_5S-CH_2-CCl-O\ (\text{ring: } H_2C-C=O) \xrightarrow{H_2O} F_5S-CH_2-C(=O)-CH_3$$

$$F_5S-CBr=C=O + \text{cyclohexene} \longrightarrow \text{bicyclic ketone with Br and } SF_5$$

Scheme 10.7

It is noteworthy that the aldehyde exists not only as such but also in an hydrated form, and as a trimer (trioxolane).[90] Acetals of this aldehyde are reported also.[48,90]

SF_5-substituted acetone is formed directly from F_5SCl and $H_2C{=}C(CH_3)OAc$,[45] or by hydrolysis of the product obtained by F_5SCl addition to diketene.[47] Aromatic (Friedel–Crafts) acylation with Cl—CO—CH_2—SF_5 affords various SF_5-acetophenones.[135] A cyclic ketone is obtained from 2+2 cycloaddition reaction of F_5S—CBr=C=O with cyclohexene.[136]

The most important compound in this chapter is certainly the acid F_5S—CH_2—COOH and its numerous derivatives. The first preparation from monomeric ketene and SF_5Cl is still the best method.[45,46,47,137] The theme can be varied.[58,138]

$$SF_5Cl + CH_2{=}C{=}O \longrightarrow F_5S{-}CH_2{-}COCl \xrightarrow{H_2O} F_5S{-}CH_2{-}COOH$$

$$SF_5Cl + H_3C{-}C{\equiv}C{-}OR \longrightarrow F_5S{-}C(CH_3){=}C(Cl)OR$$

$$\longrightarrow F_5S{-}CH(CH_3){-}COCl \longrightarrow F_5S{-}C(H)CH_3{-}COOH$$

$$SF_5Cl + H{-}C{\equiv}C{-}OR \longrightarrow F_5S{-}CH{=}C(Cl)OR$$

$$\longrightarrow F_5S{-}C(H)Br{-}CClBrOR \longrightarrow F_5S{-}C(H)Br{-}COCl$$

$$\longrightarrow F_5S{-}C(H)Br{-}COOH$$

Scheme 10.8

Perfluorinated acids $SF_5(CF_2)_n$—COOH, obtained via direct fluorination of sulfur compounds, are even commercially available. Clearly the most simple derivatives of these acids are the ketenes.[108,136,139]

The higher-functionalized ketene $F_5S(FSO_2)C{=}C{=}O$ is easily obtained from the sultone F_5S—CH_2—CF_2—O—SO_2.[111,112,140]

The stability of these ketenes allows the multitude of reactions known for ketenes, which involve in most cases nucleophilic attacks on the carbonyl atom,[88,111,136,139–142] including the dimerization.[108] One interesting intramolecular reaction is the formation of $F_4S{=}CH$—COF from F_5S—CH=C=O.[108]

$$F_5S{-}CH{=}C{=}O \rightarrow F_4S{=}CH{-}COF \tag{10.1}$$

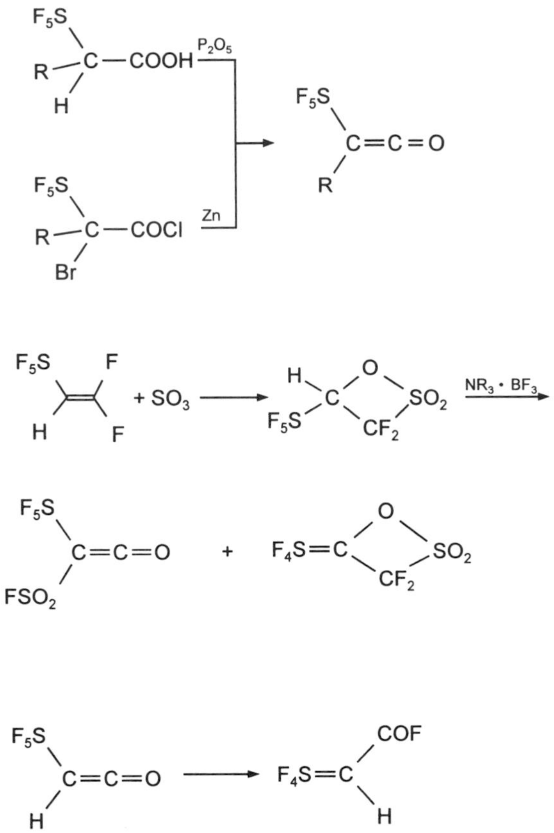

Scheme 10.9

Another alkylidene sulfurtetrafluoride is observed in the preparation sequence of $F_5S(FSO_2)C{=}C{=}O$ (Scheme 10.9).[111]

The derivatives of SF_5-carboxylic acids are numerous; there are among others: esters,[45,48,91,140,141] acyl halides,[45–47,50,58,85,98,107,111,136–138,142,143] oxadiazoles,[143] hydrazides,[143] and amides.[45,88,140,143]

Acids or acid derivatives serve for the preparation of ketones,[135,136] trifluoromethylated compounds[138,144] (by means of SF_4), bromides (by means of the Hundsdiecker reaction[46,47,58,137]), alcohols by reduction,[50,107] and of $(F_5S)_2CH_2$ by decarboxylation.[98]

10.4.7 SF_5-Substituted Nitrogen Compounds

F_5S—CN is a fascinating molecule. It was first claimed to be formed in the direct fluorination of CH_3—S—CN. Later it was found that only the isomer

CF_3—N=SF_2 is formed, which is readily available by reaction of SF_4 and varying substrates. If $(SCN)_2$ is reacted with diluted F_2 at −20°, however, SF_5CN is obtained.[6] Also F_3S—CN can be directly fluorinated to SF_5—CN.[7] Even the isocyanide SF_5NC has become known,[8] which is, of course, outside the topic of this chapter. Its structure has been elucidated by low-temperature X-ray diffraction.[145] The known chemistry of SF_5—CN is scarce thus far. A Lewis-acid–base adduct with AsF_5 has been characterized spectroscopically.[7] Reaction of SF_5—CN with $[SNS]AsF_6$ yields crystalline F_5S—$CNSNS^+AsF_6^-$ of known structure. This can be reduced to the F_5S—CNSNS radical. The structure of SF_5CN has been established by electron diffraction[82] and X-ray crystallography.[83] If amide derivatives of SF_5-carboxylic acids are not considered here, other SF_5—C—N compounds are scarce. The NOF addition of SF_5-alkenes yielding nitroso compounds has been reported above.[120] SF_5—CH=CH_2 will add N_2F_4 across the double bond; subsequent HF elimination affords *N*-fluoroiminonitriles:[146]

$$F_5S\text{-CH=CH}_2 + N_2F_4 \longrightarrow F_5S\text{-CH(NF}_2\text{)-CH}_2\text{(NF}_2\text{)} \xrightarrow[-3HF]{NaF} \text{syn. anti} \quad F_5S\text{-C(C}\equiv\text{N)=N-F}$$

$$(CF_2)_n\text{-cyclic } N\text{-CF=CF}_2 + SF_5Cl \longrightarrow (CF_2)_n\text{-cyclic } N\text{-CF(SF}_5\text{)-CF}_2\text{Cl}$$

Scheme 10.10

The successful addition of SF_5Cl across double bonds has been extended in one case to vinyl amines.[51] Unstable CF_3—NC will add SF_5Br in an α-addition.[73]

C_6F_5—NC and CF_3—NC will couple with SF_5—C≡SF_3, followed by a 1,3 fluorine shift, giving R_f—N=CF—C(SF_5)=SF_2[86]. Finally, CH_3—N≡C adds to SF_5—CBr=C=O, forming a cyclopentanetrione derivative.[136]

10.4.8 SF_5-Substituted Organic Sulfur Compounds

Here it becomes obvious that the SF_5 group can function as a particularly good stabilizing group for otherwise unstable molecular fragments. This should be true for other elemental compositions also. Aliphatic thioketons are often

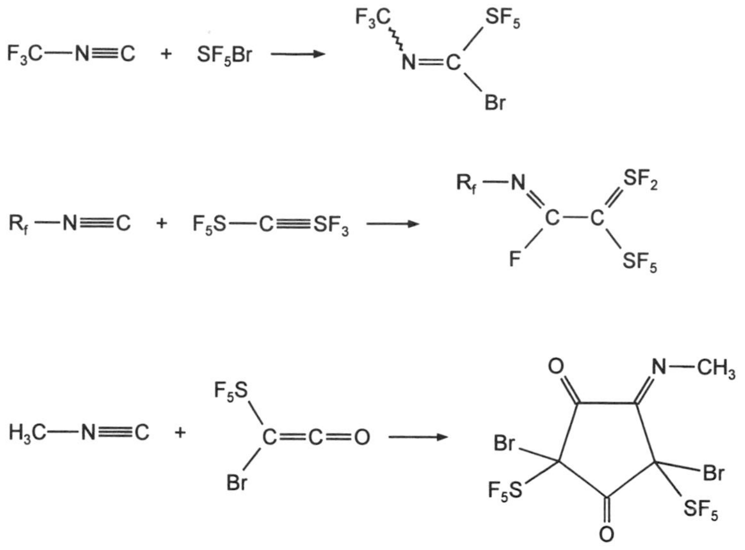

$F_3C{-}N{\equiv}C + SF_5Br \longrightarrow (F_3C)N{=}C(SF_5)Br$

$R_f{-}N{\equiv}C + F_5S{-}C{\equiv}SF_3 \longrightarrow R_f{-}N{=}C(F){-}C(SF_5){=}SF_2$

$H_3C{-}N{\equiv}C + (F_5S)(Br)C{=}C{=}O \longrightarrow$ cyclopentane ring: $C(=O)$, $C(=N{-}CH_3)$, $C(Br)(SF_5)$, $C(=O)$, $C(Br)(F_5S)$

Scheme 10.11

unstable toward dimerization, as is $(CF_3)_2C{=}S$. $SF_5(CF_3)C{=}S$ can be made from $SF_5(CF_3)C{=}SF_2$ and BBr_3 or BI_3; it is a stable, violet liquid that can be stored indefinitely.[118] Dimerization occurs only upon irradiation. The parent molecule $SF_5(CF_3)C{=}SF_2$ was the first alkylidene sulfur difluoride prepared.[123] Due to the rigidity of the S=C double bond, *E* and *Z* isomers are observed.[123] The crystal structure of $C_6F_5{-}N{=}CF{-}C(SF_5){=}SF_2$ is known;[86] $SF_5(CF_3)C{=}SF_2$ hydrolyses to stable $SF_5(CF_3)C{=}S{=}O$, coming also as *E* and *Z* isomers.[123]

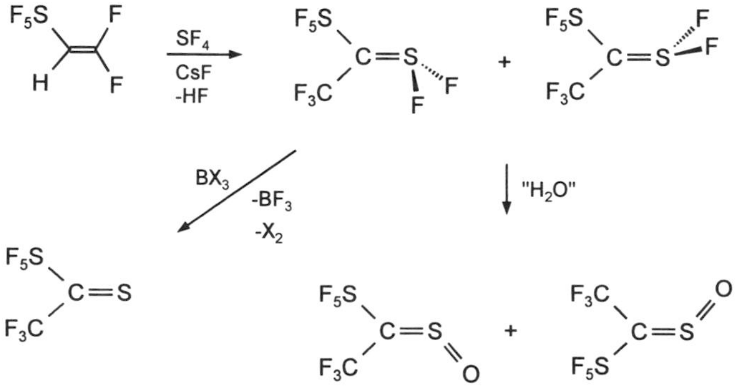

$(F_5S)HC{=}CF_2 \xrightarrow{SF_4,\ CsF,\ -HF} (F_5S)(F_3C)C{=}SF_2 + (F_5S)(F_3C)C{=}SF_2$ (two isomers)

$(F_5S)(F_3C)C{=}SF_2 \xrightarrow{BX_3,\ -BF_3,\ -X_2} (F_5S)(F_3C)C{=}S$

$(F_5S)(F_3C)C{=}SF_2 \xrightarrow{"H_2O"} (F_5S)(F_3C)C{=}S{=}O + (F_3C)(F_5S)C{=}S{=}O$

Scheme 10.12

$F_5S{-}C{\equiv}SF_3$ is the second compound containing the $-C{\equiv}SF_3$ group, made in a multistep procedure involving two SF_5Cl addition across double bonds.[84,85] It is bent at the carbon atom in the gaseous state,[84] but linear in the solid state.[85] Sterical crowding in $(SF_5)_2CH{-}CF_3$ results in a high barrier against rotation of the CF_3 group.[98] The sultone, made by 1,2 addition reaction of SO_3 on $SF_5{-}CH{=}CF_2$ (Scheme 10.9), serves as an important building block for other interesting molecules, for instance $SF_5{-}CH_2{-}SO_3H$[121] or $SF_5(FSO_2)C{=}C{=}O$.[140,141] The perfluorinated analog to the former, $SF_5{-}CF_2{-}SO_2F$, is obtained from $SF_5{-}CHF{-}SO_2F$ after elemental fluorination.[147] $SF_5{-}CHF{-}SO_2F$ in turn is obtained from the sultone made from SO_3 and $SF_5{-}CF{=}CF_2$.[148]

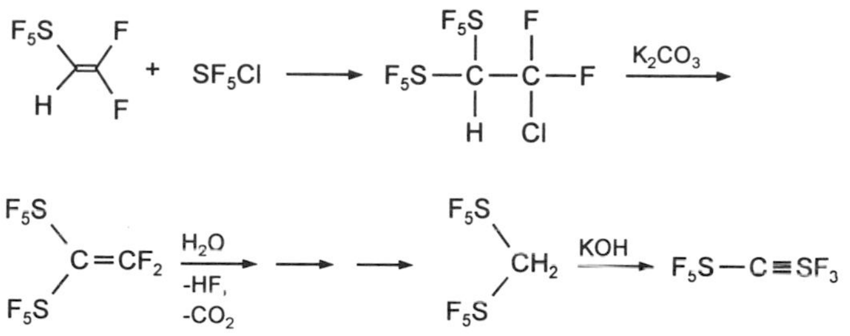

$F_5S(H)C{=}CF_2$ + SF_5Cl ⟶ $F_5S{-}C(SF_5)(H){-}CF(Cl){-}F$ $\xrightarrow{K_2CO_3}$

$(F_5S)_2C{=}CF_2$ $\xrightarrow[-HF,\ -CO_2]{H_2O}$ ⟶ ⟶ $(F_5S)_2CH_2$ $\xrightarrow{KOH}$ $F_5S{-}C{\equiv}SF_3$

Scheme 10.13

10.4.9 Organometallic Compounds Containing the λ^6-Pentafluorosulfanyl Group

The organometallic chemistry of SF_5-containing compounds is so far rather limited. Although several lithium alkyls with an SF_5 group in the α-position have been prepared by lithium-halogen exchange reactions at very low temperatures, they all tend to be unstable towards β-fluoride elimination, yielding alkylidene sulfurtetrafluoride compounds as the desired products.[46,58,74,110,114,137,138] Every attempt to use these organolithium compounds as reagents has failed thus far.[149] Attempts to prepare stable cesium derivatives by CsF addition across the C=C double bond of $F_5S{-}CH{=}CF_2$ results in isomerization, forming $F_4S{=}CH{-}CF_3$.[109] However, cesium salts containing the $F_5S{-}C(R_f){-}CF_3$ anions can be obtained by cesium fluoride addition to $F_5S{-}C(R_f){=}CF_2$, but no preparative use of these materials has been described.[77,99] In analogy to fluorinated alkenes,[96,150–152] silver fluoride adds across the double bond of $F_5S{-}CR{=}CF_2$ (R = H, F), forming moderately stable organosilver compounds which have been characterized by spectroscopic methods. A few reactions of these materials were studied including the thermal decomposition yielding the dimer $(F_5S{-}CF{-}CF_3)_2$.[153,154]

The instability of suitable reagents for transferring F_5S-alkyl moieties seems to be a limiting factor in the preparation of further organometallics containing F_5S-alkyl substituents. The only examples known so far were prepared by completely different pathways.

Arsenic pentafluoride addition across the S=C double bond of $F_4S{=}CH_2$ and $F_4S{=}CH{-}CF_3$ yields $F_5S{-}CH_2{-}AsF_4$ and $F_5S{-}CH(CF_3){-}AsF_4$ as the primary addition products. The structure of $F_5S{-}CH(CF_3){-}AsF_4$ was elucidated by X-ray crystallography. $F_5S{-}CH_2{-}AsF_4$ seemed to be unstable under the reaction conditions, since $[(F_5S{-}CH_2)_2As(F)O]{\cdot}AsF_5$ is the final product according to an X-ray crystal structure determination.[92]

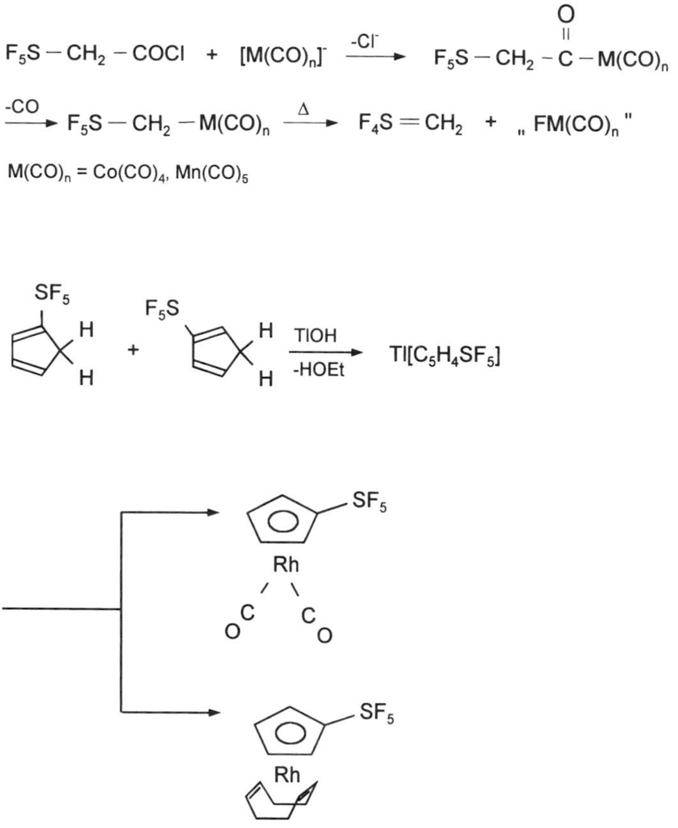

$$F_5S-CH_2-COCl \; + \; [M(CO)_n]^- \xrightarrow{-Cl^-} F_5S-CH_2-C(=O)-M(CO)_n$$

$$\xrightarrow{-CO} F_5S-CH_2-M(CO)_n \xrightarrow{\Delta} F_4S{=}CH_2 \; + \; \text{„} FM(CO)_n \text{"}$$

$M(CO)_n = Co(CO)_4$, $Mn(CO)_5$

SF5 H H + F5S H H $\xrightarrow[-HOEt]{TlOH}$ $Tl[C_5H_4SF_5]$

SF5 Rh C O C O

SF5 Rh

Scheme 10.14

The acid chloride, $F_5S{-}CH_2{-}COCl$, reacts with the metal carbonylates, $[Co(CO)_4]^-$ and $[Mn(CO)_5]^-$, giving $[F_5S{-}CH_2{-}Co(CO)_4]$ and $[F_5S{-}CH_2{-}Mn(CO)_5]$, respectively.[87] Whereas the intermediate acyl compound cannot be isolated for the cobalt derivative, it has been characterized spectroscopically for the mangenese compound. Both compounds decompose thermally, yielding $F_4S{=}CH_2$ by β-fluoride elimination.

No complex of an acyclic SF_5-substituted alkene has been reported. However, a mixture of two isomers of pentafluorosulfanylcylopentadiene was synthesized by two different multistep pathways. Deprotonation with potasium *t*-butanolate gives an unstable potasium cyclopentadienide which eliminates KF on warming above −55°C, yielding $C_5H_4{=}SF_4$. Due to the lower fluoride ion affinity of thallium(I), stable crystalline thallium(pentafluoro-λ^6-sulfanyl)-cyclopentadienylide can be isolated as THF adduct and studied by X-ray crystallography. By use of this thallium salt, two rhodium complexes, $[(\eta^5{=}C_5H_4{-}SF_5)Rh(CO)_2]$ and $[(\eta^5{-}C_5H_4{-}SF_5)Rh(cod)]$, could be obtained. The latter has been studied by X-ray crystallography demonstrating the η^5 coordination of the cyclopentadienyl ligand. Attempts to synthesize a derivative of the first row transition metals, including a ferrocene analog, failed.[58]

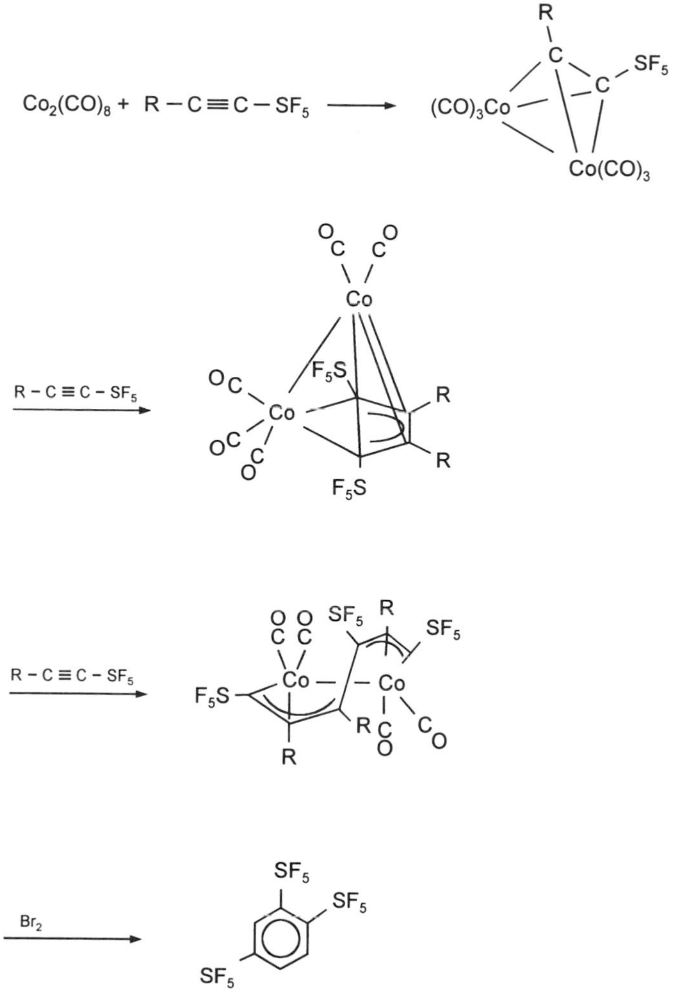

Scheme 10.15

The pentafluorosulfanyl-substituted alkynes are among the most studied organometallic derivatives. Lithium, magnesium, and mercury compounds containing the $C{\equiv}C{-}SF_5$ group can be prepared by conventional methods[129] and have been used for the preparation of the corresponding halogeno alkynes and $R_3Si{-}C{\equiv}C{-}SF_5$. The reaction of $F_5S{-}C{\equiv}C{-}CF_3$ with tetracarbonyl nickel gives a complex formulated as $[(CO)_2Ni(\eta^2\text{-}F_5S{-}C{\equiv}C{-}CF_3)]$ according to spectroscopic data. This is moderately stable in the presence of carbon monoxide but decomposes at ambient temperature, yielding SF_4, CO, and $F_3C{-}C{\equiv}C{-}C{\equiv}C{-}CF_3$ in the absence of CO. The carbonyl ligands can be easily exchanged by triphenylphosphine to yield the more stable complex $[(PPh_3)_2Ni(\eta^2\text{-}F_5S{-}C{\equiv}C{-}CF_3)]$.[155] Three different types of cobalt complexes have been prepared for several SF_5-substituted alkynes, $R{-}C{\equiv}C{-}SF_5$ ($R = H$, CF_3, C_6H_5, $SiMe_3$, SF_5), depending on the nature of the substituent R. Dimetallatetrahedrane type cluster compounds, $[(CO)_6Co_2(\mu_2\text{-}\eta^2\text{-}R{-}C{\equiv}C{-}SF_5)]$ have been isolated for all these substituents and are the only products for the more sterically demanding substituents CF_3, C_6H_5, $SiMe_3$, and SF_5[87,129,131] even under vigorous reaction conditions. In the single case of $HC{\equiv}C{-}SF_5$ a cobaltacyclopentadienyl-type complex, $[(CO)_2Co{-}C(SF_5){-}C(H){-}C(H){-}CSF_5]Co(CO)_3$, composed of two alkynes, could be isolated and structurally characterized. However, this material reacts with pentafluorosulfanylethyne, forming a so-called flyover bridge complex of the structure depicted in Figure 10.2.

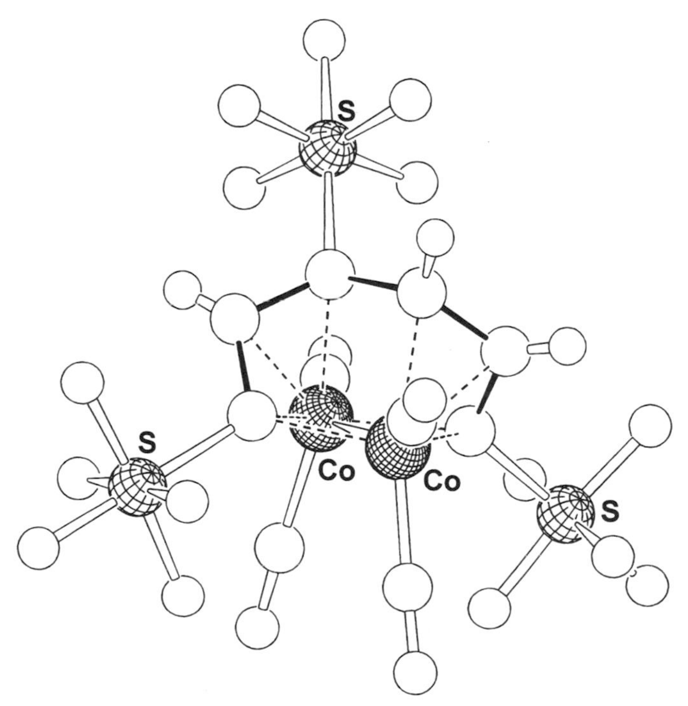

Figure 10.2 Molecular structure of $Co_2(CO)_4(F_5SCCH)_3$.

Two isomers of a similar complex were obtained on the reaction of $[(CO)_6Co_2(\mu_2\text{-}\eta^2\text{-}F_3C—C{\equiv}C—SF_5)]$ with $HC{\equiv}CSF_5$ on very long reaction times. Most importantly, $[Co_2(CO)_4(HCCSF_5)_3]$ decomposes on oxidation by bromine, yielding 1,2,4-tris(pentafluorsulfanyl)benzene. Thus the reaction $[(CO)_6Co_2(\mu_2\text{-}\eta^2\text{-}R—C{\equiv}C—SF_5)]$ complexes might offer a more systematic pathway to SF_5-substituted benzene derivatives with substituents in well-defined positions than the fluorination of disulfides. In addition, these types of alkyne complexes have been widely used in organic synthesis.[156] Pentafluorosulfanylethyne has been used among other alkynes in a regio-selective photochemical C—C bond formation with $[Fe_3(CO)_9(\mu_3\text{-}CF)_2]$ yielding $[Fe_3(CO)_8(\mu_3\text{-}CF)(\mu_3\text{-}CF—CH—CSF_5)]$.[157]

10.5. SeF_5 AND TeF_5 ORGANIC COMPOUNDS

In contrast to SF_5 organic chemistry, the total amount of published SeF_5 and TeF_5 organic chemistry is small. Although it is clear that such compounds should exist, the first notice on $C_2F_5—SeF_5$, prepared from $C_2F_5—SeF_3$ by fluorination with elemental fluorine/CsF dates back only to 1976.[158] $CF_3—SeF_3$, if fluorinated with AgF_2, gives only 3% CF_3SeF_5.[159,160] It decomposes above 70°C to CF_4 and SeF_4.[159,160] This contrasts sharply to the extreme thermal stability of CF_3SF_5, where a similar decomposition reaction is not observed before approximately 450°C.[106] Therefore it must be stated that the SeF_5 group lends a special instability if bonded to carbon in an organic compound.

$$F_3Se-C_2F_5 \xrightarrow[CsF]{F_2} F_5Se-C_2F_5$$

$$F_3Se-CF_3 \xrightarrow{AgF_2} F_5Se-CF_3$$

$$F_5Se-CF_3 \xrightarrow{>70°C} CF_4 + SeF_4$$

$$F_2C{=}CH_2 + TeF_5Cl \longrightarrow F_5Te-CH_2-CF_2Cl$$

$$F_5Te-CH_2-CF_2Cl \xrightarrow{r.t.} TeF_4 + CF_2Cl-CH_2F$$

Scheme 10.16

Not much better is the state of knowledge about TeF_5 organic compounds. In 1974 it was shown that a related *trans*-$C_2F_5—TeF_4Cl$ can be formed,[161] and a MO study on unknown $CH_3—TeF_5$ predicted an intermediate *cis* influence of the methyl group, as compared to known $HOTeF_5$, $H_2N—TeF_5$, $ClTeF_5$ (smaller *cis* influence), and $BrTeF_5$ (larger *cis* influence).[162] The one and only report on an alkyl TeF_5 compound is that of $CF_2Cl—CH_2—TeF_5$, which is formed by $TeClF_5$ addition to $CF_2{=}CH_2$ in 6% yield as colorless liquid, decomposing at room temperature to TeF_4 (and most likely CF_2Cl—CH_2F).[54]

Pentafluorophenyltellurium can be prepared from $PhTeF_3$ and XeF_2 as well as from PhTeTePh and XeF_2.[163] Hydrolyzation of $PhTeF_5$ yields $PhTeF_4OH$ as a mixture of the *cis* and *trans* isomer in a ratio of approximately 1 : 3,[164] whereas the reaction of $PhTeF_5$ with MeOH, $MeOSiMe_3$ or Me_2NSiMe_3 yields only *cis*-$PhTeF_4X$ (X = OMe, $OSMe_2$).[165] The fluorination of alkenes by $PhTeF_5$, $PhSeF_3$, and $PhSeF_5$ was described very recently, however, no details on the chemical and physical properties of $PhSeF_5$ were given.[166]

10.6 CONCLUSION

Almost any SF_5-substituted organic compound can now be made, if starting materials are at hand, and multistep procedures are taken into account. Although there is already a large body of possible applications (the large patent literature on this subject was not the topic of this chapter), further applications still wait to be explored. In contrast to this, the SeF_5 and TeF_5 organic compounds are and probably will always remain of purely academic interest.

REFERENCES

1. R. Winter, G. Gard in *Inorganic Fluorine Chemistry*, J. S. Thrasher, S. H. Strauss, Eds (ACS Symposium Series 555, Washington, D.C., 1994), p. 128.
2. C. Markowsky, V. Pashimik, *Rev. Heteroatom Chem.* **2**, 112 (1989).
3. G. A. Silvey, G. H. Cady, *J. Am. Chem. Soc.* **72**, 3624 (1950).
4. L.E. A. Tyczkowski, L. A. Bigelow, *J. Am. Chem. Soc.* **75**, 3523 (1953).
5. J. A. Attaway, R. H. Groth, L. A. Bigelow, *J. Am. Chem. Soc.* **81**, 3599 (1959).
6. O. Lösking, H. Willner, *Angew. Chem.* **101**, 1283 (1989); *Angew. Chem. Int. Ed. Engl.* **28**, 1255 (1989).
7. J. Jacobs, S. E. Ulic, H. Willner, G. Schatte, J. Passmore, S. V. Sereda, T. S. Cameron, *J. Chem. Soc., Dalton Trans.* 383 (1996).
8. J. S. Thrasher, K. V. Madappat, *Angew. Chem* **101**, 1284 (1989); *Angew. Chem. Int. Ed. Engl.* **28**, 1256 (1989).
9. L. A. Shimp, R. J. Lagow, *Inorg. Chem.* **16**, 2974 (1977).
10. A. Waterfeld, R. Mews, *J. Fluorine Chem.* **23**, 325 (1983).
11. H.-N. Huang, H. Roesky, R. J. Lagow, *Inorg. Chem.* **30**, 789 (1991).
12. D. Viets, A. Waterfeld, R. Mews, I. Weiss, H. Oberhammer, *Chem. Ber.* **125**, 535 (1992).
13. T. Abe, J. M. Shreeve, *J. Fluorine Chem.* **3**, 187 (1973/74).
14. G. A. Silvey, G. H. Cady, U.S. patent 2, 697, 726 (1954); *Chem. Abstr.* **49**, 5794a (1955).
15. R. A. Brandt, H. J. Emeléus, R. N. Haszeldine, *J. Chem. Soc.* 2198 (1952).

16. H. J. Emeléus, B. W. Tattershall, *J. Inorg. Nucl. Chem.* **28**, 1823 (1966).
17. W. Gombler, *J. Fluorine Chem.* **9**, 233 (1977).
18 W. A. Sheppard, *J. Am. Chem. Soc.* **84**, 3064 (1962).
19. T. L. St. Clair, A. K. St. Clair, J. S. Thrasher, U.S. Patent, 5,220,070 (1993); *Chem. Abstr.* **117**, 70558w (1992).
20. A. G. Williams, N. R. Foster, PCT Int. Appl. WO 94 22, 817 (1994); *Chem. Abstr.* **122**, 58831a (1995).
21. G. A. Silvey, G. H. Cady, *J. Am. Chem. Soc.* **74**, 5792 (1952).
22. A. F. Clifford, H. K. El-Shamy, H. J. Emeléus, R. N. Haszeldine, *J. Chem. Soc.* 2372 (1953).
23. R. N. Haszeldine, F. Nyman, *J. Chem. Soc.* 2684 (1956).
24. R. D. Dresdner, J. A. Young, *J. Am. Chem. Soc.* **81**, 574 (1959).
25. F. W. Hoffmann, T. C. Simmons, R. B. Beck, H. V. Holler, T. Katz, R. J. Koshar, E. R. Larsen, J. E. Mulvaney, F. E. Rogers, B. Singleton, R. S. Sparks, *J. Am. Chem. Soc.* **79**, 3424 (1957).
26. R. Dresdner, *J. Am. Chem. Soc.* **79**, 69 (1957).
27. R. D. Dresdner, T. M. Reed III, T. E. Taylor, J. A. Young, *J. Org. Chem.* **25**, 1464 (1960).
28. M. Schmeisser, F. Huber, *Z. Naturforsch.* **21b**, 285 (1966).
29. M. Hisasue, N. Watanabe, S. Yoshizawa, *Asahi Garasu Kenkyu Hokoku* **15**, 139 (1965); *Chem. Abstr.* **64**, 17034d (1966).
30. T. Abe, S. Nagase, H. Baba, *Bull. Chem. Soc. Jpn.* **46**, 3845 (1973).
31. H. Baba, K. Kodaira, S. Nagase, T. Abe, *Bull. Chem. Soc. Jpn.* **51**, 1891 (1978).
32. H. Baba, K. Kodaira, S. Nagase, T. Abe, *Bull. Chem. Soc. Jpn.* **50**, 2809 (1977).
33. R. Czerepinski, G. H. Cady, *J. Am. Chem. Soc.* **90**, 3954 (1968).
34. H. L. Roberts, *J. Chem. Soc.* 3183 (1962).
35. M. Tremblay, *Can. J. Chem.* **43**, 219 (1965).
36. J. Hutchinson, *J. Fluorine Chem.* **3**, 429 (1973/74).
37. M. R. Féliz, H. J. Schumacher, *J. Photochem.* **35**, 23 (1986).
38. A cheap and reliable method for the preparation of SF_5Cl: U. Jonethal, K. Seppelt, *J. Fluorine Chem.* **88**, 3, 1998.
39. J. R. Case, N. H. Ray, H. L. Roberts, *J. Chem. Soc.* 2066 (1961).
40. N. H. Ray, H. L. Roberts, *Brit. Patent* 891,552 (1962); *Chem. Abstr.* **57**, 8439d (1962); N. H. Ray, *Brit. Patent* 905,006 (1962); *Chem. Abstr.* **58**, 7831h (1962).
41. J. R. Case, N. H. Ray, H. L. Roberts, *J. Chem. Soc.* 2070 (1961); J. R. Case, *German Patent* 1,142,862 (1963); *Chem. Abstr.* **59**, 1484h (1963).
42. R. E. Banks, R. N. Haszeldine, W. D. Morton, *J. Chem. Soc.* (C), 1947 (1969).
43. R. A. de Marco, W. B. Fox, *J. Fluorine Chem.* **17**, 137 (1978).
44. A. D. Berry, W. B. Fox, *J. Fluorine Chem.* **6**, 175 (1975).
45. D. D. Coffman, C. W. Tullock, *U.S. Patent* 3,102,903 (1963); *Chem. Abstr.* **60**, 1599e (1964).
46. G. Kleemann, K. Seppelt, *Angew. Chem.* **90**, 547 (1978); *Angew. Chem. Int. Ed. Engl.*, **17**, 516 (1978).

47. G. Kleemann, K. Seppelt, *Chem. Ber.* **112**, 1140 (1979).
48. R. Winter, G. L. Gard, *J. Fluorine Chem.* **66**, 109 (1994).
49. E. F. Witucki, *J. Fluorine Chem.* **20**, 803 (1982).
50. J. Wessel, G. Kleemann, K. Seppelt, *Chem. Ber.* **116**, 2399 (1983).
51. A. Vij, R. L. Kirchmeier, J. M. Shreeve, T. Abe, H. Fukaya, E. Hayashi, Y. Hayakawa, T. Ono, *Inorg. Chem.* **32**, 5011 (1993).
52. A. Klauck, K. Seppelt, *Angew. Chem* **106**, 98 (1994); *Angew. Chem. Int. Ed. Engl.* **33**, 93 (1994).
53. H. W. Sidebottom, J. M. Tedder, J. C. Walton, *Trans. Faraday Soc.* **65**, 2103 (1969).
54. T. Grelbig, T. Krügerke, K. Seppelt, *Z. anorg. allg. Chem.* **544**, 74 (1987).
55. J. Kay, F. M. Page, *Trans. Faraday Soc.* **60**, 1042 (1964); E. C. M. Chen, L.-R. Shuie, E. Desai D'sa, C. F. Batten, W. E. Wentworth, *J. Chem. Phys.* **88**, 4711 (1988).
56. F. W. Hoover, D. D. Coffman; *J. Org. Chem.* **29**, 3567 (1964).
57. H. L. Roberts, *Brit. Patent* 907,648 (1962); *Chem. Abstr.* **58**, 4427d (1962).
58. B. Pötter, K. Seppelt, *Inorg. Chem.* **21**, 3147 (1982).
59. K. Seppelt, T. Drews, unpublished results.
60. B. Cohen, A. G. McDiarmid, *Inorg. Chem.* **4**, 1782 (1965).
61. K. O. Christe, E. C. Curtis, C. J. Schack, A. Roland, *Spectrochim. Acta* **33A**, 69 (1977).
62. R. J. Terjeson, J. A. Canich, G. L. Gard, *Inorg. Synth.* **27**, 329 (1990); R. Winter, R. J. Terjeson, G. L. Gard, *J. Fluorine Chem.* **89**, 105 (1998).
63. J. Steward, L. Kegley, H. F. White, G. L. Gard, *J. Org. Chem.* **34**, 760 (1969).
64. J. Mohtasham, R. J. Terjeson, G. L. Gard, *Inorg. Synth.* **29**, 33 (1992).
65. Q. C. Mir, R. Debuhr, C. Haug, H. F. White, G. L. Gard, *J. Fluorine Chem.* **16**, 373 (1980).
66. R. J. Terjeson, G. L. Gard, *J. Fluorine Chem.* **35**, 653 (1987).
67. A. D. Berry, W. B. Fox, *J. Fluorine Chem.* **7**, 449 (1976).
68. J. A. M. Canich, M. M. Ludvig, W. W. Paudler, G. L. Gard, J. M. Shreeve, *Inorg. Chem.* **24**, 3668 (1985).
69. A. D. Berry, R. A. De Marco, W. B. Fox, *J. Am. Chem. Soc.* **101**, 737 (1979).
70. Q. C. Wang, H. F. White, G. L. Gard, *J. Fluorine Chem.* **13**, 455 (1979).
71. T. A. Kovacina, R. A. De Marco, A. W. Snow, *J. Fluorine Chem.* **21**, 261 (1982).
72. A. D. Berry, W. B. Fox, *J. Org Chem.* **43**, 365 (1978).
73. D. Lentz, H. Oberhammer, *Inorg. Chem.* **24**, 4665 (1985).
74. K. Seppelt, *Angew. Chem.* **103**, 399, 1991; *Angew. Chem. Int. Ed. Engl.* **30**, 361 (1991).
75. L. F. Drullinger, J. F. Griffiths, *Spectrochim. Acta* **A27**, 1793 (1971).
76. K. O. Christe, E. C. Curtis, C. J. Schack, D. Pilipovich, *Inorg. Chem* **11**, 1679 (1972).
77. J. Bittner, J. Fuchs, K. Seppelt, *Z. Anorg. Allg. Chem.* **557**, 182 (1988).

78. I. Weiss, H. Oberhammer, G. L. Gard, R. Winter, K. Seppelt, *J. Mol. Struct.* **269**, 197 (1992).

79. K. D. Gupta, R. Mews, A. Waterfeld, J. M. Shreeve, H. Oberhammer, *Inorg. Chem.* **25**, 275 (1986).

80. C. J. Marsden, D. Christen, H. Oberhammer, *J. Mol. Struct.* **131**, 299 (1985).

81. A. G. Császár, K. Hedberg, R. J. Terjeson, G. L. Gard, *Inorg. Chem.* **26**, 955, (1987).

82. J. Jacobs, G. S. McGrady, H. Willner, D. Christen, H. Oberhammer, P. Zylka, *J. Mol. Struct.* **245**, 275 (1991).

83. J. Jacobs, D. Lentz, P. Luger, G. Perpetuo, H. Willner, to be published.

84. I. Weiss, H. Oberhammer, R. Gerhardt, K. Seppelt, *J. Am. Chem. Soc.* **112**, 6839 (1990).

85. R. Gerhardt, T. Grelbig, J. Buschmann, P. Luger, K. Seppelt, *Angew. Chem.* **100**, 1592 (1988); *Angew. Chem. Int. Ed. Engl.* **27**, 1534 (1988).

86. J. Buschmann, R. Damerius, R. Gerhardt, D. Lentz, P. Luger, R. Marschall, D. Preugschat, K. Seppelt, A. Simon, *J. Am. Chem.* Soc. **114**, 9465 (1992).

87. R. Damerius, D. Leopold, W. Schulze, K. Seppelt, *Z. Anorg. Allg. Chem.* **578**, 110 (1989).

88. D. A. Keszler, R. Winter, G. L. Gard, *Eur. J. Solid State Inorg. Chem.* **29**, 835 (1992).

89. H. Wessolowski, G.-V. Röschenthaler, R. Winter, G. L. Gard, G. Pon, R. Willett, *Eur. J. Solid State Inorg. Chem.* **29**, 1173 (1992).

90. R. Winter, R. D. Willett, G. L. Gard, *Inorg. Chem.* **28**, 2499 (1989).

91. R. Winter, G. L. Gard, R. Mews, M. Noltemeyer, *J. Fluorine Chem.* **60**, 109 (1993).

92. R. Kuschel, K. Seppelt, *J. Fluorine Chem.* **61**, 23 (1993).

93. L. S. Bartell, W. F. Bradford, *J. Mol. Struct.* **37**, 113 (1977).

94. A. L. Andreassen, S. H. Bauer, *J. Chem. Phys.* **56**, 3802 (1972)

95. R. L. Hilderbrandt, A. L. Andreassen, S. H. Bauer, *J. Phys. Chem.* **74**, 1586 (1970).

96. R. R. Burch, J. C. Calabrese, *J. Am. Chem. Soc.* **108**, 5359 (1986).

97. Cambridge crystal structure data file.

98. R. Gerhardt, K. Seppelt, *Chem. Ber.* **122**, 463 (1989).

99. J. Bittner, R. Gerhardt, K. Moock, K. Seppelt, *Z. anorg. allg. Chem.* **602**, 89 (1991).

100. P. Brant, A. D. Berry, R. A. DeMarco, F. L. Carter, W. B. Fox, J. A. Hashmall, *J. Electron Spectrosc. Relat. Phenom.* **22**, 119 (1981).

101. N. Müller, P. C. Lauterbur, G. F. Svatos, *J. Am. Chem. Soc.* **79**, 1043 (1957).

102. C. I. Merril, S. M. Williamson, G. H. Cady, D. F. Eggers Jr., *Inorg. Chem.* **1**, 215 (1962); M. T. Rogers, J. D. Graham, *J. Am. Chem. Soc.* **84**, 3666 (1962); D. R. Eaton, W. A. Sheppard, *J. Am. Chem. Soc.* **85**, 1310 (1963); N. Boden, J. W. Emsley, J. Feeney, L. H. Sutcliffe, *Trans. Faraday Soc.* **59**, 620 (1963).

103. In older publications the chemical shift sign convention of ^{19}F chemical shifts is opposite to the IUPAC rules.

104. H. Wessolowski, G.-V. Röschenthaler, R. Winter, G. L. Gard, *Z. Naturforsch.* **46b**, 126 (1991).
105. J. C. Gage, *Brit. J. Industr. Med*, **27**, 1 (1970).
106. R. Dresdner, *J. Am. Chem. Soc.* **77**, 6633 (1955).
107. G. Kleemann, K. Seppelt, *Angew. Chem.* **93**, 1096 (1981); *Angew. Chem. Int. Ed. Engl.* **20**, 1037 (1981).
108. T. Krügerke, J. Buschmann, G. Kleemann, P. Luger, K. Seppelt, *Angew. Chem.* **99**, 808 (1987); *Angew. Chem. Int. Ed. Engl.* **26**, 799 (1987).
109. T. Grelbig, B. Pötter, K. Seppelt, *Chem. Ber.* **120**, 815 (1987).
110. B. Pötter, K. Seppelt, *Angew. Chem.* **96**, 138 (1984); *Angew. Chem. Int. Ed. Engl.* **23**, 150 (1984).
111. R. Winter, D. H. Peyton, G. L. Gard, *Inorg. Chem.* **28**, 3766 (1989).
112. R. Winter, G. L. Gard, *Inorg. Chem.* **29**, 2386 (1990).
113. R. J. Terjeson, K. D. Gupta, R. M. Sheets, G. L. Gard, J. M. Shreeve, *Rev. Chimie minérale* **23**, 604 (1986).
114. B. Pötter, G. Kleemann, K. Seppelt, *Chem. Ber.* **117**, 3255 (1984).
115. R. Debuhr, J. Howbert, J. M. Canich, H.F. White, G. L. Gard, *J. Fluorine Chem.* **20**, 515 (1982).
116. E. F. Witucki, *J. Fluorine Chem.* **20**, 807 (1982).
117. R. A. De Marco, W. B. Fox, *J. Org. Chem.* **47**, 3772 (1982).
118. R. Kuschel, K. Seppelt, *Inorg.Chem.* **32**, 3568 (1993).
119. S. A. Kinkead, J. M. Shreeve, *Inorg.Chem.* **23**, 3109 (1984).
120. S. A. Kinkead, J. M. Shreeve, *Inorg.Chem.* **23**, 4174 (1984).
121. R. J. Terjeson, J. Mohtasham, G. L. Gard, *Inorg. Chem.* **27**, 2916 (1988).
122. G. L. Gard, C. Woolf, *J. Fluorine Chem.* **1**, 487 (1971/72).
123. R. Damerius, K. Seppelt, J. S. Thrasher, *Angew. Chem.* **101**, 783 (1989); *Angew. Chem. Int. Ed. Engl.* **28**, 769 (1989).
124. N. H. Ray, *J. Chem. Soc.*, 1440 (1963).
125. R. Winter, G. L. Gard, *J. Fluorine Chem.* **50**, 141 (1990).
126. A. W. Marcellis, R. E. Eibeck, *J. Fluorine Chem.* **5**, 71 (1975).
127. A. Lork, G. Gard, M. Hare, R. Mews, W.-D. Stohrer, R. Winter, *J. Chem. Soc., Chem. Commun.* 898 (1992).
128. H. Wessolowski, G.-V. Röschenthaler,R. Winter, G. L. Gard, *Z. Naturforsch.*, **46b**, 123 (1991).
129. J. Wessel, H. Hartl, K. Seppelt, *Chem. Ber.* **119**, 453 (1986).
130. J. Hutchinson, R. D. Chambers, M. P. Greenhall, 211th ACS National Meeting, New Orleans 1996.
131. T. Henkel, A. Klauck, K. Seppelt, *J. Organomet. Chem.* **501**, 1 (1995).
132. K. Seppelt, *Angew. Chem.* **89**, 325 (1977); *Angew. Chem. Int. Ed. Engl.* **16**, 322 (1977).
133. G. Klöter, K. Seppelt, *J. Am. Chem. Soc.* **101**, 347 (1979).
134. K. Seppelt, *Angew. Chem.* **88**, 56 (1976); *Angew. Chem. Int. Ed. Engl.* **15**, 44 (1976).

135. T. Henkel, T. Krügerke, K. Seppelt, *Angew. Chem.* **102**, 1171 (1990); *Angew. Chem. Int. Ed. Engl.* **29**, 1128 (1990).
136. J. Bittner, K. Seppelt, *Chem. Ber.* **123**, 2187 (1990).
137. G. Kleemann, K. Seppelt, *Chem. Ber.* **116**, 645 (1983).
138. B. Pötter, G. Kleemann, K.Seppelt, *Chem. Ber.* **117**, 3255 (1984).
139. T. Krügerke, K. Seppelt, *Chem. Ber.* **121**, 1977 (1988).
140. R. Winter, G. L. Gard, *Inorg. Chem.* **27**, 4329 (1988).
141. R. Winter, G. L. Gard, *J. Fluorine Chem.* **52**, 57 (1991).
142. R. Winter, G. L. Gard, *J. Fluorine Chem.* **52**, 73 (1991).
143. M. E. Sitzmann, *J. Fluorine Chem.* **70**, 31 (1995).
144. B. Pötter, K. Seppelt, A. Simon, E.-M. Peters, B. Hettich, *J. Am. Chem. Soc.* **107**, 980 (1985).
145. D. Lentz, P. Luger, G. Perpetuo, D. Preugschat, J. S. Thrasher, H. Wölk, to be published.
146. A. L. Logothetis, G. N. Sausen, *J. Org. Chem.* **31**, 3689 (1966).
147. L. Gard, A. Waterfeld, R. Mews, J. Mohtasham, R. Winter, *Inorg. Chem.* **29**, 4588 (1990).
148. J. M. Canich, M. M. Ludvig, G. L. Gard, J. M. Shreeve, *Inorg. Chem.* **23**, 4403 (1984).
149. G. Kleemann, Dissertation, Freie Universität Berlin, 1982.
150. W. T. Miller, R. J. Burnard, *J. Am. Chem. Soc.* **90**, 7367 (1968).
151. W. T. Miller, K. H. Snider, R. J. Hummel, *J. Am. Chem. Soc.* **91**, 6532 (1969).
152. K. K. Sun, W. T. Miller, *J. Am. Chem. Soc.* **92**, 6985 (1970).
153. R. E. Noftle, W. B. Fox, *J. Fluorine Chem.* **9**, 219 (1977).
154. H. F. Efner, R. Kirk, R. E. Noftle, M. Uhrig, *Polyhedron* **1**, 723 (1982).
155. A. D. Berry, R. A. De Marco, *Inorg. Chem.* **21**, 457 (1982).
156. G. G. Melikyan, K. M. Nicholas in *Modern Acetylene Chemistry*, P. J. Stang, F. Diederich, Eds (VCH Publishers, Weinheim, 1995), p. 99.
157. D. Lentz, H. Michael-Schulz, *Inorg. Chem.* **29**, 4396 (1990).
158. C. Lau, J. Passmore, *J. Fluorine Chem.* **7**, 261 (1976).
159. A. Haas, *J. Fluorine Chem.* **32**, 415 (1986).
160. A. Haas, H.-U. Weiler, *Chem. Ber.* **118**, 943 (1985).
161. C. D. Desjardins, C. Lau, J. Passmore, *Inorg. Nucl. Chem. Lett.* **10**, 151 (1974).
162. D. R. Armstrong, G. W. Fraser, G. D. Meikle, *Inorg. Chim. Acta* **15**, 39 (1975).
163. K. Alam, A. F. Janzen, *J. Fluorine Chem.* **27**, 467 (1985).
164. A. F. Janzen, K. Alam, M. Jang, B. J. Blackburn, A. S. Secco, *Can. J. Chem.* **66**, 1308 (1988).
165. A. F. Janzen, K. Alam, B. J. Blackburn, *J. Fluorine Chem.* **42**, 173 (1989).
166. S. A. Lermontov, S. J. Zavorin, I. V. Bakhtin, A. N. Pushin, N. S. Zefirov, P. J. Stang, *J. Fluorine Chem.* **87**, 75 (1998).

11 Polycoordinate Iodine Compounds

VIKTOR V. ZHDANKIN[1] and PETER J. STANG[2]
[1]Department of Chemistry, University of Minnesota–Duluth, Minnesota, USA
[2]Department of Chemistry, University of Utah, Salt Lake City, Utah, USA

11.1 INTRODUCTION

Iodine differs in many respects from the other common halogens. Due to the large atom size and the relatively low ionization potential, it can easily form stable polycoordinate, multivalent compounds. Unlike other halogens, iodine occurs naturally not only in the oxidation state of -1 as iodide, but also in polyvalent states as inorganic iodates. One of the major commercial sources of iodine is Chilean saltpeter, which contains from 0.05 to 1% iodine in the form of the minerals lautarite, $Ca(IO_3)_2$, and dietzeite, $7Ca(IO_3)_2{\cdot}8CaCrO_4$.[1-7] Various inorganic derivatives of polyvalent iodine in oxidation states +3, +5, and +7 were prepared as early as the beginning of the 19th century. For example, iodine trichloride was first discovered by Gay Lussac as the result of treating warm iodine or iodine monochloride with an excess of chlorine.[8] In the same paper,[8] the preparation of potassium iodate by the action of iodine on hot potash lye was described. The inorganic chemistry of polyvalent iodine has been reviewed in a number of well-known texts.[1-7]

Iodine is an essential trace element for humans and plays an important role in many biological organisms. In the human body, iodine is mainly concentrated in the thyroid gland in the form of thyroxine, a metabolism-regulating hormone. In natural organic compounds, iodine occurs exclusively in the monovalent state. The first polyvalent organic iodine compound, $PhICl_2$, was prepared by the German chemist Willgerodt in 1886.[9] This was rapidly followed by the preparation of many others, so that when Willgerodt published his comprehensive treatise in 1914 on polyvalent organoiodine species[10] nearly 500 such compounds were known. The most-investigated species were

Chemistry of Hypervalent Compounds, edited by Kin-ya Akiba.
ISBN 0-471-24019-2

diaryliodonium salts, which are used today in industry in lithography and as polymerization initiators. Since the early 1980s, interest in polyvalent organic iodine compounds has experienced a renaissance. This resurgence of interest is due to several factors: the realization that the chemical properties and reactivity of iodine(III) species are similar to those of Hg(II), Tl(III), and Pb(IV) but without the toxic and environmental problems of these heavy metal congeners; the recognition of the similarities (reductive elimination, ligand exchange, etc.) between organic transition metal complexes and polyvalent main group compounds such as organoiodine species; the timely publication of a number of key reviews and surveys (*vide infra*) in the early 1980s and finally the commercial availability of key precursors such as $PhI(OAc)_2$.

It is the purpose of this review to acquaint the reader with the main structural types of polycoordinated iodine compounds. Emphasis is placed on the post-1970s literature, as many hundreds of references have appeared in just the last dozen years. A number of previous reviews and comprehensive monographs by Varvoglis dealing mainly with organic polyvalent iodine derivatives have been published.[11-31]

11.2 GENERAL ASPECTS

In terms of the Martin–Arduengo *N*-X-*L* designation,[32] six structural types of polyvalent iodine species (**1-6**) are the most common. The first two species, 8-I-2 (**1**) and 10-I-3 (**2**), also called iodanes, are conventionally considered as derivatives of trivalent iodine, whereas the next two, 10-I-4 (**3**) and 12-I-5 (**4**) periodanes, represent the most typical structural types of pentavalent iodine. Structural types **5** and **6** are the most typical of heptavalent iodine; only inorganic compounds of iodine(VII), such as iodine(VII) fluoride, oxyfluorides, and derivatives of periodic acid, are known.

8-I-2	10-I-3	10-I-4	12-I-5	14-I-6	14-I-7
1	**2**	**3**	**4**	**5**	**6**

Species **1-4** are especially important for organic chemistry. The structure of these compounds was thoroughly discussed previously in Koser's reviews[16] and in the book by Varvoglis.[24a] The most important structural features of the organic derivatives of iodine(III), available from numerous X-ray data, may be summarized as follows:

1. The 10-I-3 species (**2**) have an approximately T-shaped structure with a collinear arrangement of the most electronegative ligands. Including the nonbonding electron pairs, the geometry about iodine is a distorted

trigonal bipyramid with the most electronegative groups occupying the axial positions and the least electronegative group and both electron pairs residing in equatorial positions.

2. The I—C bond lengths in both iodoso derivatives (**2**) and iodonium salts (**1**) are approximately equal to the sum of the covalent radii of iodine and carbon, ranging generally from 2.00 to 2.10 Å.
3. For 10-I-3 species (**2**) with two heteroligands of the same electronegativity, both I—L bonds are longer than the sum of the appropriate covalent radii, but shorter than purely ionic bonds. For example, the I—Cl bond lengths in $PhICl_2$ are 2.45 Å,[33] and the I—O bond lengths in $PhI(OAc)_2$ are 2.15–2.16 Å,[34] whereas the sum of the covalent radii of I and O is 1.99 Å.
4. The 8-I-2 species (iodonium salts) (**1**) generally have a typical distance between iodine and the nearest anion of 2.6–2.8 Å and are usually considered as ionic compounds with pseudotetrahedral geometry about the central iodine atom. However, with consideration of the anionic part of the molecule, the overall experimentally determined geometry is T-shaped, similar to the 10-I-3 species. For example, available X-ray data on alkynyliodonium triflate salts clearly indicate a T-shaped geometry.[25,30]

The structural features of iodine(III) compounds are generally explained by the hypervalent bonding model, thoroughly discussed by Musher[35] in 1969 and more recently employed by Martin[36] in his work on hypervalent halogen and sulfur compounds. In the hypervalent model, only non-hybridized 5*p* orbitals of iodine are involved in bonding. The least electronegative ligand is bound by a normal covalent bond to the singly occupied equatorial 5*p* orbital, whereas the other two ligands are attached, one to each lobe, of the axial doubly occupied 5*p* orbital of the iodine atom. As a result, a linear three-center–four-electron (3c–4e) bond is formed. Such bonds, termed "hypervalent," are weaker and longer than covalent bonds.

Recently, Reed and Schleyer provided a general theoretical description of chemical bonding in hypervalent molecules in terms of the dominance of ionic bonding and negative hyperconjugation over *d*-orbital participation.[37]

In contrast to iodine(III), only a few single crystal X-ray analyses have been reported for the iodine(V) compounds.[38–46] Depending on the ligands and with the consideration of secondary bonding, the overall observed geometry for the structural types **3** and **4** can be square pyramidal, pseudo-trigonal bipyramidal, and pseudooctahedral.

The bonding in iodine(V) compounds, IL_5, with a square pyramidal structure may be described in terms of a normal covalent bond between iodine and the ligand in an apical position, and two orthogonal, hypervalent 3c–4e bonds, accommodating four ligands. The carbon ligand and unshared electron pair in this case should occupy the apical positions, with the most electronegative ligands residing at equatorial positions.

The typical structures of iodine(VII) involve a distorted octahedral configuration (**5**) about iodine in most periodates[5] and oxyfluoride, IOF_5,[47] and the heptacoordinated, pentagonal bipyramidal species **6** for the IF_7 and IOF_6^- anions.[48,49] The pentagonal bipyramidal structure can be described as two covalent, collinear, axial bonds between iodine and the ligands in the apical positions and a coplanar, hypervalent 6c–10e bond system for the five equatorial bonds.[49]

11.3 IODINE(III) COMPOUNDS

A great variety of iodanes (**1** and **2**) have been reported in the literature. Most of these are organic derivatives with one or two carbon ligands at the iodine. Only several inorganic compounds of iodine(III), mainly oxygen-bonded derivatives, are known. Compounds of iodine(III) with one carbon ligand are represented by organic iodosyl compounds (RIO, where R usually is aryl) and their derivatives (RIX_2, where X is any noncarbon ligand). The second class includes various iodonium salts ($R_2I^+X^-$). The overwhelming majority of known, stable organic compounds of polyvalent iodine belong to these two classes. The third class, derivatives of iodine(III) with three carbon ligands (R_3I), in general are less investigated because of their low thermal stability.

11.3.1 Inorganic Iodine(III) Derivatives

Iodine(III) halides generally are unstable. Iodine trifluoride, IF_3, disproportionates to IF_5 and IF or I_2 even at low temperatures.[5] According to recent *ab initio* calculations,[50] IF_3 has a distorted T-shaped geometry with the axial I—F bond distance of 1.971 Å, the equatorial I—F bond distance of 1.901 Å, and an F_{ax}—I—F_{eq} angle of 81.7°. The chemical properties of IF_3 are almost unknown. The ligand-exchange reaction of IF_3 with trifluoroacetic anhydride leading to iodine(III) trifluoroacetates has been reported.[51] Iodine trichloride, ICl_3, is more stable than the fluoride; however, it easily decomposes to ICl and Cl_2 at elevated temperatures.[3] In the solid state, iodine trichloride exists only as a dimer, I_2Cl_6 (structure **7**), with a planar structure containing two bridging I—Cl—I bonds (I—Cl distances 2.68 and 2.72 Å) and four terminal I—Cl bonds (2.38—2.39 Å).[52] Iodine trichloride forms a number of addition products which can be regarded as salts of the acid $HICl_4$. The salts of alkali metals and ammonia are best prepared by adding chlorine into an aqueous solution of the respective iodide.[1] The potassium salt $KICl_4$ can be used as a reagent for iodochlorination of alkenes.[53]

The parent iodine(III) oxide, I_2O_3, is unknown. However, a number of its inorganic derivatives OIOR or $I(OR)_3$ have been reported in the literature. Historically, the first of these derivatives was iodosyl sulfate, $(IO)_2SO_4$, which was first isolated as early as 1844.[54] The X-ray crystal data for iodosyl sulfate shows a polymeric structure with infinite $(\text{—O—I}^{+}\text{—O—})_n$ spiral chains linked together by SO_4 tetrahedra.[55] Iodosyl sulfate and the selenate, $(IO)_2SeO_4$, are best prepared by the interaction of I_2 with I_2O_5 in concentrated sulfuric or selenic acid.[56] More recently, a similar method was used for the preparation of iodosyl fluorosulfate, $OIOSO_2F$, and triflate, OIOTf.[57] Raman and infrared spectra of these compounds indicate a polymeric structure analogous to iodosyl sulfate.[57] Both iodosyl triflate and fluorosulfate can be used as reagents for the synthesis of diaryliodonium salts by reaction with arenes or arylsilanes.[58–60] Iodine tris(fluorosulfate), $I(OSO_2F)_3$, and tris(triflate), $I(OTf)_3$, are also known.[61,62] $I(OSO_2F)_3$ can be prepared by the reaction of iodine with peroxydisulfuryl difluoride. Salts such as $KI(OSO_2F)_4$ have also been prepared.[61] $I(OTf)_3$ was recently prepared from iodine tris(trifluoroacetate) and trifluoromethanesulfonic acid.[62] Several iodine(III) tris(carboxylate) derivatives $I(O_2CR)_3$ (R = CH_3, CH_2Cl, CF_3) have been reported in the literature.[62] These compounds are best synthesized by oxidation of iodine with fuming nitric acid in the presence of the appropriate carboxylic acid and acetic anhydride. Birchall and co-workers recently reported an X-ray crystal and molecular structure of $I(OAc)_3$.[62] The geometry about iodine in this compound consists of primary bonds to the three acetate groups (I—O distances 2.159, 2.023, and 2.168 Å) and two strong intramolecular secondary bonds (I···O distances 2.463 and 2.518 Å) to two of the acetate groups, forming a pentagonal-planar arrangement **8**.

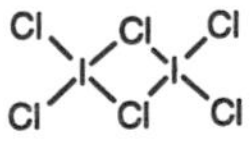

7

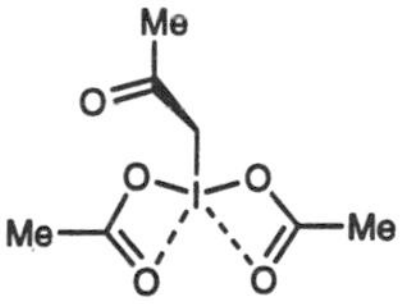

8

Iodine tris(trifluoroacetate) found some synthetic application in organic chemistry as a reagent for oxidation of alkanes, arenes, alkenes, and cyclic ethers. The organic reactions of $I(O_2CCF_3)_3$ were recently reviewed by Moriarty and Kosmeder.[63]

11.3.2 Iodine(III) Species with One Carbon Ligand

In the traditional, conventional classification scheme compounds of iodine(III) with one carbon ligand, RIX_2, are regarded as derivatives (salts or esters) of the parent iodosyl compounds, RIO, or nonisolable acids $RI(OH)_2$.[11,21,24] This classification is justified by the facile conversion of RIO into various RIX_2 by the action of the respective acids HX. The reverse reaction leading to RIO can be generally carried out by the hydrolysis of RIX_2 in the presence of a strong base.

A broad variety of stable iodosyl derivatives have been reported in the literature. The overwhelming majority of these compounds have a benzene ring as a carbon ligand linked to the iodine atom. Derivatives of polyvalent iodine with an alkyl substituent at iodine generally are highly unstable and can exist only as short-lived reactive intermediates in the oxidations of alkyliodides. However, introduction of an electron-withdrawing substituent in the alkyl moiety may lead to significant stabilization of the molecule. Typical examples of such stabilized compounds with I—C_{sp^3} bonding are perfluoroalkyl iodosyl derivatives.

The two heteroatom ligands X attached to iodine in RIX_2 are commonly represented by fluorine, chlorine, O- and N-substituents. In general, only derivatives bearing the most electronegative substituents X are sufficiently stable.

11.3.2.1 ***Iodosylarenes*** Iodosylbenzene, PhIO, is the most important member of the family of iodosyl compounds. It is best prepared by hydrolysis of $PhI(OAc)_2$ with aqueous NaOH.[64] Iodosylbenzene is a yellowish amorphous powder, which can not be recrystallized due to its polymeric nature. Heating or extended storage at room temperature results in disproportionation of iodosylbenzene to PhI and colorless, explosive iodylbenzene, $PhIO_2$.

Recent X-ray powder diffraction and EXAFS analyses of amorphous PhIO clearly indicated a linear polymeric, asymmetrically bridged structure in the solid state having the expected T-shaped geometry around the iodine centers (**9**).[65] The I—O bond lengths in this structure are distinctly different, with a longer one of 2.37 Å and a shorter one of 2.06 Å.

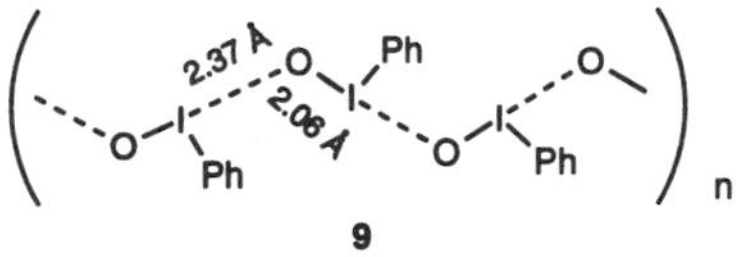

A comprehensive bibliography on the preparation and properties of a variety of other known iodosylarenes was published by Beringer and Gindler.[12] More recent examples include modified procedures for the preparation of a variety of *o, m, p*-monosubstituted iodosylbenzenes from the respective iododiacetates,[66] and the preparation of *p*-bis(iodosyl)benzene by alkaline hydrolysis of the corresponding tetrachloride.[67]

Iodosylbenzene can be easily converted into various $PhIX_2$ derivatives **10** under mild, aprotic conditions by the action of the respective trimethylsilylated reagents (Eq. 11.1):[68–78]

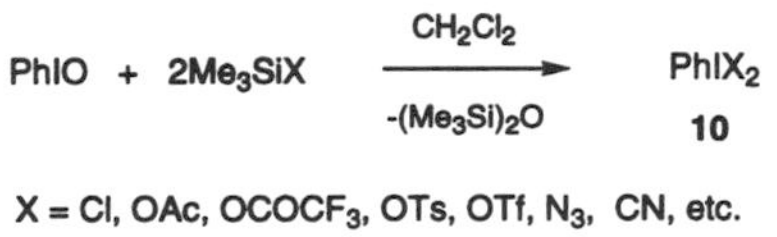

$$PhIO + 2Me_3SiX \xrightarrow[-(Me_3Si)_2O]{CH_2Cl_2} \underset{\mathbf{10}}{PhIX_2} \tag{11.1}$$

X = Cl, OAc, $OCOCF_3$, OTs, OTf, N_3, CN, etc.

The analogous sequential reaction of PhIO with two different trimethylsilylated reagents, one of which is a trimethylsulfonate, $TMSOSO_2R$, leads to mixed iodonium sulfonates **11** according to Eq. 11.2:[79]

$$\text{PhIO} + \text{Me}_3\text{SiOSO}_2\text{R} + \text{Me}_3\text{SiX} \xrightarrow[-30 \text{ to } -5\,^\circ\text{C}]{\text{CH}_2\text{Cl}_2} \underset{\mathbf{11}}{\text{PhI(OSO}_2\text{R)X}} \tag{11.2}$$

X = OAc, NHAc, NCO, CN
R = *p*-Tol, CF_3, C_4F_9

This reaction is especially useful for the preparation of cyano(phenyl)iodonium triflate (**11**, X = CN, R = CF_3), which is a highly efficient iodonium transfer reagent in the syntheses of various alkynyl and alkenyl(phenyl)iodonium salts.[80–88]

Acids or acid anhydrides also react with PhIO with the formation of $PhIX_2$ or similar derivatives. Addition of triflic anhydride to two equivalents of PhIO at 0°C gives a bright yellow precipitate of μ-oxobis[(trifluoromethanesulfonyloxy)iodobenzene] **12** (Zefirov's reagent) (Eq. 11.3), a useful reagent for the preparation of iodonium salts:[89]

$$2\text{PhIO} + \text{Tf}_2\text{O} \xrightarrow[0\,^\circ\text{C}]{\text{CH}_2\text{Cl}_2} \underset{\mathbf{12}}{\text{Ph(TfO)I–O–I(OTf)Ph}} \tag{11.3}$$

A similar reaction of PhIO with triflic acid or one equivalent of triflic anhydride at room temperature yields bisiodonium triflate **13** as the principal product (Eq. 11.4).[90–92]

$$\text{PhIO} + 2\text{TfOH} \xrightarrow[94\%]{\text{CH}_2\text{Cl}_2,\ \text{r.t., 4 h}} \mathbf{13};\quad \text{PhIO} + \text{Tf}_2\text{O} \xrightarrow[76\%]{\text{CH}_2\text{Cl}_2,\ \text{r.t., 12 h}} \mathbf{13}\ \ [\text{Ph(TfO)I–C}_6\text{H}_4\text{–I(OTf)OH}] \tag{11.4}$$

Sulfur trioxide forms two different adducts with PhIO depending on the ratio of reactants (Eq. 11.5).[93] Both sulfates **14** and **15** are hygroscopic solids, highly reactive toward nucleophilic organic substrates.

$$\underset{\mathbf{14}}{[\text{PhI(O)}]_2\text{–O}_2\text{SO}_2\ \text{(cyclic)}} \xleftarrow{1/2\text{SO}_3,\ -50\,^\circ\text{C}} \text{PhIO} \xrightarrow{\text{SO}_3,\ -50\,^\circ\text{C}} \underset{\mathbf{15}}{\overset{+}{\text{PhI}}\text{–OSO}_3^-} \tag{11.5}$$

Among other synthetically important transformations of PhIO, are the reactions with HF or SF_4 leading to $PhIF_2$[94] and the formation of isolable, but hygroscopic $PhI(OMe)_2$ in methanol solution.[95]

11.3.2.2 Halides Iodoaryl halides, $ArIX_2$, can generally be prepared either by the reaction of iodosylarenes, ArIO, with the respective acids HX, or by halogenation of aryliodides. The overwhelming majority of known compounds of this structural type are chlorides and fluorides. Only two examples of a stable bromoiodane have been reported in the literature.[96]

(Difluoroiodo)arenes, $ArIF_2$, are best prepared by reactions of iodosylarenes or their derivatives with hydrofluoric acid or sulfur tetrafluoride. The most common method involves the reaction of (dichloroiodo)arenes with 48% HF and yellow HgO.[97] However, methods based on the use of hydrofluoric acid have several disadvantages. Firstly, (difluoroiodo)arenes often are hygroscopic and highly hydrolizable, which make their separation from aqueous hydrofluoric acid and crystallization extremely difficult. Secondly, the high acidity of the reaction media complicates reactions of basic and acid sensitive substrates, such as heterocycles. Reaction of iodosylarenes with SF_4 usually gives better results.[94] In this method, SF_4 is bubbled at −20°C through a suspension of the iodosyl compound in dichloromethane. All the byproducts in this reaction are volatile, so evaporation of the solvent under anhydrous conditions affords (difluoroiodo)arenes of high purity.[94]

Fuchigami and Fujita recently reported a convenient procedure for the electrochemical preparation of (difluoroiodo)arenes.[98] In this method, the electrosynthesis of *p*-substituted $ArIF_2$ was accomplished by anodic oxidation of the respective iodoarenes with $Et_3N{\cdot}3HF$ in anhydrous acetonitrile. This procedure works especially well for the preparation of *p*-$NO_2C_6H_4IF_2$, which precipitates from the electrolytic solution in pure form during the electrolysis.

(Dichloroiodo)benzene, $PhICl_2$, was first prepared by Willgerodt in 1886 by the reaction of iodobenzene with ICl_3 or, preferably, with chlorine.[9] Reactions of aryliodides with chlorine in chloroform is still the most general approach to (dichloroiodo)arenes.[99] (Dichloroiodo)arenes are generally yellow solids, which are easy to handle and can be purified by recrystallization. Most of these compounds have relatively low thermal stability and can be stored only for short periods after preparation. X-ray structures were reported for $PhICl_2$[39] and for a highly hindered $ArICl_2$ (Ar = 2,6-bis(3,5-dichloro-2,4,6-trimethylphenyl)benzene).[100] (Dichloroiodo)arenes have found broad synthetic application as reagents for chlorination and oxidation of various organic substrates.

(Dibromoiodo)arenes with the general formula $ArIBr_2$ usually lack stability and cannot be isolated as individual compounds. However, mono-bromoiodanes, stabilized due to the involvement of the iodine in a five-membered ring, have been reported in the literature.[96] Bromide **17** can be prepared by addition of bromine to aryliodide **16** (Eq. 11.6), in the form of a stable, crystalline solid with a melting point of 173°C:[96]

11.3.2.3 Carboxylates [Bis(acyloxy)iodo]arenes, $ArI(O_2CR)_2$, are the most important, well-investigated, and practically useful organic derivatives of iodine(III). Two of them, (diacetoxyiodo)benzene (DIB) and [bis(trifluoro-

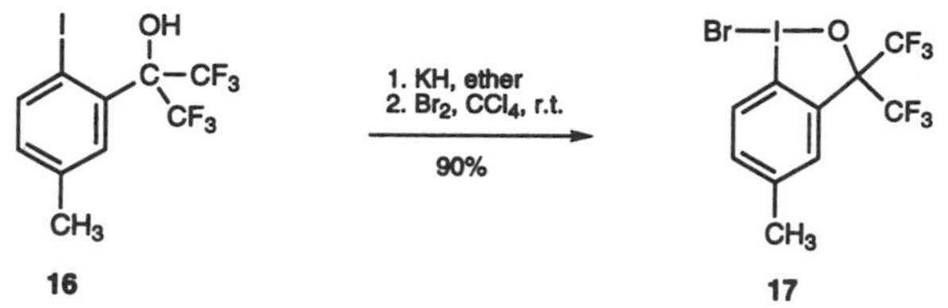

(11.6)

acetoxy)iodo]benzene (BTI), are commercially available or can be easily prepared by oxidation of iodobenzene with the respective peracid.[101] A similar peroxide oxidation of substituted iodobenzenes can be used for the preparation of other [bis(acyloxy)iodo]arenes.[102–106] For example, various mono- and di-[bis(trifluoroacetoxy)iodo] substituted arenes were synthesized in 70–85% yield from the respective iodoarenes and peroxytrifluoroacetic acid.[102–104] [Bis(acetoxy)iodo]arenes can be prepared in excellent yields by oxidation of iodoarenes with sodium perborate in acetic acid.[105–107] A broad variety of other carboxylates are easily accessible by the ligand-exchange reaction of acetates with the appropriate carboxylic acid.[21,108]

[Bis(acyloxy)iodo]arenes are generally colorless, stable microcrystalline solids which can be easily recrystallized and stored for extended periods of time without significant decomposition. Single crystal X-ray structural data for $ArI(O_2CR)_2$ indicate the expected T-shaped geometry around the iodine centers with almost collinear acyloxy groups. The I–O bond distances in $ArI(O_2CR)_2$ are considerably longer than the sum of their covalent radii (1.99 Å) and may vary in a broad range between 2.14 and 2.98 Å.[24,34,109,110]

[Bis(acyloxy)iodo]arenes are widely used as readily available precursors to a variety of other iodosyl derivatives, $ArIX_2$, and iodonium salts, $Ar_2I^+X^-$.

In the last 10–15 years [bis(acyloxy)iodo]arenes have seen widespread application as general, universal oxidizing reagents.

11.3.2.4 Derivatives of Strong Acids In contrast to aryl iodosyl dihalides and dicarboxylates, the analogous disubstituted derivatives of strong acids, $ArIX_2$ (X = RSO_3, ClO_4, NO_3, etc.), are generally unstable and extremely moisture sensitive. Koser and Wettach reported some evidence for the intermediate formation of the ditosylate $PhI(OTs)_2$ in the reaction of $PhICl_2$ with silver tosylate, but an attempt to isolate this compound failed.[111] Similarly, triflate $PhI(OTf)_2$,[68] sulfate $PhISO_4$,[93] and nitrate $PhI(ONO_2)_2$[112] lack stability and typically can only be generated *in situ*, at low temperature, under absolutely dry conditions.

The most likely, stable structural types of the iodosyl derivatives of strong acids are represented by hydroxy-substituted (**18**) and μ-oxo-bridged (**19**) iodanes which are products of partial hydrolysis of the initial $ArIX_2$.

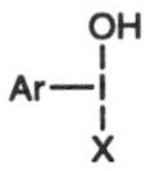

18

X = OSO_2R, $OP(O)(OR)_2$; Ar = Ph, $4\text{-}FC_6H_4$, $4\text{-}ClC_6H_4$, $4\text{-}BrC_6H_4$, $4\text{-}IC_6H_4$, $4\text{-}MeC_6H_4$, $4\text{-}NO_2C_6H_4$, $4\text{-}PhC_6H_4$, $2\text{-}PhCH_2C_6H_4$, $2\text{-}(MeNHCO)C_6H_4$, $2\text{-}(PhCH_2NHCO)C_6H_4$, $2\text{-}MeC_6H_4$, 2-naphthyl, $3\text{-}MeC_6H_4$, $2\text{-}FC_6H_4$, $3\text{-}FC_6H_4$, C_6F_5

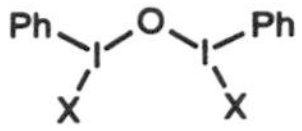

19

X = OSO_2OH, OTf, $OClO_3$, ONO_2, O_2CCF_3, BF_4, SbF_6, PF_6

Both types generally can be prepared by the reaction of iodosyl arenes or dicarboxylates with the appropriate aqueous acid. Hydroxy derivatives **18** are colorless, stable microcrystalline solids which can be recrystallized and stored for extended periods of time without significant decomposition. The most important representative of **18**, [hydroxy(tosyloxy)iodo]benzene, PhI(OH)OTs, is commercially available. Single crystal X-ray structural data for PhI(OH)OTs indicates the expected T-shaped geometry around the iodine center with almost collinear O-ligands and two different I—O bonds of 2.47 Å (I—OTs) and 1.94 Å (I—OH) consistent with the ionic character of this compound.[113] The first preparation of [hydroxy(tosyloxy)iodo]benzene (HTIB) from (diacetoxyiodo)benzene and *p*-toluenesulfonic acid monohydrate was reported in 1970 by Neiland and Karele.[114] A few years later, Koser and co-workers discovered that HTIB is an efficient reagent for the synthesis of various iodonium salts and oxytosylations, so in the contemporary literature HTIB is often referred to as "Koser's Reagent."

A variety of other sulfonate iodosyl derivatives of type **18** have been reported in the literature. [Hydroxy(methanesulfonyloxy)iodo]benzene (**18**, Ar = Ph, X = OSO_2Me) can be prepared from (diacetoxyiodo)benzene and methanesulfonic acid and is very similar to HTIB in its chemical reactivity.[115–117] A chiral analog of this reagent, [hydroxy((+)–10-camphorsulfonyloxy)iodo]benzene, has been recently reported by Varvoglis and co-workers.[118]

[Methoxy(tosyloxy)iodo]benzene, PhI(OMe)OTs, can be prepared by the treatment of HTIB with trimethyl orthoformate.[119] The methoxy-ligand in PhI(OMe)OTs can be substituted with a chiral menthyloxy group by treatment with menthol.[120]

Several sulfonates **18** with the aryl substituent other than phenyl have been reported in the literature.[121,122] For example, [hydroxy(tosyloxy)iodo]pentafluorobenzene and the analogous mesylate derivative were prepared by treatment of [bis(trifluoroacetoxy)iodo]pentafluorobenzene in acetonitrile with *p*-toluenesulfonic or methanesulfonic acid, respectively.[122]

[Hydroxy(bisphenylphosphoryloxy)iodo]benzene (**18**, X = $OP(O)(OPh)_2$) can be prepared as a stable, crystalline solid by the reaction of (diacetoxy-iodo)benzene with diphenyl phosphate in aqueous acetonitrile.[123]

μ-Oxo-bridged iodanes **19** are the most typical iodosyl derivatives of strong inorganic acids (sulfuric, nitric, perchloric, trifluoromethanesulfonic, etc.);

however, compounds of this type are also known for carboxylic acids. A distinctive feature of all μ-oxo-bridged iodanes **19** is the bright yellow color due to the $^{+}I—O—I^{+}$ bridging fragment. The available X-ray structural data for **19** (X = ONO_2,[124] and OCOR[34,125]) is also consistent with the ionic character of the I—X bonds.

Triflate **12** and perchlorate **20** are prepared from (diacetoxyiodo)benzene and aqueous perchloric or trifluoromethanesulfonic acid (Eq. 11.7).[126] In a more convenient procedure triflate **12** can be generated *in situ* from PhIO and triflic anhydride (Eq. 11.3).[89]

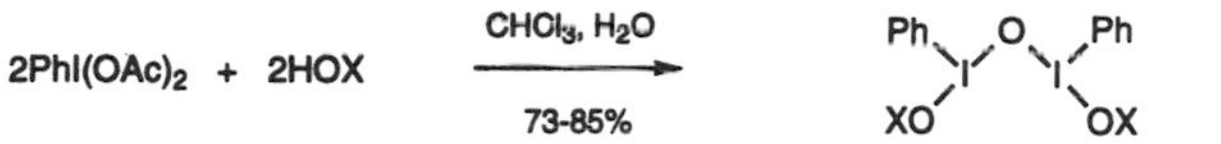

$$2PhI(OAc)_2 + 2HOX \xrightarrow[73\text{-}85\%]{CHCl_3,\ H_2O} \text{Ph(XO)I–O–I(OX)Ph} \quad (11.7)$$

12, X = Tf; **20**, X = ClO_3

The newest representatives of μ-oxo-bridged iodanes **19** derivatives of fluoroboric, fluoroantimonic, and fluorophosphoric acids, can be prepared as yellow, microcrystalline solids by the treatment of $PhI(OAc)_2$ with aqueous solutions of the respective acids.[127]

In the last dozen years, iodosyl derivatives of strong acids and, especially, iodosyl sulfonates, have found wide synthetic application as effective oxidizers and starting materials in the preparation of numerous iodonium salts. The application of iodosyl sulfonates in organic synthesis was recently reviewed by Moriarty, Vaid, and Koser.[22]

11.3.2.5 Five-Membered Iodine(III) Heterocycles The overwhelming majority of known heterocyclic iodanes are derivatives of benziodoxole (**21**), with iodine and oxygen incorporated in the five-membered ring. A few examples of heterocyclic iodanes with other elements in five- or six-membered rings are represented by benziodazoles **22**,[128,129] benziodathiazoles **23**,[130] benziodoxathioles **24**,[131] cyclic phosphonate **25**[132] and phosphate **26**.[133]

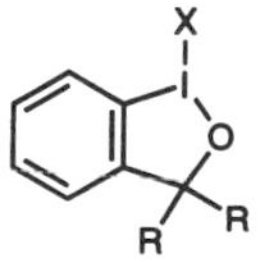

21

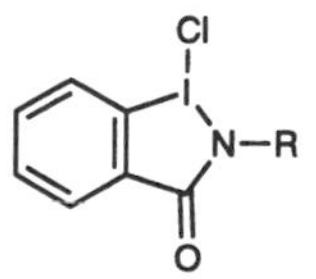

22, R = Cl, CH_2CO_2Me

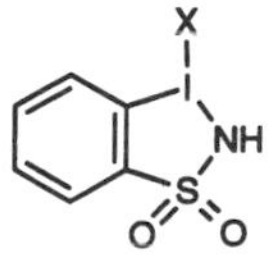

23, X = Cl, OH, OAc

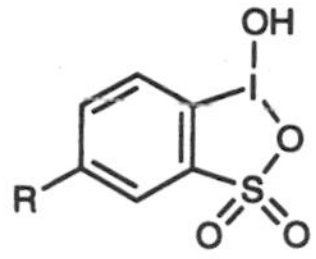

24, R = H, Me

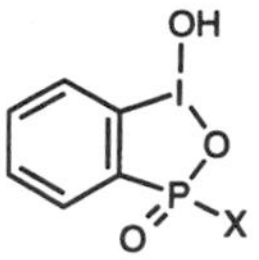

25, X = Me, OH

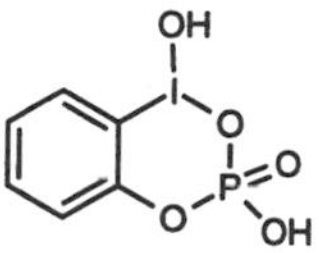

26

X-ray molecular structures were reported for a number of benziodoxole derivatives **21**,[134–140] as well as for benziodazoles **22**,[128,129] benziodoxathioles **24**,[131] and cyclic phosphonate **25** (X = Me).[132] The structural features of heterocyclic iodanes were thoroughly discussed in Koser's review[16a] and in the book by Varvoglis.[24a] In general, the five-membered ring in heterocyclic iodanes is highly distorted with almost linear alignment of the two electronegative ligands. The I—O bond length in benziodoxoles **21** varies in a wide range from 2.11 Å in carboxylates (**21**, X = *m*-$ClC_6H_4CO_2$) to 2.48 Å in the phenyl derivative (**21**, X = Ph), which indicates considerable changes in the ionic character of this bond. The endocyclic C—I—O bond angle is typically around 80°, which is a significant deviation from the expected angle of 90° for the normal T-shaped geometry of hypervalent iodine.

The distinctive feature of cyclic iodanes is a considerably higher stability than that of their acyclic analogs. This stabilization is usually explained by the bridging of an apical and an equatorial position by a five-membered ring,[96] and also by better overlap of the lone pair electrons on the iodine atom with the p-orbitals of the benzene ring in **21**.[16a,135] The greater stability of heterocyclic iodanes enables the isolation of otherwise unstable iodine(III) derivatives with I—Br, I—N_3, I—CN, I—OOR and other bonds.

The most important and best-investigated heterocyclic iodane is 1-hydroxy-1,2-benziodoxole-3(1*H*)-one **28**, the cyclic tautomer of 2-iodosylbenzoic acid **27**. The tautomeric form **28** gives the best representation for the actual structure of this compound, as indicated by its unusually low acidity (pK_a 7.25 against 2.85 for the hypothetical 2-iodosylbenzoic acid **27**) and confirmed by X-ray analysis of **28** in the crystalline state.[134a] The I—O bond distance of 2.3 Å in the five-membered ring of **28** is significantly longer than the computed covalent I—O bond length of 1.99 Å, which indicates the highly ionic nature of this bond.

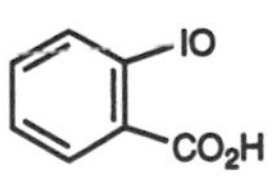

IO
CO2H
27

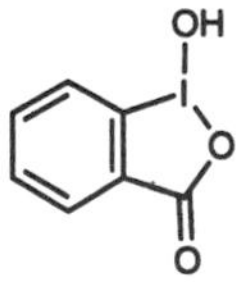

OH
I
O
O
28

1-Hydroxy-1,2-benziodoxole-3(1*H*)-one **28** is commercially available or can be easily prepared by direct oxidation of 2-iodobenzoic acid or by basic hydrolysis of the respective iododichloride. In the last 10 years, benziodoxole **28** and its derivatives have attracted considerable research interest, due to their excellent catalytic activity in the cleavage of toxic phosphates and reactive esters.[24,31]

The stabilization effect due to the incorporation of the hypervalent iodine into a five-membered heterocycle made possible the isolation of several otherwise unstable iodine(III) derivatives, such as 1-(alkylperoxy)benziodoxoles,[136,137] 1-bromobenziodoxoles,[96] 1-azidobenziodoxoles,[138] and 1-cyanobenziodoxoles.[139,140]

11.3.2.6 Derivatives with I—N Bonds Iodanes with I—N bonds are generally less common than those with I—O bonds. Most of these compounds lack stability and are sensitive to moisture. In addition to the previously discussed benziodazoles **22** and benziodathiazoles **23**, four more structural types of iodine(III) derivatives with one carbon ligand and one or two I—N bonds are known (structures **29–32**).

NR_2 = phthalimidate, succimidate, saccharinate; R^1, R^2 = alkyl, aryl; X = OTs, OCOR, etc.; Ar = Ph, 2-$O_2CC_6H_4$

The first two structures (**29,30**) are amidoiodanes, the third (**31**) an imidoiodane, and the fourth (**32**) an azide.

The only known amidoiodanes with two *N*-ligands at iodine (**29**) are derivatives of cyclic imides, such as phthalimide, succinimide, glutarimide, and saccharine.[141]

Iminoiodanes, ArI=NR, are formal analogs of iodosylbenzene. Several examples of [imino(iodo)]arenes were reported in the literature: *N*-tosyl-,[142] *N*-methanesulfonyloxy-,[143] and *N*-(trifluoroacetyl)-derivatives.[144] All known iminoiodanes were prepared by the reaction of (diacetoxyiodo)arenes with the respective amides under basic conditions; *N*-trifluoroacetyl- and *N*-methanesulfonyl-iminoiodanes are relatively unstable and explosive; however, the tosylate derivatives are stable, crystalline compounds which can be stored for extended periods.

Single crystal X-ray structural data have recently been reported for several *N*-tosyliminoiodanes, namely, PhI=NTs,[142b] 2,4,6-$Me_3C_6H_2I$=NTs,[142b] and 2-MeC_6H_4I=NTs.[145] Similar to iodosylarenes (structure **9**), iminoiodanes have a linear polymeric, asymmetrically bridged structure with the expected T-shaped geometry around the iodine centers.[142b,145]

Azidoiodanes, $PhI(N_3)X$ (X = OAc, Cl, OTMS, etc.) or $PhI(N_3)_2$, were proposed as reactive intermediates in azidonation reactions involving the combination of PhIO or $PhI(OAc)_2$ with trimethylsilyl azide or NaN_3.[146–152] Attempts to isolate these intermediates always resulted in fast decomposition at –25 to 0°C with the formation of iodobenzene and dinitrogen; however, low-temperature spectroscopy and the subsequent chemical reactions *in situ* provided some experimental evidence for the existence of these species.

The structure of the heterocyclic azidoiodane **33**, prepared from the respective 1-hydroxybenziodoxole and trimethylsilyl azide,[153] was unambiguously established by a single-crystal X-ray analysis.[138] The structural data revealed the expected hypervalent iodine, distorted T-shaped geometry, with an N1—I—O bond angle of 169.5 (2) degrees. The lengths of the bonds to the iodine atom, I—N (2.18 Å), I—O (2.13 Å), and I—C (2.11 Å), all have similar values, and generally are within the range of typical single covalent bonds in organic derivatives of polyvalent iodine (structure **33**).

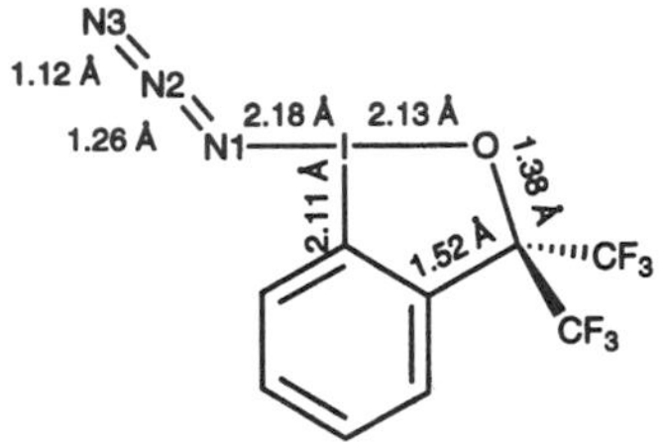

33

11.3.2.7 Iodine(III) Species with One sp^3***-Carbon Ligand*** Derivatives of polyvalent iodine with an alkyl (R) substituent at iodine, RIX_2, generally are highly unstable and can exist only as short-lived reactive intermediates in the oxidations of alkyliodides. However, the thermal stability of alkyliodosyl derivatives can be substantially increased by steric or electronic modification of the alkyl moiety, preventing decomposition of the molecule by either elimination or nucleophilic substitution pathways. A typical example of such a stabilization due to structural features is illustrated by the preparation of 4-(iodosyl)tricyclene derivatives by Morris and co-workers.[154] Difluoride **35** was prepared in the form of a pale yellow solid by treatment of a solution of 4-iodotricyclene **34** in carbon tetrachloride with an excess of xenon difluoride, followed by removal of solvent (Eq. 11.8).[154b]

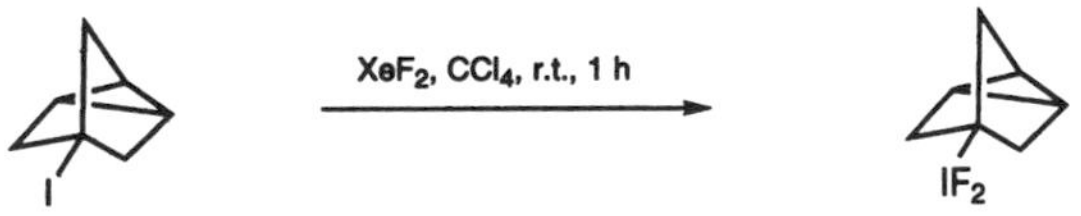

$$\mathbf{34} \xrightarrow{XeF_2,\ CCl_4,\ r.t.,\ 1\ h} \mathbf{35} \qquad (11.8)$$

Similarly, reaction of 4-iodotricyclene with *m*-CPBA affords the respective 4-[bis-(*m*-chlorobenzoyloxy)iodo]tricyclene as a stable, microcrystalline compound.[154a]

Historically, the first example of a stable alkyliodosyl derivative was represented by (dichloroiodo)methylsulfones, $RSO_2CH_2ICl_2$, the stability of which was explained by the strong electron-withdrawing properties of the sulfonyl group combined with its bulk.[155] More recent examples of alkyliodosyl derivatives, stabilized due to electron-withdrawing substituents, include the derivatives of 1-iodo-1*H*,1*H*-perfluoroalkanes, $R_fCH_2IX_2$,[156–159] and 1-iodoperfluoroalkanes, R_fIX_2.[160–163]

Three types of derivatives of 1-iodo-1*H*,1*H*-perfluoroalkanes are known: bis(trifluoroacetates) $R_fCH_2I(CO_2CF_3)_2$,[157a,158] difluorides $R_fCH_2IF_2$,[157b] dichlorides $R_fCH_2ICl_2$,[156] and (hydroxy)sulfonates, $R_fCH_2I(OH)OSO_2R$.[159] Bis(trifluoroacetates) **37** are prepared in almost quantitative yield by the oxidation of 1-iodo-1*H*,1*H*-perfluoroalkanes **36** with peroxytrifluoroacetic acid (Eq. 11.9).[157a,158]

$$C_nF_{2n+1}CH_2{-}I \xrightarrow{CF_3CO_3H,\ CF_3CO_2H,\ 0\ \text{to}\ 25\ ^\circ C,\ 24\ h} C_nF_{2n+1}CH_2{-}I(OCOCF_3)_2 \qquad (11.9)$$

36 $n = 1,2,3,7$, etc. **37**

Trifluoroacetates **37** can be converted to sulfonates $R_fCH_2I(OH)OSO_2R$ by treatment with the appropriate sulfonic acid.[159] In contrast to the starting trifluoroacetates **37**, sulfonates have a substantially higher thermal stability and are not water-sensitive.

Montanari and DesMarteau recently reported the synthesis of stable 1-(dichloroiodo)-1*H*,1*H*-perfluoroalkanes by the direct chlorination of the respective iodides.[156]

Another important structural class of stable alkyliodosyl compounds is represented by the derivatives of 1-iodosylperfluoroalkanes, R_fIX_2.[160–163] Four types of these compounds are known: difluorides R_fIF_2,[94,163] bis(trifluoroacetates) $R_fI(CO_2CF_3)_2$,[160,162] (hydroxy)sulfonates $R_fI(OH)OSO_2R$,[162] and iodosylperfluoroalkanes R_fIO.[94,164a]

Bis(trifluoroacetates), $R_fI(CO_2CF_3)_2$, were originally prepared by Yagupolskii and co-workers by oxidation of 1-iodoperfluoroalkanes with peroxytrifluoroacetic acid.[160] These compounds have found practical application for the preparation of perfluoroalkyl(phenyl)iodonium salts, which are used as electrophilic fluoroalkylating reagents (FITS reagents).[165] Bis(trifluoroacetates) can be converted to difluorides R_fIF_2 by reaction with SF_4, or hydrolyzed to iodosylperfluoroalkanes.[94,164a]

[Hydroxy(sulfonyloxy)iodo]perfluoroalkanes, $R_fI(OH)OSO_2R$, are conveniently prepared by treatment of trifluoroacetates with the appropriate sulfonic acid.[162]

Recently, a new example of stable alkyliodosyl derivatives, [bis(trifluoroacetoxy)iodo]methylsulfones, $RSO_2CH_2I(CO_2CF_3)_2$, was reported.[164b]

11.3.3 Iodine(III) Species with Two Carbon Ligands

According to conventional classification, iodonium salts are defined as positively charged 8-I-2 species with two carbon ligands and a negatively charged counter ion, $R_2I^+X^-$. A large variety of such compounds are known and an overwhelming majority of them have at least one aryl group as a ligand. The most common and well-investigated class of these compounds is diaryliodonium salts, known for over 100 years and extensively covered in previous reviews.[10–13,16] In the last 10–15 years several new structural classes of iodonium species have been synthesized; the most important of these are the aryliodonium derivatives bearing cyano-, alkynyl-, alkenyl-, or fluoroalkyl groups as the second ligand.

11.3.3.1. ***Cyanoiodonium Salts*** Three structural types of cyanoiodanes are known: (dicyano)iodonium triflate, $(CN)_2IOTf$,[166,167] [cyano(organosulfonyloxy)iodo]arenes, $ArI(CN)OSO_2R$,[79,102,168] and 1-cyanobenziodoxoles.[169] The thermally unstable and air-sensitive (dicyano)iodonium triflate **39** can be prepared by the reaction of iodosyl triflate **38** with cyanotrimethylsilane in methylene chloride (Eq. 11.10).[166,167]

$$\underset{\mathbf{38}}{O{=}I{-}OTf} + 2\,NCSiMe_3 \xrightarrow[40\text{-}60\%]{CH_2Cl_2,\ -78\ \text{to}\ -20\ ^\circ C} \underset{\mathbf{39}}{NC{-}\overset{+}{I}(CN)\ \ ^-OTf} \tag{11.10}$$

[Cyano(organosulfonyloxy)iodo]arenes **40** are generally prepared by reactions of ArIO or $ArI(OAc)_2$ with trimethylsilyl triflate and cyanotrimethylsilane (Eq 11.11). Cyanides **40** are relatively stable, white microcrystalline solids that decompose over several days at room temperature, but may be stored for extended periods in a refrigerator without change.[79,102,168]

$$ArIO + Me_3SiOTf + NCSiMe_3 \xrightarrow[89\%]{CH_2Cl_2,\ -20\ \text{to}\ 0\ ^\circ C} \underset{\mathbf{40}}{Ar{-}\overset{+}{I}(CN)\ \ ^-OTf} \tag{11.11}$$

Cyanobenziodoxoles **42** represent the most stable structural type of cyanoiodanes. They can be prepared in one step by the reaction of cyanotrimethylsilane with the respective 1-hydroxybenziodoxoles **41** (Eq. 11.12):[169]

$$\mathbf{41} \xrightarrow[74\text{-}94\%]{Me_3SiCN,\ CH_3CN,\ \text{r.t.},\ 3\ h} \mathbf{42} \tag{11.12}$$

41, $R = CH_3$, CF_3 or $2R = O$ **42**

All three adducts **42** were isolated as thermally stable white, microcrystalline solids. The structure of one of them was unambiguously established by a single-crystal X-ray analysis.[169b]

11.3.3.2 Alkynyliodonium Salts Alkynyl(phenyl)iodonium salts are relatively recent members of the family of polyvalent organic iodine species.[16,25,29,30] They are best prepared by interaction of a sila- or tin-acetylene with an electrophilic iodine reagent.[80–88,170,171] Key reagents are PhIO, HTIB, the μ-oxo-bridged species **12** and the cyano(phenyl)iodonium triflate **40**.

Interaction of a large variety of β-functionalized alkynylstannanes with **40** provides ready access to diverse β-functionalized alkynyliodonium salts **43** (Eq. 11.13) in 42–89% isolated yields.[86]

$$\mathrm{YC{\equiv}CSnR_3} + \underset{\textbf{40}}{\mathrm{PhI(CN)OTf}} \xrightarrow[-42^\circ \text{ to } 20^\circ\,\mathrm{C}]{\mathrm{CH_2Cl_2}} \mathrm{YC{\equiv}C\overset{+}{I}Ph\ \overset{-}{O}Tf} + \mathrm{R_3SnCN} \tag{11.13}$$

43: Y = CN, Cl, $ArSO_2$, RC(O)-, $R_2NC(O)$-, etc.

This procedure also allows the preparation of the novel bis-iodonium and diynyl(phenyl)iodonium triflates salts.[80,81,83,88]

Several X-ray molecular structure determinations for alkynyliodonium salts have been reported.[25,29] The data are all consistent with the pseudo-trigonal bipyramidal, or T-shaped geometry, of 8-I-2 species. In all known cases, the aryl group occupies an equatorial position, whereas the alkynyl moiety and the counter ion occupy apical positions. The alkynyl—I bond length is 2.0± 0.01 Å. The I—O distances vary from 2.34 to 2.69 Å. The C_{sp}—I—O bond angles vary from 166° to 172° and the C_{sp^2}—I—C_{sp} bond angles are between 90° and 95°.

The preparation, properties, and chemistry of alkynyliodonium salts were recently summarized in several comprehensive reviews.[16b,25,29]

11.3.3.3 Alkenyliodonium Salts Several examples of alkenyliodonium salts have been known for many years, but only in the last decade have these compounds became readily available and found some synthetic application. The first general method of synthesis of alkenyliodonium species was developed by Ochiai in the mid-1980s.[172] This method is based on the reaction of silylated alkenes with iodosylbenzene in the presence of Lewis acids, leading to the stereoselective formation of various alkenyliodonium tetrafluoroborates in good yield.

A very general and mild procedure for stereospecific synthesis of alkenyliodonium triflates **45** involves phenyl(cyano)iodonium triflate **40** as an iodonium transfer reagent in reactions with stannylated alkenes **44** (Eq. 11.14).[173] This method was also applied to the preparation of the parent vinyliodonium triflate from tributyl(vinyl)tin.[87]

$$\underset{\textbf{44}}{\mathrm{R^1R^2C{=}CH{-}SnBu_3}} + \underset{\textbf{40}}{\mathrm{PhI(CN)OTf}} \xrightarrow[60\text{-}90\%]{\mathrm{CH_2Cl_2},\ -23\,^\circ\mathrm{C}} \underset{\textbf{45}}{\mathrm{R^1R^2C{=}CH{-}\overset{+}{I}Ph\ \overset{-}{O}Tf}} \tag{11.14}$$

R^1, R^2 = H, Alkyl, Ph

A variety of alkenyliodonium salts were synthesized via electrophilic addition of aryliodosyl derivatives to alkynes or allenes.[67,77,90–92]

Z-β-Substituted vinyliodonium salts can be prepared by a Michael-type nucleophilic addition of nucleophiles to alkynyliodonium salts.[174]

Several single crystal X-ray structures of alkenyliodonium salts have been reported in the literature.[77,81,172a,175] These structures generally can be considered as ionic, with a typical distance between iodine and the nearest anion of 2.8–3.0 Å. However, with consideration of the anionic part of the molecule, the overall experimentally determined geometry is T-shaped, similar to the 10-I-3 species. The C=C double bonds in alkenyliodonium salts have normal lengths of 1.31–1.34 Å, the C_{sp^2}—I distances can vary from 2.07 to 2.13 Å, and the C—I—C angles are in the range 91°–99° degrees.

11.3.3.4 Aryl- and Heteroaryliodonium Salts Aryl-, as well as heteroaryliodonium salts, belong to the most common, stable, and well investigated class of polyvalent iodine compounds. The chemistry of aryliodonium salts was extensively covered in previous reviews.[5–13,16] Aryliodonium salts are usually prepared by reaction of arenes or silylated arenes with electrophilic aryliodosyl derivatives.[5–13,16]

Inorganic iodosyl derivatives, such as iodosyl trifluoromethanesulfonate and iodosyl fluorosulfate, react with arenes or trimethylsilylarenes under mild conditions to afford iodonium triflates or hydrosulfates.[58,59] Naumann and coworkers used a similar procedure for the preparation of bis(pentafluorophenyl)iodonium triflate from iodine tris(trifluoroacetate), pentafluorobenzene, and the corresponding sulfonic acid.[176]

A very mild and general method for the preparation of aryl- and heteroaryliodonium triflates is based on iodonium transfer reactions of iodine(III) cyanides with the respective aryl- or heteroarylstannanes.[102,104,166,167] (Dicyano)iodonium triflate **46** generated *in situ* reacts with tributyltin derivatives of aromatic compounds **47**, affording the corresponding diaryliodonium salts **48** (Eq. 11.15).[167]

$$\underset{\mathbf{46}}{(NC)_2\overset{+}{I}\cdot{}^{-}OTf} + \underset{\mathbf{47}}{2ArSnBu_3} \xrightarrow[45\text{-}74\%]{CH_2Cl_2,\ -40\ \text{to}\ 20\ ^\circ C} \underset{\mathbf{48}}{Ar_2\overset{+}{I}\ {}^{-}OTf} \tag{11.15}$$

Ar = Ph, 3-MeOC$_6$H$_4$, 4-MeOC$_6$H$_4$, or heteroaryl

Since (dicyano)iodonium triflate **46** does not possess any pronounced electrophilic or oxidizing properties and is not acidic, it is especially useful for the synthesis of various bis(heteroaryl)iodonium salts from acid-sensitive substrates.[166]

Another iodonium transfer reagent, β-(chlorovinyl)iodonium dichloride $ClCH{=}CHICl_2$, is especially useful for the preparation of aryliodonium salts from lithiated arenes.[177] Recently, this compound was applied as the iodonium transfer reagent in the synthesis of hybrid, iodonium–transition-metal macrocyclic squares.[178]

A number of X-ray molecular structure determinations have been reported for aryl- and heteroaryliodonium salts with the structural features of the iodonium moiety generally similar to that previously described for the alkynyl and alkenyl species.[179]

11.3.3.5 Alkyl- and Fluoroalkyliodonium Salts Iodonium salts with one or two non-substituted aliphatic alkyl groups generally lack stability. However, several examples of these unstable species were generated and investigated by nuclear magnetic resonance (NMR) spectroscopy at low temperatures and some of them even were isolated in the form of labile crystalline salts.[16a,24]

The presence of electron-withdrawing groups in the alkyl group of iodonium salts has a pronounced stabilizing effect. The most stable and important derivatives of this type are perfluoroalkyl(aryl)iodonium salts. These salts as chlorides were first prepared in 1971 by Yagupolskii and co-workers[180] and later widely applied as electrophilic fluoroalkylating reagents by Umemoto and co-workers.[165] The triflate salts, $PhIC_nF_{2n+1}OTf$, originally were synthesized by the reaction of [bis(trifluoroacetoxy)iodo]perfluoroalkanes with benzene in the presence of triflic acid.[161] A recent and more general method for the preparation of various perfluoroalkyl(aryl)iodonium sulfonates involves the reaction of [hydroxy(sulfonyloxy)iodo]perfluoroalkanes with arylsilanes under Lewis acid catalysis.[162]

In a similar manner, 1*H*, 1*H*-perfluoroalkyl(aryl)iodonium triflates **50** are best prepared by reaction of triflates **49** with trimethylsilylarenes under mild conditions (Eq. 11.16).[159]

$$C_nF_{2n+1}CH_2{-}I(OH)(OSO_2CF_3)\ (\mathbf{49}) + ArSiMe_3 \xrightarrow[62\text{-}85\%]{CH_2Cl_2,\ -30\ \text{to}\ 0\ ^\circ C,\ 2\ h} C_nF_{2n+1}CH_2{-}\overset{+}{I}{-}Ar\ \ ^{-}OSO_2CF_3\ (\mathbf{50})$$

n = 1,2; Ar = Ph, 4-MeC_6H_4, 2-MeC_6H_4, 4-$Me_3SiC_6H_4$, 4-$MeOC_6H_4$

(11.16)

Fluoroalkyl(aryl)iodonium triflates have found practical application as electrophilic fluoroalkylating reagents toward a variety of organic substrates, such as arenes, carbanions, alkynes, alkenes, carbonyl compounds, amines, phosphines, and sulfides.

Recently, the preparation, X-ray crystal structure, and chemistry of [(arylsulfonyl)methyl](phenyl)iodonium triflates, $RSO_2CH_2IPh{\cdot}OTf$, was reported.[164b]

11.3.3.6 Iodonium Ylides The first preparation of an iodonium ylide by the reaction of dimedone and $PhIF_2$ was reported by Neiland and co-workers[181] in 1957. Since then a large number of stable iodonium ylides have appeared, and many synthetic applications have emerged. The chemistry of iodonium ylides was reviewed by Koser,[16a] and more recently it was extensively covered in the book of Varvoglis.[24]

There is significant current interest in phenyliodonium ylides stabilized by two organosulfonyl groups, $PhIC(SO_2R)_2$.[182,183] Zhu and Chen first reported the preparation of phenyliodonium bis(perfluoroalkanesulfonyl)methide **51** from DIB and bis(perfluoroalkanesulfonyl)methane (Eq. 11.17).[182a]

$$PhI(OAc)_2 + CH_2(SO_2R)_2 \xrightarrow[40\text{-}75\%]{CH_2Cl_2,\ 20\text{ to }40\ ^\circ C,\ 12\text{-}24\ h} Ph\overset{+}{I}{-}^{-}C(SO_2R)_2 \quad (11.17)$$

$R = CF_3, C_4F_9$ or $2R = (CF_2)_3$ **51**

Ylides **51** have unusually high thermal stability, they can be stored without decomposition at room temperature for several months. The X-ray structural analysis for the trifluoromethanesulfonyl derivative **51** shows a structure with the C—I bond length about 1.9 Å and the C—I—C bond angle 98°.[182c]

Hackenberg and Hanack reported the synthesis of similar, nonsymmetric cyano(perfluoroalkanesulfonyl) substituted ylides by an analogous reaction.[183]

Phenyliodonium bis(benzenesulfonyl)methide, $PhIC(SO_2Ph)_2$ and the dicarbonyl derivative $PhIC(COR)_2$ are generally prepared by a reaction of $PhI(OAc)_2$ with $H_2C(SO_2R)_2$ or $H_2C(COR)_2$ under basic conditions.[181,182,184] These two types of iodonium ylides usually have relatively low thermal stability.

11.3.4 Iodine(III) Species with Three Carbon Ligands

In general, iodine(III) species with three carbon ligands are unstable at room temperature. Only a few examples of isolable compounds of this type were reported in the earlier literature, namely, triphenyl iodine[185] and 5-aryl-5*H*-dibenziodoles.[186] All triaryl iodanes are highly unstable and air-sensitive.

A more recent example of a relatively stable, isolable compound of this type is (dicyanoiodo)benzene, $PhI(CN)_2$, which can be prepared in good yield by the reaction of iodosylbenzene with trimethylsilyl cyanide in methylene chloride.[69]

11.4 PENTACOORDINATE IODINE COMPOUNDS

11.4.1 Inorganic Iodine(V) Derivatives

Iodine pentafluoride, IF_5, is the only known binary interhalogen compound of iodine(V).[1–7] This compound was first prepared in 1871 by heating of I_2 with AgF.[3] It is a colorless liquid with a boiling point of 98°C and a freezing point of 9.6°C.[3] According to microwave and electron-diffraction measurements, the

IF_5 molecule has a tetrahedral pyramidal geometry with four equatorial fluorine atoms and one axial fluorine atom with bond distances F_{eq}—I = 1.869 Å and F_{ax}—I = 1.844 Å (structure **52**).[187] The X-ray crystal structure of the molecular addition compound $XeF_2 \cdot IF_5$ with more-or-less discrete IF_5 molecules has also been reported.[188] The salts $M^+IF_6^-$ can be prepared by the reaction of IF_5 with the respective fluorides, MF. Crystal structures of $(CH_3)_4N^+IF_6^-$ and $NO^+IF_6^-$ were recently reported.[189] Reaction of IF_5 with I_2O_5 gives stable, solid oxyfluorides IOF_3 and IO_2F.[3,5]

Various oxygen-bonded iodine(V) derivatives are known.[1–7] Iodine pentoxide, the most important and thermally stable of the iodine oxides, is best prepared by dehydration of iodic acid at 200°C in the form of a hygroscopic, white solid. It readily absorbs water from the atmosphere, giving the hydrate, HI_3O_8. On dissolution in water, I_2O_5 regenerates HIO_3.[5]

Iodine pentoxide has a polymeric structure (**53**) with primary I—O bonds of 1.77 to 1.95 Å in the distinguishable molecular I_2O_5 units and intermolecular I···O distances as short as 2.23 Å.[190]

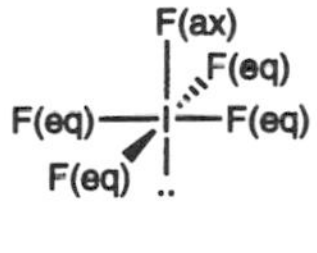

F(ax)
F(eq)
F(eq)—I—F(eq)
F(eq)

52

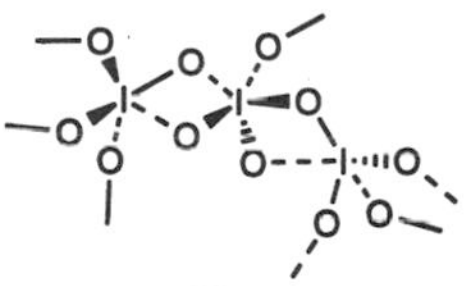

53

Several mixed iodine(III/V) oxides, such as I_2O_4 and I_4O_9, are also known.[5] Recently, a structure of a mixed iodine(V/VII) oxide, I_4O_{12}, was reported.[191]

A number of iodyl derivatives (IO_2X) of strong inorganic acids have been synthesized and characterized. The earliest reports on iodyl derivatives of sulfuric or polysulfuric acids date well back into the last century.[1,2] More recently, a crystal structure of a mixed iodine(III/V) sulfate, $(I_3O_6)(HSO_4)$, was reported.[192] The fluorosulfate IO_2OSO_2F[193] and the triflate IO_2OSO_2F[194] are also known. Raman and IR spectra of these compounds indicate the presence of discrete IO_2—groups with relatively weak secondary I···O bonding.[194]

11.4.2 Organic Iodine(V) Derivatives

The most common structural class of organic iodine(V) compounds is represented by iodylarenes, $ArIO_2$. Iodylarenes are generally prepared by direct oxidation of iodoarenes with strong oxidants, or by disproportionation of iodosylarenes. Iodylbenzene is a colorless, microcrystalline solid, which detonates violently upon heating. X-ray structural analysis of $PhIO_2$ revealed a distorted octahedral arrangement with three primary bonds to iodine with bond distances in the range 1.92–2.01 Å and three secondary I—O bonds of 2.57–2.73 Å.[44]

There is substantial current interest in the cyclic iodylarenes, which have better stability due to the incorporation of the pentavalent iodine into a five-membered ring.[43,46,195,196] Similar to 2-iodosylbenzoic acid, 2-iodylbenzoic acid has the actual structure of the cyclic hydroxyiodane oxide (**54**), as determined by X-ray structural analysis.[42,43] It is most commonly prepared by oxidation of 2-iodobenzoic acid with potassium bromate in sulfuric acid (Eq. 11.18).[45,196,197]

$KBrO_3$, H_2SO_4, 68 °C, 4 h; 93% → **54** (11.18)

Oxide **54** was reported to be explosive under excessive heating or impact,[198] but Dess and Martin suggested that the explosive properties of some samplcs **54** were due to the presence of bromate or other impurities.[45] An alternative preparation of iodylbenzoic acids involve oxidation of the respective iodobenzoic acids with excess peracetic acid or aqueous sodium hypochlorite.[195a]

Benziodoxole oxide **54** is widely used as a starting material for the preparation of triacetate **55**, the so-called Dess–Martin periodane.[45,195] Recently, Ireland and Liu reported an improved procedure for its preparation by the reaction of oxide **54** with acetic anhydride in the presence of *p*-toluenesulfonic acid (Eq. 11.19).[196]

54 → Ac_2O, 0.5% TsOH, 80 °C, 2 h; 91% → **55** (11.19)

In the early 1990s Martin and co-workers reported the synthesis, structure and properties of several other, stable, cyclic periodanes **56**–**58**.[45,46,195b]

56: R = H, Me; X = F, OAc, $OCOCF_3$, etc.

57: R = Me, *t*-Bu; 2X = O or X= F, OTf, $OCOCF_3$, etc.

58: X = F, Cl, Br, O^-, $NHCMe_3$, OTf, $OCOCF_3$, etc.

The Dess–Martin periodane and other cyclic iodylarenes have received some practical application in organic chemistry as selective oxidizers.

11.5 IODINE(VII) COMPOUNDS

Only inorganic derivatives of iodine(VII) are known. Iodine(VII) fluoride, IF_7, has been known since 1931[199] and many papers dealing with its properties and structure have been published. It is best prepared[200] by heating IF_5 with F_2 at 150°C in the form of a colorless gas with a musty odor. On cooling, colorless crystals are formed which melt under slight pressure at 6.5°C. IF_7 is a unique, neutral, simple binary compound in which heptacoordination is shown. Numerous attempts of crystal structure determination of solid IF_7 were inconclusive due to disorder problems. Nevertheless, important structural information was obtained from ^{19}F NMR, microwave, vibrational spectra, and gas-phase electron diffraction data.[48] The isolated IF_7 molecule (structure **59**) has the form of a pentagonal bipyramid belonging to the symmetry group D_{5h}, in which the bonds of the axial F—I—F unit (1.786 Å) are shorter than those of the more congested equatorial IF_5 unit (1.858 Å).

Two oxyfluorides, IO_2F_3[201,202] and IOF_5,[47] were reported in the literature. IO_2F_3 can be synthesized by the reaction of tetrafluoroorthoperiodic acid, $HOIOF_4$, and SO_3 in the form of a yellow sublimable solid with a melting point of 41°C.[201] According to ^{19}F NMR and Raman spectroscopy, in the liquid state IO_2F_3 exists as a cyclic trimer with *cis*-oxygen bridges, in the boat conformation.[202] The second oxyfluoride, IOF_5, can be prepared by the interaction of IF_7 with SiO_2.[200] The high symmetry, C_{4v}, of IOF_5 renders its structure determination very difficult. Nevertheless, on the basis of a combined electron diffraction-microwave study[203] and *ab initio* calculations,[47] structure **60** was determined for IOF_5. In this structure, the axial and the equatorial I—O bonds are of comparable lengths (about 1.826 Å), the I—O bond distance is 1.725 Å, and the O—I F$_{eq}$ bond angle is close to 97.2°.[47] IOF_5 reacts with tetramethylammonium fluoride to give a stable, crystalline salt, $(CH_3)_4N^+IOF_6^-$.[49] X-ray analysis of this salt showed the expected, pentagonal bipyramidal structure for the IOF_6^- anion with the oxygen atom occupying one of the axial positions (structure **61**). The I—O bond in this compound is relatively short (1.75—1.77 Å), which indicates significant double-bond character. The axial I—F bonds (1.823 Å) and the equatorial I—F bonds (average 1.88 Å) in IOF_6^- are significantly longer than the corresponding bonds in IF_7, which can be attributed to greater I–F bond polarities due to the formal negative charge on IOF_6^-.[49]

Iodine(VII) oxide, I_2O_7, is unknown. A mixed iodine(V/VII) oxide, I_4O_{12}, was recently reported.[204] The crystal structure of this oxide includes the basic molecular unit I_4O_{12} with two octahedrally coordinated iodine(VII) atoms and two trigonal pyramidal iodine atoms in the +V oxidation state.

Periodic acid and periodates are the most important, commercially available compounds of iodine(VII). Periodates were first prepared in 1833 by oxidation of $NaIO_3$ by chlorine in alkaline solutions.[1] The industrial preparation of periodates is based on electrolytic oxidation of iodates.[7] Aqueous solutions of periodic acid are best obtained by treating barium paraperiodate with concentrated nitric acid. In solution it exists as a fairly weak orthoperiodic acid, H_5IO_6, which can be dehydrated to HIO_4 by heating to 100°C *in vacuo*.[5] Structural investigations indicate an octahedral configuration about iodine in most periodates.[5]

Periodic acid and periodates have found wide application in organic chemistry as powerful oxidizers.[205] The well-known periodate glycol oxidation involves the cyclic periodate esters as key intermediates; however, none of the organic periodate esters has been isolated as a stable compound.

REFERENCES

1. (a) J. W. Mellor, *A Comprehensive Treatise on Inorganic and Theoretical Chemistry*. (Longmans, Green and Co, London, 1922), Vol. 2. (b) *Supplement to Mellor's Comprehensive Treatise on Inorganic and Theoretical Chemistry* (Longmans, Green and Co, London, 1956), Supplement 2, Part 1.
2. *Gmelins Handbuch der Anorganischen Chemie*, 8 Auflage (Verlag Chemie, Berlin, 1933), "Jod", System-nummer 8.
3. R. C. Brasted, in *Comprehensive Inorganic Chemistry*, M. C. Sneed, J. L. Maynard, R. C. Brasted, Eds (D. Van Norstrand Co., Inc., Princeton, New Jersey, 1954), Vol. 3, Ch. 4, p. 78.
4. *Halogen Chemistry*, V. Gutmann, Ed. (Academic Press, New York, 1967), Vols 1–3.
5. A. J. Downs, C. J. Adams, in *Comprehensive Inorganic Chemistry*, J. C. Bailar, Jr., H. J. Emeleus, R. Nyholm, A. F. Trotman-Dickenson, Eds (Pergamon Press, Oxford, 1973), Vol. 2, pp 1107–1594.
6. N. N. Greenwood, A. Earnshaw, *Chemistry of the Elements* (Pergamon Press, Oxford, 1984), pp. 920–1041.
7. J. Shamir, in *Encyclopedia of Inorganic Chemistry*, R. B. King, Ed. (Wiley, Chichester, New York, 1994), Vol. 2, pp. 646–660.
8. J. L. Gay Lussac, *Ann. Chim.* **91**, 5 (1814).
9. C. Willgerodt, *J. Prakt. Chem.* **33**, 154 (1886).

10. C. Willgerodt, *Die Organischen Verbindungen mit Mehrwertigen Jod* (Ferdinand Enke Verlag, Stuttgart, 1914).
11. R. B. Sandin, *Chem. Rev.* **32**, 249 (1943).
12. F. M. Beringer, E. M. Gindler, *Iodine Abstract Rev.* **3**, 70 (1956).
13. D.F. Banks, *Chem. Rev.* **66**, 243 (1966).
14. A. Varvoglis, *Chem. Soc. Rev.* **10,** 7099 (1981).
15. T. Umemoto, *Yuki Gosei Kagaku Kyokai Shi* **41**, 251 (1983).
16. (a) G. F. Koser, in *The Chemistry of Functional Groups, Suppl. D*, S. Patai, Z. Rappoport, Eds (Wiley-Interscience, Chichester, 1983), Chs 18 and 25, pp. 721–811 and 1265–1351; (b) G. F. Koser, in *The Chemistry of Halides, Pseudo-Halides and Azides, Suppl. D2*, S. Patai, Z. Rappoport, Eds (Wiley-Interscience, Chichester, 1995), Ch. 21, pp. 1173–1274.
17. T. T. Nguyen, J. C. Martin, in *Comprehensive Heterocyclic Chemistry*, A. R. Katritzky, C. W. Rees, Eds (Pergamon Press, Oxford, 1984), Vol. 1, p. 563.
18. A. Varvoglis, *Synthesis*, 7099 (1984).
19. M. Ochiai, Y. Nagao, *Yuki Gosei Kagaku Kyokai Shi* **44,** 660 (1986).
20. R. M. Moriarty, O. Prakash, *Acc. Chem. Res.* **19**, 244 (1986).
21. E. B. Merkushev, *Russian Chem. Rev.* **56**, 826 (1987).
22. R. M. Moriarty, R. K. Vaid, G. F. Koser, *Synlett*, 365 (1990).
23. R. M. Moriarty, R. K. Vaid, *Synthesis*, 431 (1990).
24. (a) A. Varvoglis, *Hypervalent Iodine in Organic Synthesis* (Academic Press, New York, 1996); (b) A. Varvoglis, *The Organic Chemistry of Polycoordinated Iodine.* (VCH Publishers, Inc., New York, 1992).
25. P. J. Stang, *Angew. Chem. Int. Ed. Engl.* **31**, 274 (1992).
26. O. Prakash, N. Saini, P. K. Sharma, *Synlett*, 221 (1994).
27. O. Prakash, N. Saini, P. K. Sharma, *Heterocycles* **38**, 409 (1994).
28. O. Prakash, S. P. Singh, *Aldrichimica Acta* **27**, 15 (1994); O. Prakash, *Aldrichimica Acta* **28**, 63 (1995).
29. P. J. Stang, in *The Chemistry of Triple-Bonded Functional Groups, Supplement C2*, S. Patai, Ed. (John Wiley & Sons, Ltd., Chichester, 1994), Vol. 2, Ch. 20, pp 1164–1182.
30. P. J. Stang, in *Modern Acetylene Chemistry*, P.J . Stang, F. Diederich, Eds (VCH Publishers, Weinheim, 1995), Ch. 3, pp. 67–98.
31. P. J. Stang, V. V. Zhdankin, *Chem. Rev.* **96**, 1123 (1996).
32. C. W. Perkins, J. C. Martin, A. J. Arduengo III, W. Law, A. Alegria, J. K. Kochi, *J. Am. Chem. Soc.* **102**, 7753 (1980).
33. E. M. Archer, T. G. D. van Schalkwyk, *Acta. Cryst.* **6**, 88 (1953).
34. N. W. Alcock, R. M. Countryman, S. Esperas, J. F. Sawyer, *J. Chem. Soc., Dalton Trans.*, 854 (1994).
35. J. I. Musher, *Angew. Chem. Int. Ed. Engl.* **8**, 65 (1969).
36. J. C. Martin, *Science* **221**, 509 (1983).
37. A. E. Reed, P. v. R. Schleyer, *J. Am. Chem. Soc.* **112**, 1434 (1990).
38. R. D. Burbank, G. R. Jones, *Inorg. Chem.* **13**, 1071 (1974).

39. A. J. Edwards, P. Taylor, *J. Chem. Soc., Dalton Trans.*, 2174 (1975).
40. D. B. Dess, J. C. Martin, *J. Am. Chem. Soc.* **104**, 902 (1982).
41. A. P. Bozopoulos, P. J. Rentzeperis, *Acta Crystallogr.* **C43**, 142 (1987).
42. J. Gougoutas, *Cryst. Struct. Comm.* **10**, 489 (1981).
43. A. R. Katritzky, J. P. Savage, G. J. Palenic, K. Quian, Z. Zhang, H. D. Durst, *J. Chem. Soc., Perkin Trans. 2*, 1657 (1994).
44. N. W. Alcock, J. F. Sawyer, *J. Chem. Soc., Dalton Trans.*, 115 (1980).
45. D. B. Dess, J. C. Martin, *J. Am. Chem. Soc.* **113**, 7277 (1991).
46. D. B. Dess, J.C. Martin, *J. Am. Chem. Soc.* **115**, 2488 (1993).
47. K. O. Christe, E. C. Curtis, D. A. Dixon, *J. Am. Chem. Soc.* **115**, 9655 (1993).
48. K. O. Christe, E. C. Curtis, D. A. Dixon, *J. Am. Chem. Soc.* **115**, 1520 (1993).
49. K. O. Christe, D. A. Dixon, A. R. Mahjoub, H. P. A. Mercier, J. C. P. Sanders, K. Seppelt, G. J. Schrobilgen, W. W. Wilson, *J. Am. Chem. Soc.* **115**, 2696 (1993).
50. A. I. Boldyrev, V. V. Zhdankin, J. Simons, P. J. Stang, *J. Am. Chem. Soc.* **114**, 10569 (1992).
51. M. Schmeisser, D. Naumann, R. Scheele, *J. Fluorine Chem.* **1**, 369 (1972).
52. K. H. Basijk, E. H. Wiebenga, *Acta Crystallogr.* **7**, 417 (1954).
53. N. S. Zefirov, G. A. Sereda, S. E. Sosonuk, N. V. Zyk, T. I. Likhomanova, *Synthesis* 1359 (1995).
54. M. E. Millon, *Ann. Chim. Phys.* **12**, 345 (1844).
55. S. Furuseth, K, Selte, H. Hope, A. Kjekshus, B. Klewe, *Acta Chem. Scand.* **A28**, 71 (1974).
56. G. Daehlie, A. Kjekshus, *Acta Chem. Scand.* **18**, 144 (1964).
57. J. R. Dalziel, H. A. Carter, F. Aubke, *Inorg. Chem.* **15**, 1247 (1976).
58. P. J. Stang, V. V. Zhdankin, R. Tykwinski, N. S. Zefirov, *Tetrahedron Lett.* **32**, 7497 (1991).
59. N. S. Zefirov, T. M. Kasumov, A .S. Koz'min, V. D. Sorokin, P. J. Stang, V. V. Zhdankin, *Synthesis*, 1209 (1993).
60. T. M. Kasumov, V. K. Brel, A. S. Koz'min, N. S. Zefirov, *Synthesis*, 775 (1995).
61. R. J. Gillespie, J. B. Milne, *Inorg. Chem.* **5**, 1236 (1966).
62. T. Birchall, C. S. Frampton, P. Kapoor, *Inorg. Chem.* **28**, 636 (1989).
63. R. M. Moriarty, J. W. Kosmeder II, in *Encyclopedia of Reagents for Organic Synthesis* L.A. Paquette, Ed. (John Wiley & Sons, Ltd., Chichester, England, 1995), Vol. 4, pp. 2821–2822.
64. H. Saltzman, J. G. Sharefkin, *Org. Syn.* **43**, 60 (1963); *Org. Syn., Coll.Vol. V*, 658 (1973).
65. C. J. Carmalt, J. G. Crossley, J. G. Knight, P. Lightfoot, A. Martin, M. P. Muldowney, N. C. Norman, A. G. Orpen, *J. Chem. Soc., Chem. Commun.* 2367 (1994).
66. Y. Gao, X. Jiao, W. Fan, J.-K. Shen, Q. Shi, F. Basolo, *J. Coord. Chem.* **29**, 349 (1993).
67. P. J. Stang, V. V. Zhdankin, N. S. Zefirov, *Mendeleev Commun.*, 159 (1992).

68. N. S. Zefirov, V. V. Zhdankin, S. O. Safronov, A. A. Kaznacheev, *Zh. Org. Khim.* **25**, 1807 (1989).

69. V. V. Zhdankin, R. Tykwinski, B. L. Williamson, P. J. Stang, N. S. Zefirov, *Tetrahedron Lett.* **32**, 733 (1991).

70. P. Magnus, C. Hulme, W. Weber, *J. Am. Chem. Soc.* **116**, 4501 (1994).

71. P. Magnus, J. Lacour, W. Weber, *J. Am. Chem. Soc.* **115**, 9347 (1993).

72. P. Magnus, J. Lacour, *J. Am. Chem. Soc.* **114**, 3993 (1992).

73. P. Magnus, J. Lacour, *J. Am. Chem. Soc.* **114**, 767 (1992).

74. M. Arimoto, H. Yamaguchi, E. Fujita, M. Ochiai, Y. Nagao, *Tetrahedron Lett.* **28**, 6289 (1987).

75. R. M. Moriarty, R. K. Vaid, V. T. Ravikumar, B. K. Vaid, T. E. Hopkins, *Tetrahedron*, **44**, 1603 (1988).

76. R. M. Moriarty, R. K. Vaid, T. E. Hopkins, B. K. Vaid, A. Tuncay, *Tetrahedron Lett.* **30**, 3019 (1989).

77. D. A. Gately, T. A. Luther, J. R. Norton, M. M. Miller, O. P. Anderson, *J. Org. Chem.* **57**, 6496 (1992).

78. R. M. Moriarty, S. M. Tuladhar, R. Penmasta, A. K. Awasthi, *J. Am. Chem. Soc.* **112**, 3228 (1990).

79. V. V. Zhdankin, C. M. Crittell, P. J. Stang, N. S. Zefirov, *Tetrahedron Lett.* **31**, 4821 (1990).

80. P. J. Stang, V. V. Zhdankin, *J. Am. Chem. Soc.* **112**, 6437 (1990).

81. P. J. Stang, V. V. Zhdankin, *J. Am. Chem. Soc.* **113**, 4571 (1991).

82. P. J. Stang, B. L. Williamson, V. V. Zhdankin, *J. Am. Chem. Soc.* **113**, 5870 (1991).

83. P. J. Stang, R. Tykwinski, V. V. Zhdankin, *J. Org. Chem.* **57**, 1861 (1992).

84. B. L. Williamson, R. Tykwinski, P. J. Stang, *J. Am. Chem. Soc.* **116**, 93 (1994).

85. R. R. Tykwinski, P. J. Stang, *Tetrahedron* **49**, 3043 (1993).

86. B. L. Williamson, P. J. Stang, A. M. Arif, *J. Am. Chem. Soc.* **115**, 2590 (1993).

87. P. J. Stang, J. Ullmann, *Angew. Chem. Int. Ed. Engl.* **30**, 1469 (1991).

88. P. J. Stang, J. Ullmann, *Synthesis*, 1073 (1991).

89. R. T. Hembre, C. P. Scott, J. R. Norton, *J. Org. Chem.* **52**, 3650 (1987).

90. T. Kitamura, R. Furuki, K. Nagata, L. Zheng, H. Taniguchi, *Synlett*, 193 (1993).

91. T. Kitamura, R. Furuki, K. Nagata, L. Zheng, H. Taniguchi, P. J. Stang, *J. Org. Chem.* **57**, 6810 (1992).

92. T. Kitamura, R. Furuki, H. Taniguchi, P. J. Stang, *Tetrahedron* **48**, 7149 (1992).

93. N. S. Zefirov, V. D. Sorokin, V. V. Zhdankin, A. A. Koz'min, *Zh. Org. Khim.* **22**, 450 (1986).

94. V. V. Lyalin, V. V. Orda, L. A. Alekseeva, L. M. Yagupolskii, *Zh. Org. Khim.* **6**, 329 (1970).

95. B. C. Schardt, C. L. Hill, *Inorg. Chem.* **22**, 1563 (1983).

96. R. L. Amey, J. C. Martin, *J. Org. Chem.* **44**, 1779 (1979).

97. W. Carpenter, *J. Org. Chem.* **31**, 2688 (1966).

98. T. Fuchigami, T. Fujita, *J. Org. Chem.* **59**, 7190 (1994).

99. H. J. Lucas, E. R. Kennedy, *Org. Syn., Coll.Vol. III*, 482 (1955).
100. J. D. Protasiewicz, *J. Chem. Soc., Chem. Commun.*, 1115 (1995).
101. J. G. Sharefkin, H. Saltzman, *Org. Syn., Coll.Vol. V*, 660 (1973).
102. V. V. Zhdankin, M. C. Scheuller, P. J. Stang, *Tetrahedron Lett.* **34**, 6853 (1993).
103. P. J. Stang, V. V. Zhdankin, *J. Am. Chem. Soc.* **115**, 9808 (1993).
104. P. M. Gallop, M. A. Paz, P. J. Stang, V. V. Zhdankin, R. Tykwinski, *J. Am. Chem. Soc.* **115**, 11702 (1993).
105. A. McKillop, D. Kemp, *Tetrahedron* **45**, 3299 (1989).
106. M. Ochiai, K. Oshima, T. Ito, Y. Masaki, M. Shiro, *Tetrahedron Lett.* **32**, 1327 (1991).
107. M. Ochiai, Y. Takaoka, Y. Masaki, Y. Nagao, M. Shiro, *J. Am. Chem. Soc.* **112**, 5677 (1990).
108. D. G. Ray III, G.F. Koser, *J. Org. Chem.* **57**, 1607 (1992).
109. N. S. Zefirov, S. I. Kozhushkov, V. V. Zhdankin, S. O. Safronov, *Zh. Org. Khim.* **25**, 1109 (1989).
110. C. K. Lee, T. C. W. Mak, W. K. Li, J. F. Kirner, *Acta Crystallogr.* **B33**, 1620 (1977); M. Bardan, T. Birchall, C. S. Frampton, P. Kapoor, *Can. J. Chem.* **67**, 1878 (1989); H. J. Frohn, J. Helber, A. Richter, *Chem. Ztg.* **107**, 169 (1983).
111. G. F. Koser, R. H. Wettach, *J. Org. Chem.* **42**, 1476 (1977).
112. M. Schmeisser, E. Lehmann, D. Naumann, *Chem. Ber.* **110**, 2665 (1977).
113. G. F. Koser, R. H. Wettach, J. M. Troup, B. A. Frenz, *J. Org. Chem.* **41**, 3609 (1976).
114. O. Neiland, B. Karele, *J. Org. Chem. USSR (Engl. Transl.)* **6**, 889 (1970).
115. N. S. Zefirov, V. V. Zhdankin, Yu. V. Dan'kov, A. S. Koz'min, O. S. Chizhov, *Zh. Org. Khim.* **21**, 2461 (1985).
116. P. J. Stang, B. W. Surber, Z-C. Chen, K. A. Roberts, A. G. Anderson, *J. Am. Chem. Soc.* **109**, 228 (1987).
117. J. S. Lodaya, G. F. Koser, *J. Org. Chem.* **53**, 210 (1988).
118. E. Hatzigrigoriou, A. Varvoglis, M. Bakola-Christianopoulou, *J. Org. Chem.* **55**, 315 (1990).
119. G. F. Koser, R. H. Wettach, *J. Org. Chem.* **45**, 4988 (1980).
120. D. G. Ray III, G. F. Koser, *J. Am. Chem. Soc.* **112**, 5672 (1990).
121. A. J. Margida, G. F. Koser, *J. Org. Chem.* **49**, 4703 (1984); C. S. Carman, G. F. Koser, *J. Org. Chem.* **48**, 2534 (1983); A. J. Margida, G. F. Koser, *J. Org. Chem.* **49**, 3643 (1984).
122. R. M. Moriarty, R. Penmasta, I. Prakash, *Tetrahedron Lett.* **28**, 877 (1987).
123. G. F. Koser, J. S. Lodaya, D. G. Ray III, P. B. Kokil, *J. Am. Chem. Soc.* **110**, 2987 (1988).
124. G. W. Bushnell, A. Fischer, P. N. Ibrahim, *J. Chem. Soc., Perkin Trans. 2*, 1281 (1988).
125. J. Gallos, A. Varvoglis, N. W. Alcock, *J. Chem. Soc., Perkin Trans. 1*, 757 (1985); S. C. Kokkou, C. J. Cheer, *Acta Crystallogr.* **C42**, 1748 (1986).
126. N. S. Zefirov, A. S. Koz'min, V. V. Zhdankin, *Izv. AN SSSR, ser. khim.* 1682 (1983).

127. V. V. Zhdankin, R. Tykwinski, M. Mullikin, B. Berglund, R. Caple, N. S. Zefirov, A. S. Koz'min, *J. Org. Chem.* **54**, 2609 (1989).
128. T. M. Balthazar, D. E. Godaz, B. R. Stults, *J. Org. Chem.* **44**, 1447 (1979).
129. D. G. Naae, J. Z. Gougoutas, *J. Org. Chem.* **40**, 2129 (1975).
130. H. Jaffe, J. E. Leffler, *J. Org. Chem.* **40**, 797 (1975).
131. G. F. Koser, G. Sun, C. W. Porter, W. J. Youngs, *J. Org. Chem.*, **58**, 7310 (1993).
132. T. M. Balthazar, J. A. Miles, B. R. Stults, *J. Org. Chem.* **43**, 4538 (1978).
133. J. E. Leffler, H. Jaffe, *J. Org. Chem.* **38**, 2719 (1973).
134. (a) E. Shefter, W. Wolf, *J. Pharm. Sci.* **54**, 104 (1965); (b) M. C. Etter, *J. Am. Chem. Soc.* **98**, 5326, 5331 (1976); J. Z. Gougoutas, L. Lessinger, *J. Solid State Chem.* **9**, 155 (1974).
135. M. Ochiai, Y. Masaki, M. Shiro, *J. Org. Chem.* **56**, 5511 (1991).
136. M. Ochiai, T. Ito, Y. Masaki, M. Shiro, *J. Am. Chem. Soc.* **114**, 6269 (1992).
137. M. Ochiai, T. Ito, M. Shiro, *J. Chem. Soc., Chem. Commun.*, 218 (1993).
138. V. V. Zhdankin, A. P. Krasutsky, C. J. Kuehl, A. J. Simonsen, J. K. Woodward, B. Mismash, J. T. Bolz, *J. Am. Chem. Soc.* **118**, 5192 (1996).
139. V. V. Zhdankin, C. J. Kuehl, A. M. Arif, P. J. Stang, *Mendeleev Commun.*, 50 (1996).
140. S. Akai, T. Okuno, M. Egi, T. Takada, H. Tohma, Y. Kita, *Heterocycles* **42**, 47 (1996).
141. L. Hadjiarapoglou, S. Spyroudis, A. Varvoglis, *Synthesis*, 207 (1983); M. Papadopoulou, A. Varvoglis, *J. Chem. Res. (S)*, 166 (1984).
142. (a) Y. Yamada, T. Yamamoto, M. Okawara, *Chem. Lett.*, 361 (1975); (b) A. K. Mishra, M. M. Olmstead, J. J. Ellison, P. P. Power, *Inorg. Chem.* **34**, 3210 (1995).
143. R. A. Abramovitch, T. D. Bailey, T. Takaya, V. Uma, *J. Org. Chem.* **39**, 340 (1974).
144. D. Mansuy, J. P. Mahy, A. Dureault, G. Bedi, P. Battioni, *J. Chem. Soc., Chem. Commun.*, 1161 (1984).
145. R. L. Cicero, D. Zhao, J. D. Protasiewicz, *Inorg. Chem.* **35**, 275 (1996).
146. F. Cech, E. Zbiral, *Tetrahedron* **31**, 605 (1975); J. Ehrenfreund, E. Zbiral, *Liebigs Ann. Chem.*, 290 (1973).
147. R. M. Moriarty, J. S. Khosrowshahi, *Tetrahedron Lett.* **27**, 2809 (1986).
148. Y. Kita, H. Tohma, K. Hatanaka, T. Takada, S. Fujita, S. Mitoh, H. Sakurai, S. Oka, *J. Am. Chem. Soc.* **116**, 3684 (1994); Y. Kita, H. Tohma, T. Takada, S. Mitoh, S. Fujita, M. Gyoten, *Synlett*, 427 (1994).
149. M. Tingoli, M. Tiecco, D. Chianelli, R. Balducci, A. Temperini, *J. Org. Chem.* **56**, 6809 (1991).
150. S. Czernecki, D. Randriamandimby, *Tetrahedron Lett.* **34**, 7915 (1993).
151. P. Magnus, M. B. Roe, C. Hulme, *J. Chem. Soc., Chem. Commun.* 263 (1995); Magnus, P.; Hulme, C. *Tetrahedron Lett.* **35**, 8097 (1994).
152. A. Kirschning, S. Domann, G. Dräger, L. Rose, *Synlett*, 767 (1995).
153. A. P. Krasutsky, C. J. Kuehl, V. V. Zhdankin, *Synlett*, 1081 (1995); V. V. Zhdankin, C. J. Kuehl, A. P. Krasutsky, M. S. Formaneck, J. T. Bolz, *Tetrahedron Lett.* **35**, 9677 (1994).

154. (a) D. G. Morris, A. G. Shepperd, *J. Chem. Soc., Chem. Commun.*, 1250 (1981); (b) G. W. Bradley, J. H. Holloway, H. J. Koh, D. G. Morris, P. G. Watson, *J. Chem. Soc., Perkin Trans. 1*, 3001 (1992).

155. J. L. Cotter, L. J. Andrews, R. M. Keefer, *J. Am. Chem. Soc.* **84**, 4692 (1962).

156. V. Montanari, G. Resnati, D. D. DesMarteau, *J. Org. Chem.* **59**, 6093 (1994).

157. (a) V. V. Lyalin, V. V. Orda, L. A. Alekseeva, L. M. Yagupolskii, *J. Org. Chem. USSR* **8**, 1027 (1972); (b) A. A. Mironova, I. V. Soloshonok, I. I. Maletina, V. V. Orda, L. M. Yagupolskii, *Zh. Org. Khim.* **24**, 593 (1988).

158. T. Umemoto, Y. Gotoh, *Bull. Chem. Soc. Jpn.* **60**, 3307 (1987).

159. V. V. Zhdankin, C. J. Kuehl, A. J. Simonsen, *Tetrahedron Lett.* **36**, 2203 (1995).

160. L. M. Yagupolskii, I. I. Maletina, N. V. Kondratenko, V. V. Orda, *Synthesis*, 835 (1978).

161. T. Umemoto, Y. Kuriu, H. Shuyama, O. Miyano, S. Nakayama, *J. Fluorine Chem.* **31**, 37 (1986).

162. V. V. Zhdankin, C. J. Kuehl, *Tetrahedron Lett.* **35**, 1809 (1994); V. V. Zhdankin, C. J. Kuehl, J. T. Bolz, N. S. Zefirov, *Mendeleev Commun.*, 165 (1994); C. J. Kuehl, J. T. Bolz, V. V. Zhdankin, *Synthesis*, 312 (1995).

163. W. Tyrra, D. Naumann, *Can. J. Chem.* **69**, 327 (1991).

164. (a) V. V. Zhdankin, C. J. Kuehl, A. J. Simonsen, *Main Group Chemistry* **1**, 349 (1996); (b) V. V. Zhdankin, S. A. Erickson, K. J. Hanson, *J. Am. Chem. Soc.* **119**, 4775 (1997).

165. T. Umemoto, *Tetrahedron Lett.* **25**, 81 (1984); T. Umemoto, Y. Gotoh, *Bull. Chem. Soc. Jpn.* **64**, 2008 (1991); T. Umemoto, *Chem. Rev.* **96**, 1757 (1996).

166. P. J. Stang, R. Tykwinski, V. V. Zhdankin, *J. Heterocyclic Chem.* **29**, 815 (1992).

167. P. J. Stang, V. V. Zhdankin, R. Tykwinski, N. S. Zefirov, *Tetrahedron Lett.* **33**, 1419 (1992).

168. P. J. Stang, V. V. Zhdankin, *J. Am. Chem. Soc.* **113**, 4572 (1991).

169. (a) V. V. Zhdankin, C. J. Kuehl, J. T. Bolz, M. S. Formaneck, A. J. Simonsen, *Tetrahedron Lett.* **35**, 7323 (1994); V. V. Zhdankin, C. J. Kuehl, A. P. Krasutsky, J. T. Bolz, B. Mismash, J. K. Woodward, A. J. Simonsen, *Tetrahedron Lett.* **36**, 7975 (1995); (b) V. V. Zhdankin, C. J. Kuehl, A. M. Arif, P. J. Stang, *Mendeleev Commun.*, 50 (1996).

170. M. D. Bachi, N. Bar-Ner, C. M. Crittell, P. J. Stang, B. L. Williamson, *J. Org. Chem.* **56**, 3912 (1991); P. J. Stang, A. M. Arif, C. M. Crittell, *Angew. Chem. Int. Ed. Engl.* **29**, 287 (1990).

171. M. Ochiai, M. Kunishima, K. Sumi, Y. Nagao, E. Fujita, *Tetrahedron Lett.* **26**, 4501 (1985); M. Ochiai, T. Ito, Y. Takaoka, Y. Masaki, M. Kunishima, S. Tani, Y. Nagao, *J. Chem. Soc., Chem. Commun.*, 118 (1990).

172. (a) M. Ochiai, K. Sumi, Y. Takaoka, M. Kunishima, Y. Nagao, M. Shiro, E. Fujita, *Tetrahedron* **44**, 4095 (1988); (b) M. Ochiai, K. Sumi, Y. Nagao, E. Fujita, *Tetrahedron Lett.* **26**, 2351 (1985).

173. R. J. Hinkle, G. T. Poulter, P. J. Stang, *J. Am. Chem. Soc.* **115**, 11626 (1993); R. J. Hinkle, P. J. Stang, *Synthesis*, 313 (1994).

174. M. Ochiai, K. Uemura, K. Oshima, Y. Masaki, M. Kunishima, S. Tani, *Tetrahedron Lett.* **32**, 4753 (1991).

175. P. J. Stang, H. Wingert, A. M. Arif, *J. Am. Chem. Soc.* **109**, 7235 (1987).
176. W. Tyrra, H. Butler, D. Naumann, *J. Fluorine Chem.* **60**, 79 (1993).
177. (a) F. M. Beringer, R. Nathan, *J. Org. Chem.* **34**, 685, 2095 (1969); (b) P. J. Stang, B. Olenyuk, K. Chen, *Synthesis*, 937 (1985).
178. P. J. Stang, K. Chen, *J. Am. Chem. Soc.* **117**, 1667 (1995); P. J. Stang, K. Chen, A. M. Arif, *J. Am. Chem. Soc.* **117**, 8793 (1995).
179. W. B. Wright, E. A. Meyers, *Cryst. Struct. Commun.* **1**, 95 (1972); V. N. Petrov, C. V. Tindeman, Yu. T. Struchkov, F. A. Chungtai, V. A. Budylin, Yu. G. Bundel, *Dokl. Akad. Nauk SSSR* **269**, 614 (1983); S. W. Page, E. P. Mazzola, A. D. Mighell, V. L. Himes, C. R. Hubbard, *J. Am. Chem. Soc.* **101**, 5858 (1979); C. R. Hubbard, V. L. Himes, A. D. Mighell, *Acta Crystallogr.* **B36**, 2819 (1987); R. M. Moriarty, O. Prakash, W. A. Freeman, *J. Am. Chem. Soc.* **106**, 6082 (1984); N. W. Alcock, R. M. Countryman, *J. Chem. Soc., Dalton Trans.*, 217 (1997); T. L. Khotsyanova, T. A. Babushkina, V. V. Saatsazov, T. P. Tolstaya, I. N. Lisichkina, G. K. Semin, *Koord. Khim.* **2**, 1567 (1976); N. W. Alcock, R. M. Countryman, *J. Chem. Soc., Dalton Trans.*, 193 (1987); A. P. Bozopoulos, P. J. Rentzeperis, *Acta Crystallogr.* **C43**, 914 (1987).
180. V. V. Lyalin, V. V. Orda, L. A. Alekseyeva, L. M. Yagupolskii, *J. Org. Chem. USSR* **7**, 1524 (1971).
181. E. Gudrinietse, O. Neiland, G. Vanag, *Zh. Obsch. Khim.* **27**, 2737 (1957).
182. (a) S. Zhu, Q. Chen, *J. Chem. Soc., Chem. Commun.*, 1459 (1990); (b) S. Zhu, Q. Chen, W. Kuang, *J. Fluorine Chem.* **60**, 39 (1993); (c) S. Zhu, *Heteroatom Chemistry* **5**, 9 (1994).
183. J. Hackenberg, M. Hanack, *J. Chem. Soc., Chem. Commun.*, 470 (1991).
184. L. Hadjiarapoglou, S. Spyroudis, A. Varvoglis, *J. Am. Chem. Soc.* **107**, 7178 (1985); L. Hadjiarapoglou, A. Varvoglis, N. W. Alcock, G. A. Pike, *J. Chem. Soc., Perkin Trans. 1*, 2839 (1988).
185. G. Wittig, M. Rieber, *Ann. Chem.* **562**, 187 (1949).
186. F. M. Beringer, L. L. Chang, *J. Org. Chem.* **36**, 4055 (1971); H. J. Reich, C. S. Cooperman, *J. Am. Chem. Soc.* **95**, 5077 (1973).
187. A. G. Robiette, R. H. Bradley, P. N. Brier, *J. Chem. Soc., Chem. Commun.*, 1567 (1971).
188. G. R. Jones, R. D. Burbank, N. Bartlett, *Inorg. Chem.* **9**, 2264 (1970).
189. A. R. Mahjoub, K. Seppelt, *Angew. Chem. Int. Ed. Engl.* **30**, 323 (1991)
190. K. Selte, A. Kjekshus, *Acta Chem. Scand.* **24**, 1912 (1970).
191. T. Kraft, M. Jensen, *J. Am. Chem. Soc.* **117**, 6795 (1995).
192. A. Rehr, M. Jensen, *Angew. Chem. Int. Ed. Engl.* **31**, 274 (1992).
193. H. A. Carter, F. Aubke, *Inorg. Chem.* **10**, 2296 (1971).
194. J. R. Dalziel, H. A. Carter, F. Aubke, *Inorg. Chem.* **15**, 1247 (1976).
195. (a) A. R. Katritzky, J. P. Savage, J. K. Gallos, H. D. Durst, *J. Chem. Soc., Perkin Trans. 2*, 1515 (1990); A. R. Katritzky, J. P. Savage, J. K. Gallos, H. D. Durst, *Org. Prep. Proc. Int.* **21**, 157 (1989); (b) L. Weclas-Henderson, T. T. Nguyen, R. A. Hayes, J. C. Martin, *J. Org. Chem.* **56**, 6565 (1991); (c) D. A. Evans, S. W. Kaldor, T. K. Jones, J. Clardy, T. J. Stout, *J. Am. Chem. Soc.* **112**, 7001 (1990); S. W. Bailey, R. Y. Chandrasekaran, J. E. Ayling, *J. Org. Chem.* **57**, 4470 (1992); M.

Frigerio, M. Santagostino, *Tetrahedron Lett.* **35**, 8019 (1994); M. J. Batchelor, R. J. Gillespie, J. M. C. Golec, C. J. R. Hedgecock, *Tetrahedron Lett.* **34**, 167 (1993); X. Deruyttere, L. Dumortier, J. Van der Eycken, M. Vandewalle, *Synlett*, 51 (1992); D. S. Yamashita, V. P. Rocco, S. Danishefsky, *Tetrahedron Lett.* **32**, 6667 (1991); M. J. Robins, V. Samano, M. D. Johnson, *J. Org. Chem.* **55**, 410 (1990); R. J. Linderman, D. M. Graves, *J. Org. Chem.* **54**, 661 (1989); J. P. Burkhart, N. P. Peet, P. Bey, *Tetrahedron Lett.* **29**, 3433 (1988).

196. R. E. Ireland, L. Liu, *J. Org. Chem.* **58**, 2899 (1993).
197. D. B. Dess, J. C. Martin, *J. Org. Chem.* **48**, 4155 (1983).
198. J. B. Plumb, D. J. Harper, *Chem. Eng. News* **68**, 3 (1990).
199. O. Ruff, R. Keim, *Z. Anorg. Chem.* **201**, 245 (1931).
200. C. J. Schack, D. Pilipovich, S. N. Cohz, D. F. Sheehan, *J. Phys. Chem.* **72**, 4697 (1968).
201. A. Engelbrecht, P. Peterfly, *Angew. Chem. Int. Ed. Engl.* **8**, 65 (1969).
202. R. J. Gillespie, J. P. Krasznai, *Inorg. Chem.* **15**, 1251 (1976).
203. L. S. Bartell, F. B. Clippard, E. J. Jacob, *Inorg. Chem.* **15**, 3009 (1976).
204. T. Kraft, M. Jensen, *J. Am. Chem. Soc.* **117**, 6795 (1995).
205. L. Fieser, M. Fieser, *Reagents for Organic Synthesis* (Wiley-Interscience, New York, 1967), Vol. 1, pp. 809–819; Vol. 2, pp. 311–315.

12 Organic Synthesis Using Hypervalent Organoiodanes

MASAHITO OCHIAI
Faculty of Pharmaceutical Sciences, University of Tokushima, Tokushima, Japan

Hypervalent iodine(III) compounds IL_3 have a 10-I-3 T-shaped structure,[1] and are known as iodanes.[2] Including the lone pair electrons, iodanes have a geometry of a trigonal bipyramid. A compound of the general structure of R_2IX **1** (R: *sp* and/or sp^2 carbon atom ligands, X: heteroatom ligands such as halogens, OTs, BF_4, OCOR, etc.) is usually called an iodonium salt of 8-I-2 structure, which means such compounds are not hypervalent compounds. The term onium salts refers to compounds with tetrahedral geometry whose octet structure has eight electrons in the valence shell of the positively charged atom. However, X-ray structural data of **1** show it to have T-shaped geometry with a C—I—C angle of nearly 90° and with good linearity for the axial triad X—I—C.[2,3] The I—X bonds of **1** are presumed to be partially ionic in nature, because the I—X bond distances are much longer than the computed covalent bond lengths but considerably shorter than the sum of van der Waals radii. Although the structure in solution is unknown, the preferred name for **1** seems to be an iodane rather than an iodonium salt, given that the overall experimentally determined geometry of **1** is a pseudotrigonal bipyramid rather than pseudotetrahedral. In this chapter, therefore, the compound R_2IX **1** is termed an iodane, which has a hypervalent X—I—C bond of partially ionic character.

X
R—I
R

1

Chemistry of Hypervalent Compounds, edited by Kin-ya Akiba.
ISBN 0-471-24019-2

The high reactivity of allyl(aryl)-, alkenyl(aryl)-, and alkynyl(aryl)iodanes toward nucleophilic species is the focus of this chapter. The very high leaving group ability of aryliodanyl groups makes these iodanes a versatile reagent in organic synthesis.[4]

12.1 HYPERLEAVING ABILITY OF ARYLIODANYL GROUPS

Solvolysis of (1-cyclohexenyl)aryliodanes **2a**, prepared by BF_3-catalyzed silicon– or tin–iodane exchange reaction of vinylsilane or stannane with iodosylarenes, proceeds at a reasonable rate in aqueous alcohol solutions even at room temperature, and generates the vinyl cation **3a** with liberation of iodoarenes.[5] Intermediacy of the cyclohexenyl cation **3a** was established by the observation of carbocation rearrangement during the solvolysis of **2b**, in which an initially generated bent vinyl cation **3b** rearranges to a more stable linear vinyl cation **3c**.[6] The observed rate constants for the solvolysis shown in Table 12.1 imply that the phenyliodanyl group $Ph(BF_4)I$— is a remarkably good nucleofuge with a leaving group ability about 10^6 times greater than triflate, a so-called "superleaving" group.[7] Relative leaving group abilities are listed in Table 12.2,[8] showing the aryliodanyl groups to be the most efficient leaving groups that have ever been determined quantitatively.

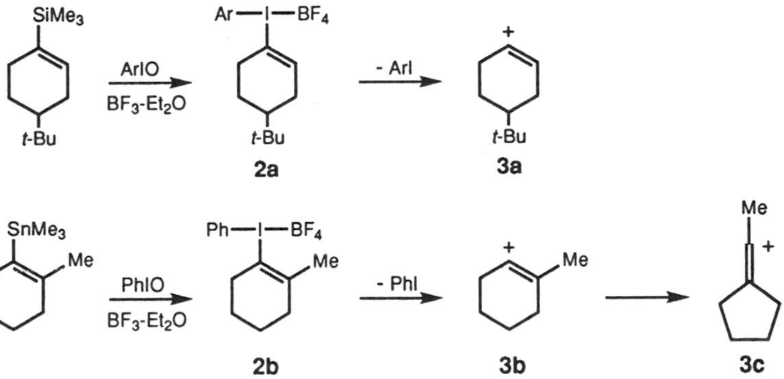

A nucleofuge such as the aryliodanyl group is termed a *hyperleaving group*. The hyperleaving group must show a leaving group ability higher than that of a superleaving group such as TfO, and also be a hypervalent leaving group. As shown in the following equation, the leaving process of a hyperleaving group must involve an energetically preferable reduction of the hypervalent atom to the normal valency of octet structure, which is the origin of the high leaving group ability. This process is called reductive elimination. The positively charged dimethylsulfonio group with tetrahedral geometry, and hence with no hypervalency, shows poor leaving group ability (Table 12.2).

TABLE 12.1 Rate Constants ($10^4 k_{obsd}s^{-1}$) for Solvolysis of 2 in 60:40 Ethanol–Water

Iodane (Ar)	Temp(°C) 25	35	50	69	$\Delta H^{\ddagger}$ (kcal mol^{-1})	$\Delta S^{\ddagger}$ (cal mol^{-1}°C^{-1})
2a (Ph)		0.229	2.32	26.9	28.7	13.3
2a (*p*-MePh)		0.114	1.23	14.0	28.5	11.8
2a (*p*-ClPh)		0.594	5.74	60.3	27.9	12.8
2b	11.4					
1-*c*-C_6H_9OTf[a]			2.8×10^{-6}			

[a]Ref. 6.

TABLE 12.2 Relative Leaving Group Abilities

Nucleofuge	k_{rel}	Nucleofuge	k_{rel}
AcO	1.4×10^{-6}	I	9.1×10
F	9.0×10^{-6}	MsO	3.0×10^{4}
Me_2S^+	5.3×10^{-2}	TsO	3.7×10^{4}
Cl	1.0	TfO	1.4×10^{8}
F_3CCO_2	2.5	*p*-MePh(BF_4)I	6.2×10^{13}
NO_3	7.2	Ph(BF_4)I	1.2×10^{14}
Br	1.4×10	*p*-ClPh(BF_4)I	2.9×10^{14}

$$-\overset{|}{\underset{|}{C}}-\underset{\underset{Ph}{|}}{I}-X \longrightarrow -\overset{|}{\underset{|}{C}}^{+} + Ph-I + X^{-}$$

[10-I-3] → [8-I-1]

Hammett substituent constants of iodanyl and periodanyl groups have been estimated by ^{19}F nuclear magnetic resonance (NMR) spectroscopy of *m*- and *p*-substituted fluorobenzenes.[9] As expected, the phenyliodanyl group, Ph(BF_4)I—, is an inductively strong electron-withdrawing group with large σ_I (1.34) and small σ_R (0.03) values.

12.2 REACTION OF ALLYLIODANES

BF_3-catalyzed reactions of allylsilane, -germane, and -stannane with iodosylbenzene undergo metal–iodane exchange at −20°C and generate allyliodanes **4**. They serve as a reactive allyl cation equivalent and undergo Friedel–Crafts monoallylation of electron-rich arenes at that temperature, because of a hyperleaving ability of phenyliodanyl group.[10] Oxygen and nitrogen nucleophiles such as alcohols, carboxylic acids, and trimethylsilyl azide react with **4**, yielding allyl ethers, esters, and azides, respectively.[11]

$$CH_2{=}CHCH_2M \xrightarrow[BF_3\text{-}Et_2O]{PhIO} CH_2{=}CHCH_2I(Ph)X \ (\mathbf{4}) \xrightarrow[-\,PhI]{ArH} CH_2{=}CHCH_2Ar$$

M: Me_3Si, Me_3Ge, *n*-Bu_3Sn

12.3 REACTION OF ALKENYLIODANES

12.3.1 Generation of Iodine(III) Ylides

The observed large deshieldings of α-vinylic protons of alkenyl(phenyl)iodanes in ^{1}H NMR chemical shifts compared to those of alkenyl iodides (Table 12.3)[12] indicate that the α-C—H bonds of alkenyliodanes are quite acidic. The PhI^+ group has been reported to increase the C–H acidity (pK_a in aqueous solution) of malonic ester by eight orders of magnitude.[13]

TABLE 12.3 ^{1}H NMR Chemical Shift (δ, ppm) in $CDCl_3$

Vinylic Proton	*E*-5	*E*-6	*Z*-5	*Z*-6
H_α	5.97	6.78	ca. 6.2	7.00
H_β	6.51	6.98	ca. 6.2	6.61

Weak bases such as amines could abstract the acidic α-vinylic hydrogens of alkenyliodanes, generating iodine(III) ylide.[14] Reaction of *E*-β-ethoxyvinyl-(phenyl)iodane **7** with triethylamine in D_2O—THF at room temperature undergoes deuterium exchange of the α-vinylic proton, indicating the generation of the vinyliodine(III) ylide **8** via α-proton abstraction.

12.3.2 Generation of Alkylidenecarbenes

Because of the hyperleaving group ability of aryliodanyl groups, the vinyl(phenyl)iodine(III) ylides readily undergo reductive elimination of iodobenzene to generate reactive alkylidenecarbenes, which in turn can undergo regioselectively an intramolecular 1,5-C—H insertion yielding cyclopentenes. This α-elimination of alkenyl(phenyl)iodane **10** and the subsequent 1,5-C—H insertion of **11** proceed under mild conditions (Et_3N/25°C) to give the bicyclo-[3.3.0]octene **12**.[15] These processes are compatible with an OH group, which makes possible the 1,5-O—H insertion of **17**, yielding the 2,3-dihydrofuran **18**.[16]

Me I(Ph)BF_4 **10** Et_3N THF, RT 84% Me **11** H Me H **12**

O Me I(Ph)BF_4 **13** *t*-BuOK THF, -78 °C 61% O Me **14** H O Me H **15**

HO $PhSO_2$ I(Ph)BF_4 **16** Et_3N THF, RT 61% O H $PhSO_2$ **17** O SO_2Ph **18**

Reaction of vinyl(phenyl)iodane **19** with Et_3N at 0°C leads to formation of the terminal alkyne quantitatively. Mechanistic studies with α- or β-deuterated **19** indicate that the alkyne-forming reaction predominantly involves generation of alkylidenecarbenes via α-elimination and their 1,2-hydrogen shift, but not direct syn-β-elimination.[15] The absence of formation of cyclopentenes in these reactions is in good agreement with the result that 1,5-C—H insertion of alkylidenecarbenes cannot compete with 1,2-α-hydrogen migration.[17]

n-C_8H_{17} D I(Ph)BF_4 **19-*αd*** (100%D) Et_3N 0 °C *n*-C_8H_{17} H → *n*-C_8H_{17}—≡—H 10%D

n-C_8H_{17} D I(Ph)BF_4 **19-*βd*** (92%D) Et_3N 0 °C *n*-C_8H_{17} D → *n*-C_8H_{17}—≡—D 89%D

α-(Phenylsulfenyl) and α-(phenylsulfinyl) groups of alkylidenecarbenes **21** tend to show the excellent migratory aptitude, and their reactions lead exclusively to formation of the alkynes **23**, presumably via the sulfur ylide intermediate **22**. In marked contrast, [α-(phenylsulfonyl)alkylidene]carbene **25** predominantly undergoes 1,5-C—H insertions, yielding synthetically useful 1-sulfonylcyclopentene **26**. The tendency of α-sulfur groups to migrate appears to depend to some extent on the availability of lone pair electrons at the sulfur atom.[16] In contrast to the reaction of α-chloroalkylidenecarbene **29** (X = Cl), which gives a mixture of the 1,5-C—H insertion product **30** (X = Cl) and the rearranged alkyne **31** (X = Cl), exclusive formation of the bromoalkyne **31** (X = Br) from α-bromocarbene **29** (X = BR) is attributable to the migratory aptitude of α-bromine atom being higher than that of α-chlorine atom.[18]

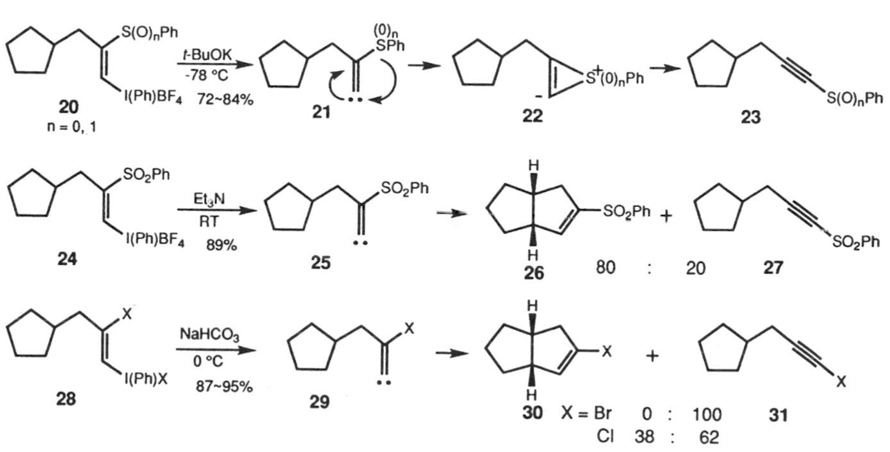

Base-induced α-elimination of the phenyliodanyl group with hypernucleofugalities from Z-(β-halovinyl)iodanes **32** would directly generate α-haloalkylidenecarbenes. However, the following mechanistic alternative should be considered for the generation of the α-haloalkylidenecarbenes: (1) stereoelectronically preferable anti-β-elimination of hydrogen halides by a base, yielding alkynyliodanes **33**; (2) Michael-type addition of the halide ions; (3) reductive elimination of the phenyliodanyl group.

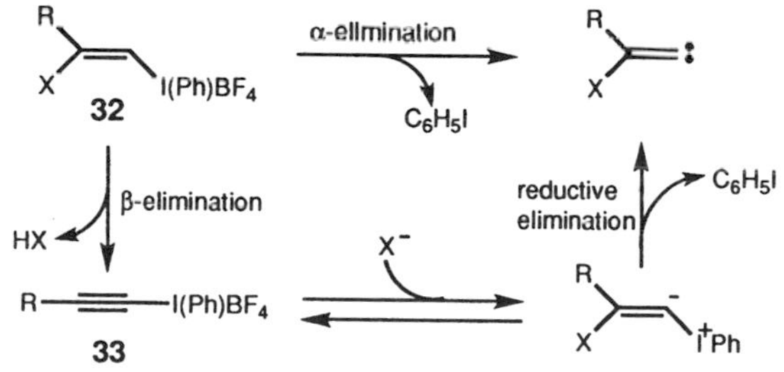

Formation of both the vinyliodane-derived product and the alkynyliodane-derived product in a ratio of 70:30 from the intermolecular crossover reaction of a 1:1 mixture of the Z-(β-bromovinyl)iodane **34** and the alkynyliodane **35** with $NaHCO_3$ clearly demonstrates that the generation of α-bromoalkylidenecarbenes involves not only α-elimination of phenyliodanyl groups but also anti-β-elimination of hydrogen bromide. These results were further supported by the intramolecular version of the crossover experiments using **36**.[18]

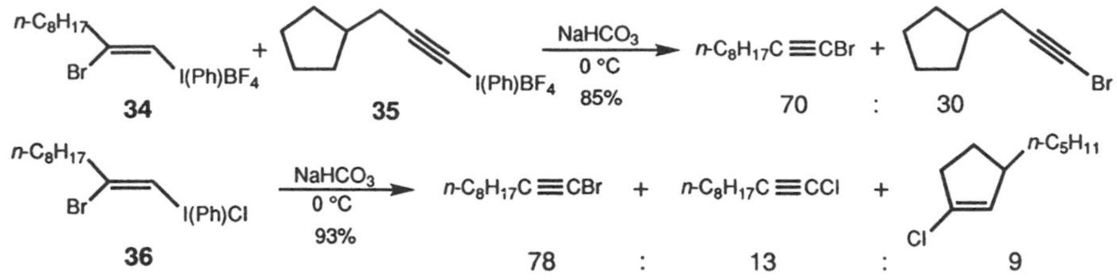

When alkylidenecarbenes are geneated in ethereal solvents, formation of the solvent–alkylidenecarbene complex (i.e., oxonium ylide), is observed.[19] The alkylidenecarbene **42** generated by Et_3N-induced α-elimination of **37** in THF undergoes regioselective 1,5-C—H insertions, 1,2-shifts of the butyl group, and electrophilic attack on the tertiary amine, followed by protonation, to give **38**, **39**, and **40**, respectively. In addition to these carbene-derived products, the reaction affords the three-component coupling product **41** produced through nucleophilic attack of THF on **42** generating the oxonium ylide **43**, followed by protonation with subsequent ring-opening of the resulting oxonium salt **44** by nucleophilic attack of Et_3N. The reactions were observed to be temperature dependent, as reflected in variations in the product profiles (Table 12.4): lowering the reaction temperature tended to decrease the yields of alkylidenecarbene-derived products **38–40** and to increase those of the vinyloxonium ylide-derived product **41**.[20] This temperature dependence is explained in terms of reversible oxonium ylide formation. However, no evidence was observed to suggest the existence of an equilibrium between the free alkylidenecarbene and the sulfonium ylide in the reaction in tetrahydrothiophene.

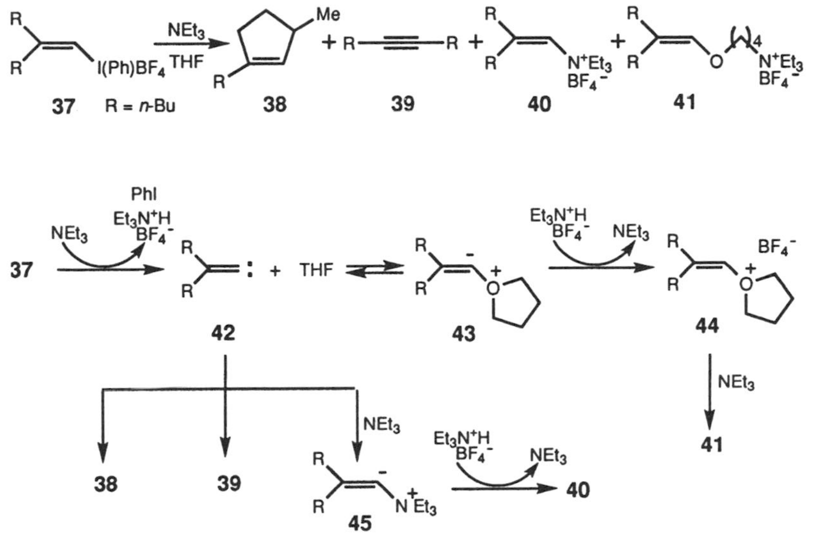

TABLE 12.4 Effects of Temperature on Reaction of Vinyliodane 37 with Et_3N in THF

	Product (% yield)					Ratio
Conditions	**38**	**39**	**40**	**41**	Total[a]	(**38** + **39** + **40**):**41**[b]
60°C (10 h)	74	14	< 1	2	90	98:2
40°C (10 h)	75	11	< 1	9	95	90:10
20°C (10 h)	75	8	1	15	99	85:15
0°C (10 C)	71	3	1	23	98	77:23
−20°C (2 d), 0°C (10 h)	48	2	1	27	78	65:35
−40°C (2 d), 0°C (10 h)	44	2	1	40	87	54:46
−60°C (2 d), 0°C (10 h)	44	3	3	42	92	54:46

[a]Total yields of **38–41**.
[b]Ratios of the alkyidenecarbene-derived products **38–40** versus the vinyloxonium ylide-derived product **41**.

The alkylidenecarbene derived from the vinyliodane **46** undergoes addition reaction to olefins to give methylenecyclopropanes **47**.[21] The addition to *cis*- and *trans*-4-methyl-2-pentenes is stereospecific with retention of the olefin geometry in more than 99% selectivity, implying that the alkylidenecarbene generated is a singlet electronic state.

Me, Me, I(Ph)BF_4 **46** — $ArCH{=}CH_2$, *t*-BuOK or Et_3N, 3 °C, CH_2Cl_2 → Me, Me, Ar **47**

To gain some insight into the freeness of the vinyliodane-derived carbenic species, both stereoisomers of **48** were allowed to react with Et_3N or *tert*-BuOK.[15] It seems reasonable to assume that if there is any association with the leaving group, that is, iodobenzene in the transition state for 1,5-C—H insertions such as **52**, there should be a memory effect that reflects the stereochemistry of **48** in the product ratios. The results revealed that the product ratios of **49** and **50** were about 1:1, regardless of the stereochemistry of **48** and the reaction conditions. The lack of regioselectivity for the intramolecular insertion of carbenes derived from the vinyliodane clearly indicates that the loss of stereochemistry of the vinyliodane precedes the intramolecular insertion. Thus, these results suggest the involvement of the free alkylidenecarbene **51** rather than the carbenoid **52**.

Generation of alkylidenecarbenes in the presence of sulfides undergoes onium transfer reactions to give vinylsulfonium salts. The involvement of free alkylidenecarbenes was also suggested by the stereochemical outcome of

E-**48** $\xrightarrow{t\text{-BuOK}}$ **49** + **50** (1 : 1) $\xleftarrow{t\text{-BuOK}}$ *Z*-**48**

51

52: B = base

this onium transfer reaction.[20] Stereoconvergence with *E*-vinylsulfonium salt *E*-**54** as a major product was observed in the reactions of both *E*- and *Z*-vinyliodanes **53** with diiosopropylethylamine in the presence of diphenyl sulfide. Sincc reactions of alkylidenecarbenoids with nucleophiles have been shown to proceed in favor of inversion of configuration, the high degree of stereoconvergence of the olefin geometry in this onium transfer reaction that we observed strongly indicates intermediacy of the free alkylidenecarbene. This stereoselectivity is most likely a consequence of the steric differences between the two α-substituents of the free alkylidenecarbene generated.

53 $\xrightarrow[Ph_2S]{i\text{-}Pr_2NEt}$ *E*-**54** + *Z*-**54**

	E-**54**		*Z*-**54**
E-**53**	69	:	31
Z-**53**	71	:	29

Relative reactivities toward 2-methyl-l-propenylidene derived from **46** of ring-substituted styrenes were measured and the magnitudes of ρ values obtained by a Hammett correlation are summarized in Table 12.5.[21] A free alkylidenecarbene has been known to be mildly electrophilic in nature and shows a relatively small negative ρ value, whereas a carbenoid shows a very large degree of electrophilicity and a large negative ρ valuc. The small ρ values of −0.56 (Et_3N) and −0.55 (*tert*-BuOK) for **46** indicate that the vinyliodane-derived alkylidenecarbenes are mildly electrophilic and are free carbenes.

Strong bases, like alkyllithiums, or drastic reaction conditions are required to generate carbenic species from vinyl halides, thus precluding the presence of many functional groups in the substrate. The vinyliodane method produces

TABLE 12.5 ρ Values for the Reaction of 2-Methyl-1-propenylidene with Styrenes

Carbene Precursor	Reaction Conditions	ρ
46	$Et_3N/CH_2Cl_2/3°C$	−0.56 (σ)
46	$KOBu^t/CH_2Cl_2/3°C$	−0.55 (σ)
$Me_2C{=}CN_2$	$KOBu^t/CH_2Cl_2/-78°C$	−0.51 (σ)
$Me_2C{=}CHOTf$	$KOBu^t/-20°C$	−0.75 (σ)
$Me_2C{=}C(OTf)TMS$	$R_4N^+F^-/0°C$	−0.44 (σ)
$Me_2C{=}CHN_2Ts$	0°C	−0.71 (σ)
$Me_2C{=}CHBr$	$KOBu^t/-10°C$	−4.3 (σ^+)
$Me_2C{=}CBr_2$	MeLi/−40°C	−4.3 (σ^+)
N-nitrosooxazolidone	$LiOCH_2CH_2OEt/40°C$	−3.4 (σ^+)

free alkylidenecarbenes under mild conditions, making the reaction compatible with a variety of functional groups.

12.3.3 Nucleophilic Vinylic Substitution of Alkenyliodanes

Alkenyl(phenyl)iodanes are highly reactive in nucleophilic vinylic substitutions because of a hyperleaving group ability of aryliodanyl groups. The representative transformation shown for the cyclic vinyliodane **2a** (Ar = Ph) illustrates the broad range of utility of the iodane.[12a,22] A variety of substituted olefins including α-cyano and α-nitro olefins, vinyl sulfides, vinyl halides, and α, β-unsaturated esters were prepared under mild conditions (at low or room temperature). It is essential to the success of these reactions that the hetero ligand (BF_4) on iodine(III) is non-nucleophilic in character, so as to avoid competition with nucleophiles.

Nucleophilic substitutions at vinylic carbons of moderately activated olefins usually proceed with predominant retention of configuration, thought to proceed via addition–elimination in most cases.[23] Alternatively, highly activated olefins afford products with partial or complete stereoconvergence via a multistep process. Exclusive inversion in nucleophilic vinylic substitutions has not been observed for simple systems. Very interestingly, nucleophilic vinylic substitution of *E*-β-alkylvinyliodane **19** with tetrabutylammonium halides proceeds with exclusive inversion of configuration at room temperature.[24]

Nucleophilic vinylic substitutions of **19** with n-Bu_4NCl, Br, and I gave *Z*-alkenyl halides **55** with complete inversion, along with formation of a small amount of 1-decyne. However, n-Bu_4NF afforded only the alkyne. The selectivity for substitution over elimination decreases in the order of $n\text{-}Bu_4NI > n\text{-}Bu_4NBr > n\text{-}Bu_4NCl \gg n\text{-}Bu_4NF$, reflecting the decreasing softness of halide ions. Surprisingly, primary kinetic deuterium isotope effects and the observed deuterium content of the product 1-decyne in the reaction of α- and β-deuterated **19** with n-Bu_4NCl clearly indicate that 1-decyne was predominantly produced as the result of syn-β-elimination. Generation of an alkylidenecarbene by α-elimination, which occurred in the reaction of **19** with triethylamine as mentioned, was not involved. α-Elimination generating the alkylidenecarbene, however, was involved in the reaction with n-Bu_4NF.

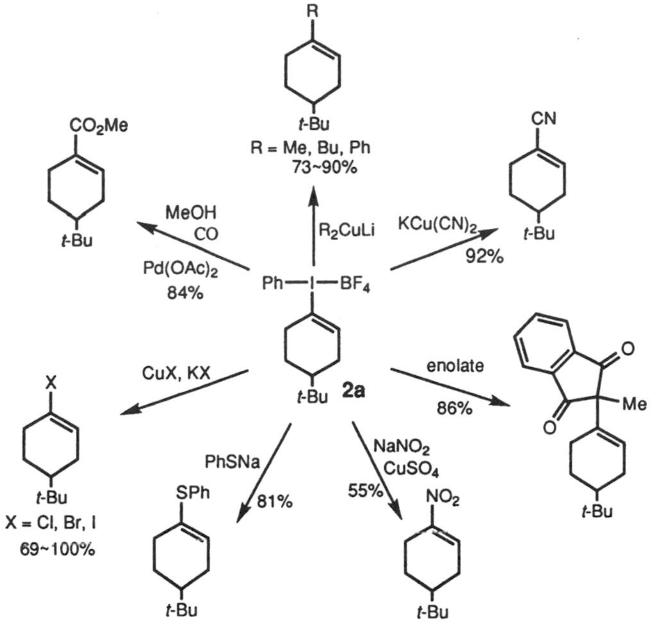

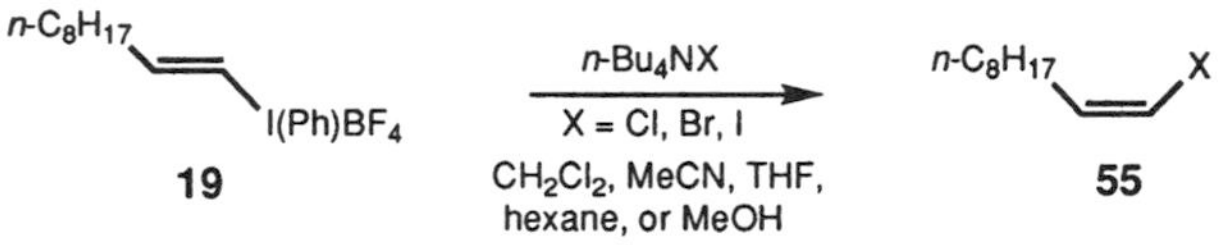

The fact that rapid ligand exchange of **19** with halides takes place to form vinyliodanes **56** was established by ^1H NMR. Pseudorotation on the iodine(III) leads to the formation of **57**.[25] Direct vinylic S_N2 displacement of **57** by attack of another halide produces *Z*-**55** via the transition state **58**. Alternatively, intramolecular *cis*-β-proton abstraction by the ligand X of **56** gives 1-alkyne. The hyperleaving group ability of the phenyliodanyl group would be the origin of this unusual inversion of configuration in this nucleophilic vinylic substitution.

In marked contrast, nucleophilic substitutions of **19** with a combination of cuprous halides and potassium halides lead to complete retention of stereochemistry, which probably involves oxidative addition of cuprates, ligand coupling at the iodine(III) of **59**, and then ligand coupling at the copper(III) of **60**.[24] Reaction of alkenyliodanes with arylcuprates also occurs with retention of configuration.[12a,26]

Retention of stereochemistry is observed in the reactions of Z-β-(phenylsulfonyl)alkenyliodanes **61** and Z-β-haloalkenyliodanes with tetrabutylammonium halides or sodium benzenesulfinate.[27] An addition–elimination mechanism involving the intermediacy of α-sulfonyl-stabilized carbanion **63** by attack of the nucleophiles to the π^* orbital of **61** at the C_α atom is compatible with the retention of stereochemistry. Negative hyperconjugation-directed internal 60° rotation of **63**, followed by reductive elimination of PhI, gives **62**.

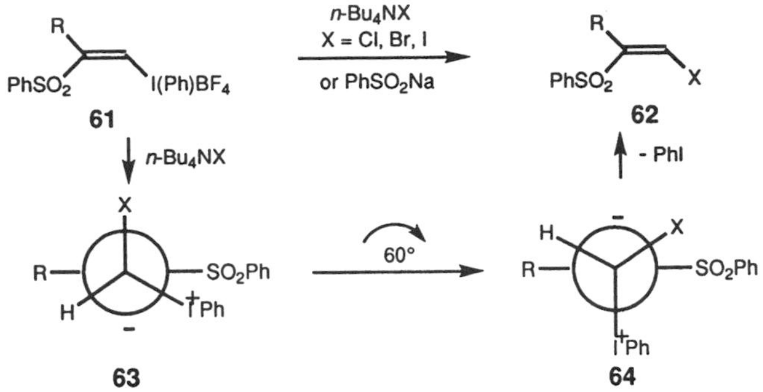

The first example of a Michael-type addition of nucleophiles to alkenyliodanes at the C_β atom, which is a common mode of reactions for alkynyliodanes, was reported recently.[28] Nucleophilic vinylic substitutions of *Z*-β-haloalkenyliodanes **32** with sodium benzenesulfinate (2.5 equiv.) afforded the retained *Z*-1,2-bis(phenylsulfonyl)alkenes stereoselectively: this reaction involves a Michael-type addition of benzenesulfinate anion as the first step of the reaction, yielding *Z*-β-(phenylsulfonyl)alkenyliodanes **65** with retention of stereochemistry.

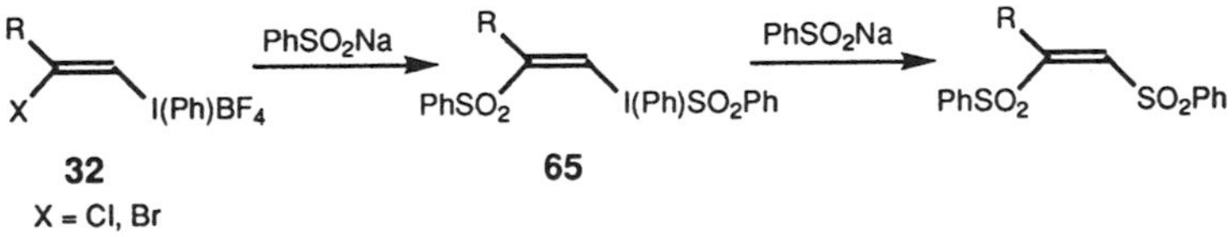

12.3.4 Friedel–Crafts Vinylation

Direct vinylations of aromatic compounds by electrophilic aromatic substitution are very limited because of the low reactivity of vinylic compounds such as vinyl halides and esters, and the instability of the products under the reaction conditions. Intramolecular Friedel–Crafts vinylation of alkenyliodanes **66** occurs on gentle heating at 40–60°C in MeOH, CHC_3, acetone, or MeCN.[29] The reaction could be applied to the synthesis of a chromene derivative. An intermolecular version of this Friedel–Crafts vinylation has also been demonstrated (**2a**→**68**). An addition–elimination mechanism for this Friedel–Crafts vinylation of alkenyliodanes has been proposed. Interestingly, heating of these alkenyliodanes in DMSO gives carbonyl compounds.

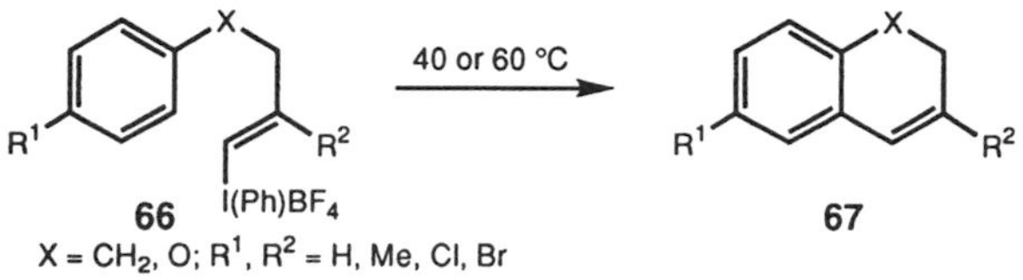

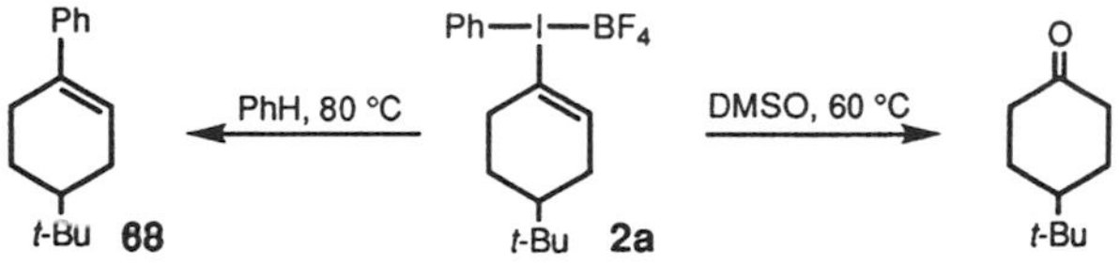

X = CH_2, O; R^1, R^2 = H, Me, Cl, Br

12.3.5 Palladium-Catalyzed Reaction

Palladium-catalyzed methoxycarbonylation of alkenyliodane **69** occurs at room temperature to give the α, β-unsaturated ester stereospecifically with retention of configuration.[12a] Moriarty and co-workers reported an important Heck-type synthesis of conjugated dienes.[30] The palladium-catalyzed coupling of **70** with olefins proceeds at room temperature with high stereoselectivity, yielding the diene **71**. Shibasaki and co-workers demonstrated the asymmetric version of this Heck-type olefination with a chiral phosphine ligand (*R*)-2, 2′-bis(diphenylphosphino)-1,1′-binaphthyl [(*R*)-BINAP].[31] Alkenyliodanes undergo palladium-catalyzed cross-coupling with allyl-, alkenyl-, and alkynylstannanes.[32]

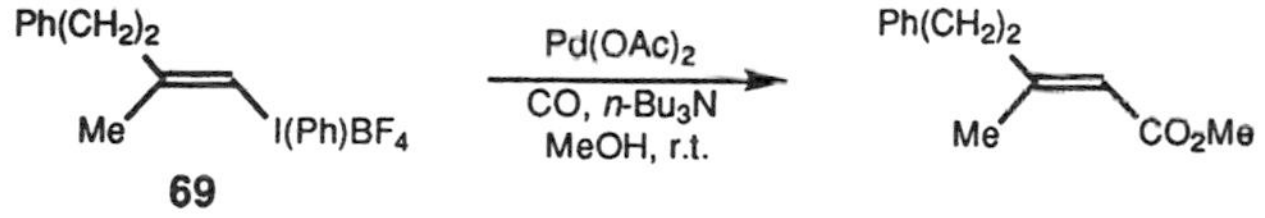

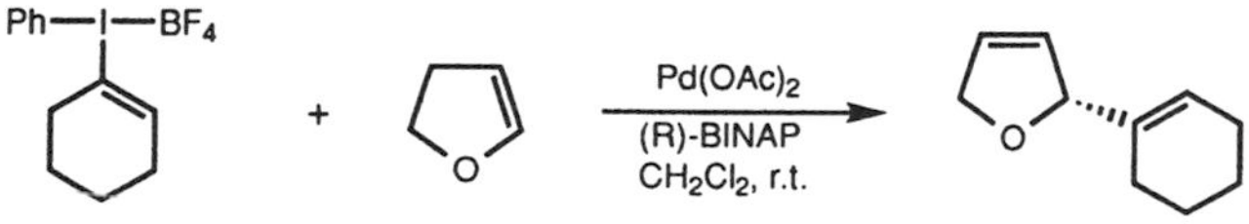

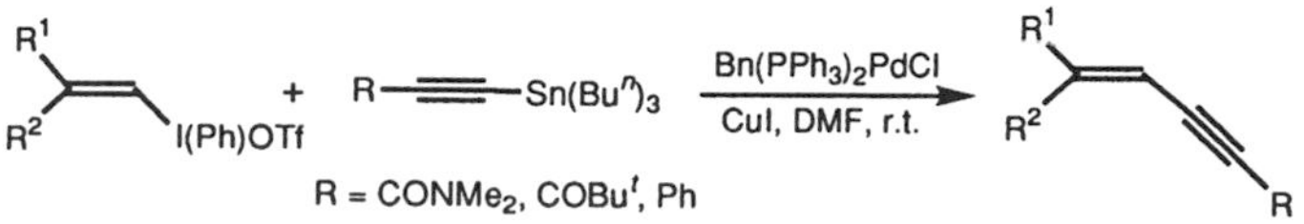

R = COMe, CO_2Me, Ph

R = $CONMe_2$, $COBu^t$, Ph

12.3.6 Claisen Rearrangement of Allenyl(aryl)iodanes

Allenyliodanes are a new class of hypervalent organoiodanes and could be generated by the reaction of propargylsilanes with aryliodanes. Reaction of propargylsilanes, germanes, and stannanes **73**, with (diacetoxyiodo)benzene **72** in the presence of BF_3—Et_2O undergoes reductive propargylation of the aryliodane to give *o*-propargyliodoarenes **74** in good yields.[33] This reductive propargylation of aryliodanes proceeds regioselectively at the ortho position under mild conditions (−20°C) and always involves reduction of iodanes to univalent iodides. Furthermore, the C—C bond formation takes place at the propargylic carbon atom attached to the metals of **73**.

AcO—I—OAc
72 + R—C≡C—CH₂M (73, M = SiMe₃, GePh₃, SnPh₃) → (BF₃-Et₂O, CH₂Cl₂, -20 °C) → 74

A mechanism which is compatible with these results involves electrophilic attack of the positively charged iodine of **72**, activated with BF_3, to **73** via an S_E2' process, generating hypervalent allenyl(aryl)iodanes **75**, which, in turn, undergo [3,3]-sigmatropic rearrangement providing **76**. Reductive elimination of **76** produces **74**. The lack of crossover products argues for the intramolecularity of the rearrangement of **75**. [3,3]-Sigmatropic rearrangement of **75** takes place under much milder conditions than the Claisen rearrangement and its nitrogen and sulfur variants, which usually require heating at about 150–250°C. The lower activation energy associated with the rearrangement of **75** can be interpreted in terms of the small bond energy needed to break the apical C—I(III) bond. In general, aryliodanes $ArIX_2$ adopt a T-shaped geometry, the hypervalent I(III)-X bonds being well overlapped with the aromatic π bond. This favorable orbital interaction could facilitate the rearrangement of **75**.

72 + 73 → (BF_3, S_E2') → 75 → ([3, 3]-Claisen) → 76 → (reductive elimination, −AcOH) → 74

When both ortho positions of aryliodanes were occupied with alkyl substituents, the reductive Claisen rearrangement afforded meta substitution products. This is the first demonstration that meta Claisen rearrangement occurs preferentially even when a free para position is available.

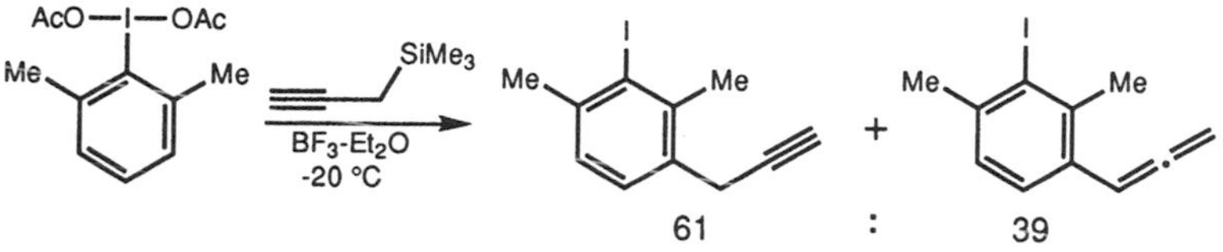

AcO—I—OAc
Me Me
SiMe$_3$
BF$_3$-Et$_2$O
-20 °C
I Me Me + I Me Me
61 : 39

In marked contrast, the allenyliodane **79**, generated from *p*-methoxyiodane **77** by the reaction with propargylsilane, undergoes deiodinative ipso Claisen rearrangement selectively, yielding ipso-substituted propargylarene **78**.[34] The transition state of 1,2-rearrangement of ortho propargyl group of **80** to the ipso site leading to the cation **81** will be efficiently stabilized by the π-donor *p*-methoxy group; subsequent deiodination of **81** affords **78**.

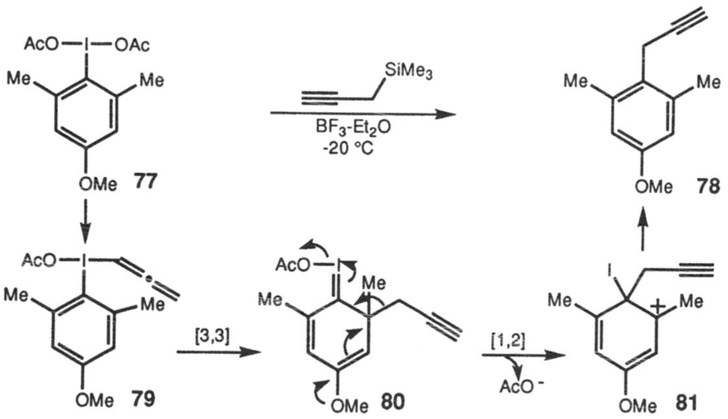

AcO—I—OAc
Me Me
OMe **77**
SiMe$_3$
BF$_3$-Et$_2$O
-20 °C
Me Me
OMe **78**
AcO—I
Me Me
OMe **79**
[3,3]
AcO—I
Me Me
OMe **80**
[1,2]
AcO$^-$
I
Me + Me
OMe **81**

The allenyl(aryl)iodanes **84** with an electron-withdrawing nitro group at the meta position show no tendency to undergo reductive [3,3]-sigmatropic rearrangement, but instead act as propargyl cation-equivalent species. They undergo nucleophilic substitutions with alcohols, carboxylic acids, and nitriles regioselectively to give propargyl ethers, esters, and amides **83**, presumably via collapse to propargyl cation **85** with reductive elimination of *m*-nitroiodobenzene.[35] The leaving ability of *m*-nitrophenyliodanyl group can be evaluated from the Hammett ρ value for solvolysis of cyclohexenyliodanes **2a** to be 16 times greater than that of phenyliodanyl group.[5] This very high leaving ability of the *m*-nitrophenyliodanyl group would be responsible for the selective collapse of allenyl(*m*-nitrophenyl)iodane **84** to propargyl cation **85**.

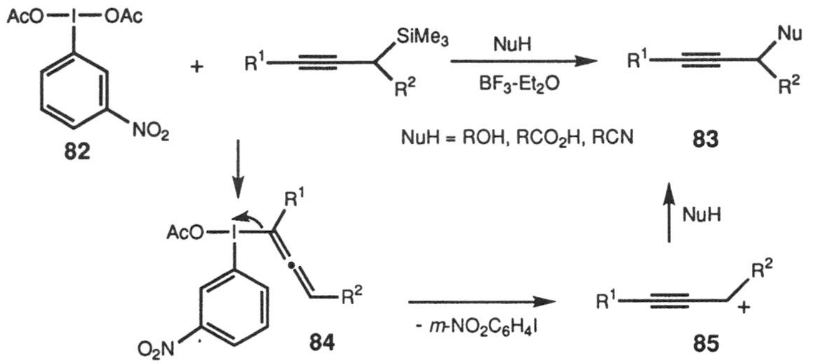

The BF_3-catalyzed reaction of the cyclic hydroxyiodane **86** with propargylsilane affords 2-iodo-3-propargylbenzoic acid **87** via the reductive iodane-Claisen rearrangement. The reaction with bis(trimethylsilyl)butyne **88**, however, led to the unexpected formation of an (alkylperoxy)iodane **89**.[36]

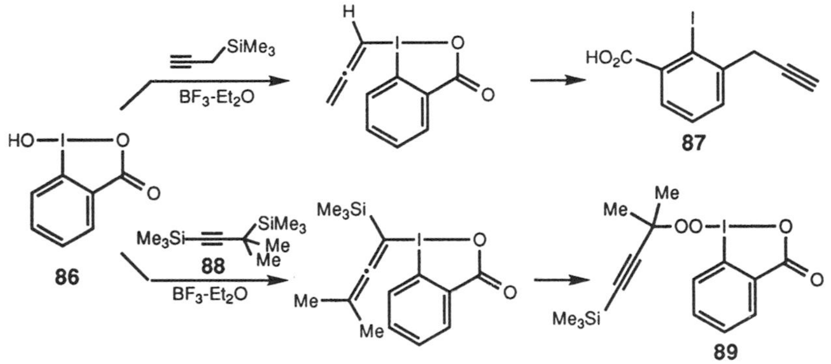

(Alkylperoxy)aryliodanes generally show a high tendency to decompose. This ready decomposition can be attributed to the small dissociation energy of the apical hypervalent peroxy-iodine(III) bond, and will be facilitated by conjugative overlap of the breaking hypervalent bond with π-orbitals of the aromatic nucleus. Fixation of an apical hetero ligand and an equatorial aromatic ligand on iodine(III) by the formation of five-membered heterocycles such as an iodoxolone, which makes this destabilizing orbital interaction very difficult, appears to lead to the enhanced stability of the (alkylperoxy)aryliodanes. The BF_3-catalyzed ligand exchange of the cyclic hydroxyiodane **86** with *tert*-butyl hydroperoxide gives the crystalline (*tert*-butylperoxy)iodane **90**, which is very stable in the solid state and can be safely stored at room temperature for an indefinite period of time.[37] In solution, however, the cyclic peroxyiodane **90** can generate the *tert*-butylperoxy radical even at room temperature via homolytic bond cleavage of the hypervalent iodine(III)-peroxy bond and can oxidize benzyl and allyl ethers to the esters at room temperature in the presence of alkali metal carbonates, which involves generation of α-oxy carbon-centered radicals.

86 —(t-BuOOH, BF_3-Et_2O, 0 °C)→ **90** (t-BuOO—I benziodoxolone) —(RO–CH$_2$Ph or RO–CH$_2$CH=CH$_2$, K_2CO_3)→ RO–C(O)Ph or RO–C(O)CH=CH$_2$

12.4 REACTION OF ALKYNYLIODANES

In 1981, Koser and co-workers reported the first general method for the synthesis of alkynyl(tosyloxy)iodanes **92** by the reaction of terminal alkynes with hydroxy(tosyloxy)iodane **91**.[38] The BF_3-catalyzed silicon, germanium, tin–iodane exchange reaction, developed for the generation of allyliodanes **4**, provides an efficient method for the synthesis of alkynyliodanes **33**.[39] Stang and co-workers developed a very useful procedure for the preparation of diverse β-functionalized alkynyliodane **94**, which involves a reaction of alkynylstannanes with cyanoiodane **93**.[40]

R—≡ —(PhI(OH)OTs **91**)→ R—≡—I(Ph)OTs **92**

R—≡—MR′$_3$ —(PhIO, BF_3-Et_2O or PhIO, $Et_3O^+BF_4^-$)→ R—≡—I(Ph)BF_4 **33**

MR′$_3$ = $SiMe_3$, $GeMe_3$, $SnMe_3$

R—≡—SnR′$_3$ —(PhI(CN)OTf **93**)→ R—≡—I(Ph)OTf **94**

R = CN, Cl, SO_2Ar, COR, $CONR_2$, CO_2Me
R′ = Me, Et, n-Bu

This cyanoiodane procedure allows the preparation of the interesting alkynyliodane **96** from the stannane **95**.[41] Use of dicyanoiodane **97** gives the dialkynyliodane **98**.[42] Taniguchi and Kitamura reported the synthesis of the novel alkynyliodane **99**.[43]

Bu_3Sn—≡—$SnBu_3$ (**95**) —PhI(CN)OTf **93**→ TfO(Ph)I—≡—I(Ph)OTf (**96**)

R—≡—$SnBu_3$ —$(NC)_2IOTf$ **97**→ $(R—≡)_2IOTf$ (**98**)

R—≡—$SiMe_3$ —PhIO, 2 TfOH→ R—≡—I(OTf)—C_6H_4—I(Ph)OTf (**99**)

12.4.1 Michael-Carbene Insertion (MCI) Reactions

Reaction of the alkynyliodane **100** with hard carbanions like 2-lithiofuran occurs at the hypervalent iodine atom and undergoes ligand exchange, yielding diaryliodanes.[44]

t-Bu—≡—I(Ar)OTs (**100**) + 2-furyl-Li ⟶ 2-furyl-I(Ar)OTs

It has been generally accepted that one of the most common modes of the reactions of alkynyliodanes involves Michael addition of nucleophiles to the electron-deficient acetylenic β-carbon atom. In 1986, we found that this reaction constitutes a key step of a highly versatile cyclopentene annulation of alkynyliodanes via the tandem Michael-carbene insertion (MCI) reaction.[45] Examples of this MCI reactions are shown in the following equations.

n-C_8H_{17}—≡—$I(Ph)BF_4$ (**101**) + 2-phenylindane-1,3-dione —t-BuONa [5+0]→ spiro product (Ph, n-C_5H_{11})

cyclohexyl—$(CH_2)_2$—≡—$I(Ph)BF_4$ + 2-phenylindane-1,3-dione —t-BuONa [5+0]→ product (Ph)

cyclopentyl—CH_2—≡—$I(Ph)BF_4$ + Meldrum's acid derivative (Me, Me, Me) —t-BuOK [5+0]→ product (H, H, Me, Me, Me)

Me—≡—$I(Ph)BF_4$ + 2-acetyl-3-(n-C_4H_9)cyclohexanone (Me) —t-BuOK [2+3]→ product ($COCH_3$, Me, n-C_3H_7)

The anti-Michael addition of soft carbanions, β-dicarbonyl enolates (Nu^-), to the alkynyliodanes **33** generates the labile iodine(III) ylide **102**, which, because of a hyperleaving group ability of a phenyliodanyl group, undergoes reductive elimination of iodobenzene to produce the reactive alkylidenecarbene **103a**. The intramolecular regioselective 1,5-C—H insertion gives the cyclopentene **104a**. Since all carbon atoms of the cyclopentene ring of **104a** come from the alkynyliodanes **33**, the reaction is termed a [5 + 0] cyclopentene annulation. The MCI reaction also becomes valuable as a [2 + 3] cyclopentene annulation, in which two sp^2 and three sp^3 carbon atoms of the cyclopentene ring of **104b** originate from acetylenic carbons of the alkynyliodanes **33** and the carbon nucleophiles, respectively.

R—≡—I(Ph)BF4 (**33**) —Nu⁻→ **102** → **103**

[5+0] MCI reaction: R = ...; **103a** → **104a**

[2+3] MCI reaction: Nu = ...; **103b** → **104b**

The MCI reaction provides a route to polysubstituted furans.[45] The formation of the furan **106** can be interpreted in terms of an intramolecular 1,5-insertion of the enolized alkylidenecarbene **105** into the enolic O—H bond. Exclusive formation of the furan **106** implies that the intramolecular 1,5-insertion into C—H bonds of methylene groups cannot compete with that into O—H bonds of enols.

n-C_8H_{17}—≡—I(Ph)BF4 (**101**) —PhCOCH2SO2Ph, t-BuONa→ **105** → **106**

Me—≡—I(Ph)BF4 —PhCOCH2CN, t-BuOK→ furan (NC, Me, Ph)

The MCI reactions of the alkynyliodane **99** with oxygen nucleophiles were reported by Taniguchi and Kitamura.[46] The α-oxyalkylidenecarbene **107**, generated by the reaction with phenoxide anion, shows a high selectivity for 1,5-insertion to the aromatic over the aliphatic C—H bonds to give benzofuran **108**. This nice reaction leads to the synthesis of furotropones and furanonaphthoquinones.[47]

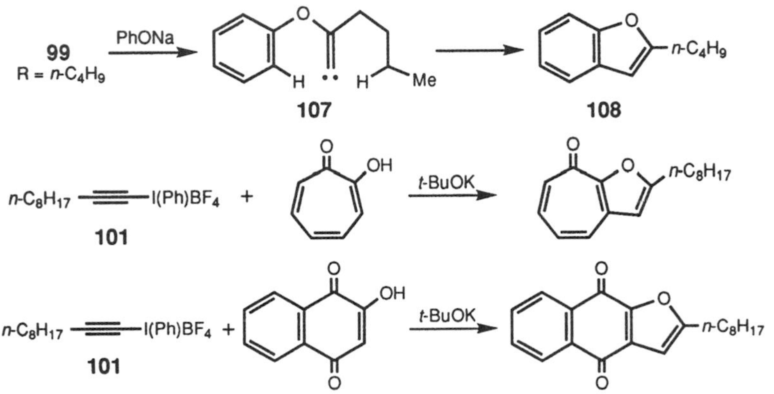

Reaction of the alkynyliodane **101** with benzenesulfinate anion in THF affords directly 1-(phenylsulfonyl)cyclopentene **109** via the [5 + 0] MCI reaction.[16] However, when the reaction was carried out under sufficiently acidic conditions ($PhSO_2H$/MeOH), protonation to the iodine(III) ylide **110** produced by anti-Michael addition is faster than the elimination of iodobenzene leading to an alkylidenecarbene and, therefore, gives *Z*-vinyliodane **111** stereoselectively.[16] These results combined with the formation of 1-sulfonylcyclopentene **26** from the vinyliodane **24** provide firm evidence for the tandem MCI pathway.

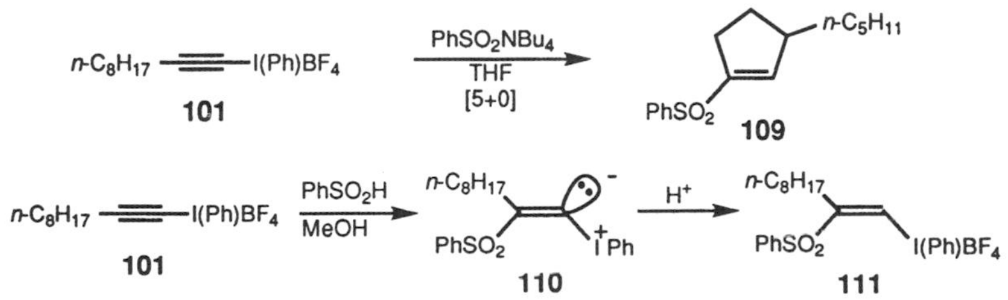

An excellent procedure for the synthesis of functionalized cyclopentenones and γ-lactams was developed by Stang and co-workers.[48] The MCI reaction of β-ketoethynyl- and β-amidoethynyliodanes with sulfinate anion affords **112**–**114**. They further demonstrated that the bis(alkynyliodane) **115** makes possible the double [5 + 0] MCI reaction yielding bis(cyclopentene) products.[49]

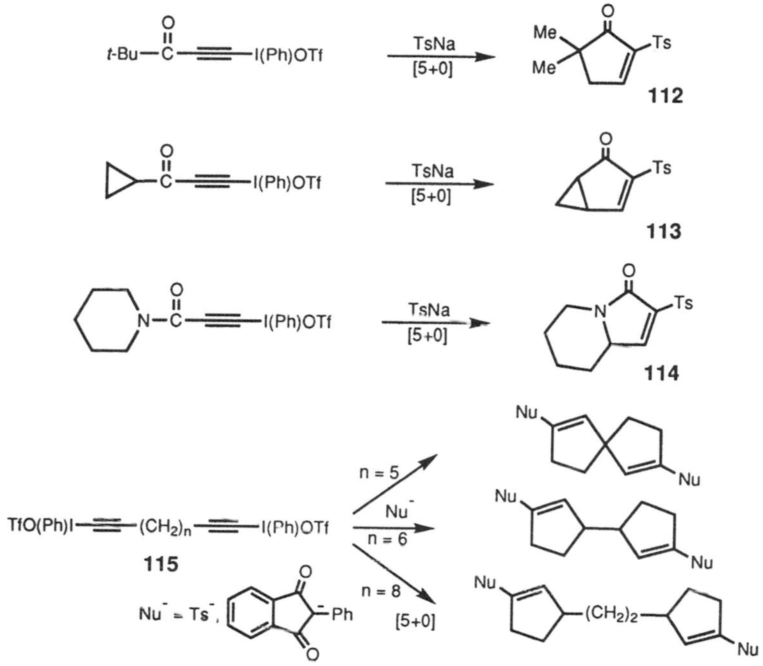

The [2 + 3] MCI reaction employing tosylamide as nitrogen nucleophiles was reported by Feldman and co-workers.[50] They developed a nicely designed intramolecular version of the MCI reaction.[51] Treatment of the tosylamide-bearing alkynylstannane **116** with the cyanoiodane **93** gave a labile alkynyliodane **117**, which, on exposure to a base, undergoes an intramolecular Michael addition to afford **119**. Recently, an intramolecular 1,4-C—H insertion of alkylidenecarbenes yielding the cyclobutene **120** was reported.[52]

12.4.2 Michael-Carbene Rearrangement (MCR) Reactions

In 1965, Beringer and Galton reported an alkynylation reaction of the anion of phenylindandione with (phenylethynyl)iodane **121**, yielding the alkyne **122**.[53] They proposed two types of mechanisms for this alkynylation: (1) direct nucleophilic displacement on the acetylenic α-carbon atom, and (2) an electron-transfer free-radical mechanism. Twenty years later, however, we demonstrated that this reaction might proceed via Michael addition.[45] Selective formation of the ^{13}C-labeled alkyne **124** with ^{13}C predominantly at the α-carbon atom of the alkynyl group in the reaction of **123** provides evidence for a Michael addition pathway, followed by 1,2-phenyl rearrangement of the alkylidenecarbene **125**. Apparently, the migratory aptitude of a phenyl group is so high that the 1,5-C—H insertion of **125** cannot compete with the 1,2-phenyl migration.

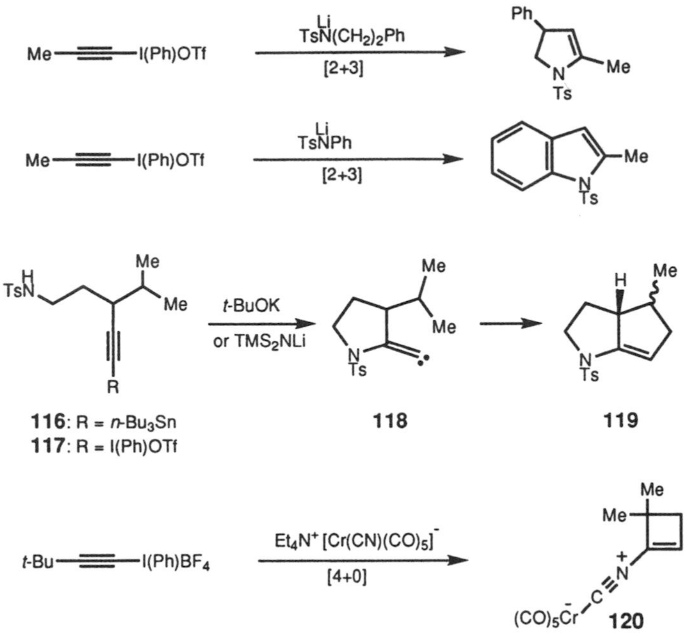

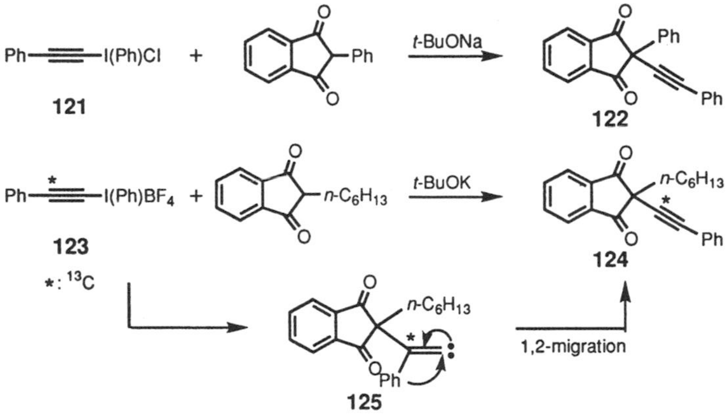

Similarly, ethynylation of β-dicarbonyl enolates via the tandem Michael-carbene rearrangement (MCR) pathway occurs smoothly by the reaction with ethynyliodane **126**, because of the high migratory aptitude of α-hydrogen of alkylidenecarbenes.[54] Regioselective ethynylation of monocarbonyl enolates with **126** affords the alkyne **127** in a good yield.[55]

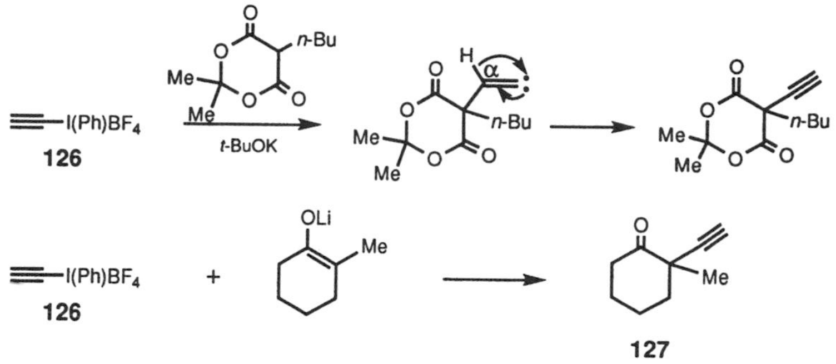

Nucleophiles with high tendency to migrate undergo this MCR reaction preferentially, providing a useful route for the synthesis of substituted alkynes. Examples of the MCR reaction with little or no tendency to compete with the MCI pathway are summarized in the following equation. It appears likely that, because of the electron-deficient nature of the carbenic center of alkylidenecarbenes, the electron-rich substituent, especially when the migrating atoms have lone pair electrons, tends to migrate.

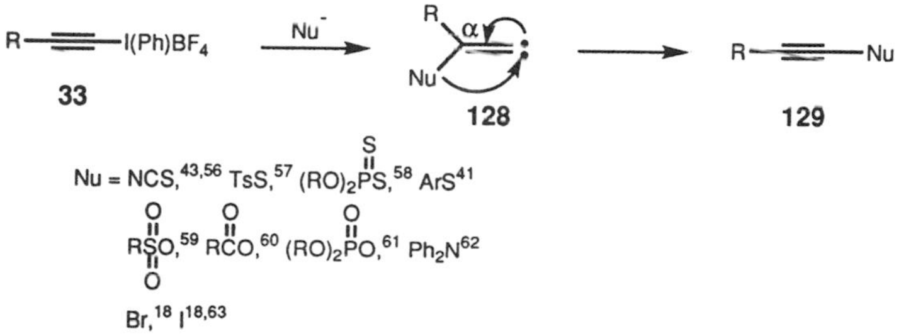

In some cases, the MCI pathway competes with the MCR reaction. Reaction of the alkynyliodane **101** with benezenesulfinate anion in H_2O leads to a mixture of the MCI product **109** and the MCR product **130**, because of a moderate migratory aptitude of arylsulfonyl groups.[16] When the alkynyliodanes have no available C—H bonds for the 1,5-insertions, the MCR products such as the alkynyl sulfone **132** are formed exclusively.[64]

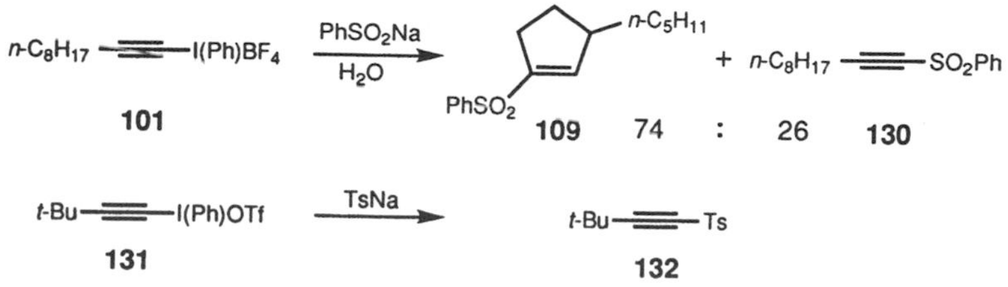

Phosphorous nucleophiles furnish substituted alkynes. Reaction with Ph_3P and $(MeO)_3P$ gives the alkynylphosphonium salts **133**[65] and the alkynylphosphonates **134**,[66] respectively. Both the ligand-coupling mechanism on iodine(III) and the MCR pathway may lead to formation of these alkynes. Similarly, reaction of alkynyliodanes with Ph_3As affords the alkynylarsonium salts **135**.[67]

$$R-C\equiv C-I(Ph)BF_4 \; (\mathbf{33}) \xrightarrow[\text{(sunlight)}]{Ph_3P} R-C\equiv C-\overset{+}{P}Ph_3\; BF_4^- \; (\mathbf{133})$$

$$\mathbf{33} \xrightarrow{(MeO)_3P} R-C\equiv C-P(=O)(OMe)_2 \; (\mathbf{134})$$

$$\mathbf{33} \xrightarrow{Ph_3As} R-C\equiv C-\overset{+}{As}Ph_3\; BF_4^- \; (\mathbf{135})$$

Because of the limited pages available for this review, only Michael-type addition reactions of alkynyliodanes are described. For other very useful reactions of alkynyliodanes involving Diels–Alder reactions, 1,3-dioplar cycloadditions, and reactions with transition metal complexes, see the excellent reviews of Koser[3b] and Stang.[4i,l]

ACKNOWLEDGMENTS

Our own work described here was supported by grants from the Ministry of Education, Science, Culture, and Sports, the Iwatani Naoji Foundation's Research Grant, and the Asahi Glass Foundation.

REFERENCES

1. (a) C. W. Perkins, J. C. Martin, A. J. Auduengo, W. Law, A. Alegria, J. K. Kochi, *J. Am. Chem. Soc.* **102**, 7753 (1980). (b) J. C. Martin, *Science* **221**, 509 (1983).
2. A. Varvoglis, *The Chemistry of Polycoordinated Iodine* (VCH Publishers, New York, 1992).
3. (a) G. F. Koser, in *The Chemistry of Functional Groups, Supplement D*, S. Patai and Z. Rappoport, Eds (Wiley, New York, 1983), Chapter 18 and 25. (b) G. F. Koser, in *The Chemistry of Functional Groups, Supplement D2*, S. Patai and Z. Rappoport, Eds (Wiley, New York, 1995), Chapter 21.
4. For reviews of organoiodanes, see: (a) D. F. Banks, *Chem. Rev.* **66**, 243 (1966). (b). A. Varvoglis, *Synthesis*, 709 (1984). (c) M. Ochiai, Y. Nagao, *J. Synth. Org. Chem., Jpn.* **44**, 660 (1986) (d) R. M. Moriarty, O. Prakash, *Acc. Chem. Res.* **19**, 244 (1986). (e) E. B. Merkushev, *Russian Chemical Reviews* **56**, 826 (1987). (f) M. Ochiai, *Revs. Heteroatom Chem.* **2**, 92 (1989). (g) R. M. Moriarty, R. K. Vaid, *Synthesis*, 431

(1990). (h) R. M. Moriarty, R. K. Vaid, G. F. Koser, *Synlett.* 365 (1990) (i) P. J. Stang, *Angew. Chem., Int. Ed. Engl.* **31**, 274 (1992). (j) Y. Kita, H. Tohma, T. Yakura, *Trends in Organic Chemistry*, **3**, 113 (1992). (k) T. Kitamura, *J. Synth. Org. Chem., Jpn.* **53**, 893 (1995). (l) P. J. Stang, V. V. Zhdankin, *Chem. Rev.* **96**, 1123 (1996).

5. T. Okuyama, T. Takino, T. Sueda, Ochiai, *J. Am. Chem. Soc.* **117**, 3360 (1995).
6. M. Hanack, K.-A. Fuchs, C. J. Collins, *J. Am. Chem. Soc.* **105**, 4008 (1983).
7. W. D. Pfeifer, C. A. Bahn, P. v. R. Schleyer, S. Bocher, C. E. Harding, K. Hummel, M. Hanack, P. J. Stang, *J. Am. Chem. Soc.* **93**, 1513 (1971).
8. (a) D. S. Noyce, J. A. Virgilio, *J. Org. Chem.* **37**, 2643 (1972). (b) P. J. Stang, M. Hanack, L. R. Subramanian, *Synthesis*, 85 (1982).
9. A. A. Mironova, I. I. Maletina, S. V. Iksanova, V. V. Orda, L. M. Yagupolskii, *Zh. Org. Chem.* **25**, 306 (1989).
10. M. Ochiai, E. Fujita, M. Arimoto, H. Yamaguchi, *Chem. Pharm. Bull.* **33**, 41 (1985).
11. (a) M. Ochiai, E. Fujita, M. Arimoto, H. Yamaguchi, *J. Chem. Soc., Chem. Commun.* 1108 (1982). (b) M. Ochiai, E. Fujita, M. Arimoto, H. Yamaguchi, *Tetrahedron Lett.* **24**, 777 (1983). (c) M. Ochiai, E. Fujita, M. Arimoto, H. Yamaguchi, *Chem. Pharm. Bull.* **33**, 989 (1985). (d) M. Arimoto, H. Yamaguchi, E. Fujita, M. Ochiai, Y. Nagao, *Tetrahedron Lett.* **28**, 6289 (1987).
12. (a) M. Ochiai, K. Sumi, Y. Takaoka, M. Kunishima, Y. Nagao, M. Shiro, E. Fujita, *Tetrahedron*, **44**, 4095 (1988). (b) M. Ochiai, K. Oshima, Y. Masaki, *J. Chem. Soc., Chem. Commun.* 869 (1991).
13. O. Neiland, B. Karele, *Zu. Org. Chem.* **7**, 1611 (1971).
14. M. Ochiai, M. Kunishima, K. Fuji, M. Shiro, Y. Nagao, *J. Chem. Soc., Chem. Commun.* 1076 (1988). (b) P. J. Stang, H. Wingert, A. M. Arif, *J. Am. Chem. Soc.* **109**, 7235 (1987).
15. M. Ochiai, Y. Takaoka, Y. Nagao, *J. Am. Chem. Soc.* **110**, 6565 (1988).
16. M. Ochiai, M. Kunishima, S. Tani, Y. Nagao, *J. Am. Chem. Soc.* **113**, 3135 (1991).
17. J. C. Gilbert, D. H. Giamalva, U. Weerasooriya, *J. Org. Chem.* **48**, 5251 (1983).
18. M. Ochiai, K. Uemura, Y. Masaki, *J. Am. Chem. Soc.* **115**, 2528 (1993).
19. (a) J. C. Gilbert, U. Weerasooriya, *Tetrahedron Lett.* **21**, 2041 (1980) (b) A. Oku, T. Harada, Y. Nozaki, Y. Yamaura, *J. Am. Chem. Soc.* **107**, 2189 (1985). (c) A. Oku, K. Kimura, S. Ohwaki, *Acta Chem. Scand.* **47**, 391 (1993).
20. T. Sueda, T. Nagaoka, S. Goto, M. Ochiai, *J. Am. Chem. Soc.* **118**, 10141 (1996).
21. M. Ochiai, T. Sueda, K. Uemura, Y. Masaki, *J. Org. Chem.* **60**, 2624 (1995).
22. M. Ochiai, K. Sumi, Y. Nagao, E. Fujita, *Tetrahedron Lett.* **26**, 2351 (1985).
23. Z. Rappoport, *Recl. Trav. Chim. Pays-Bas* **104**, 309 (1985).
24. M. Ochiai, K. Oshima, Y. Masaki, *J. Am. Chem. Soc.* **113**, 7059 (1991).
25. M. Ochiai, Y. Takaoka, Y. Masaki, Y. Nagao, M. Shiro, *J. Am. Chem. Soc.* **112**, 5677 (1990).
26. (a) M. Ishikura, M. Terashima, *J. Chem. Soc. Chem. Commun.* 727 (1989). (b) M. Ishikura, M. Terashima, *Heterocycles*, **27**, 2619 (1988). (c) P. J. Stang, T. Blume, V. V. Zhdankin, *Synthesis*, 35 (1993).

27. M. Ochiai, K. Oshima, Y. Masaki, *Tetrahedron Lett.* **32**, 7711 (1991) (b). M. Ochiai, K. Oshima, Y. Masaki, M. Kunishima, S. Tani, *Tetrahedron Lett.* **34**, 4829 (1993). (c) M. Ochiai, K. Oshima, Y. Masaki, *Chem. Lett.* 871 (1994).
28. M. Ochiai, Y. Kitagawa, M. Toyonari, K. Uemura, *Tetrahedron Lett.* **35**, 9407 (1994).
29. M. Ochiai, Y. Takaoka, K. Sumi, Y. Nagao, *J. Chem. Soc., Chem. Commun.* 1382 (1986).
30. (a) R. M. Moriarty, W. R. Epa, A. K. Awasthi, *J. Am. Chem. Soc.* **113**, 6315 (1991). (b) S.-K. Kang, K.-Y. Jung, C.-H. Park, S.-B. Jang, *Tetrahedron Lett.* **36**, 8047 (1995). (c) S.-K. Kang, H.-W. Lee, S.-B. Jang, P.-S. Ho, *J. Chem. Soc., Chem. Commun.* 835 (1996).
31. Y. Kurihara, M. Sodeoka, M. Shibasaki, *Chem. Pharm. Bull.* **42**, 2357 (1994).
32. (a) R. M. Moriarty, W. R. Epa, *Tetrahedron Lett.* **33**, 4095 (1992). (b) R. J. Hinkle, G. T. Poulter, P. J. Stang, *J. Am. Chem. Soc.* **115**, 11626 (1993).
33. M. Ochiai, T. Ito, Y. Takaoka, Y. Masaki, *J. Am. Chem. Soc.* **113**, 1319 (1991).
34. (a) M. Ochiai, T. Ito, Y. Masaki, *J. Chem. Soc., Chem. Commun.* 15 (1992) (b) M. Ochiai, T. Ito, *J. Org. Chem.* **60**, 2274 (1995).
35. M. Kida, T. Sueda, S. Goto, T. Okuyama, M. Ochiai, *J. Chem. Soc., Chem. Commun.* 1933 (1996).
36. M. Ochiai, T. Ito, M. Shiro, *J. Chem. Soc., Chem. Commun.* 218 (1993).
37. (a) M. Ochiai, T. Ito, Y. Masaki, M. Shiro, *J. Am. Chem. Soc.* **114**, 6269 (1992) (b) M. Ochiai, T. Ito, H. Takahashi, A. Nakanishi, M. Toyonari, T. Sueda, S. Goto, M. Shiro, *J. Am. Chem. Soc.* **118**, 7716 (1996).
38. (a) G. F. Koser, L. Rebrovic, R. H. Wettach, *J. Org. Chem.* **46**, 4324 (1981). (b) L. Rebrovic, G. F. Koser, *J. Org. Chem.* **49**, 4700 (1984).
39. (a) M. Ochiai, M. Kunishima, K. Sumi, Y. Nagao, E. Fujita, *Tetrahedron Lett.* **26**, 4501 (1985). (b) M. Ochiai, M. Kunishima, Y. Nagao, K. Fuji, E. Fujita, *J. Chem. Soc., Chem. Commun.* 1708 (1987). (c) M. Kunishima, PhD. Dissertation, Institute for Chemical Research, Kyoto University, 1989.
40. (a) P. J. Stang, B. L. Williamson, V. V. Zhdankin, *J. Am. Chem. Soc.* **113**, 5870 (1991). (b) B. L. Williamson, P. J. Stang, A. M. Arif, *J. Am. Chem. Soc.* **115**, 2590 (1993).
41. (a) P. J. Stang, V. V. Zhdankin, *J. Am. Chem. Soc.* **112**, 6437 (1990). (b) P. J. Stang, V. V. Zhdankin, *J. Am. Chem. Soc.* **113**, 4571 (1991).
42. (a) P. J. Stang, V. V. Zhdankin, R. Tykwinski, *Tetrahedron Lett.* **33**, 1419 (1992). (b) P. J. Stang, V. V. Zhdankin, A. M. Arif, *J. Am. Chem. Soc.* **113**, 8997 (1991).
43. (a) T. Kitamura, R. Furuki, L. Zheng, T. Fujimoto, H. Taniguchi, *Chem. Lett.* 2241 (1992). (b) T. Kitamura, R. Furuki, L. Zheng, K. Nagata, T. Fukuoka, Y. Fujiwara, H. Taniguchi, *Bull. Chem. Soc. Jpn.* **68**, 3637 (1995).
44. A. J. Margida, G. F. Koser, *J. Org. Chem.* **49**, 4703 (1984).
45. M. Ochiai, M. Kunishima, Y. Nagao, K. Fuji, M. Shiro, E. Fujita, *J. Am. Chem. Soc.* **108**, 8281 (1986).
46. T. Kitamura, L. Zheng, H. Taniguchi, M. Sakurai, R. Tanaka, *Tetrahedron Lett.* **34**, 4055 (1993).
47. T. Shu, D.-W. Chen, M. Ochiai, *Tetrahedron Lett.* **37**, 5539 (1996).

48. B. L. Williamson, R. R. Tykwinski, P. J. Stang, *J. Am. Chem. Soc.* **116**, 93 (1994).
49. R. R. Tykwinski, P. J. Stang, N. E. Persky, *Tetrahedron Lett.* **35**, 23 (1994).
50. K. S. Feldman, M. M. Bruendl, K. Schildknegt, *J. Org. Chem.* **60**, 7722 (1995).
51. K. Schildknegt, A. C. Bohnstedt, K. S. Feldman, A. Sambandam, *J. Am. Chem. Soc.* **117**, 7544 (1995).
52. R. Kunz, W. P. Fehlhammer, *Angew. Chem. Int. Ed. Engl.* **33**, 330 (1994).
53. F. M. Beringer, S. A. Galton, *J. Org. Chem.* **30**, 1930 (1965).
54. (a) M. Ochiai, T. Ito, Y. Takaoka, Y. Masaki, M. Kunishima, S. Tani, Y. Nagao, *J. Chem. Soc., Chem. Commun.* 118 (1990). (b) T. Suzuki, Y. Uozumi, M. Shibasaki, *J. Chem. Soc., Chem. Commun.* 1593 (1991). (c) M. D. Bachi, N. Bar-Ner, P. J. Stang, B. L. Williamson, *J. Org. Chem.* **58**, 7923 (1993).
55. M. Ochiai, T. Ito, unpublished observation.
56. D. R. Fischer, B. L. Williamson, P. J. Stang, *Synlett.* 535 (1992).
57. B. L. Williamson, P. Murch, D. R. Fischer, P. J. Stang, *Synlett.* 858 (1993).
58. Z.-D. Liu, Z.-C. Chen, *J. Org. Chem.* **58**, 1924 (1993).
59. (a) P. J. Stang, B. W. Surber, *J. Am. Chem. Soc.* **107**, 1452 (1985). (b) P. J. Stang, B. W. Surber, Z.-C. Chen, K. A. Roberts, A. G. Anderson, *J. Am. Chem. Soc.* **109**, 228 (1987).
60. (a) P. J. Stang, M. Boehshar, J. Lin, *J. Am. Soc.* **108**, 7832 (1986). (b) P. J. Stang, M. Boehshar, H. Wingert, T. Kitamura, *J. Am. Chem. Soc.* **110**, 3272 (1988).
61. P. J. Stang, T. Kitamura, M. Boehshar, H. Wingert, *J. Am. Chem. Soc.* **111**, 2225 (1989).
62. P. Murch, B. L. Williamson, P. J. Stang, *Synthesis*, 1255 (1994).
63. M. Ochiai, K. Uemura, K. Oshima, Y. Masaki, M. Kinishima, S. Tani, *Tetrahedron Lett.* **32**, 4753 (1991).
64. (a) R. R. Tykwinski, B. L. Williamson, D. R. Fischer, P. J. Stang, A. M. Arif, *J. Org. Chem.* **58**, 5235 (1993). (b) Z.-D. Liu, Z.-C. Chen, *Synth. Commun.* **22**, 1997 (1992).
65. (a) M. Ochiai, M. Kunishima, Y. Nagao, K. Fuji, E. Fujita, *J. Chem. Soc., Chem. Commun.* 1708 (1987). (b) A. Schmidpeter, P. Mayer, J. Stocker, K. A. Roberts, P. J. Stang, *Heteroatom Chem.* **2**, 569 (1991). (c) P. J. Stang, C. M. Crittell, *J. Org. Chem.* **57**, 4305 (1992). (d) K. K. Laali, M. Regitz, M. Birkel, P. J. Stang, C. M. Crittell, *J. Org. Chem.* **58**, 4105 (1993).
66. J. S. Lodaya, G. F. Koser, *J. Org. Chem.* **55**, 1513 (1990).
67. T. Nagaoka, T. Sueda, M. Ochiai, *Tetrahedron Lett.* **36**, 261 (1995).

13 Hypervalent Compounds of Xenon

VALERY K. BREL AND NIKOLAI S. ZEFIROV
Institute of Physiologically Active Compounds, Russian Academy of Science, Chernogolovka, Moscow Region, Russia

13.1 INTRODUCTION

The first stable xenon compounds were obtained by N. Bartlett in 1962.[1] He synthesized some salts containing xenon as a cation from elementary xenon and platinum hexafluoride (Eq. 13.1):

$$Xe + PtF_6 \longrightarrow XePtF_6 \quad (13.1)$$

Less than a year after this report, xenon tetrafluoride was prepared by reacting xenon and fluorine at 400°C.[2] Later other oxidizers were shown to be applicable in reactions with elementary xenon. For example, elementary xenon oxidation by silver difluoride in hydrogen fluoride also leads to xenon difluoride,[3] and the same treatment in arsenic pentafluoride leads to a derivative of dixenon cation (Eqs. 13.2, 13.3): [4–6]

$$2AgF_2 + 2BF_3 + Xe \longrightarrow 2AgBF_4 + XeF_2 \quad (13.2)$$

$$4AgF_2 + 5AsF_5 + 2Xe \longrightarrow 4AgAsF_6 + [Xe_2F_3]^+AsF_6^- \quad (13.3)$$

These discoveries led to the beginning of an absolutely unexpected field of chemistry – the chemistry of noble gases. Nowadays there exist more than 500 articles concerning the chemistry of hypervalent xenon derivatives,[1] and this number grows continuously. These works have developed in two directions. The first one concerns application of the most useful compound – xenon difluoride (XeF_2) – as a reagent in chemistry and, in particularly, in organic

Chemistry of Hypervalent Compounds, edited by Kin-ya Akiba.
ISBN 0-471-24019-2

synthesis. XeF_2 possesses clear oxidative, electrophilic, and fluorinating properties, and therefore is widely applied in organic chemistry.[7]

The second direction concerns the development of xenon chemistry itself, that is, a synthesis of variable derivatives of xenon of different valences (Xe(II), Xe(IV), Xe(VI)), which often possess amazing structures. It is the goal of the present chapter to present a short overview of this area of xenon chemistry. In other words, we have tried to depict briefly an area of hypervalent xenon derivatives, and their interconversions and reactions.

13.2 POLYFLUORINATED DERIVATIVES OF XENON

13.2.2 Synthesis of Polyfluorinated Derivatives of Xenon

Synthesis of xenon derivatives can be divided into two categories: (1) direct interaction of elementary xenon with different oxidizing systems, and (2) interconversions of xenon derivatives (usually, synthesis of various derivatives from fluorides).

Xenon tetrafluoride was the first stable xenon derivative, obtained by N. Bartlett and, the same year, by direct interaction of elements Xe and F_2 at high temperatures.[7] Later, efforts were concentrated on creation of more convenient and reliable methods of synthesis of xenon fluorides. Xenon difluoride, tetrafluoride, and hexafluoride were obtained in satisfactory yields[8–11] due to optimal choice of reaction conditions (pressure, temperature) and xenon-to-fluorine ratios. UV irradiation of a xenon–fluorine mixture was found to lead to a formation of xenon fluorides.[12–16] Being very simple, the photochemical method was the most popular and accessible one. This method allowed one to obtain xenon difluoride of 99% purity. Besides elementary fluorine, other fluorination agents were found to be applicable as well. For example, available chlorine trifluoride was successfully used for large scale-up synthesis[17] of xenon difluoride, and O_2F_2 was used for synthesis of xenon tetrafluoride and hexafluoride.[18]

Methods of synthesis of xenon fluorides[1] are not limited to those mentioned above. Other examples of such oxidations are: (1) irradiation of crystalline xenon with a flow of fluorine atoms, obtained by a thermocatalytic method;[19–21] (2) γ-irradiation of fluorine/xenon mixture;[22] (3) electrochemical fluorination.[23] However, these methods are not widely applied, being mostly illustrations of the variety of elementary xenon chemistry methods. There exist data concerning other xenon halides, for instance xenon chlorides,[24,25] which have not been studied thoroughly due to their low stability.

13.2.2 Complexing Ability of Polyfluorinated Derivatives of Xenon

One of the main features of xenon fluorides is their ability for complexation with strong aprotic Lewis acids. These reactions were studied for xenon fluorides of different valences. In particular, they were studied in detail for Xe(II) and Xe(VI). For example, if Lewis acids react with xenon difluoride, the fluoroxenonium-type salts $[XeF]^+[MF_m]^-$ or $[Xe_2F_3]^+[MF_m]^-$ are formed, these salts being very effective oxidizing systems for many substrates. Interaction of XeF_2 with HF, BF_3, or $BF_3 \cdot Et_2O$ as Lewis acid takes place, and probably leads to a formation of complexes with the polarization of $FXe^{\delta+}F \cdots BF_3^{\delta-}$ type. The formation of this complex explains the so-called electrophilic fluorination of organic substrates using xenon difluoride in the presence of Lewis acids.

Syntheses of stable complexes of XeF_2 with fluorides of different metals are extensively reported. For example, the formation of complexes with the following compositions was proved: $XeF_2 \cdot 2SnF_4$,[26] $3XeF_2 \cdot 4SnF_4$,[26] $2XeF_2 \cdot SnF_4$,[26] $XeF_2 \cdot VF_5$,[27] $XeF_2 \cdot NbF_5$,[27] $XeF_2 \cdot TaF_5$,[27] $XeF_2 \cdot BiF_5$,[27–29] $XeF_2 \cdot SbF_5$,[27–29] $XeF_2 \cdot MnF_4$,[30–32] $XeF_2 \cdot 2MnF_4$,[30,32] $XeF_2 \cdot CrF_4$,[33–35] and $[Ag(XeF_2)_2]AsF_6$.[36] Also, the formation of $XeF_2 \cdot XeF_4$,[37] $XeF_2 \cdot IF_5$,[38,39] and $XeF_2 \cdot XeOF_4$[40] complexes has been mentioned. It is difficult to cover in this chapter all the reports concerning the synthesis of salts via the reaction of xenon difluoride with metal fluorides. However, it should be emphasized that xenon difluoride usually acts as a donor of F^- anions, forming salts of the type $[XeF]^+[MF_m]^-$. These structures were clearly proved by IR and Raman spectroscopy and in some cases by X-ray diffraction analysis. Complexes of xenon difluoride with alkali metals could not be obtained directly,[41,42] but at high temperatures the salts of type M_2XeF_6 (M = Na, K, Rb, Cs) were synthesized with Xe(VI) being the anion $[XeF_6]^{-2}$ (Eqs. 13.4, 13.5):[43]

$$2MF + 2XeF_2 \longrightarrow M_2XeF_6 + Xe \qquad (13.4)$$

$$2MF + XeF_4 \longrightarrow M_2XeF_6 \quad M = K, Rb, Cs \qquad (13.5)$$

This result shows, indirectly, that xenon difluoride cannot act as an F^- anion acceptor.

By analogy to xenon difluoride, xenon hexafluoride can form complexes with fluorides of different elements. Being a donor of fluoride anions, XeF_6 forms two types of cations: (1) XeF_5^+ and (2) $Xe_2F_{11}^+$, having a bridge structure. So far the complexes of XeF_6 with various salts of elements,[44] including fluorides of lead,[45] tin,[45] zirconium,[46] hafnium,[46] vanadium,[47] manganese,[48] nickel,[49,50] niobium,[51] silver,[52] gold,[52] germanium,[53] and a majority of the lantanides[54] have been obtained.

13.3 OXYGEN-CONTAINING DERIVATIVES OF XENON

As was mentioned above, xenon fluorides are the starting compounds for obtaining other xenon derivatives. Their interaction with oxygen-containing substrates leads to a formation of compounds with covalent bonds Xe—O or Xe=O. Nowadays this class of compounds is one of the most diverse. A formation of an Xe=O bond usually occurs during xenon fluoride hydrolysis, the hydrolysis of xenon difluoride being studied particularly thoroughly. Appelman's[55,56] suggestion concerning the transient formation of Xe=O at the first stage of hydrolysis, which, in turn, dissociates to xenon and hydrogen peroxide, was later proved by other authors.[57–61] These species exhibit strong oxidizing properties. In fact, XeF_2 aqueous solutions were able to oxidize different inorganic[55–57] and organic[58–61] compounds. It should be taken into account that in some cases the interaction of xenon difluoride with alcohols occurs with the formation of electrophilic alkoxyxenon fluorides, [ROXeF], which can react with unsaturated compounds.[62]

Xenon hexafluoride hydrolysis by two equivalents of water (precise amount!) leads to xenon oxytetrafluoride, $XeOF_4$,[63] whereas use of an excess of water may lead to extremely explosive xenon trioxyde,[64,65] probably via intermediacy of xenon dioxydifluoride.

Xenon dioxydifluoride can react with arsenic pentafluoride with the formation of both cationic structures and, untypically, structures of the bridged type. The salt type depends upon the ratio of the applied initial compounds (Scheme 13.1).[65]

$$XeO_2F_2 + AsF_5 \xrightarrow{HF} [XeO_2F]^+AsF_6^-$$

$$\downarrow -AsF_5$$

$$2XeO_2F_2 + AsF_5 \xrightarrow{HF} [FO_2XeFXeO_2F]^+AsF_6^-$$

Scheme 13.1

Xenon tetrafluoride formation is also possible in the reaction of hexafluoride oxidation by SeO_2F_2,[66] $NaNO_3$,[67–69] and TeO_3F.[70] However, these methods have some limitations due to a toxicity of reagents or to the side formation of explosive xenon trioxide. To overcome these problems, the reaction should be carried out with phosphorus fluorooxide (Eq. 13.6):[71]

$$XeF_6 + O{=}PF_3 \longrightarrow XeOF_4 + PF_5 \qquad (13.6)$$

Phosphorus fluorooxide was found to be a good oxidizer of xenon hexafluoride and a possibility of interchange between phosphorus and xenon ligands was demonstrated. The reaction occurs with almost quantitative yield without XeO_3 formation.

Xenon tetrafluorooxide, $XeOF_4$, can form complexes with fluorides of alkali metals in different ratios[72,73] and salts of type $[XeOF_5]^-Me^+$.[74,75] These species can possess a very peculiar molecular geometry. Thus, Seppelt has found that $XeOF_5^-$ anion in $NO^+XeOF_5^-$ salt has a pentagonal pyramidal structure with the oxygen (but not fluorine!) atom in apical position and five fluorine atoms in one basal plane (Eq. 13.7). This geometrical arrangement is an extremely rare one.[76]

$$XeOF_4 + O{=}NF \longrightarrow NO^+XeOF_5^- \tag{13.7}$$

Xenon (VIII) derivatives – perxenonates (xenon peroxide compounds) – could be obtained according to Scheme 13.2.[77]

$$XeF_6 \xrightarrow{HOP(O)F_2} XeO_3 \xrightarrow{NaOH} Na_4XeO_6 + Xe$$

Scheme 13.2

This class of compounds is of interest due to the fact that perxenonates are the initial compounds for synthesis of perxenates[78,79] and hydroxenates[80] of various compositions.

Nowadays perxenonates of almost all alkaline, alkaline earth, and actinide elements[81] can be synthesized. The oxidizing potential of these compounds in acidic media is very high; in some cases, it reaches 3.0 V.[82]

Elementary xenon fluorides can react with different oxygen-containing acids with the formation of rather stable compounds. The following reactions were studied in detail for xenon difluoride (Eqs. 13.8, 13.9):

$$XeF_2 + HOR \longrightarrow ROXeF + HF \tag{13.8}$$

$$FXeOR + HOR \longrightarrow (RO)_2Xe + HF \tag{13.9}$$

Xenon fluorides react with most oxygen-containing acids, but a strict selection of ligands —OR is necessary to obtain stable products of reactions. To form one or several covalent bonds Xe—O, the selected ligand should be an effective electron-withdrawing group and, to some extent, be able to mimic properties of the substituted fluorine ligand. Nowadays, compounds with the following substituents are known among two-valent xenon derivatives: OSO_2F,[83,84] OSO_2CF_3,[85] $OC(O)CF_3$,[85–87] $OP(O)F_2$,[88] $OIOF_4$,[89,90] $OClO_3$,[83,85] $OSeF_5$,[91] $OTeF_5$,[86,92] and ONO_2.[84] The synthesis is usually carried out by reaction of corresponding acids with xenon difluoride. However, another way of forming Xe—O bonds is also possible, namely by an addition of FXe and F moieties to element=O bonds. This approach was realized in the case of interaction of xenon difluoride with sulfur trioxide.[93] Depending on the components ratio, xenon mono- or difluorosulfate were obtained (Scheme 13.3).

$$Xe(OSO_2F)_2 \xleftarrow{2SO_3} XeF_2 \xrightarrow{SO_3} FXeOSO_2F$$

Scheme 13.3

Analogously, xenon difluoride reacts with iodine trifluorodioxide as shown in Scheme 13.4.[90]

$$IO_2F_3 + XeF_2 \rightleftharpoons FXeOIOF_4$$

$$IO_2F_3 + FXeR \rightleftharpoons RXeOIOF_4$$
$R = OIOF_5, OSO_2F.$

Scheme 13.4

This example demonstrates the possibility of synthesis of xenon derivatives, having different ligands.

Substitution of one or more *non*-fluorine ligands by others is possible, in principle, though it has not been thoroughly studied. Syvret and Schrobilgen illustrated this possibility with the following example (Eq. 13.10):[90]

$$Xe(OTeF_5)_2 + 2HOIOF_4 \longrightarrow Xe(OIF_4O)_2 + 2HOTeF_5 \qquad (13.10)$$

Reaction of xenon difluoride with *cis*-$(HO)_2TeF_4$ leads to the formation of a stable polymer having the composition $(XeO_2TeF_4)_n$. This polymer has a melting point of around 80°C.

In some cases, problems connected with formation of side-products (e.g., HF) occurred during these syntheses. Application of boric acid esters, $B(OR)_3$ ($R = SO_2CF_3, C(O)CF_3$),[94] as the reagents gives a nice solution to such problems.

Even stricter demands as to ligand selection are raised for the synthesis of stable four- and six-valent xenon derivatives, having covalent bonds Xe—O. Ligands $OTeF_5$, $OSeF_5$ were experimentally shown to be the most suitable ones due to their electron-withdrawing character and the best stabilizing effect,[95] as well as their similarity to the F-ligand (Scheme 13.5).[96–98] For example, Seppelt carried out an extensive series of synthetic and structural studies concerning Xe(IV)[98–100] and Xe(VI)[99–101] derivatives bearing one or more $OTeF_5$ ligands.

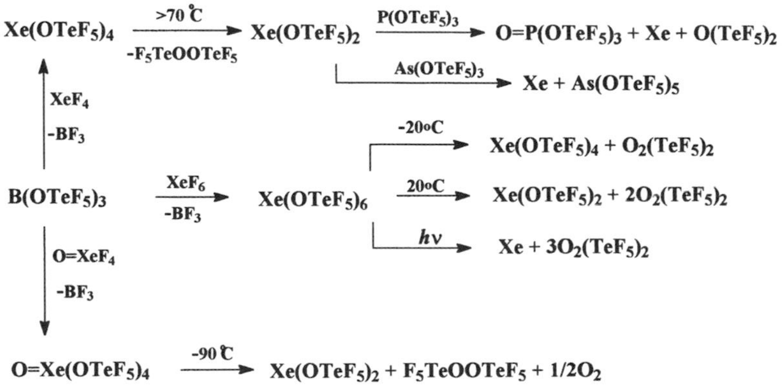

$$Xe(OTeF_5)_4 \xrightarrow[-F_5TeOOTeF_5]{>70\,^\circ C} Xe(OTeF_5)_2 \xrightarrow{P(OTeF_5)_3} O{=}P(OTeF_5)_3 + Xe + O(TeF_5)_2$$

$$Xe(OTeF_5)_2 \xrightarrow{As(OTeF_5)_3} Xe + As(OTeF_5)_5$$

$$B(OTeF_5)_3 \xrightarrow[-BF_3]{XeF_4} Xe(OTeF_5)_4$$

$$B(OTeF_5)_3 \xrightarrow[-BF_3]{XeF_6} Xe(OTeF_5)_6$$

$$Xe(OTeF_5)_6 \xrightarrow{-20\,^\circ C} Xe(OTeF_5)_4 + O_2(TeF_5)_2$$

$$Xe(OTeF_5)_6 \xrightarrow{20\,^\circ C} Xe(OTeF_5)_2 + 2O_2(TeF_5)_2$$

$$Xe(OTeF_5)_6 \xrightarrow{h\nu} Xe + 3O_2(TeF_5)_2$$

$$B(OTeF_5)_3 \xrightarrow[-BF_3]{O{=}XeF_4} O{=}Xe(OTeF_5)_4$$

$$O{=}Xe(OTeF_5)_4 \xrightarrow{-90\,^\circ C} Xe(OTeF_5)_2 + F_5TeOOTeF_5 + 1/2O_2$$

Scheme 13.5

Boric acid esters could be successfully used for the synthesis of xenon(VI) dioxydi(pentafluorotellurate) and xenon(VI) oxytetra(pentafluorotellurate) (Eqs 13.11, 13.12).[102]

$$XeO_2F_2 + 2B(OTeF_5)_3 \longrightarrow 3(F_5TeO)_2XeO_2 + 2BF_3 \quad (13.11)$$

$$3XeOF_4 + 4B(OTeF_5)_3 \longrightarrow 3O{=}Xe(OTeF_5)_4 + 4BF_3 \quad (13.12)$$

As was mentioned above, xenon fluorides react with oxygen-containing acids under rather mild conditions with almost quantitative yields to give xenon derivatives of different structures. Nowadays two-valent compounds, obtained from xenon difluoride and different acids, are the most thoroughly studied ones. Depending on the reagents ratio, it is possible to synthesize compounds of two types: F—Xe—OZ and ZO—Xe—OZ (where ZO is an anion of the corresponding acid). These compounds are unstable and degradable under moderate conditions due to high polarization of Xe—O bond. Decomposition mechanism can include both radical and ionic pathways. For example, both mechanisms[103] (ionic – path (**a**), and radical – path (**b**)) could be realized at the decomposition of xenon fluorocarboxylates (Scheme 13.6).

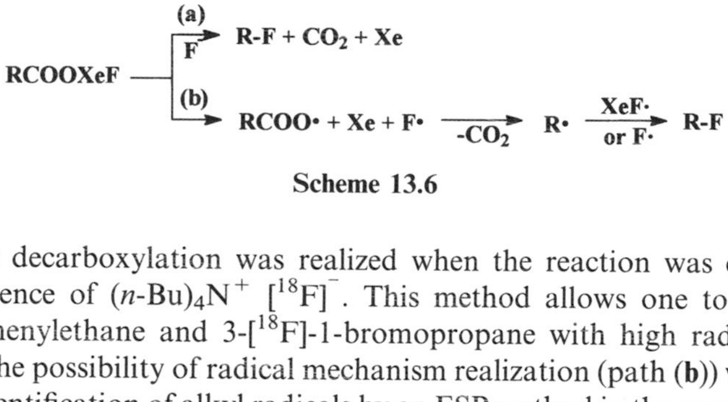

RCOOXeF → (a) F^- → R-F + CO_2 + Xe

RCOOXeF → (b) → RCOO• + Xe + F• $\xrightarrow{-CO_2}$ R• $\xrightarrow[\text{or F}\cdot]{\text{XeF}\cdot}$ R-F

Scheme 13.6

Path (**a**) decarboxylation was realized when the reaction was carried out in the presence of $(n\text{-Bu})_4N^+$ $[^{18}F]^-$. This method allows one to synthesize 2-$[^{18}F]$-1-phenylethane and 3-$[^{18}F]$-1-bromopropane with high radiochemical yields.[103] The possibility of radical mechanism realization (path (**b**)) was proved by direct identification of alkyl radicals by an ESP method in the course of study of the reaction of xenon difluoride with heptenoic acid (Scheme 13.7).[104]

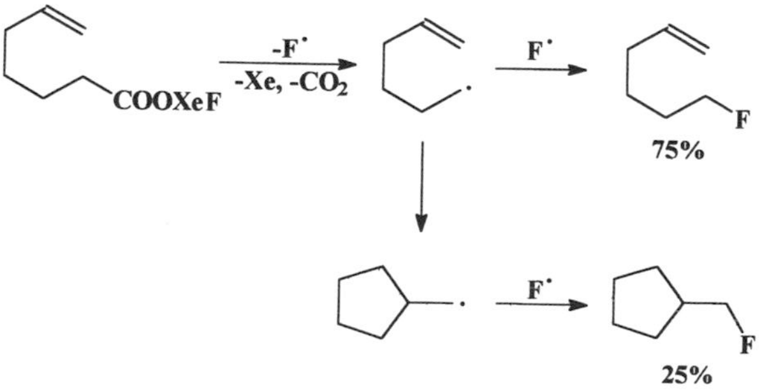

COOXeF $\xrightarrow[\text{-Xe, -CO}_2]{\text{-F}^\cdot}$ (radical) $\xrightarrow{F^\cdot}$ (fluoride) 75%

(cyclopentylmethyl radical) $\xrightarrow{F^\cdot}$ (fluoromethylcyclopentane) 25%

Scheme 13.7

Radical generation in the course of decomposition of the corresponding xenon carboxylates is nowadays used for preparative purposes to introduce substituents of different structures into organic substrates, for instance into heterocyclic rings (Scheme 13.8).[105]

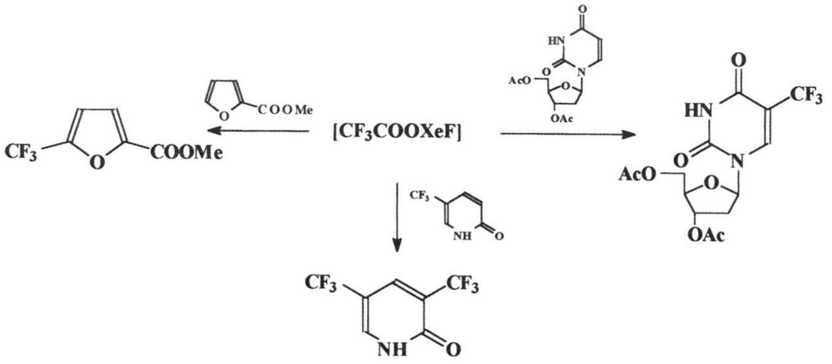

Scheme 13.8

Decomposition of fluorine-containing carboxylates of xenon was studied in detail.[105–111] Xenon carboxylates were obtained *in situ* without preliminary isolation. Thus, in the reaction of perfluoroglutaric acid with two equivalents of xenon difluoride, the corresponding xenon derivative is formed *in situ*, with subsequent decomposition to give the biradical species, which, in turn, can be trapped by aromatics or halogens (Scheme 13.9).[107] The latter case is an interesting variant of Hunsdiecker–Borodin reaction.[112]

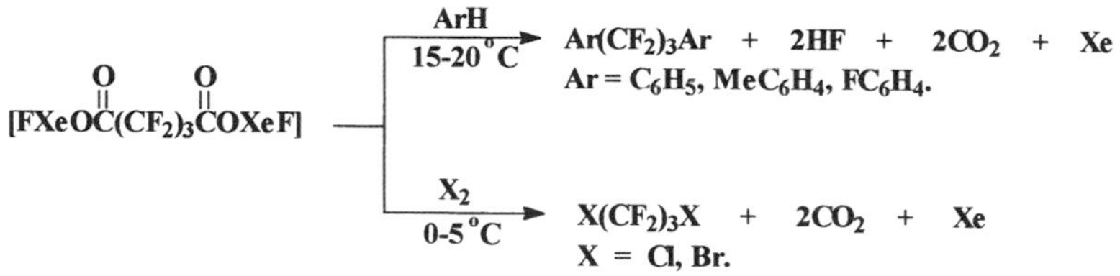

$$[FXeOC(O)(CF_2)_3C(O)OXeF] \xrightarrow[15\text{-}20\,^\circ C]{ArH} Ar(CF_2)_3Ar + 2HF + 2CO_2 + Xe$$

$Ar = C_6H_5, MeC_6H_4, FC_6H_4.$

$$[FXeOC(O)(CF_2)_3C(O)OXeF] \xrightarrow[0\text{-}5\,^\circ C]{X_2} X(CF_2)_3X + 2CO_2 + Xe$$

$X = Cl, Br.$

Scheme 13.9

Xenon carboxylates possess high oxidative potential. For example, xenon bis(trifluoroacetate) can react with aryliodides. In particular, xenon bis(trifluoroacetate), obtained *in situ* from xenon difluoride and trifluoroacetic acid in the presence of trifluoroacetic acid anhydride, reacts with aryliodides with the formation of bis(trifluoroacetoxy)iodine arenes. If the reaction is carried out in the absence of trifluoroacetic acid anhydride, it leads to a synthesis of l-oxo-bis[trifluoroacetate](aryl)iodides, their structure being determined by X-ray diffraction (Scheme 13.10).[113]

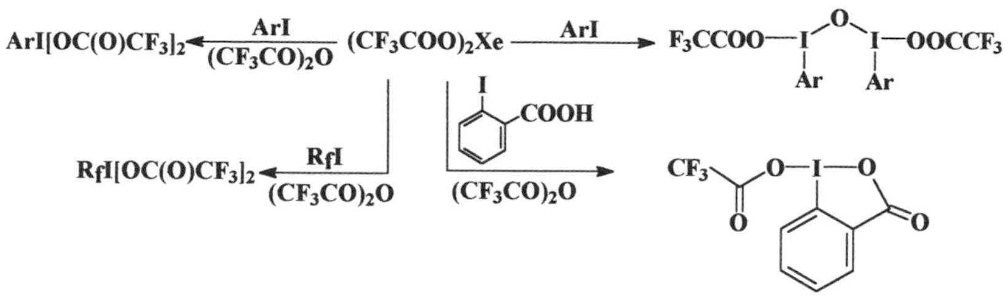

$$ArI[OC(O)CF_3]_2 \xleftarrow[(CF_3CO)_2O]{ArI} (CF_3COO)_2Xe \xrightarrow{ArI} F_3CCOO{-}I(Ar){-}O{-}I(Ar){-}OOCCF_3$$

$$RfI[OC(O)CF_3]_2 \xleftarrow[(CF_3CO)_2O]{RfI} (CF_3COO)_2Xe \xrightarrow[(CF_3CO)_2O]{\text{2-iodobenzoic acid}} \text{(trifluoroacetoxy)benziodoxolone}$$

Scheme 13.10

Xenon-monosubstituted compounds (e.g., fluorotriflate and fluoromesilate) also possess a strong oxidative tendency towards aryliodides, oxidizing them to fluoro(aryl)iodonium salts (Scheme 13.11).[114,115]

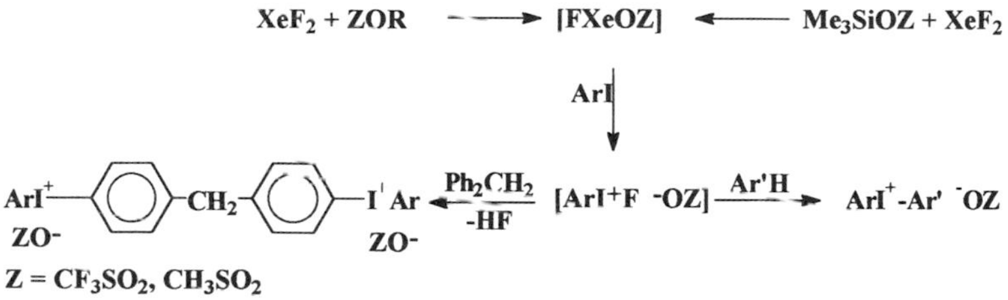

$$XeF_2 + ZOR \longrightarrow [FXeOZ] \longleftarrow Me_3SiOZ + XeF_2$$

$$[FXeOZ] \xrightarrow{ArI} [ArI^{+}F\ ^{-}OZ] \xrightarrow{Ar'H} ArI^{+}\text{-}Ar'\ ^{-}OZ$$

$$[ArI^{+}F\ ^{-}OZ] \xrightarrow[-HF]{Ph_2CH_2} ArI^{+}{-}C_6H_4{-}CH_2{-}C_6H_4{-}I^{+}Ar\ \ (ZO^{-})_2$$

$Z = CF_3SO_2, CH_3SO_2$

Scheme 13.11

Xenon monosulfates (e.g., fluoro(fluorosulfate)xenon(II)) possess electrophilic properties and their interaction with unsaturated substrates leads to electrophilic Ad_E-reactions.[93,116,117] Analysis of the adduct compositions reveals that the reactions proceed in accordance with Markovnikov rule. The strong electrophilic properties of xenon fluoro(fluorosulfates) and xenon fluoro(triflates) allow one to carry out the reaction with poor electron-donating olefins (e.g., polyhalogenated ones) under rather mild conditions (Scheme 13.12).[118]

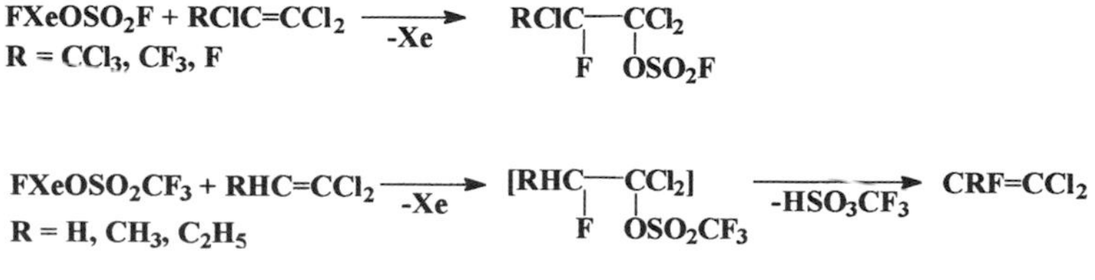

$$FXeOSO_2F + RClC{=}CCl_2 \xrightarrow[-Xe]{} RClC(F){-}CCl_2(OSO_2F)$$

$R = CCl_3, CF_3, F$

$$FXeOSO_2CF_3 + RHC{=}CCl_2 \xrightarrow[-Xe]{} [RHC(F){-}CCl_2(OSO_2CF_3)] \xrightarrow[-HSO_3CF_3]{} CRF{=}CCl_2$$

$R = H, CH_3, C_2H_5$

Scheme 13.12

13.5 NITROGEN-CONTAINING XENON DERIVATIVES

As compared with the oxygen-containing xenon derivatives, compounds with Xe—N bonds are much less known, due to the smaller variety of strong N-electron-acceptor ligands necessary for the synthesis of stable xenon compounds. The synthesis of the derivatives discussed was first carried out by DesMarteau who studied the interaction of xenon difluoride with imidobis (sulfuryl fluoride).[119–121] Depending on the reagents ratio, fluoro[imidobis(sulfuryl fluoride)]xenon(II) or bis[imidobis(sulfuryl fluoride)]xenon(II) were obtained (Scheme 13.13).

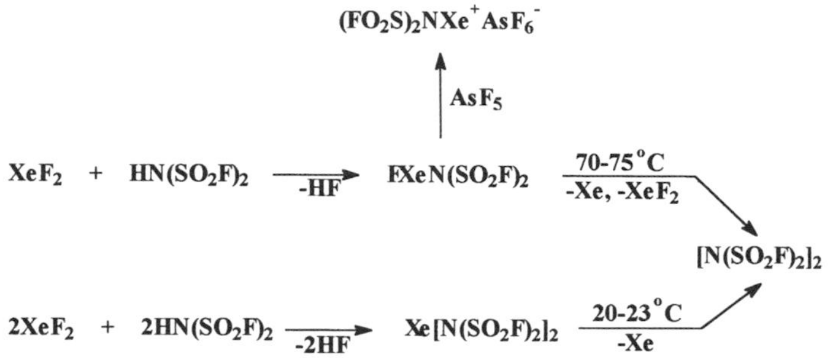

$$(FO_2S)_2NXe^+AsF_6^- \xleftarrow{AsF_5} FXeN(SO_2F)_2$$

$$XeF_2 + HN(SO_2F)_2 \xrightarrow[-HF]{} FXeN(SO_2F)_2 \xrightarrow[-Xe,\ -XeF_2]{70\text{-}75^\circ C} [N(SO_2F)_2]_2$$

$$2XeF_2 + 2HN(SO_2F)_2 \xrightarrow[-2HF]{} Xe[N(SO_2F)_2]_2 \xrightarrow[-Xe]{20\text{-}23^\circ C} [N(SO_2F)_2]_2$$

Scheme 13.13

However, the yield of monosubstituted compound was only 80%, whereas the disubstituted derivative was obtained with 20% yield. This fact indicates a weak tendency of the monosubstituted derivative toward any future substitution of fluorine atoms. The compounds obtained are stable, but they decompose when heated or in storage, with a formation of $\cdot N(SO_2F)_2$ radicals, which were detected by an ESR method. Concerning thermostability of the compounds, it should be mentioned each that bis-derivative decomposes rapidly at 20°C, whereas destruction of monoderivative takes place at 70–75°C. The structures of the compounds obtained were thoroughly studied by various methods, including X-ray diffraction analysis.[121–123] Fluoro[imidobis(sulfuryl fluoride)]xenon(II) can appear as a donor of fluoride anions in a reaction with Lewis acids. Indeed, it reacts with arsenic pentafluoride and antimony pentafluoride with a formation of cationic structure (Scheme 13.14). Heating to 23°C in vacuum leads to a formation of a "bridge" cationic structure, which was studied by spectral methods and X-ray diffraction.[120,124,125]

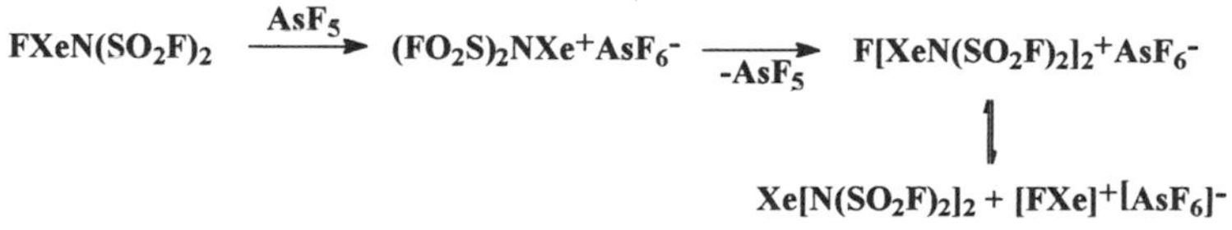

$$FXeN(SO_2F)_2 \xrightarrow{AsF_5} (FO_2S)_2NXe^+AsF_6^- \xrightarrow[-AsF_5]{} F[XeN(SO_2F)_2]_2{}^+AsF_6^-$$

$$\downarrow$$

$$Xe[N(SO_2F)_2]_2 + [FXe]^+[AsF_6]^-$$

Scheme 13.14

Bis(trifluoromethanesulfonyl)imide can also be used as a nitrogen ligand, which stabilizes the compound with Xe—N bonds. However, there are no publications concerning the direct interaction of XeF_2 with $HN(SO_2CF_3)_2$, although the alternative method of Xe—N bond formation, based on Si—N bond splitting in N-trimethylsilylbis(trifluoromethansulfonyl)imide was suggested (Eq. 13.13).

$$XeF_2 + 2Me_3SiN(SO_2CF_3)_2 \longrightarrow Xe[N(SO_2CF_3)_2]_2 + 2Me_3SiF \qquad (13.13)$$

The Si—N bond is easily split by xenon difluoride, giving a stable bis[bis (trifluoromethylsulfonyl)imido]xenon. This decomposes at around 72°C with a formation of the $\cdot N(SO_2CF_3)_2$ radical.[126] Thus the $N(SO_2CF_3)_2$ ligand has a higher stabilizing effect than its structural analog $N(SO_2F)_2$. The elaborated method of synthesis of xenon derivatives containing an Xe—N bond was applied to $(CF_3)_2NH$ and HNF_2. The authors could not isolate any derivatives containing Xe—N bonds; however, the formation of $F_2N—NF_2$ in this reaction seemed to provide indirect evidence of the intermediacy of the compounds of $FXe—NF_2$ type.[121]

Another approach to the formation of Xe–N bonds can be realized via application of XeF_2 salts, obtained by interaction with Lewis acids. In this case, the XeF^+ cation reacts with an atom possessing an electron lone pair to form a new bond with this cation, which is stabilized by a counter ion (e.g. AsF_5^-). This approach was realized in the interaction of $XeF^+AsF_6^-$ salt with hydrocyanic acid,[127,128] nitriles,[127] perfluoropyridine,[129] and perfluorotriazine (Scheme 13.15).[130]

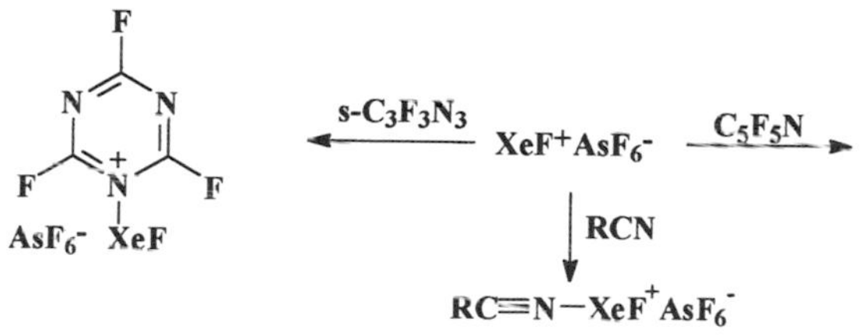

R = H, CH_3, CH_2F, C_2H_5, C_2F_5, C_3F_7, C_6F_5

Scheme 13.15

The Xe—N bond in the $RCN—XeF^+$ cation may be considered as a classical Lewis acid–base donor–acceptor bond. The theoretical description of this Xe—N bond indicates a considerable degree of ionic character, which appears to be a dominant feature of stability of such $RCN—XeF^+$ cations.[128]

Data concerning (IV)-, (VI)-, and (VIII)-valent xenon compounds with Xe–F bonds are absent from the literature. There exist, however, several reports concerning the possibility of formation of the following salts: $NO^+XeF_5^-$ and $N(CH_3)_4{}^+XeF_5^-$, which were obtained by reaction of xenon tetrafluoride with nitrosyl fluoride and tetramethylammonium fluoride, correspondingly.[131] The XeF_5^- anion of these salts was established to have a pentagonal, exactly flat structure with identical valent angles (F—Xe—F around 72°) and similar bond lengths.

Formation of $NF_4^+XeF_7^-$ was mentioned[132] in the reaction of xenon hexafluoride with $[NF_4^+ \; HF_2^-]$.

13.5 CARBON-CONTAINING XENON COMPOUNDS

At the end of the 1970s, a synthesis of the first stable xenon-organic compound, $Xe(CF_3)_2$, was reported, the compound being obtained by xenon difluoride interaction with $\cdot CF_3$ radicals.[133] This compound is stable at low temperatures. Raising the temperature up to about 20°C leads to its complete decomposition, with the formation of xenon difluoride and perfluoroalkane mixture. For a long time this publication remained the only example of the compounds having Xe—C bonds, though speculations concerning the possible intermediacy of compounds with Xe—C bonds have occasionally appeared in the literature.[134–137] These proposals were based on the structures of the isolated compounds, which can be rationalized only by accepting xenon fixation by a C—element bond.

Only in 1989, Naumann and Tyrra[138] and Frohn and Jakobs[139] made simultaneous claims of successful syntheses of stable xenon-organic compounds of cationic type. The suggested method was based on C—B bond splitting.[138–141] In particular, the compounds of $[ArXe]^+ \; [BF_4]^-$ type were obtained by interaction of Ar_3B with xenon difluoride in the presence of BF_3 (Eq. 13.14):

$$Ar_3B + 3XeF_2 + 2BF_3 \longrightarrow 3[XeAr]^+[BF_4]^- \qquad (13.14)$$
$$Ar = C_6H_5;\ 2,4,6\text{-}F_3C_6H_2.$$

These xenon compounds having Xe—C covalent bonds are rather stable, their structure being determined by ^{129}Xe, ^{19}F, ^{13}C, and ^{1}H NMR spectroscopy.[138,140] Some chemical properties of the obtained xenon derivatives were studied (Eqs. 13.15–13.17).

$$[XeAr]^+[BF_4]^- + Te(C_6F_5)_2 \longrightarrow [(C_6F_5)_2TeAr]^+[BF_4]^- + Xe \qquad (13.15)$$

$$[XeAr]^+[BF_4]^- + C_6F_5I \longrightarrow [C_6F_5IAr]^+[BF_4]^- + Xe \qquad (13.16)$$

$$[XeAr]^+[BF_4]^- + KCl \longrightarrow ArCl + KBF_4 + Xe \qquad (13.17)$$

It should be mentioned that $[XeAr]^+[BF_4]^-$ compounds have moderate reactivity; for instance the reaction with C_6F_5I proceeds only at 60°C. $[XeC_6F_5]^+[BF_4]^-$ readily reacts with alkali metal halides within a few seconds to yield the corresponding halogen-substituted pentafluorobenzenes, whereas 1-chloro-2,4,6-trifluorobenzene is formed by the interaction of $[XeC_6H_2F_3]^+[BF_4]^-$ with KCl only after several days of reaction. The remarkable stability of $[XeC_6F_5]^+[BF_4]^-$ and $[XeC_6H_2F_3]^+[BF_4]^-$ is also proved by their relative inertness in the presence of atmospheric oxygen and moisture. Storage in dry air for 21 days did not lead to decomposition; and hydrolysis in an aqueous acetonitrile solution was complete in seven days. Arylxenonium cations having hydrogen atoms in the aromatic ring were synthesized by difluoride interaction with arylboranes (Eqs. 13.18, 13.19): [138,140]

$$(m\text{-}CF_3C_6H_4)_2BF + XeF_2 \longrightarrow [m\text{-}CF_3C_6H_4Xe]^+[m\text{-}CF_3C_6H_4BF_3]^- \quad (13.18)$$

$$(p\text{-}FC_6H_4)_3B + XeF_2 \longrightarrow [p\text{-}FC_6H_4Xe]^+[(p\text{-}FC_6H_4)_2BF_2]^- \quad (13.19)$$

Analogous synthesis of xenon-organic compounds can be performed in the absence of boron trifluoride;[139,141] the compounds were obtained in methylene chloride at around −50°C with good yields.

Some chemical properties of these Xe-organic compounds were investigated. In fact, the above-mentioned para-fluoro-substituted compound was found to be rather unstable and its spectral characteristics were not studied. However, the structure of this compound was proved by its reaction with $[NEt_4]^+I^-$, which leads to the formation of p-FC_6H_4I. More data were obtained in the case of relatively stable polyfluorinated derivatives. For example, the reaction with benzyl cyanide proceeds as follows (Eq. 13.20):

$$[C_6F_5Xe]^+[(C_6H_5)_2BF_2]^- + C_6H_5CH_2CN \longrightarrow C_6H_5(C_6F_5)CHCN + HF + (C_6F_5)_2BF + Xe \quad (13.20)$$

The reaction of triphenylmethane with $[C_6F_5Xe]^+$ cation led to the formation of tetrarylmethanes (Eq. 13.21):

$$[C_6F_5Xe]^+[(C_6F_5)_2BF_2]^- + (C_6H_5)_3CH \longrightarrow \underset{\text{major}}{(C_6H_5)_3CC_6F_5} + \underset{\text{minor}}{C_6F_5\text{-}C_6H_4(C_6H_5)_2C\text{-}C_6F_5} \quad (13.21)$$

(13.21)

This reaction demonstrates that both activated aliphatic C—H bonds and aromatic C—H bonds can be substituted by an electrophilic aryl group. Actually the synthetic potential of this processes is worthy of a thorough study.

The $[C_6F_5Xe]^+$ cation can react with the $Cd(C_6F_5)_2$ (Eq. 13.22):

$$[C_6F_5Xe]^+[(C_6F_5)_2BF_2]^- + (C_6F_5)_2Cd \longrightarrow \underset{\text{major}}{C_6F_5\text{-}C_6F_5} + \underset{\text{minor}}{C_6HF_5} \quad (13.22)$$

The products obtained are the same as in the reaction of XeF_2 with one equivalent of $Cd(C_6F_5)_2$. Both reactions probably go via transient formation of

$Xe(C_6F_5)_2$, which is unstable and undergoes Xe° elimination. The formation of C_6HF_5 results from homolytic cleavage of Xe—C bond and the attack of the forming $\cdot C_6F_5$ radical on the H-containing solvent.[137,139] The structure of the $[C_6H_5Xe]^+$ cation was clearly determined by the X-ray diffraction method.[142]

Perfluoroarylxenonium salts can be a starting material for the synthesis of another class of xenon-organic compounds – alkenylxenon(II) derivatives.[143–145] It is possible to introduce two or four fluorine atoms into an aromatic ring of initial xenonium salt, using xenon difluoride as a fluorinating reagent (Scheme 13.16).

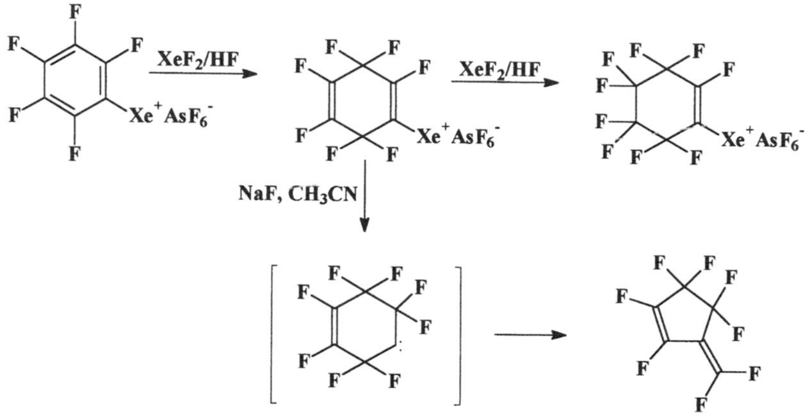

Scheme 13.16

These alkylxenonium salts are colorless compounds, stable at room temperature. Cyclopenten derivative was isolated as a product of reaction of (heptafluoro-1,4-cyclohexadien-1-yl) xenonium salt with F^- anions (using the system NaF/MeCN). There is a good probability that the reaction proceeds *via* the carbene formation, which then rearranges to cyclopenten.[146]

Finally, another class of xenon-organic compounds, namely alkynylxenonium(II) salts containing ϵC—Xe bond, was recently discovered. These salts were synthesized by splitting of ϵC—B or C—Si bonds in the proper acetylene substrates by xenon difluoride (Scheme 13.17).[147]

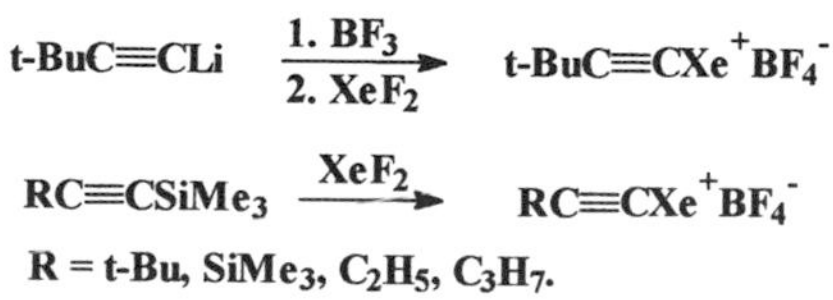

$$t\text{-BuC}\equiv\text{CLi} \xrightarrow[2.\ XeF_2]{1.\ BF_3} t\text{-BuC}\equiv\text{CXe}^+BF_4^-$$

$$RC\equiv CSiMe_3 \xrightarrow{XeF_2} RC\equiv CXe^+BF_4^-$$

R = t-Bu, $SiMe_3$, C_2H_5, C_3H_7.

Scheme 13.17

Analysis of the above mentioned data allows one to conclude that xenon-organic chemistry does exist nowadays and is definitely a developing area.

13.6 OTHER XENON DERIVATIVES

Besides the above-mentioned examples, there exist several reports concerning compounds having a covalent bond of xenon with some other elements. In particular, the reaction between xenon and O_2BF_4 was observed to lead to oxygen, fluorine, and a white solid, which decomposed at 243K into an equimolar mixture of Xe and BF_3.[148] The authors suggested the structure FXe—BF_2 for this solid (Scheme 13.18).

$$Xe + O_2BF_4 \longrightarrow F\text{-}Xe\text{-}BF_2 + [FO_2] \longrightarrow O_2 + F_2$$

Scheme 13.18

Recently, the possibility of Xe—Si bond formation was declared.[149] The gaseous trifluorosilylxenon cation, F_3SiXe^+, a stable species with Si—Xe bond, can be obtained under mass-spectrometric conditions, from the nucleophilic displacement of HF by Xe from protonated SiF_4 (Scheme 13.19).[149]

$$F_3Si\text{-}FH^+ + Xe \longrightarrow F_3SiXe^+ + HF$$

Scheme 13.19

13.7 CONCLUSION

Nowadays we face the fascinating and developed area of the chemistry of inert gases (xenon, in particular). Unusual compounds, having in some cases unique geometry, strong oxidative and electrophilic properties as well as fluorinating properties have been synthesized. The time is coming to accept and to utilize these unusual properties in practice, in particular, in organic synthesis. The material presented in this chapter gives evidence that the true synthetic potential of the hypervalent xenon compounds and their structural variability is ready for wide development.

REFERENCES

1. N. Bartlett, *Proc. Chem. Soc.*, 218 (1962).
2. H. H. Classen, H. Selig, J. G. Malm, *J. Am. Chem. Soc.*, **84**, 3593 (1962).
3. B. Zemva, R. Hagiwara, W. J. Casteel, Jr., K. Lutar, A. Jesih, N. Bartlett, *J. Am. Chem. Soc.*, **112**, 4846 (1990).
4. L. Stein, J. R. Norris, *J. Chem. Soc. Chem. Commun.*, 502 (1978).
5. D. R. Brown, M. J. Clegg, R. C. Fowler, A.R. Minihan, J. R. Norris, L. Stein, *Inorg. Chem.*, **31**, 5041 (1992).

6. L. Stein, W. W. Henderson, *J. Am. Chem. Soc.*, **102**, 2856 (1980).
7. (a) R. Filler, *Israel J. Chem.*, **17**, 71 (1978); (b) V. V. Bardin, Yu. L. Yagupol'skii, in *New Fluorinating Agents in Organic Synthesis,* L. German and S. Zemskov, Eds (Springer-Verlag, New York and Berlin, 1989), p. 63; (c) M. A. Tius, *Tetrahedron,* **51**, 6605 (1995).
8. C. L. Chernick, J. L. Malm, *Inorg. Synth.,* **8**, 258 (1966).
9. W. E. Falconer, W. A. Sunder, *J. Nucl. Chem.,* **29**, 1380 (1967).
10. A. B. Neding, V. B. Sokolov, *Usp. Khim.*, **43**, 2146 (1974).
11. F. Schreiner, G. N. McDonald, C. L. Chernich, *J. Phys. Chem.,* **72**, 1162 (1968).
12. J. L. Weeks, C. L. Chernick, M. S. Matheson, *J. Am. Chem. Soc.,* **84**, 4612 (1962).
13. L. V. Streng, A. B. Streng, *Inorg. Chem.,* **4**, 1370 (1965).
14. S. M. Williamson, *Inorg. Synth.,* **11**, 147 (1968).
15. J. H. Holloway, *J. Chem. Soc. Chem. Commun.,* 22 (1966).
16. A. Smalc, K. Lutar, J. Slivnik, *J. Fluorine Chem.*, **8**, 95 (1976).
17. V. N. Mit'kin, S. V. Zemskov, *Izv. Akad. Nauk. SSSR, Inorg. Materials*, **17**, 1897 (1981).
18. J. B. Nielsen, S. A. Kinkead, J. D. Purson, P. G. Eller, *Inorg. Chem.,* **29**, 1779 (1990).
19. V. N. Bezmel'nizyn, V. A. Legasov, S. N. Spirin, B. B. Chaivanov, *Dokl. Akad. Nauk SSSR,* **262**, 1153 (1982).
20. A. V. Eletzkyi, V. A. Legasov, S. N. Spirin, E. V. Stepanov, B. B. Chaivanov *Dokl. Akad. Nauk SSSR*, **280**, 909 (1985).
21. V. N. Bezmel'nizyn, V. A. Legasov, B. B. Chaivanov, *Dokl. Akad. Nauk SSSR,* **235**, 96 (1977).
22. D. R. Mackenzie, R. H. Wiswall, *Noble Gas Compounds* (University of Chicago Press, Chicago and London, 1963), p. 81.
23. A. D. Kirshenbaum, L. V. Streng, A. G. Streng, A. V. Grosse, *J. Am. Chem. Soc.,* **85**, 360 (1963).
24. R. D. Willett, S. D. Petersen, B. A. Coyle, *J. Am. Chem. Soc.,* **99**, 8202 (1977).
25. C. R. Bieler, K. C. Janda, *J. Am. Chem. Soc.* **112**, 2033 (1990).
26. B. Druzina, B. Zemva, *J. Fluorine Chem.,* **34**, 233 (1986).
27. A. I. Popov, V. F. Sukhoverkhov, N. A. Tchumaevsky, *Zhur. Neorganich. Khim.,* **35**, 1111 (1990).
28. A. I. Popov, A. V. Scharabarin, V. F. Sukhoverkhov, N. A. Tchumaevsky, *Z. Anorg. Allg. Chem.,* **576**, 242 (1989).
29. A. Zalkin, *Inorg. Chem.,* **17**, 1318 (1978).
30. I. G. Zaytcheva, A. I. Popov, Ya. M. Kiselev, *Zhur. Neorganich. Khim.,* **33**, 928 (1988).
31. B. Zemva, J. Zupan, J. Slivnik, *J. Inorg. Nucl. Chem.,* **33**, 3953 (1971).
32. M. Bohinc, J. Grannec, J. Slivnik, B. Zemva, *J. Inorg. Nucl. Chem.,* **38**, 75 (1976).
33. K. Lutar, I. Leban, T. Ogrin, B. Zemva, *Eur. J. Solid State Inorg. Chem.,* **29**, 713 (1992).
34. J. Slivnik, B. Zemva, *Z. Anorg. Allg. Chem.,* **385**, 137 (1971).

35. B. Zemva, J. Zupan, J. Slivnik, *J. Inorg. Nucl. Chem.,* **35**, 3941 (1973).
36. R. Hagiwara, F. Hollander, C. Maines, N. Bartlett, *Eur. J. Solid State Inorg. Chem.,* **28**, 855 (1991).
37. J. H. Burns, R. D. Ellison, H. A. Levy, *Acta Crystallogr.,* **18**, 5316 (1968).
38. F. O. Sladky, N. Bartlett, *J. Chem. Soc., (A),* 2188 (1969).
39. G. R. Jones, R. D. Burbank, N. Bartlett, *Inorg. Chem.,* **9**, 2264 (1970).
40. N. Bartlett, M. Wechsberg, *Z. Anorg. Allg. Chem.,* **385**, 5 (1971).
41. N. Bartlett, *Chem. Eng. News,* **41**, 36 (1963).
42. J. H. Holloway, *Noble Gas Chemistry* (Pergamon Press, London, 1968).
43. V. I. Spirin, Yu. M. Kiselev, N. E. Fadeeva, A. I. Popov, N. A. Tchumaevsky, *Z. Anorg. Allg. Chem.,* **559**, 171 (1988).
44. B. Zemva, *Croat. Chem. Acta,* **61**, 163 (1988).
45. B. Zemva, A. Jesih, *J. Fluorine Chem.,* **24**, 281 (1984).
46. B. Zemva, S. Milicev, J. Slivnik, *J. Fluorine Chem.,* **11**, 545 (1978).
47. A. Jesih, B. Zemva, J. Slivnik, *J. Fluorine Chem.,* **19,** 221 (1982).
48. B. Zemva, J. Slivnik, *J. Fluorine Chem.,* **17**, 375 (1981).
49. A. Jesih, K. Lutar, I. Leban, B. Zemva, *Inorg. Chem.,* **28**, 2911 (1989).
50. A. Jesih, K. Lutar, I. Leban, B. Zemva, *Eur. J. Solid State Inorg. Chem.,* **28**, 829 (1991).
51. B. Zemva, J. Slivnik, *J. Fluorine Chem.,* **8**, 369 (1976).
52. K. Lutar, A. Jesih, I. Leban, B. Zemva, N. Bartlett, *Inorg. Chem.,* **28**, 3467 (1989).
53. T. E. Mallouk, B. Desbat, N. Bartlett, *Inorg. Chem.,* **23**, 3160 (1984).
54. M. F. Beuermann, S. Milicev, K. Lutar, B. Zemva, *Eur. J. Solid State Inorg. Chem.,* **31**, 545 (1994).
55. Å. H. Appelman, *Inorg. Chem.,* **6**, 1305 (1967).
56. Å. H. Appelman, *J. Am. Chem. Soc.,* **90**, 1900 (1968).
57. V. A. Khalkin, Yu. V. Norseev, V. D. Nefedov, M. A. Toropova, V. I. Kirin, *Dokl. Akad. Nauk SSSR,* **195**, 623 (1970).
58. T. H. Dunning, P. J. Hay, *J. Chem. Phys.*, **66**, 3767 (1977).
59. A. A. Goncharov, Yu. N. Kozlov, A. P. Purmal', *Zh. Phyz. Khim.*, **52**, 2870 (1978).
60. A. A. Goncharov, Yu. N. Kozlov, *Zh. Phyz. Khim.*, **52**, 945 (1978).
61. A. A. Goncharov, Yu. N. Kozlov, A. P. Purmal', *Zh. Phyz. Khim.*, **55**, 1633 (1981).
62. D. F. Shellhamer, S. L. Carter, R. H. Dunham, S. N. Graham, M. P. Spitsbergen, V. Heasley, R. D. Chapman, M. L. Druelinger, *J. Chem. Soc. Perkin Trans II,* 159 (1989).
63. D. F. Smith, *Science,* **140**, 899 (1963).
64. C. L. Chernick, H. H. Classen, J. G. Malm, P. L. Plurien, *Noble Gas Compounds* (University of Chicago Press, Chicago and London, 1963), p. 287.
65. K. O. Christe, W. W. Wilson, *Inorg. Chem.,* **27**, 2714 (1988).
66. K. Seppelt, H.H. Rupp, *Z. Anorg. Allg. Chem.,* **409**, 331 (1974).
67. W. W. Wilson, K. O. Christe, *Inorg. Chem.,* **26**, 916 (1987).

68. K. O. Christe, W. W. Wilson, *Inorg. Chem.,* **27**, 1296 (1988).
69. K. O. Christe, W. W. Wilson, *Inorg. Chem.,* **27**, 3763 (1988).
70. H. P. A. Mercier, G. J. Schrobilgen, *Inorg. Chem.,* **32**, 145 (1993).
71. J. B. Nielsen, S. A. Kinkead, P. G. Eller, *Inorg. Chem.,* **29**, 145 (1990).
72. H. Selig, *Inorg. Chem.,* **5**, 183 (1966).
73. G. J. Moody, H. Selig, *Inorg. Nucl. Chem. Lett.,* **2**, 319 (1966).
74. G. J. Schrobilgen, D. Martin-Rovet, P. Charpin, M. Lance, *J. Chem. Soc., Chem. Commun.,* 894 (1980).
75. J. H. Holloway, V. Kaucic, D. Martin-Rovet, D. R. Russell, G. J. Schrobilgen, H. Selig, *Inorg. Chem.,* **24**, 678 (1985).
76. A. Ellern, K. Seppelt, *Angew. Chem. Int. Ed. Engl.,* **34**, 1586 (1995).
77. J. Foropoulos, D. D. DesMarteau, *Inorg.Chem.,* **21**, 2503, (1982).
78. L. D. Shustov, N. S. Tolmacheva, Sh. Sh. Nabiev, E. K. Il'in, V. D. Klimov, V. P. Yshakov, *Zh. Neorganich. Khim.,* **34**, 1673 (1989).
79. V. K. Isupov, A. V. Oleinikov, N. N. Aleinikov, *Zh. Neorganich. Khim.,* **34**, 2080 (1989).
80. N. N. Aleinikov, S. A. Kashtanov, I. A. Pomytkin, A. M. Sipyagin, *Izv. Akad. Nauk SSSR, Ser. Khim.*, 184 (1995).
81. J. G. Malm, E. H, Appelman, *Atom Energy Revs.,* **7**, 3 (1969)
82. V. A. Legasov, *Vest. Akad. Nauk SSSR*, 3, (1976).
83. N. Bartlett, M. Wechsberg, F. O. Sladky, P. A.Bulliner, G. R. Jones, R. D. Burbank, *J. Chem. Soc. Chem. Commun.,* 703 (1969).
84. M. Eisenberg, D. D. DesMarteau, *Inorg. Nucl. Chem. Lett.,* **6**, 29 (1970).
85. M. Wechsberg, P. A. Bulliner, F. O. Sladky, R. Mews, N. Bartlett, *Inorg. Chem.,* **11**, 3063 (1972).
86. F. O. Sladky, *Monatsh. Chem.,* **101**, 1571 (1970); *Angew. Chem.*, **81**, 536 (1969).
87. J. I. Musher, *J. Am. Chem. Soc.,* **90**, 7371 (1968).
88. M. Eisenberg, D. D. DesMarteau, *Inorg. Chem.,* **11**, 1901 (1972).
89. R. G. Syvret, G. J. Schrobilgen, *J. Chem. Soc. Chem. Commun.,* 1529 (1985).
90. R. G. Syvret, G. J. Schrobilgen, *Inorg. Chem.,* **28**, 1564 (1989).
91. K. Seppelt, *Angew. Chem. Int. Ed. Engl.,* **11**, 723 (1972).
92. F. O. Sladky, *Monatsh. Chem.,* **101**, 1559 (1970); *Angew. Chem.*, **81**, 330 (1969).
93. V. K. Brel, A. S. Koz'min, V. I. Uvarov, N. S. Zefirov, V. V. Zhdankin, P. J. Stang, *Tetrahedron Lett.,* **31**, 5225 (1990).
94. B. Cremer-Lober, H. Butler, D. Naumann, W. Tyrra, *Z. Anorg. Allg. Chem.,* **607** 34 (1992).
95. K. Seppelt, D. Lenz, *Progress in Inorganic Chemistry* (Wiley & Sons, New York, 1982), Vol. 29, p. 167.
96. F. Sladky, H. Kropshofer, *Inorg. Nucl. Chem. Lett.,* **8**, 195 (1972).
97. T. Birchall, R. D. Myers, H. de Waard, G. J. Schrobilgen, *Inorg. Chem.,* **21**, 1068 (1982).
98. D. Lentz, K. Seppelt, *Angew. Chem. Int. Ed. Engl.,* **17**, 356 (1978).
99. E. Jacob, D. Lentz, K. Seppelt, A. Simon, *Z. Anorg. Allg. Chem.,* **472**, 7 (1981).

100. L. Turowsky, K. Seppelt, *Z. Anorg. Allg. Chem.*, **609**, 153 (1992).
101. D. Lentz, K. Seppelt, *Angew. Chem. Int. Ed. Engl.*, **18**, 66 (1979).
102. G. A. Schumacher, G. J. Schrobilgen, *Inorg. Chem.*, **23**, 2923 (1984).
103. T. B. Patrick, K. K. Johri, D. H. White, W. S. Bertrand, R. Mokhtar, M. R. Kilbourn, M. J. Welch, *Can. J. Chem.*, **64**, 138 (1986).
104. T. B. Patrick, S. Khazaeli, S. Nadji, K. Hering-Smith, D. Reif, *J. Org. Chem.*, **58**, 705 (1993).
105. Y. Tanabe, N. Matsuo, N. Ohno, *J. Org. Chem.*, **53**, 4582 (1988).
106. A. M. Sipyagin, I. N. Pomytkin, S. V. Kartzev, N. N. Aleinikov, V. G. Kartzev, *Dokl. Akad. Nauk SSSR*, **311**, 1137 (1990).
107. V. K. Brel, V. I. Uvarov, N. S. Zefirov, P. J. Stang, R. Caple, *J. Org. Chem.*, **58**, 6922 (1993).
108. I. V. Martynov, V. K. Brel, V. I. Uvarov, I. A. Pomytkin, N. N. Aleinikov, S. A. Kashtanov, *Izv. Akad., Nauk SSSR, Ser. Khim.*, 466 (1988).
109. I. V. Martynov, V. K. Brel, V. I. Uvarov, I. A. Pomytkin, N. N. Aleinikov, S. A. Kashtanov, *Izv. Akad., Nauk SSSR, Ser. Khim.*, 2639 (1988).
110. V. I. Uvarov, V. K. Brel, I. V. Martynov, N. N. Aleinikov, S. A. Kashtanov, *Izv. Akad., Nauk SSSR, Ser. Khim.*, 2650 (1988).
111. V. K. Brel, A. S. Koz'min, I. V. Martynov, V. I, Uvarov, N. S. Zefirov, V. V. Zhdankin, P.J. Stang, *Tetrahedron Lett.*, **31**, 4799 (1990).
112. (a) A. Borodin, *Lieb. Ann.*, **119**, 121 (1861); (b) C. Wilson, *Organic Reactions* (John Wiley, London, 1957), Vol. 9, p. 332.
113. (a) T. M. Kasumov, V. K. Brel, Y. K. Grishin, N. S. Zefirov, P. J. Stang, *Tetrahedron*, **53**, 1145-1150 (1997); (b) A. N. Chekhlov. T. M. Kasumov, V. K. Brel, N. S. Zefirov, *Zh. Struk. Khim.*, **37**, 939 (1996).
114. T. M. Kasumov, V. K. Brel, A. S. Koz'min, N. S. Zefirov, K. A. Potekhin, P. J. Stang, *New J. Chem.*, **21**, 1342 (1998).
115. T. M. Kasumov, V. K. Brel, A. S. Koz'min, E. V. Balashova, K. A. Potekhin, Yu.A. Struchkov, N. S. Zefirov, *Dokl. Akad. Nauk.*, **349**, 634 (1996).
116. V. K. Brel, V. I. Uvarov, N. S. Zefirov, *J. Fluorine Chem.*, **58**, 317 (1992).
117. N. S. Zefirov, A. A. Gakh, V. V. Zhdankin, P. J. Stang, *J. Org. Chem.*, **56**, 1416 (1991).
118. V. I. Uvarov, V. K. Brel, *Izv. Akad. Nauk., Ser. Khim.*, 764 (1994).
119. R. D. LeBlond, D. D. DesMarteau, *J. Chem. Soc. Chem. Commun.*, 555 (1974).
120. D. D. DesMarteau, *J. Am. Chem. Soc.*, **100**, 6270 (1978).
121. D. D. DesMarteau, R. D. LeBlond, S. F. Hossain, D. Nothe, *J. Am. Chem. Soc.*, **103**, 7734 (1981).
122. J. F. Sawyer, G. J. Schrobilgen, S. J. Sutherland, *J. Chem. Soc. Chem. Commun.*, 210 (1982).
123. J. F. Sawyer, G. J. Schrobilgen, S. J. Sutherland, *Inorg. Chem.*, **21**, 4064 (1982).
124. G. A. Schumacher, G. J. Schrobilgen, *Inorg. Chem.*, **22**, 2178 (1983).
125. R. Faggiani, D. K. Kennepohl, C. J. Lock, G. J. Schrobilgen, *Inorg. Chem.*, **25**, 563 (1986).
126. J. Foropoulos, D. D. DesMarteau, *J. Am. Chem. Soc.*, **104**, 4260 (1982).

127. A. A. A. Emara, G. J. Schrobilgen, *J. Chem. Soc., Chem. Commun.,* 1644 (1987).
128. A. A. A. Emara, G. J. Schrobilgen, *Inorg. Chem.,* **31**, 1323 (1992).
129. A. A. A. Emara, G. J. Schrobilgen, *J. Chem. Soc., Chem. Commun.,* 257 (1988).
130. G. J. Schrobilgen, *J. Chem. Soc. Chem. Commun.,* 1506 (1988).
131. K.O. Christe, E. C. Curtis, D. A. Dixon, H. P. Mercier, J. C. P. Sanders, G. J. Schrobilgen, *J. Am. Chem. Soc.,* **113**, 3351 (1991).
132. K. O. Christe, W. W. Wilson, *Inorg. Chem.,* **21**, 4113 (1982).
133. L. J. Turbini, R. E. Aikman, R. J. Lagow, *J. Am. Chem. Soc.,* **101**, 5834 (1979).
134. V. V. Bardin, G. G. Furin, G. G. Yakobson, *Zh. Org. Khim.,* **18**, 604 (1982).
135. K. P. Butin, Ya. M. Kiselev, T. V. Magdesieva, O. A. Reutov, *Izv. Akad Nauk SSSR, Ser. Khim.,* 716 (1982).
136. R. G. Bulgakov, G. Ya. Maistrenko, G. A. Tolstikov, B. N. Yakovlev, V. P. Kazakov, *Izv. Akad Nauk SSSR., Ser. Khim.,* 2644 (1984).
137. W. Tyrra, D. Naumann, *Can. J. Chem.,* **67**, 1949 (1988).
138. D. Naumann, W. Tyrra, *J. Chem. Soc. Chem. Commun.,* 47 (1989).
139. H. J. Frohn, S. Jakobs, *J. Chem. Soc. Chem. Commun.,* 625 (1989).
140. H. Butler, D. Naumann, W. Tyrra. *Eur. J. Solid State Inorg. Chem.,* **29**, 739 (1992).
141. H. J.Frohn, S. Jakobs, C. Rossbach. *Eur. J. Solid State Inorg.Chem.,* **29**, 729 (1992).
142. H. J.Frohn, S. Jakobs, G. Henkel, *Angew.Chem.,* **101**, 1534 (1989).
143. H. J. Frohn, V. V. Bardin, *J. Chem. Soc. Chem. Commun.,* 1072 (1993).
144. H. J. Frohn, A. Klose, V. V. Bardin, A. J. Kruppa, T. V. Leshina, *J. Fluorine Chem.,* **70**, 147 (1995).
145. H. J. Frohn, A. Klose, V. V. Bardin, *J. Fluorine Chem.,* **64**, 201 (1993).
146. H. J. Frohn, V. V. Bardin, *Mendeleev Commun.,* 115 (1995).
147. V. V. Zhdankin, P. J. Stang, N. S. Zefirov, *J. Chem. Soc. Chem. Commun.,* 578 (1992).
148. C. T. Goetschel, K. R. Loos, *J. Am. Chem. Soc.,* **94**, 3018 (1972).
149. R. Cipollini, F. Grandinetti, *J. Chem. Soc. Chem. Commun.,* 773 (1995).

INDEX

NOTE: Italic page numbers indicate illustrations; page numbers followed by t indicate tables.